Handbook on Data Envelopment Analysis in Business, Finance, and Sustainability

Recent Trends and Developments

Transformations in Banking, Finance and Regulation

Print ISSN: 2752-5821
Online ISSN: 2752-583X

Presently, the banking and finance sectors evolve within a globalized and highly uncertain environment. Each business and financial decision requires careful consideration of country-specific and international factors due to high levels of shock transmission and volatility spillovers. Furthermore, it would be remiss to omit the COVID-19 health crisis, which caused harmful effects to all areas of the real economy and made profound changes in the way our firms, economies, and markets used to function.

Many other already well-known aspects, including the repercussions of quantitative easing policies, disruptive technologies in finance (e.g., FinTech, Big Data, Data Analytics, Artificial Intelligence, and Blockchains), as well as innovative financing instruments for startups (e.g., ICOs and crowdfunding) are currently revolutionizing banking and finance and posing numerous challenges to regulatory bodies and policymakers. Climate change emergencies also put pressure on governments and firms worldwide to improve policy on governance, corporate social responsibility, and sustainability practices.

This book series "Transformations in Banking, Finance and Regulation" attempts to address these issues by focusing on transformative perspectives in banking and finance. Accordingly, it provides evidence-based guidance, recommendations, and pathways to assist businesses and policy decision-makers through an interdisciplinary and in-depth understanding of ongoing changes in the behavior of economic agents.

Published

For the complete list of volumes in this series, please visit
www.worldscientific.com/series/tbfr

Transformations in Banking, Finance and Regulation
Volume 16

Handbook on Data Envelopment Analysis in Business, Finance, and Sustainability

Recent Trends and Developments

Editors

Sabri Boubaker
EM Normandie Business School, France

Thanh Ngo
Massey University, New Zealand
VNU University of Economics and Business, Vietnam

World Scientific

NEW JERSEY • LONDON • SINGAPORE • BEIJING • SHANGHAI • HONG KONG • TAIPEI • CHENNAI • TOKYO

Published by

World Scientific Publishing Europe Ltd.

57 Shelton Street, Covent Garden, London WC2H 9HE

Head office: 5 Toh Tuck Link, Singapore 596224

USA office: 27 Warren Street, Suite 401-402, Hackensack, NJ 07601

Library of Congress Cataloging-in-Publication Data
Names: Boubaker, Sabri, editor. | Ngo, Thanh, editor.
Title: Handbook on data envelopment analysis in business, finance, and sustainability : recent trends and developments / [edited by] Sabri Boubaker, EM Normandie Business School, France, Thanh Ngo, Massey University, New Zealand & VNU University of Economics and Business, Vietnam.
Description: New Jersey : World Scientific, [2024] | Series: Transformations in banking, finance and regulation, 2752-5821 ; volume 16 | Includes bibliographical references and index.
Identifiers: LCCN 2023057869 | ISBN 9781800615779 (hardcover) | ISBN 9781800615786 (ebook) | ISBN 9781800615793 (ebook other)
Subjects: LCSH: Data envelopment analysis.
Classification: LCC HA31.38 .H36 2024 | DDC 519.7/2--dc23/eng/20240202
LC record available at https://lccn.loc.gov/2023057869

British Library Cataloguing-in-Publication Data
A catalogue record for this book is available from the British Library.

For any available supplementary material, please visit
https://www.worldscientific.com/worldscibooks/10.1142/Q0465#t=suppl

Desk Editors: Nimal Koliyat/Shi Ying Koe

Typeset by Stallion Press
Email: enquiries@stallionpress.com

https://doi.org/10.1142/9781800615786_fmatter

Foreword

I was the examiner for Thanh Ngo's PhD thesis nearly a decade ago and, like most examiners for PhD students, hoped that this would lead to a successful career in academia. His publications and this handbook provide ample evidence of his success, and I am delighted to write this foreword. His co-editor, Sabri Boubaker, has a very impressive record of academic publications in many of the top finance and management journals. So, between the two of them, I expected this handbook to be of high quality and usefulness. A quick glance through the authors reveals a diversity of their disciplinary backgrounds, including economics, accounting, finance, business analytics, industrial engineering, health, banking, and information management. Such variety leads to multiple application areas, ranging from emergency services to local government to supply chains, in just the first three chapters. These different disciplinary backgrounds also lead to a variety of data envelopment analysis (DEA) methods, such as network DEA, simulations, meta-frontiers, two-stage models, and inverted DEA.

The handbook is split into three parts, covering business, banking and finance, and sustainability. Having an interest in health and the public sector, I particularly like the topics in Part I, especially the use of DEA to assess the efficiency of contagion control and medical treatment for COVID-19 as well as optimizing the locations of emergency medical centers. Part II has a nice mix of applications, ranging from insurance firms to pension funds as well as banking. Sustainability is the major theme of Part III, which includes a diverse collection of papers around different aspects such as sustainability indicators and undesirable outputs.

The chapters are a mix of applications of DEA in various settings and literature reviews on specific topics. The latter are also a mix of literature surveys and descriptions of the measures and techniques used. For example, Chapter 10 provides a systematic review of microfinance research, while another study provides details of investigations on the efficiencies of countries mobilizing resources and fostering economic integration. As the authors' preface states, the book encompasses both theoretical developments in DEA and empirical studies.

The handbook is substantial, with 19 chapters and just under 600 pages, and forms a valuable resource given that it covers multiple areas and methods. As such, it provides researchers with useful references on particular topics as well as ideas for future research. I am pleased to recommend it to the DEA community.

Paul Rouse
The University of Auckland
New Zealand

https://doi.org/10.1142/9781800615786_fmatter

Preface

Data Envelopment Analysis (DEA) has gained popularity since its introduction in 1978 (Charnes *et al.*, 1978) as a tool for measuring the efficiency, productivity, and performance of decision-making units (DMUs) such as firms or entities (Sherman and Gold, 1985; Boubaker *et al.*, 2018b, 2022a, 2023; Le *et al.*, 2023; Mai *et al.*, 2023). DEA has been widely used to rank DMUs, identify inefficiencies, evaluate management practices, assess policy effectiveness, and allocate resources (Berger and Humphrey, 1997; Vidal-García *et al.*, 2018; Boubaker *et al.*, 2022b, 2022c). There has been exponential growth in DEA research in recent years, with an average of nearly 700 publications per year since 2004 (Emrouznejad and Yang, 2018). These publications have appeared in highly respected journals, covering a wide range of fields, including banking and finance, healthcare, agriculture, transportation, and education (Murillo-Zamorano, 2004; Liu *et al.*, 2013; Kaffash *et al.*, 2019; Daraio *et al.*, 2020). With more research interests on the topics of eco-efficiency, environmental efficiency, and sustainable development under advanced methodologies such as network DEA, bootstrap DEA, slack-based DEA, inverse DEA, and stochastic DEA (Simar and Wilson, 2007; Tone and Tsutsui, 2009; Olesen and Petersen, 2016; Nguyen *et al.*, 2019; Ngo and Tsui, 2021; Hammami *et al.*, 2022), it is expected that the DEA literature continue to grow. The recent developments of artificial intelligence (AI) and machine learning (ML) (Nandy and Singh, 2021; Thaker *et al.*, 2021; Zhu, 2022; Boubaker *et al.*, 2023) in the *Industry 4.0* (Wen, 2015; Azadi *et al.*, 2023; Ngo *et al.*, 2023) also contribute to the above scenario.

A key advantage of DEA is its non-parametric nature, which means that it does not require any *a priori* assumptions about the underlying production function of the DMUs (Reinhard *et al.*, 2000). This makes DEA particularly suitable for situations with multiple inputs and outputs, which is common in many business, finance, and accounting studies (Botti *et al.*, 2014; Vidal-García *et al.*, 2019; Boubaker *et al.*, 2022c, 2021; Ho *et al.*, 2021). DEA has also been widely used in corporate finance, corporate governance, and corporate social responsibility (Boubaker *et al.*, 2012, 2018a; Boubaker and Nguyen, 2012, 2018). Therefore, it is important to examine the recent trends and developments of DEA in these fields.

This book provides a valuable opportunity to explore the recent trends and developments in DEA research within the context of business, finance, and accounting. It offers a comprehensive collection of insightful studies, including reviews, methodological advancements, and empirical research conducted in various country environments and organizations. The 19 chapters of this book are divided into three parts. Part I focuses on DEA studies in the field of business, covering areas such as healthcare, supply chain management, and governmental organizations. The seven chapters in this section offer valuable insights into the application of DEA in different business contexts. Part II contains seven studies that examine the use of DEA in banking and finance. These chapters explore various aspects of the banking sector, insurance firms, and pension funds, providing a deeper understanding of the efficiency and performance of these financial institutions. Part III, with five chapters, concentrates on the applications of DEA in sustainability. This section explores topics such as sustainability indicators, resource mobilization, food production, and farming, shedding light on the role of DEA in promoting sustainable practices.

By encompassing a wide array of topics and applications, this book significantly contributes to the advancement of DEA research in the fields of business, finance, and accounting. Its comprehensive coverage offers valuable insights to researchers, practitioners, and policymakers alike interested in gaining a deeper understanding of organizational efficiency and performance across various sectors. Researchers will find the book a valuable resource as it not only presents the latest theoretical developments in DEA but also showcases empirical studies that demonstrate its practical utility.

Practitioners, including managers and consultants, can benefit from the book's insights by incorporating DEA techniques into their decision-making processes to drive organizational improvements and enhance efficiency. Moreover, policymakers involved in areas such as public policy and regulation can derive valuable insights from the book's analysis of DEA applications. Understanding the factors contributing to efficiency and performance in different sectors can inform the design and implementation of policies that promote productivity and sustainable development.

In summary, this book's comprehensive and interdisciplinary approach, coupled with its diverse case studies, provides a rich resource for advancing DEA research and its practical applications to business, finance, and accounting. It serves as a valuable reference for a wide range of stakeholders seeking to enhance organizational efficiency and performance in various sectors.

Sabri Boubaker & Thanh Ngo

References

Azadi, M., Moghaddas, Z., Cheng, T.C.E., & Farzipoor Saen, R. (2023). Assessing the sustainability of cloud computing service providers for Industry 4.0: A state-of-the-art analytical approach. *International Journal of Production Research,* 61(12), 4196–4213. https://doi.org/10.1080/00207543.2021.1959666.

Berger, A.N. & Humphrey, D.B. (1997). Efficiency of financial institutions: International survey and directions for future research. *European Journal of Operational Research,* 98(2), 175–212.

Botti, L., Boubaker, S., Hamrouni, A., & Solonandrasana, B. (2014). Corporate governance efficiency and Internet financial reporting quality. *Review of Accounting and Finance,* 13(1), 43–64. https://doi.org/10.1108/RAF-11-2012-0117.

Boubaker, S. & Nguyen, D.K. (Eds.). (2012). *Board Directors and Corporate Social Responsibility.* London, UK: Palgrave Macmillan.

Boubaker, S. & Nguyen, D.K. (Eds.). (2018). *Corporate Governance and Corporate Social Responsibility: Emerging Markets Focus.* London, UK: World Scientific.

Boubaker, S., Cumming, D., & Nguyen, D.K. (Eds.). (2018a). *Research Handbook of Finance and Sustainability.* Cheltenham, UK: Edward Elgar Publishing.

Boubaker, S., Le, T.D.Q., & Ngo, T. (2022a). Managing bank performance under COVID-19: A novel inverse DEA efficiency approach. *International Transactions in Operational Research,* 30(2023), 2436–2452. https://doi.org/10.1111/itor.13132.

Boubaker, S., Manita, R., & Mefteh-Wali, S. (2022b). Foreign currency hedging and firm productive efficiency. *Annals of Operations Research,* 313(2), 833–854. https://doi.org/10.1007/s10479-020-03730-5.

Boubaker, S., Manita, R., & Rouatbi, W. (2021). Large shareholders control contestability and firm productive efficiency. *Annals of Operations Research,* 296(1), 591–614. https://doi.org/10.1007/s10479-019-03402-z.

Boubaker, S., Nguyen, B.D., & Nguyen, D.K. (Eds.). (2012). *Corporate Governance: Recent Developments and New Trends.* Berlin, Germany: Springer.

Boubaker, S., Do, D.T., Hammami, H., & Ly, K.C. (2022c). The role of bank affiliation in bank efficiency: A fuzzy multi-objective data envelopment analysis approach. *Annals of Operations Research,* 311, 611–639. https://doi.org/10.1007/s10479-020-03817-z.

Boubaker, S., Houcine, A., Ftiti, Z., & Masri, H. (2018b). Does audit quality affect firms' investment efficiency? *Journal of the Operational Research Society,* 69(10), 1688–1699. https://doi.org/10.1080/01605682.2018.1489357.

Boubaker, S., Le, T.D.Q., Ngo, T., & Manita, R. (2023). Predicting the performance of MSMEs: A hybrid DEA-machine learning approach. *Annals of Operations Research.* https://doi.org/10.1007/s10479-023-05230-8.

Charnes, A., Cooper, W.W., & Rhodes, E. (1978). Measuring the efficiency of decision-making units. *European Journal of Operational Research,* 2(6), 429–444.

Daraio, C., Kerstens, K., Nepomuceno, T., & Sickles, R.C. (2020). Empirical surveys of frontier applications: A meta-review. *International Transactions in Operational Research,* 27(2), 709–738. https://doi.org/10.1111/itor.12649.

Emrouznejad, A. & Yang, G.-l. (2018). A survey and analysis of the first 40 years of scholarly literature in DEA: 1978-2016. *Socio-Economic Planning Sciences,* 61(1), 1–5. https://doi.org/10.1016/j.seps.2017.01.008.

Hammami, H., Ngo, T., Tripe, D., & Vo, D.-T. (2022). Ranking with a Euclidean common set of weights in data envelopment analysis: With application to the Eurozone banking sector. *Annals of Operations Research*, 311, 675–694. https://doi.org/10.1007/s10479-020-03759-6.

Ho, T.H., Nguyen, D.T., Ngo, T., & Le, T.D.Q. (2021). Efficiency in Vietnamese banking: A meta-regression analysis approach. *International Journal of Financial Studies*, 9(3).

Kaffash, S., Azizi, R., Huang, Y., & Zhu, J. (2019). A survey of data envelopment analysis applications in the insurance industry 1993–2018. *European Journal of Operational Research,* 284(3), 801–813. https://doi.org/10.1016/j.ejor.2019.07.034.

Le, T.D.Q., Ngo, T., Ho, T.H., & Nguyen, D.T. (2023). Determinants of bank performance: Revisiting the role of CEO's personality traits using graphology. *Bulletin of Monetary Economics and Banking,* 26(2), 52–87. https://doi.org/10.59091/1410-8046.2066.

Liu, J.S., Lu, L.Y.Y., Lu, W.-M., & Lin, B.J.Y. (2013). A survey of DEA applications. *OMEGA*, 41(5), 893–902.

Mai, X.T.T., Nguyen, H.T.N., Ngo, T., *et al.* (2023). Efficiency of the Islamic banking sector: Evidence from two-stage DEA double frontiers analysis. *International Journal of Financial Studies*, 11(1), 32.

Murillo-Zamorano, L.R. (2004). Economic efficiency and frontier techniques. *Journal of Economic Surveys*, 18(1), 33–77.

Nandy, A. & Singh, P.K. (2021). Application of fuzzy DEA and machine learning algorithms in efficiency estimation of paddy producers of rural Eastern India. *Benchmarking: An International Journal,* 28(1), 229–248. https://doi.org/10.1108/BIJ-01-2020-0012.

Ngo, T. & Tsui, K.W.H. (2021). Estimating the confidence intervals for DEA efficiency scores of Asia-Pacific airlines. *Operational Research,* 22, 3411–3434. https://doi.org/10.1007/s12351-021-00667-w.

Ngo, T., Guégan, D., Vo, D.-T., *et al.* (Eds.). (2023). *The Financial Industry 4.0.* Basel, Switzerland: MDPI.

Nguyen, H.-D., Ngo, T., Le, T.D.Q., *et al.* (2019). The role of knowledge in sustainable agriculture: Evidence from rice farms' technical efficiency in Hanoi, Vietnam. *Sustainability*, 11(9), 2472.

Olesen, O.B. & Petersen, N.C. (2016). Stochastic data envelopment analysis — A review. *European Journal of Operational Research*, 251(1), 2–21. https://doi.org/10.1016/j.ejor.2015.07.058.

Reinhard, S., Lovell, C.A.K., & Thijssen, G.J. (2000). Environmental efficiency with multiple environmentally detrimental variables; estimated with SFA and DEA. *European Journal of Operational Research,* 121(2), 287–303.

Sherman, D.H. & Gold, F. (1985). Bank branches operating efficiency: Evaluation with Data Envelopment Analysis. *Journal of Banking & Finance*, 9, 297–315.

Simar, L. & Wilson, P.W. (2007). Estimation and inference in two-stage, semi-parametric models of production processes. *Journal of Econometrics*, 136, 31–64.

Thaker, K., Charles, V., Pant, A., & Gherman, T. (2021). A DEA and random forest regression approach to studying bank efficiency and corporate governance. *Journal of the Operational Research Society,* 73(6), 1258–1277. https://doi.org/10.1080/01605682.2021.1907239.

Tone, K. & Tsutsui, M. (2009). Network DEA: A slacks-based measure approach. *European Journal of Operational Research*, 197, 243–252.

Vidal-García, J., Vidal, M., Boubaker, S., & Hassan, M. (2018). The efficiency of mutual funds. *Annals of Operations Research,* 267(1), 555–584. https://doi.org/10.1007/s10479-017-2429-z.

Vidal-García, J., Vidal, M., Boubaker, S., & Manita, R. (2019). Idiosyncratic risk and mutual fund performance. *Annals of Operations Research*, 281(1), 349–372. https://doi.org/10.1007/s10479-018-2794-2.

Wen, M. (2015). *Uncertain Data Envelopment Analysis.* Berlin, Germany: Springer.

Zhu, J. (2022). DEA under big data: Data-enabled analytics and network data envelopment analysis. *Annals of Operations Research,* 309, 761–783. https://doi.org/10.1007/s10479-020-03668-8.

https://doi.org/10.1142/9781800615786_fmatter

About the Editors

Sabri Boubaker is a professor of finance at EM Normandie Business School, Paris, France, and the University of Swansea, Swansea, Wales, United Kingdom. Sabri holds a PhD in finance from the University of Paris Est (2006) and a HDR degree (Habilitation for Supervising Doctoral Research) in 2010 from the same university. Professor Boubaker is the editor-in-chief of the *Journal of International Financial Management and Accounting*. Sabri Boubaker has recently published more than 100 academic papers in international refereed journals, including the *Journal of Corporate Finance*, the *Journal of Banking and Finance*, the *Journal of International Money and Finance, Financial Management,* the *Journal of Business Ethics,* the *Journal of Economic Dynamics and Control, Auditing: A Journal of Theory & Practice,* the *British Journal of Management*, and the *Journal of Financial Research.* Professor Boubaker has also edited many books on corporate finance and financial markets and serves on the editorial boards of several peer-reviewed finance journals. He is the co-founder and co-chair of the annual Paris Financial Management Conference (2013–), the Vietnam Symposium in Banking and Finance (2016–), and the Annual Meeting of the International Society for the Advancement of Financial Economics (2022–).

Thanh Ngo is a senior lecturer at the School of Aviation, Massey University, New Zealand, and the VNU University of Economics and Business, Vietnam. He holds a PhD in economics (majoring in banking) from Massey University in 2015. His work involves efficiency and productivity analysis in banking and finance, sustainability, agriculture and manufacturing sectors, aviation, and transportation economics. Dr. Ngo's research papers have been published in *Transportation Research Part A*, the *Annals of Operations Research*, *Transport Policy*, *Business Strategy and the Environment*, and the *International Journal of Managerial Finance*, among others. He is currently an associate editor for the *Asian Economic and Financial Review* and also serves on the editorial boards of several journals, such as *PLoS ONE* and the *International Journal of Financial Studies*.

https://doi.org/10.1142/9781800615786_fmatter

About the Contributors

Adel Hatami-Marbini is a professor of supply chain management & analytics at Huddersfield Business School (HBS), UK. Prior to that, he held the position of director of the Innovative Logistics and Supply Chains research cluster at Leicester Castle Business School at De Montfort University in Leicester, UK. His areas of research interest include operations and supply chain management, performance measurement, productivity analysis, decision sciences, and operational research. He holds his BSc and MSc in industrial engineering and received his PhD in economics and management sciences from the Louvain School of Management (LSM) at Université catholique de Louvain in Belgium.

He was a visiting scholar at Aston University (UK), the Technical University of Ostrava (Czech Republic), the University of Seville (Spain), the American University of Beirut (Lebanon), and Virginia Tech (US). His research profile is sustained and internationally recognized with a strong track record of published academic outputs, which include over 90 scientific peer-reviewed papers, book chapters, working papers, and over 40 conference presentations. Adel also has editorial responsibilities for several leading and top journals. In particular, he is acting as the associate editor for *OR Spectrum* (Springer, ABS3, IF:2.688), the *Journal of Intelligent Manufacturing* (Springer, IF:4.311), and *Kybernetes* (Emerald, IF:1.754).

Alishba Naseer holds a bachelor's degree in accounting and finance from Kinnaird College for Women University in Pakistan. Her research interests include corporate governance, financial markets, and corporate finance.

Amar Oukil is an associate professor of operations management at Sultan Qaboos University in Oman. Dr. Oukil graduated from Lancaster University (England) with a PhD in management science (combinatorial optimization) and holds a master's in applied mathematics (operations research) from the University of Montreal (Canada). Dr. Oukil's academic career has extended over the past three decades in diverse sociocultural environments. Dr. Oukil's research interests lie primarily in developing new computerized methodologies and algorithms to support decision-making within broad business- and industry-related fields. Dr. Oukil worked on developing stabilizing paradigms to address degeneracy issues occurring in large-scale NP-hard mathematical programming models, which are typical formulations of many real-life systems, such as road and telecommunication networks and electricity and water distribution grids. Performance analysis of systems is another strong area of expertise for Dr. Oukil. His inclination toward such a research direction stems from the wide scope of its application spectrum, which covers multiple operational and decisional contexts and beyond. Data envelopment analysis (DEA) is a major technique used throughout most pertaining research. Dr. Oukil upholds an impressive publication record by disseminating his research findings in highly ranked scientific journals, including *Computers & Operations Research*, *Computers & Industrial Engineering*, the *International Journal of Computer Applications in Technology*, the *International Journal of Intelligent Systems*, the *International Journal of Advanced Manufacturing Technology*, *Measurement*, the *International Journal of Management Science*, the *International Journal of Production Research*, and *Socio-Economic Planning Sciences*.

Ana Martins is a professor of finance at the School of Management, Hospitality, and Tourism at the University of Algarve in Portugal. She holds a PhD in economic and business sciences, with a specialization in management (2013). Currently the subdirector of this school, she also collaborates on several business and research projects in the area of financial consulting and training actions of the Certified Accountants Board. Her main research interests involve the areas of valuation and efficiency.

Andrew C. Worthington is a professor of finance in the Department of Accounting, Finance and Economics at Griffith University in Australia. His earlier career appointments include the University of Wollongong, Queensland University of Technology, and the University of New England. He holds master's degrees from the University of New South Wales and the University of New England and obtained his doctorate from the University of Queensland. Professor Worthington has researched widely in the efficiency and productivity analysis of financial institutions, including banks and other deposit-taking institutions, life and general insurance companies, and pension funds.

Ane Elixabete Ripoll-Zarraga is an auditor (IFRS, GAAP, and consolidation), a consultant with more than 20 years of experience, and a senior lecturer in accounting in degrees, postgraduate programs, and executive education. Ane Elixabete holds an industrial doctorate (Generalitat de Catalunya) in entrepreneurship and management from the Universitat Autònoma de Barcelona (2018), awarded the Extraordinary Doctorate Award (June 2022). Her research interests include operations management in air transport, using parametric (stochastic frontier analysis), and non-parametric methods (data envelopment analysis). One critical aspect is public data according to the national regulatory framework and accounting policies, such as capital measures (depreciation) or dividend policies, for benchmarking purposes requiring harmonization. Her publications have appeared in highly recognized journals, such as the *Annals of Operations Research*, the *Journal of Air Transport Management*, and *Transport Policy*. Ane Elixabete belongs to the German Aviation Research Society (GARS) to promote the liaison between the aviation industry, including politicians and regulators, and the academy to produce relevant research according to current industry challenges. She is also an organizer of the Research Day at the European Aviation Conference (EAC). Her recent research in DEA visualization applies to health economics and public services. Ane Elixabete is an experienced reviewer in highly ranked journals, supporting the quality review process by investing time and effort in order to ensure robust findings for empirical implications and decision-making.

Aparna Bhatia is an associate professor at the University School of Financial Studies at Guru Nanak Dev University, Amritsar, in Punjab, India. She holds a PhD in finance from Guru Nanak Dev University. She has published many academic papers in national and international refereed journals, including *Indian Growth and Development Review*, the *Indian Journal of Corporate Governance*, the *International Journal of Indian Culture and Business Management*, the *Journal of Asia Pacific Business*, the *International Journal of Emerging Markets*, *Management and Labor Studies*, *VISION*, the *International Journal of Behavioral Accounting and Finance*, the *International Journal of Law and Management*, the *Journal of Management Development*, *Benchmarking: An International Journal*, and the *Jindal Journal of Business Research*. She has more than 70 publications to her credit. She has also attended several conferences and chaired sessions in various seminars.

Apriani Dorkas Rambu Atahau is a professor of finance at the Department of Management, Faculty of Economics and Business, Universitas Kristen Satya Wacana in Indonesia. Apriani's specialization is in the area of banking and finance. Her research interests are mainly banking, finance, and other financial institutions, such as pension funds and microfinance. Her recent article, co-authored with Tom Cronje from Curtin University, Australia, and published by the *Economic Systems* journal, is titled "Bank Lending: Bank Ownership," focusing on the pre- and post-global financial crisis scenarios.

Brian Dollery is an emeritus professor of economics and director of the UNE Centre for Local Government at the University of New England, Armidale, NSW, in Australia, and a visiting researcher at the Faculty of International Social Sciences at Yokohama National University, Japan. He has written extensively on local governments, especially on their structure, finances, and reform. His recent books include *Local Public, Fiscal and Financial Governance* (2020), *Australian Local Government Reform* (2015), *Funding the Future* (2013), and *Councils in Cooperation* (2012).

Carlos Alberto Zúniga-González is a titular professor of agricultural economy at the National Autonomous University of Nicaragua, Leon (Nicaragua). He holds a PhD in agricultural economist, with a specialization in sustainable local development, from the American World University, USA (2008–2011). He has recently published many academic papers in international refereed journals, including *F1000Research*, *PLoS ONE*, *REICE: Revista Electrónica de Investigación en Ciencias Económicas*, *NEXO*, *Tecnología en*

March, *Revista Iberoamericana de Bioeconomia y Cambio Climático*, *Revista Cientifica de la Facultad de Ciencias Veterinarias de la Universidad del Zulia*, and *Revista electrónica calidad en la educación superior*. Professor Zúniga-González has also edited many books on agricultural economy and bioeconomy and serves on the editorial boards of several peer-reviewed journals.

Carolyn-Thi Thanh Dung Tran is currently a senior lecturer in business analytics at the Australian Institute of Higher Education (AIH), Sydney, Australia. She is also a researcher at the Center for Local Government, University of New England (Australia) and an honorary research fellow at the University of Sydney, University of Technology, Sydney, and Western Sydney University in NSW, Australia. She was formerly the director of the Scholarship of Learning and Teaching at the International College of Management, Sydney. She has written extensively on higher education and local government, especially on operational efficiency, municipality economies of scale, and productivity. Her recent publications have appeared in *Public Administration*, *Studies in Higher Education*, *Local Government Studies*, *Policy & Politics*, the *Australian Journal of Public Administration*, *Public Management Quarterly*, *Applied Economics*, and the *Singapore Economic Review*.

Damaris Chieregato Vicentin is a postdoctoral researcher at State University of Campinas (Unicamp), Limeira, Brazil. She holds a PhD degree in industrial engineering at the Federal University of Sao Carlos (UFSCar/DEP), with a split-site doctorate period at the University of Waikato, a master's degree in industrial engineering at the University of State São Paulo (UNESP/FEB), an MBA in financial management, Accounting and Auditing at Foundation Getulio Vargas (FGV), and a bachelor's degree in business administration at the Toledo Institution of Education (ITE). She has experience and research interests in quality management, with an emphasis on statistical tools, financial management, industrial engineering, logistics, supply chains, sustainability, and quality management. She has recently published academic papers in international refereed journals, including the *International Journal of Quality and Reliability Management* and the *International Journal of Advanced Manufacturing Technology*. She is also a reviewer for the *International Journal of Quality and Reliability Management*.

David Sousa is an assistant professor specializing in management and quality at the School of Health, University of Algarve, Portugal. Previously, he served as a laboratory manager at AD-ABC (Algarve Biomedical Center) in Portugal. David is currently enrolled in the economic and business sciences PhD program and holds a master's in management, with a specialization in health (2022).

Dominik Naeher is a postdoctoral research fellow at the University of Göttingen, Germany. His research focuses on a range of topics in development economics, international economics, applied economic theory, and the economics of information. His work has been published in international refereed journals, such as *Economica*, the *International Economic Review*, and the *World Bank Economic Review*. Prior to his current position, Dr. Naeher was an assistant professor at University College Dublin and a doctoral researcher at Goethe University Frankfurt. He obtained his PhD in 2018 from the Graduate School of Economics, Finance, and Management in Frankfurt after studying mathematics and economics at TU Dortmund University. He has also worked as a consultant at the World Bank and the Asian Development Bank.

Elgin Antonio Vivas-Viachica is a titular professor of agrarian economy at the National Agrarian University of Nicaragua (UNA), Nicaragua. He holds a PhD in agrarian economics from Dokuchaev State Agrarian University, former Soviet Union. He has recently published many academic papers in international refereed journals, including *PLoS ONE*, *La Calera*, and *Revista Iberoamericana de Bioeconomia y Cambio Climático*. Professor Vivas-Viachica has also edited many books and handbooks on agricultural economy and rural development and serves on the editorial boards of several peer-reviewed journals.

Farah Naz is assistant professor of Accounting & Finance at Kinnaird College for Women, Lahore, Pakistan. She holds a PhD in finance from School of Accounting & Finance, University of Central Punjab. Her research interests include microfinance, fintech, financial inclusion, financial markets and risk management. As an experienced assistant professor with a strong academic background in Accounting and Finance, she brings a wealth of experience to both teaching and research. As an educator, she is dedicated to sharing my knowledge and experience with the students and helping them develop the skills they need to excel in their careers. Beyond her academic pursuits, she committed to giving back to society and making a positive impact in the community.

She has publications in the *Journals of Economic Analysis and Policy, Journal of Teaching and Education, Journal of Contemporary Issues in Business Research, Electronic Commerce Research, A Research Journal of Commerce, Economics and Social Sciences, International Review of Economics and Finance, Resources, Policy, Journal of Management and Research, International Journal of Financial Engineering, Cogent Business and Management, WSEAS, Transactions on Business and Economics, International Journal of Business Reflections*, and *Journal of Corporate Accounting and Finance.*

Hakan Yildiz is an associate professor of global supply chain management at Wayne State University's Mike Ilitch School of Business. He has a BSc in industrial engineering from Bilkent University and an MS and a PhD in operations research from Carnegie Mellon University's Tepper School of Business. His research interests include supply chain risk management, logistics optimization, resource management in service operations, and supply chain analytics. His research has appeared in several journals, including *Decision Sciences Journal*, the *European Journal of Operational Research*, the *International Journal of Production Research*, and the *Journal of Business Logistics*. He has done research projects in collaboration with several firms, such as Bosch, Dow Chemical Company, Kelly Outsourcing & Consulting Group, and Lear Corporation.

İzzettin Özdemir is currently working in the private sector. He completed his undergraduate education in the Department of Econometrics at Karadeniz Technical University, Turkey (2016). He has a master's degree in economics from Zonguldak Bülent Ecevit University (2021). He conducts studies in the fields of economics and econometrics.

Jorge Luis Icabalceta-Mairena is a titular professor of agricultural economy at the National Autonomous University of Nicaragua, Managua (Nicaragua). He holds a PhD in agricultural economist from Louisiana State University, Baton Rouge, Louisiana, US (1995–2001). He has recently published many academic papers in international refereed journals, including *Ciencias Agrícolas* and

Tecnología y Salud, La Calera. Professor Icabalceta-Mairena has also edited many books on agricultural economy and econometric models and serves on the editorial boards of several peer-reviewed journals.

Lan Nguyen is a research fellow in the Centre for the Business and Economics of Health at the University of Queensland, Australia. She holds a master of economics degree from the National Economics University, Vietnam, and a master of commerce degree from the University of Queensland, and she obtained her doctorate from Griffith University under the supervision of her co-author, Professor Andrew C. Worthington. Dr. Nguyen's research interests include the efficiency of private health insurance funds and adverse selection and moral hazards among private health insurance policyholders.

Mario Ramón López is a titular professor of agricultural economy at the National Autonomous University of Nicaragua, Managua (Nicaragua). He holds a PhD in Economics (Faculty of Economic and Social Sciences) from the University of Zulia. Maracaibo, Zulia, VE (2012–2016). He has recently published many academic papers in international refereed journals, including *REICE: Revista Electrónica de Investigación en Ciencias Económicas*, *Revista Iberoamericana de Bioeconomia y Cambio Climático*, and *Revista Compromiso Social*. Professor López has also edited many books on agricultural economy and innovation and serves on the editorial boards of several peer-reviewed journals.

Md Aslam Mia holds a PhD in economics (with research in the area of microfinance) from the University of Malaya, Malaysia. He is currently a senior lecturer at the School of Management (SOM), Universiti Sains Malaysia (USM). Dr. Mia has published around 50 articles related to productivity and efficiency of financial institutions, market structure, microfinance, and urban economics in internationally reputed peer-reviewed journals, such as the *Annals of Public and Cooperative Economics*, *Gender in Management*, *The European Journal of Development Research*, *Business History*, *Finance Research Letters*, *Economics Letters*, the *Journal of the Asia Pacific Economy*, the *Journal of Cleaner Production*, *Social Indicators Research*, the *International Journal of Social Economics*, *Cities*, *Economic Analysis and Policy*, and *Strategic Change*. Recently, his sole-authored book on *Social Purpose, Commercialization, and Innovations in Microfinance: A Global Perspective* has been published by Springer. He has also presented his research findings at over 25 national and international conferences/seminars during the past eight years. Due to his significant contribution to research and publication, Dr. Mia received the "PhD Candidate with Highest Impact Publications" award from the University of Malaya in 2018. Dr. Mia has also conducted over 25 workshops/seminars on various research issues/techniques/tools both within and outside Malaysia.

Md Imran Hossain is a PhD candidate and an awardee of the Graduate Research Assistant (GRA) at the School of Management (SOM), Universiti Sains Malaysia (USM), Malaysia. His research interests focus on various dimensions of the sustainability of microfinance institutions (MFIs), including corporate governance and risk-taking. He has conducted several studies in the field of microfinance and attended two international conferences. He won the best paper award after presenting one of his latest research papers in the *International Post-Graduate Research Symposium 2022*, organized by the School of Management, Universiti Sains Malaysia. Mr. Hossain is also a faculty member at the Department of Finance, Jagannath University, Dhaka, Bangladesh.

Megha Mahendru is an assistant professor in the University School of Financial Studies, Guru Nanak Dev University, Amritsar (Punjab), India. She holds a PhD in finance from Guru Nanak Dev University, Amritsar. She has recently published many academic papers in international refereed journals, including the *Indian Growth and Development Review*, the *Indian Journal of Corporate Governance*, the *Eurasian Journal of Business and Economics*, the *International Journal of Behavioral Accounting and Finance*, the *International Journal of Law and Management*, the *Journal of Management Development*, *Benchmarking: An International Journal*, and the *Jindal Journal of Business Research*. She has also attended many conferences related to her areas of interest.

Mehtap Çakmak Barsbay is an associate professor of Healthcare Management at the Faculty of Health Sciences at Ankara University, Türkiye. She has MSc and PhD degrees in Public Administration from Hacettepe University (Türkiye). She has also registered nursing experience and background. Her research interests focus on evaluation of health policy and politics, and program assessment methods. She still teaches in several study programs of health care management on "Health Planning", "Comparative Health Systems" and "Health Insurance" courses. She is an active member of ISPOR — The Professional Society for Health Economics and Outcomes Research. She teaches health economics and comparative health policies at the university. She is also PhD supervisor of several thesis on burden of perinatal mental health diseases. She involved in CA18138 Rise-up PPD Project and CA18211 Perinatal Mental Health Birth-related Trauma.

Napoleon Vicente Blanco-Orozco is a titular professor of agricultural economy at the National Engineer University (Nicaragua). He holds a PhD in sciences from the Technological Institute of Costa Rica (ITCR) (2010–2015). He has recently published many academic papers in international refereed journals, including the *Journal of Agricultural Studies*, *Fuentes el Revention Energetico*, *Nexo Revista Cientifica*, *Tecnología en March*, and *Revista Iberoamericana de Bioeconomia y Cambio Climático*. Professor Blanco-Orozco has also edited many books on agricultural economy, electrical machines, and project management and serves on the editorial boards of several peer-reviewed journals.

Nilofar Varzgani is an assistant professor of business systems and analytics at La Salle University, Philadelphia, US. She holds a PhD in management science from Rutgers University, NJ, US (2017). She has recently published academic papers in international refereed journals, including the *International Journal of Management and Decision Making*, the *International Journal of Systems Science: Operations & Logistics*, and the *Decision Analytics Journal*. Professor Varzgani was recently recognized as the Lindback Emerging Scholar (2022–2024) at La Salle University. She serves as a board member for the Young PhD Scholar Advisory Board at the Hunan University of Technology (China). She also serves on the editorial boards of several peer-reviewed operations research and logistics journals.

Paskalis Seran is a lecturer in financial management at the Department of Management, Faculty of Economic and Business, Universitas Katolik Widya Mandira, Kupang, Indonesia. He earned his doctorate in financial management from Universitas Kristen Satya Wacana, Salatiga, Indonesia. His research interests are mainly related to pension funds, banks, and microfinance. His recent article published by the *International Journal of Financial Studies* is titled "The Efficiency of Indonesian Pension Funds: A Two-Stage Additive Network DEA Approach," coauthored with Usil Sis Sucahyo, Apriani D.R. Atahau, and Supramono Supramono from Universitas Kristen Satya Wacana, Indonesia.

Paulo Nocera Alves Junior is currently an assistant professor at the Universidad Católica del Norte (UCN), Escuela de Ingeniería de Coquimbo (EIC), Coquimbo, Chile. He was a postdoctoral researcher at the University of São Paulo (USP/ESALQ/ESALQ-LOG). He holds a PhD degree in industrial engineering from the University of São Paulo (USP/EESC), with a split-site doctorate period at the Universidad Adolfo Ibáñez (UAI), a master's degree in industrial engineering from the São Paulo State University (UNESP/FEB), and a bachelor's degree in industrial engineering from the University of Franca (UNIFRAN). His research interests and expertise are related to inventory control, logistics, supply chain management, sustainability, performance measurement, composite indicators, efficiency analysis, data envelopment analysis, optimal control theory, operations research, multivariate analysis, and quantitative methods, mainly developing and applying data envelopment analysis models to evaluate supply chains, logistics, and inventory control systems. Also, he has published his research outcomes in international high-impact peer-reviewed journals indexed in the Web of Science, such as the *Journal of Cleaner Production*, *Agricultural Systems*, the *Journal of Marine Science and Engineering*, *Agriculture*, *Transport Policy*, *Research in Transportation Economics*, *Promet — Traffic & Transportation*, and *Socio-Economic Planning Sciences*. Besides, he has published in journals included in the Emerging Sources Citation Index, such as *Benchmarking: An International Journal* and *Transportation Research Procedia*.

Raghavan Narayanan has over 25 years of banking experience, spanning project financing, asset management, and financial advisory services, and brings unique private sector development insights into the evaluation sciences. At the European Bank for Reconstruction and Development, he manages the Impact Assessments and Foresights division building evidence-based insights and foresights to support the EBRD's Banking activities. Prior to the EBRD, he was with the World Bank Group's oversight unit, IEG, and specialized in thematic, independent evaluations and strategic corporate initiatives. He was tasked with managing the division's flagship reports and coordinating on the Mobilizing Finance for Development (FfD) agenda. His recent research on private sector development and investor behaviors is published as a chapter in the *Global Handbook on Impact Investing* (Wiley, December 2020). Before joining the World Bank Group, Raghavan was an economist and policy advisor for the energy and extractives industry and has rotated through various positions in venture capital and private equity investment funds. He has completed executive programs at Harvard University's Kennedy School of Government (USA) on economic development and trained at Carleton University (Canada) on evaluation methodologies. He holds an MBA from Georgetown University (USA) and a BEng from Nanyang Technological University (Singapore).

Ram Pratap Sinha is an associate professor of economics with more than 32 years of teaching experience at the undergraduate and postgraduate levels. He is posted at the Government College of Engineering & Leather Technology. He has 115 journal publications to his credit and has contributed to 20 edited volumes. Besides, he has written two books. He is a reviewer for many reputed international and national journals in the areas of economics, finance, and marketing. He has also supervised several PhD and MPhil theses in the areas of economics and marketing. He has taught at numerous FDPs and workshops in the areas of data envelopment analysis and econometrics.

Seyed Mojtaba Sajadi holds a PhD in industrial engineering and systems management from Amirkabir University of Technology, Tehran, Iran. He received his MSc and BSc in industrial engineering from the University of Tehran and Sharif University of Technology, respectively. Mojtaba was working as an assistant/associate professor at the University of Tehran, Iran, and Coventry University, UK, for 12 years. He was a visiting researcher at the University of Antwerp, Belgium, in 2010. He joined Aston University in January 2023, and he is now working in the Operations and Information Department. The focus of his research is on simulation-based optimization in supply chain management and business analytics. He has extensive expertise in different sorts of simulation, including discrete-event simulation

(DES), agent-based simulation (ABS), system dynamics (SD), Monte Carlo, and hybrid simulation, and their applications in different fields of research studies, such as project management, healthcare, manufacturing systems, supply chain, disaster management, banking and insurance, ICT, and after-sales services. Besides, he has a solid background in optimization, business analytics, and using mathematical modeling, metaheuristic algorithms, and machine learning techniques for developing effective simulation-based optimization approaches in order to address complex real-world problems.

Sıdıka Ece Yılmaz is an instructor at Adana Alparslan Türkeş Science and Technology University (Türkiye). She holds a master's degree in management and organization (2018) and PhD in the Business Administration (2023). She has recently published some academic papers in national and international refereed journals. She has also taken part in the European Union projects. She also conducts career studies and career events at the Career Center of Adana Alparslan Türkeş Science and Technology University.

Supramono Supramono is a professor of finance at the Department of Management, Faculty of Economics and Business, Universitas Kristen Satya Wacana, Indonesia. His research interests include behavioral finance, capital markets, and corporate finance. He has published several articles in reputable international journals.

Tooba Lutfullah holds bachelors degree in accounting and finance. She possesses expertise in the research areas of corporate sustainability, corporate governance, financial markets, fintech and corporate finance. She achieved first position in the short course competition of Global Leadership Week (GLW) 2021, Realizing Our SDG's: Unveiling Local Agility and SME's Resilience conducted by the Faculty of Economics Universitas Negeri Malang, Indonesia. This provided high experience and polished communication skills with team members and presentation for the competition.

Her publications include Board gender diversity and the financial performance of Pakistani non-financial firms: A fuzzy-set QCA approach, published in *The Journal of Corporate Accounting and Finance* — 2023; Survival of the fittest: Do firms actively or passively learn survival?, published in *Journal of Economic and Administrative Sciences* — 2023; Volatility forecasting of crude oil, gold, and silver futures: A case of Pakistan mercantile exchange, published in

WSEAS *Transactions on Business and Economics* — 2023; International financial reporting standards and firm's value relevance in textile sector of Pakistan, published in *Asian Finance Research Journal* — 2023; Relationship between corporate sustainability and return on invested capital of Pakistani firms, published in *Pakistan Journal of Social Research* — 2022; Testing daily seasonality using value premium portfolios returns: The case of Pakistan, published in *Audit Accounting Review* — 2022; The moderating impact of political intervention on the relationship between corporate governance and firm performance, published in *Asian Finance Research Journal* — 2022.

Usil Sis Sucahyo is an assistant professor in accounting at the Department of Accountancy Faculty of Economics and Business, Universitas Kristen Satya Wacana, Indonesia. His research interests include behavioral accounting and management control systems. He has published several articles in reputable international journals.

William Chung is an associate professor of management sciences at the City University of Hong Kong. He earned his PhD in management sciences at the University of Waterloo, Canada. His personal research interests mainly focus on developing mathematical methodologies for energy-environmental policy problems, such as large-scale equilibrium models, benchmarking methods for the energy consumption performance of buildings, and decomposition analysis of energy intensity. His papers can be found in the following journals: *Operations Research*, the *European Journal of Operational Research (EJOR)*, *Computational Economics*, *Energy Economics*, *Energy Policy*, *Energy*, *Applied Energy*, and *Energy and Buildings*. In addition, he is the director and founder of the Energy and Environmental Policy Research Unit at the City University of Hong Kong.

Yong-Tong Chen is an assistant professor of information management at Beijing Normal University (Beijing, China). She holds a PhD from the Hong Kong Polytechnic University in Hong Kong (2017). She has recently published many academic papers in international refereed journals, including the *International Journal of Production Research*, *Robotics and Computer-Integrated Manufacturing*, and *Mathematics*.

https://doi.org/10.1142/9781800615786_fmatter

Contents

https://doi.org/10.1142/9781800615786_fmatter

List of Abbreviations

AE	Allocative economic
App. of Biotechnology	Biotechnology applications
BB	Bio-business
BCC	Banker, Charnes, and Cooper Model
BX	Biosolid Xolotlan
CCR	Charnes, Cooper, and Rhodes Model
CIRCB	Center for Research and Reproduction of Biological Controllers
CNU	National Council of Universities
Cordoba	Nicaraguan National Currency
CRS	Constants returns to scale
CRSTE	Constants returns to scale technical efficiency
DDD	Dichlorodiphenyldichloroethane
DDT	Dichlorodiphenyltrichloroethane
DEA	Data envelopment analysis
DEAP	Program for data envelopment analysis
DMU's	Decision-making units (farms)
DRS	Decreasing returns to scales
DS	Dose
ECAV	School of Agricultural and Veterinary Sciences
ECOOALPHAA	Private company of agricultural inputs
EE	Economic efficiency
ENACAL	Nicaraguan Company of Aqueducts and Sanitary Sewers
ES	Efficiency of scale

F.CC.MM	Faculty of Medical Sciences
FF.CC.QQ	Faculty of Chemical Sciences
FAREM	Multidisciplinary Regional Faculty
FARQ	Architecture Faculty
FDR	Rural Development Faculty
FEC	Electrical Engineering Faculty
FIQ	Chemical Engineering Faculty
FTC	Science and Technology Faculty
GL	Gallon = 4.546092 liter
IRS	Increase returns to scale
LR	Likelihood ratio
Mz	1 Manzana [Nicaragua] = 0.7044 hectare
NIRS	Non-increasing returns to scale
NPV	Nuclear polyhedrosis virus
QQ	Quintal = 100 pounds (1 pound = 0.45kg)
SFA	Stochastic frontiers analysis
TE	Technical efficiency
TFP	Total factor productivity
UNA	National Agrarian University
UNAN-León	National Autonomous University of Nicaragua, León
UNAN-Managua	National Autonomous University of Nicaragua, Managua
UNI	National Engineering University
VRS	Variable returns to scale
VRSTE	Variable returns to scale technical efficiency

Part I

Data Envelopment Analysis (DEA) in Business

https://doi.org/10.1142/9781800615786_0001

Chapter 1

Toward Healthcare Improvements Using Data Envelopment Analysis: The Case of Emergency Medical Services

Adel Hatami-Marbini[*,§], **Nilofar Varzgani**[†,¶], and **Seyed Mojtaba Sajadi**[‡,||]

[*]Department of Accounting, Finance, Logistics and Economics, Huddersfield Business School, University of Huddersfield, Huddersfield, UK

[†]Department of Business Systems and Analytics, School of Business, La Salle University, Philadelphia, PA, USA

[‡]Operations and Information Management Department, Aston Business School, Aston University, Birmingham B4 7ET, UK

[§]a.hatamimarbini@hud.ac.uk

[¶]varzgani@lasalle.edu

[||]s.sajadi@aston.ac.uk

Abstract

The healthcare system is no stranger to resource challenges in the face of unlimited demand to fulfill healthcare objectives of satisfying patients, maintaining service quality, and maximizing profit. An emergency medical services (EMSs) system plays a crucial role in stabilizing and transporting seriously injured patients to hospitals within healthcare systems. The EMS function is influenced by several criteria, such as call rate, traffic condition, setup, and operating costs. Therefore, the optimal design of EMS systems, including determining the location of emergency medical bases and allocating ambulances, helps improve service performance. This chapter explains the methodology and empirical results of a mathematical modeling and simulation-based optimization approach aimed at identifying the optimal location of emergency medical centers and assigning ambulances to the selected centers to maximize survival rates and minimize the total cost of the EMS system. A case study of the city of Isfahan in Iran is presented to demonstrate the applicability and efficacy of the proposed approach. The simulation-based optimization model was implemented in four selected municipal regions of Isfahan to obtain an appropriate design for emergency center locations and ambulances allocation with three types of patients (classified by the urgency of help required) and two types of ambulances. Six scenarios were defined to simulate the model in a dynamic environment and measure the survival rate and total cost of each scenario. In view of the survival rate and costs, data envelopment analysis (DEA) was then used to rank scenarios and select the best ones. The patient type was found to have a significant effect on the DEA rankings of the different input scenarios. An analysis across scenarios showed that adding portable stations in the regions that have the highest percentage of urgent patient calls can help increase the survival rate at a lower cost.

Keywords: Emergency medical services, simulation-based optimization, healthcare system design, service operations and improvement, survival rate, resource allocation, stochastic models, public policy efficiency, heterogenous patients, dual-objective optimization.

1. Introduction

Over the past few years, overall healthcare costs have shown a significant increase all across the globe. In the US alone, after adjustments for price inflation, annual healthcare spending on inpatient, ambulatory, retail pharmaceutical, nursing facility, emergency department, and dental care increased by $933.5 billion between 1996 and 2013, from $1.2 trillion to $2.1 trillion (Dieleman *et al.*, 2017). The complexity of healthcare systems poses the challenge of managing scarce resources to meet growing patient needs. For instance, take the example of Dublin, the capital of the Republic of Ireland, where the estimated cost of supplying emergency ambulance services was over £28 million in 1996, when handling almost 300,000 patient journeys, in which there were nine emergency ambulance fleets with 272 emergency vehicles located at 89 ambulance stations. In addition

to securing prehospital emergency medical treatment and the transportation of the utterly unwell and injured patients to the hospital, ambulance fleets transport patients between hospitals for urgent or planned treatment (Dept. of Health and Children, Ireland, 1997). In contrast, the 2015 Operational Plan for Ireland's National Ambulance Services (NAS) took place within the context of a gross budget amounting to €144 million (National Ambulance Service, 2015). In the same operational plan, they projected service activities for 2015, including (but not limited to) responding to roughly 308,000 calls, with 79,000 categorized as those with the highest acuity level, utilizing a fleet of 500 vehicles, facilitating the transportation of approximately 4,800 intermediate care patients, conducting training for 50 paramedics, up to 40 advanced paramedics, and a minimum of 17 control staff.

Emergency service providers are striving hard to meet quality standards and ensure patient safety. Such service providers are continuously monitoring a range of activities, performance indicators, and clinical outcomes and have to remain open to learning and changes in the light of performance outcomes and review feedback. It is hence essential to adopt the most efficient approaches to resource allocation to achieve this goal.

Emergency medical services (EMSs) are an indispensable part of any healthcare system that aims at providing prehospital emergency medical care. EMSs include the services of supervisors, managers, directors, administrators, and coordinators. When solving today's toughest EMS challenges amidst rising demands, we need to take into consideration not only the complexity of the model but also the tremendous value of this infrastructure to the community. On the frontlines of national disasters and health crises, EMSs play a crucial role in preventing deaths and injuries. With the shift in community needs as a result of changing lifestyles and increased life expectancy, requests for EMSs have been on the rise. The provision of high-quality EMSs is expected to be consistently delivered at the national level. Healthcare managers constantly struggle with competing budget demands and staffing challenges, making the delivery of adequate service levels even more difficult. This complexity of EMS systems calls for resource allocation approaches that are capable of optimizing productivity and efficiency.

This chapter is organized as follows: Section 2 covers the literature review of previously studied models for EMSs and the use of data envelopment analysis (DEA) in this field. In Section 3, the authors describe their proposed modeling methodology and its underlying assumptions. Section 4 presents a case study of the city of Isfahan using this proposed methodology, and Section 5 concludes this chapter with some remarks.

2. Literature Review

Some researchers have minutely studied healthcare management problems to establish the extent to which emergency ambulance services are provided efficiently and economically. Accordingly, models have been proposed to enable the analyzing and understanding of the insider problem from theoretical and practical perspectives, e.g., see White (2007).

The primary duty of EMSs is to offer urgent vital services and transport patients to a specific hospital or clinic. Many studies have been carried out to seek the optimal locations of facilities, leading to the performance improvement of the EMS system. Despite a wide variety of developed models, their key purpose is to optimally assign emergency facilities to required services so that the desired outcomes are achieved. Examples include studies that cover topics such as the location set covering problem (LSCP) developed by Toregas *et al.* (1971) and the maximal covering location problem (MCLP) proposed by Church & Revelle (1974). Gendreau *et al.* (2006) reconsidered the MCLP as a dynamic relocation strategy to seek new locations for the EMS centers to maximize the coverage level requested for ambulances in the Montreal region. Nevertheless, many related research projects have neglected the real conditions of EMS systems, resulting in their findings being inapplicable (Li *et al.*, 2011; Goldberg, 2004). Hence, to make the models more practical, the stochastic and dynamic nature of EMSs needs to be considered when modeling a system. Galvão & Morabito (2008) attempted to do so by developing stochastic models based on queueing theory, in which ambulances play the role of servers. Ingolfsson *et al.* (2008) extended a model to maximize the average coverage of requests by using the least number of ambulances allocated to patients when the time delay and travel time are assumed to be uncertain.

The existing models for EMS management pose complex, dynamic, and contemporary challenges. This necessitates analyses of the efficiency of an EMS system with distinctive properties through simulation, which is based on a computerized model (Günal & Pidd, 2010). McCormack & Coates (2015) formulated a mathematical programming model for ambulance fleet allocation and base EMS locations using a simulation model and a genetic algorithm. Compared to other techniques, simulation has certain advantages, including the consideration of real characteristics and the improvement of the model's validity. Mathematical modeling and simulation methods were originally used by Toro-Díaz *et al.* (2013) to improve the logistics performance of EMS systems. In their study, they described the application of lean principles in the logistics and healthcare industries and discussed the conceptual development of the lean-theory of constraint (TOC) approach.

With special emphasis on EMSs, emergency station location, ambulance allocation, and facility management are notable subjects for healthcare studies (Goldberg, 2004). The main objective of facility location and resource allocation for EMSs is to cover all regions of a city, along with providing high service quality at a low cost. The number of requests for EMSs depends on working days per week and working hours per day. This is due to the feasibility of improving the efficiency of the EMS system by optimally relocating the emergency stations and allotting ambulances to them. Brotcorne *et al.* (2003) reviewed the state-of-the-art models for emergency facility location and service allocation over the past three decades. Generally, location and allocation models in relation to EMSs can be categorized into two major groups: (i) deterministic models and (ii) stochastic models.

The first group lays emphasis on deterministic mathematical programming models to optimize the management of facilities in a static environment, whereas in the second group, scholars develop stochastic models with uncertain parameters under dynamic conditions to cover the largest possible demand and optimally allocate resources. One of the primary stochastic models was the maximum expected coverage locating problem (MEXCLP) proposed by Daskin (1983). Daskin's study aimed to meet as large a demand as possible with the minimum number of facilities (vehicles). Revelle & Hogan (1989) proposed an optimization model to minimize the number of required ambulances in order to serve the maximum requests with sufficient reliability. Gendreau *et al.* (2001) focused on the redeployment problem for a fleet of ambulances in the context of EMSs. The authors tried to address it by developing a dynamic method to maximize the covered regions with the minimum response time via a parallel tabu search heuristic and simulation.

Given the fluctuating demand for ambulances, Rajagopalan *et al.* (2008) developed a multi-period model to identify the minimum number of ambulances and their locations over time, in which coverage requirements are met with a predefined reliability level. The experimental findings of this study and a simulation model showed that tabu search gives fast and near-optimal solutions for maximizing coverage problems. A tabu search-based heuristic has been developed by Toro-Díaz *et al.* (2015) for large-scale EMS systems, in which the model contains both strategic (location) and operative (dispatching) decisions to balance efficiency and fairness. Rajagopalan & Saydam (2009) proposed a model for allocating a set of ambulances to minimize expected response distances using a heuristic search algorithm while fulfilling coverage requirements. Schmid & Doerner (2010) considered stochastic demand and multiple demand points that were assumed to be covered by more than one vehicle. Their multi-period model allows ambulances to be transported between stations with the objective of maximizing coverage at various points simultaneously over the planning horizon.

Incorporating dynamic and stochastic settings in the models to optimize EMS systems is clearly valuable for improving the practical implementation of these systems by offering the system providers high service quality in terms of response time and coverage level. These models are normally referred to as *NP* and *NP-hard complexity* classes (see, e.g., Saydam *et al.*, 1994; Gendteau *et al.*, 1997, 1999). To simplify this complexity, the simulation approach has been extensively used in the literature. Simulation is a computerized approach to imitate real-world operational processes and systems in a dynamic environment, and it differs from the analytical approach, in which the system analysis is theoretical. Simulation has been widely used in healthcare applications. For instance, Sah *et al.* (2017) exploited the combination of simulation and goal programming for total system improvement in an Indian hospital, leading to a reduction in delays and bottlenecks throughout the hospital processes. In their research, they specified several criteria, including wait time of patients and resource utilization, to achieve the goal of analyzing and improving the system. Simulation helped them minimize the total cost of the hospital, subject to an adequate allocation of hospital staff and beds.

Abouelјinane *et al.* (2013) provided a survey of simulation models applied to EMS problems. They classified the decisions that affect the performance of EMS systems at the design and operational levels into three categories: long-term decisions, mid-term decisions, and short-term decisions. The long-term decisions include decisions about the identification of suitable skills, the number of human resources, and the location of the central EMS station. For example, Harewood (2002) proposed a multi-objective model to identify EMS locations and the total number of ambulances required to meet the service requirement. The mid-term decisions consist of those concerned with shift scheduling, the total number of EMS stations and their resource requirements, and scheduling resources such as rescue teams and ambulances. For example, Goldberg *et al.* (1990) developed a simulation model to schedule vehicles allocated to EMS centers. The short-term decisions are those concerned with dispatching rescue teams in order to improve their efficiency, choosing a suitable hospital or clinic for every patient, and determining redeployment strategies such as EMSs relocation in order to offer better services. For example, Peleg & Pliskin (2004) presented a redeployment strategy to reduce response time. They used a geographic information system (GIS) to determine the travel time between EMS locations and demand points, as well as between demand points and hospital locations.

In the literature, EMS performance is broadly assessed based on three key factors: timeliness, survival rate, and cost. Hence, these factors have been considered in most studies for improving a system's performance. Wang *et al.* (2012) developed an agent-based simulation of response to a disaster to study the assignment policies of victims to hospitals based on available GIS and response resources, such as ambulances and hospital beds. Nogueira *et al.* (2016) applied both optimization and simulation techniques to study the EMSs of Belo Horizonte, Brazil. Their optimization model aimed to identify the locations of ambulance bases and allocate ambulances to those bases, and a simulation was run to take the dynamic behavior of the system into account.

Van Barneveld *et al.* (2016) studied the effect of ambulance relocations on the performance of ambulance service providers.

They modeled ambulance relocation from the current arrangement to the target arrangement by way of a linear bottleneck assignment problem. In their developed model, the performance of the ambulance service provider was measured using a general penalty function. Zaffar *et al.* (2016) carried out a comparative study to contrast the performance of ambulance location models based on four criteria: percentage of calls covered, survivability, average response time, and workload balance among the fleet. To this end, they used a simulation optimization approach to compare the performance of three EMS location models, namely maximum coverage, minimum average response time, and maximum survivability. They eventually showed that the maximum survivability objective is more efficient than both response time and coverage criteria. Fritze *et al.* (2018) proposed an integrated model of spatial information and integer programming for the EMS location problem. This model applies the MCLP to ensure that all residents are covered by EMSs at minimum cost. Andersson *et al.* (2020) used the maximum expected performance location problem for heterogenous regions (MEPLP-HR) developed by Leknes *et al.* (2017) to analyze both strategic (locating ambulance stations) and tactical (allocating ambulances to the stations) decisions over multiple periods.

Ünlüyurt & Tunçer (2016) employed four different mathematical models to determine EMS station locations and their maximum coverage. Additionally, they evaluated the performance of EMS location models via discrete event simulation. Eventually, using Istanbul data, they showed that a simulation-based evaluation methodology can give a fair framework to assess the effectiveness of models.

Aringhieri *et al.* (2016) reviewed recent studies of EMS systems to define ongoing challenges for future research avenues. Ahmadi-Javid *et al.* (2017) surveyed articles relevant to healthcare facility (HCF) location problems and classified the models into two categories: non-emergency and emergency facilities. Furthermore, they considered different perspectives on HCF locations, such as location and allocation model for developing health services (see, e.g., Oliveira & Bevan, 2006), ambulance location and relocation models (see, e.g., Erkut *et al.*, 2008), location of HCFs from modeling aspects (see, e.g., Syam & Côte, 2010), emergency response facility location (see, e.g., Iannoni *et al.*, 2009), methodological advancement in healthcare accessibility (see, e.g., Brooks *et al.*, 2013), home healthcare logistics (see, e.g., Bricon *et al.*, 2005), and an overview of planning and management of EMSs (see, e.g., Li *et al.*, 2011). Liu *et al.* (2019) used a robust optimization method for optimizing an EMS system. Their objective was to minimize the total cost of an EMS system based on station construction and the locations, allocation, procurement, and maintenance of ambulances. Boujemaa *et al.* (2020) proposed a multistage stochastic programming model for ambulance redeployment planning. They considered two types of ambulances and two sets of calls for requesting ambulances. Since their model possesses a high degree of complexity, the heuristic method was used for solving the developed model.

Recently, Bélanger & Edmonston (2018) provided a broad overview of studies relevant to vehicle location and relocation, as well as dispatching decisions in the context of ambulance fleet management. First, they grouped the studies on static ambulance location models into three classes:

(1) single-coverage deterministic models (see, e.g., Galvãoa & ReVelle, 1996);
(2) multiple-coverage deterministic models (see, e.g., Liu *et al.*, 2014);
(3) probabilistic and stochastic models (see, e.g., McLay, 2009), aiming to address ambulance location problems at the tactical level.

Second, they reviewed the most recent approaches by classifying them into three classes:

(1) stochastic and robust location–allocation models (see, e.g., Beraldi *et al.*, 2004);
(2) maximal survival models (see, e.g., McLay & Mayorga, 2010);
(3) equity models (see, e.g., Chanta *et al.*, 2011).

Third, they examined multi-period relocation models and dynamic relocation models in ambulance location/relocation problems (see, e.g., Başar *et al.*, 2011). Finally, they reviewed research on dispatching decisions for allocating vehicles to an emergency call.

Measuring the performance of healthcare systems is an essential process that comes with some difficulties and challenges and results in identifying weaknesses and inefficient sources. DEA is a well-known non-parametric method for assessing the relative efficiencies of a group of decision-making units (DMUs). Farrell (1957) originated many of the ideas and principles underlying DEA. After a lengthy interval, Charnes *et al.* (1978) built on the thought-provoking seminal work and introduced a powerful DEA methodology to assess the relative efficiencies of multi-input, multi-output DMUs. Since the emergence of DEA, there has been significant growth both in theoretical developments and applications (Emrouznejad *et al.*, 2008; Hatami-Marbini *et al.*, 2011; Cook & Seiford, 2009). Liu *et al.* (2013) conducted a literature survey on DEA applications, and their results show that on the whole, around two-thirds (63.6%) of DEA papers embed empirical data, while the remaining one-third is purely methodological. They also found that among the multifaceted applications, the top-five industries addressed are banking, healthcare, agriculture and farming, transportation, and education. The applications that showed the highest growth momentum recently are energy and environment, as well as finance. In finance, most recent studies have been conducted with the use of DEA for performance measures to assess productivity and efficiency on different fronts. For instance, Boubaker *et al.* (2022) evaluated the performance and efficiency of 49 Islamic banks across 10 countries during 2019–2020 to assess how those banks can preserve their performance and remain resilient in the aftermath of the COVID-19 pandemic. Using the conventional inverse DEA (InvDEA) approach, they showed that because of reductions in their outputs, 31 out of the 49 banks studied would need to reduce their inputs so that their efficiency could remain unchanged. Another example is Boubaker *et al.* (2021), who employed parametric and non-parametric frontier approaches (DEA) to estimate the productive efficiency of firms in the case of large shareholders and control contestability. They found that private benefits of control prevent large shareholders from favoring productive projects. Similarly, many recent studies have unveiled the benefits of using the DEA methodology in identifying sources of inefficiency, ranking the DMUs, evaluating management, evaluating the effectiveness of programs or policies, creating a quantitative basis for reallocating resources, etc. Consider the example of the study conducted by Tu *et al.* (2021), where they investigated the interrelationships between the development of fintech credit and the efficiency of banking systems in 80 countries from 2013 to 2017. The findings indicate a two-way relationship between them. The authors conclude that fintech credit should be encouraged by authorities around the world.

DEA has also been used to assess the different aspects of healthcare systems, such as hospital efficiency (Athanassopoulos & Gounaris, 2001), public policy efficiency (Coppola *et al.*, 2003; Miller & Adam, 1996), heart surgery efficiency (Chilingerian, 1995), and HCF efficiency (Hollingsworth, 2008; Ferrier *et al.*, 2006; Nayar & Ozcan, 2008). Golabian *et al.* (2022) conducted a study to obtain the best return strategy for ambulances to maximize the expected coverage concerning a predefined dispatch policy. They proposed a hypercube queuing model to maximize customers' coverage probability, in which the locations of busy ambulances in each state are not known and approximated based on customer arrival rates.

There are some studies which extended the analysis to either predicting demand for ambulances or making location decisions for temporary emergency service centers during disasters. Grekousis & Liu (2019) put forth a novel approach by attempting to predict the demand for ambulances in EMSs using artificial intelligence. They proposed a three-level model to first predict future demand and then applied a location–allocation model to site ambulances prior to an actual emergency occurrence. They also used a case study based on data from Athens, Greece, on actual emergency events to validate their model. Karatas & Yakıcı(2021) proposed a multi-objective facility location analytics model for determining the number and locations of temporary emergency service centers for a regional natural gas distribution company in Turkey. While their work is important in natural disasters and other extreme events, it may not directly be applicable to meeting the existing demand for EMSs outside of these events.

DEA, developed by Charnes *et al.* (1978), is an exceedingly endorsed and powerful method for measuring the performances of the public and private sectors (Eilat *et al.*, 2008; Enrouznejad *et al.* 2008). Since the mid-1980s, there has been increasing interest in the application of DEA to healthcare problems. The literature review conducted by Hollingsworth *et al.* (1999) examined 91 DEA-based studies on the efficiency and productivity of healthcare systems from both theoretical and practical perspectives. Their review is divided into two groups of studies: the deployment of DEA to measure the efficiency and productivity of hospitals, e.g., Burgess & Wilson (1995), and general healthcare, e.g., Färe *et al.* (1992). Hollingsworth (2003) built on the earlier survey and provided an overview of 188 related studies on non-parametric and parametric efficiency measurement in healthcare and health. Chilingerian & Sherman (2004) reviewed the DEA literature, focusing on the efficiency measurement of healthcare providers, such as general hospitals, academic medical centers, nursing homes, and physicians. They particularly discussed DEA models applied to healthcare and compiled a list of inputs and outputs based on the extant research literature.

Although many research studies have been done to assess EMS performance, the pertinent literature has paid less attention to system cost as a vital assessment factor. Considering real-life conditions can help us derive more practical and reliable solutions to EMS problems. Although some recent EMS studies, such as Nickel *et al.* (2016) and Boujemâa *et al.* (2019), tried to consider real conditions in their models, they neglected some important conditions, such as weather and traffic.

3. Model

This chapter presents a simulation-based optimization approach originally published by the authors (Hatami-Marbini *et al.*, 2022) built on the maximization of the survival rate and minimization of the total cost. The model was purposely developed for a project at the Emergency Management Centre in Isfahan, a central province in Iran, to improve its service performance.

Figure 1 displays the processes of patient rescue in the EMS system. This figure defines all the steps of the rescue process necessary to make a valid model for emergency locations and allocating ambulances.

Response time is defined as the time elapsed from a call received by a call center to the time when an ambulance arrives at the patient's location. Response time is a key factor and has a direct effect on the survival rate in such a way that its decrease leads to a rise in the survival rate. That is, a quick EMS response is essential to improving survival rates and EMS performance.

The model presented in this chapter was built on the work of Knight *et al.* (2012). It expands on their work by adding cost

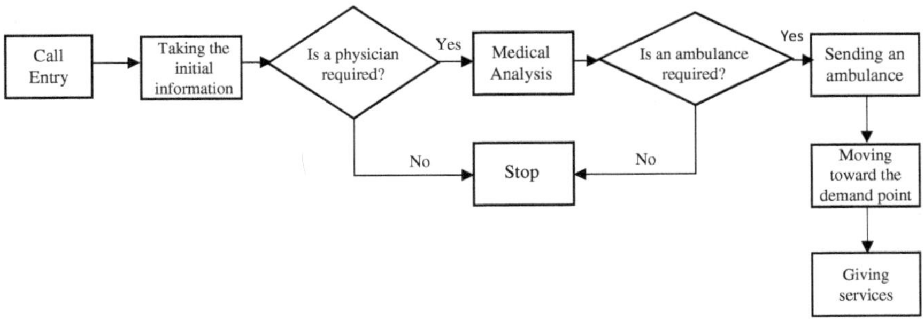

Figure 1. Rescue processes for patients.

Note: This figure displays the processes of patient rescue in the EMS system. This figure defines all the steps of the rescue process necessary to make a valid model for emergency locations and allocating ambulances.

minimization as an objective function, thus enabling managers to improve the total performance of an EMS system. Contrary to Knight *et al.* (2012), we assume two different types of ambulances and the establishment cost of EMS centers, as well as considering weather and traffic as stochastic conditions for EMS location and allocation. The results calculated from the optimization model are considered the primary scenario to formulate our simulation model. Other scenarios build on the primary scenario by altering variables such as the number and location of EMSs. In addition, the stochastic conditions of emergency locations and their allocated ambulances are considered in the simulation model. To compare and prioritize the results obtained from simulation, the DEA method is applied on the basis of two predetermined objectives: survival rate and total cost. To summarize, the theoretical and practical contributions of the framework developed in this study are threefold:

(1) A dual-objective optimization model is formulated to design an EMS system by maximizing survival rates and minimizing the total cost simultaneously. The mathematical model culminates in the optimal number and location of required EMSs as well as the optimal number of ambulances allocated to the selected EMS locations.
(2) A computer simulation analysis is applied to take the dynamic conditions of the EMS system into account in order to reach an appropriate response.

(3) Finally, we draw on the dataset from the Emergency Management Centre in Isfahan to illustrate the applicability and efficacy of the proposed framework. The results show that the implementation of our proposed framework leads to an improved survival rate.

In the following sections, the maximal expected survival location model for heterogenous patients (MESLMHP) and DEA models are reviewed and formulated (Knight *et al.*, 2012; Charnes *et al.*, 1978).

3.1. Maximal expected survival location model for heterogenous patients

The survival function plays a key role in locating the ambulance stations nearer to areas with high demand for EMSs since the possibility of survival decreases exponentially with increasing response time. As a significant contribution, Erkut *et al.* (2008) proposed EMS location models that included a survival function in the existing coverage models. They showed that more traditional coverage models suffer from certain weaknesses in practice and a survival-maximizing approach enables the decision-maker to obtain reliable and robust solutions. "A survival function is a monotonically decreasing function of the response time of an emergency medical service (EMS) vehicle to a patient that returns the probability of survival for the patient" (Erkut *et al.*, 2008, p. 42). Rather than designing the network location model, Knight *et al.* (2012) developed the MESLMHP based on Erkut *et al.* (2008) to improve the performance of an extant EMS system. In this respect, MESLMHP aims to maximize the overall expected survival probability of different types of patients. In rough terms, several different types of survival probabilities can be defined based on patient types. The patient types can be defined on the basis of time standards (for instance, type I patients are those who can be reached by an ambulance within eight minutes) or by observing various patient conditions with respect to a relative survival curve.

Assume that there are m ambulance stations and n demand locations. Patients are categorized into k different patient types on the basis of severity level and survival probability. Let λ_i^l indicate the demand of patient type l $(l = 1, 2, \ldots, k)$ from demand location

Table 1. List of notations used in mathematical formulation of the MESLMHP.

Notations		Description
Indices	i	Index for demand locations ($i = 1, 2, \dots, n$)
	l	Index for patient types ($l = 1, 2, \dots, k$)
	j	Index for ambulance stations ($j = 1, 2, \dots, m$)
	λ_i^l	Demand of patient type l from demand location i
Parameters	π_j	Average utilization of ambulance at station j
	Z	Total number of available ambulances
	ρ_{ij}	jth preferred station for the ith demand point
	t_{ij}	Travel time between station j and demand node i
	w_l	Relative importance of patient type l
	$s_l(t_{ij})$	Probability degree of survival associated with patient type l for time t_{ij}
Decision variables	x_j	Number of ambulances allocated to emergency station j

i ($i = 1, 2, \dots, n$). The average utilization of ambulance type h at station j is denoted by π_j. The mathematical notations are provided in Table 1. In view of π_j, the probability of ambulance availability is indicated by $(1 - \pi_j)$. The preference matrix $\rho = [\rho_{ij}]_{n \cdot m}$ is used to ensure that an emergency station is allocated to a demand point. We point out that if the jth emergency station has no available ambulance, the $(j + 1)$th station will be selected. Furthermore, there can be two general survival functions, where the first function is determined on the basis of severity and response time target, while the second one is based on considering different patient conditions with respect to survival probabilities. Note that a combination of both functions can be observed in some circumstances. Figure 2 displays three feasible survival functions. The two dotted lines represent the survival functions, which show that the survival probability decreases with time, so a quick response from EMSs is essential. Another function depicted by a solid line in Figure 2 is a step function that is used to consider rigid targets, such as defined service norms. Let $s_l : \mathbb{R} \to [0, 1]_{\mathbb{R}}$ be a survival function for each patient type l, and let t_{ij} be the travel time between station j and demand node i. There might be l different survival functions, which are reliant on the number of patient types, and the importance weights w_l, are allocated to the lth patient.

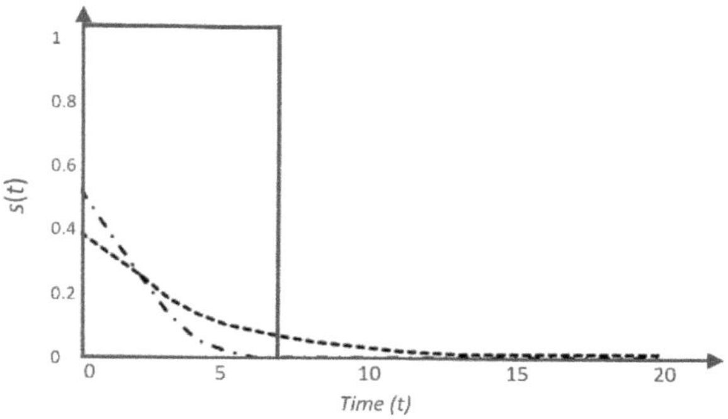

Figure 2. Survival functions.

Note: This figure shows that the survival probability decreases with time, and quick action of EMS is essential.

The MESLMHP model is given as follows:

$$\max \sum_{l=1}^{k} w_l \sum_{i=1}^{n} \lambda_i^l B_{i,\rho_{ij}}^l, \tag{1}$$

$$\text{s.t.:} \quad \sum_{j=1}^{m} x_j = Z, \quad x_j \in \mathbb{Z}^{0+}, \quad j = 1, 2, \ldots, m,$$

where $B_{i,\rho_{ij}}^l = \sum_{j=1}^m s_l(t_{i,\rho_{ij}}) \underbrace{(1 - \pi_{\rho_{ij}}^{x_{\rho_{ij}}})\Pi_{r=1}^{j-1}\pi_{\rho_{ir}}^{x_{\rho_{ir}}}}_{\phi}$ indicates the probability (expected degree) of survival of a patient of type l from demand node i, with ϕ representing the probability of at least one ambulance located at j service node being free, while ambulances located at nodes 1 to -1 are busy. The equality constraint set of model (1) shows that Z ambulances are allocated to all EMS stations to maximize the weighted sum of all demand nodes in terms of each probability function. To find the actual utilization π_j, an iterative approach is developed on the basis of queuing theory, where each EMS center can be modeled as an $M_j/M_j/x_j$ queue. The iterations are stopped after an allocation of ambulances matches the demand and the utilization.

The MESLMHP includes the following six phases (Knight *et al.*, 2012):

Phase 1: Estimate the mean service rate at station j (μ_j).
Phase 2: Set the average utilization of the jth ambulance station for the initial iteration ($\pi_j^{(0)}$).
Phase 3: Find x_j for all centers by solving the MESLMHP problem.
Phase 4: Compute the demand distribution $\Lambda_j^{(k)}$ for the jth ambulance station.
Phase 5: Obtain π_j based on the queuing model for $M_j/M_j/x_j$ which ensures the allocation calculated in Phase 3.
Phase 6: Iterate Phases 3, 4, and 5 until convergence is achieved. Since this method does not guarantee convergence, the model runs for some predetermined number of iterations or time. The steady-state graphs for cost or survival can be used to specify the convergence conditions. A steady state is a case where all state variables remain constant.

Our book chapter aims to consider the total costs of the EMS system installation, e.g., building and equipment costs. Therefore, our model is a bi-objective optimization model to design an emergency medical system in which the first objective function aims to maximize the survival rate and the second one aims to minimize the total cost of the EMS system. Notably, the simulation method is used in this chapter to seek the value of $x_j (j = 1, 2, \ldots, m)$. In addition, the demand rate of each demand point is extracted from the data calls.

3.2. *Data envelopment analysis*

DEA is a well-established approach for evaluating the relative efficiency of a group of functionally similar DMUs (e.g., institutions, banks, hospitals, and hotels) that transforms multiple inputs (resources) into multiple outputs (goods and services) (Charnes *et al.*, 1978). DEA builds on the foundation of "total weights flexibility". The original DEA model proposes that the efficiency of a DMU is given by the maximum of the ratio of weighted outputs to weighted inputs, subject to the constraint that the same ratio for all the DMUs must be at most equal to one. Consider n DMUs, in which each DMU consumes m various inputs to produce s various outputs.

Let $x_{ij}, i = 1, \ldots, m$ and $y_{rj}, r = 1, \ldots, s$ denote the input and output vectors of DMU_j, $j = 1, \ldots, n$, respectively. Charnes *et al.* (1978) proposed the first DEA-CCR model under constant returns to scale (CRS), followed by the DEA-BCC model (Banker *et al.*, 1984) that hypothesizes variable returns to scale (VRS). This chapter's focus is on the following VRS model to measure the technical efficiency of the DMU_o:

$$\max \rho = \sum_{r=1}^{s} u_r y_{ro} + u_o, \tag{2}$$

$$\begin{aligned} \text{s.t.:} \quad & \sum_{i=1}^{m} v_i x_{io} = 1, \\ & \sum_{r=1}^{s} u_r y_{rj} - \sum_{i=1}^{m} v_i x_{ij} + u_o \leq 0, \quad j = 1, 2, \ldots, n, \\ & v_i, u_r \geq 0, \quad i = 1, 2, \ldots, m, \quad r = 1, 2, \ldots, s, \end{aligned}$$

where u_r and v_i are the importance weights associated with the rth output and ith input, respectively, and ρ represents the efficiency measure of DMU_o. Note that the free-in-sign variable, u_o, is associated with the convexity condition in the dual model that leads to a convex hull. If the optimal objective value is equal to 1, $\rho^* = 1$, then DMU_o is called *efficient*. Otherwise, DMU_o is called *inefficient*.

There have been various developments in the application of DEA over the years. Some of these developments have been effective in drawing valid inferences and improving statistical efficiency in two-stage procedures. One such model is the single and double bootstrap procedure proposed by Simar & Wilson (2007). Another example, which was also discussed in the literature review, is the InvDEA approach adopted by Boubaker *et al.* (2022). In one of the latest applications of DEA, Hamami *et al.* (2022) provide a new method to define a Euclidean common set of weights (ECSW) in DEA that allows ranking both efficient and inefficient firms and is more realistic in terms of the determination of weights. Their method was tested to show that the model generates rankings for banks consistent with their credit ratings.

3.3. *The resulting model*

In healthcare, fixed expenditures associated with buildings and equipment play a pivotal role in healthcare management and absorb more than 80% of the total cost. Although increasing the number of ambulances, stations, and crews on duty can improve the ability of EMSs to respond to emergency calls, the system may not necessarily become more efficient because the rise in resources would result in long periods of inactivity for crews between calls and a massive fixed cost. In this chapter, the objective is to design an EMS system that minimizes the total cost (the costs of preparation and construction of stations and the purchasing cost of ambulance vehicles) while maximizing the overall expected survival probability of patients. In reality, there are several types of ambulances in EMS systems. For instance, there are four types of ambulances in the US. Types I and III ambulances feature a square patient compartment installed onto the chassis. The difference between Types I and III lies in their chassis: Type I is attached to a truck-like chassis, while Type III is attached to a cutaway van chassis. Types I and III ambulances are often equipped to be used in locations such as airports, chemical plants, and oil refineries, as well as being ready for advanced life support. Type II ambulances with a van-type chassis are commonly used by hospitals and health departments when patients require basic life support. Type IV is classified as mini ambulances, and their compact design enables them to maneuver in regions where other types of ambulances may not, leading to a great reduction in the overall response time.

Contrary to the MESLMHP, without loss of generality, it is assumed that two types of ambulances ($h = 1, 2$) operate in EMSs, the so-called *general* and *advanced* ambulances. General ambulances are equipped with primary and emergency facilities, while advanced ambulances have more facilities, including devices to treat cardiac patients. According to the above-mentioned ambulance types, Types I and II are classified as general ambulances, and Type III is classified as advanced ambulances.

Let Z_h denote the maximum number of the hth ambulance type which are available to respond to any emergency. Here, c_1 and c_2 represent the purchasing cost of general and advanced ambulance types, respectively. The fixed activation and construction cost of each station is assumed to be identical for all stations and denoted by C, and the maximum number of stations which can be in operation is denoted by D. We have two zero-one variables, y_j and x_j^h, where y_j indicates whether or not station j is in operation and x_j^h denotes the number of ambulances of Types I and II allocated to emergency station j. Given the definitions of parameters and variables mentioned above, we propose the following bi-objective network location model to find the optimal number of necessary stations and allocated ambulances in terms of their types:

$$\min \quad \sum_{j=1}^{m} Cy_j + \sum_{j=1}^{m}\sum_{h=1}^{2} c_h x_j^h, \qquad (3.1)\ (3)$$

$$\max \sum_{h=1}^{2}\sum_{l=1}^{k} w_l \sum_{i=1}^{n} \lambda_i^l \sum_{j=1}^{m} s_l(t_{i,\rho_{ij}}) \left(1 - \pi_{\rho_{ij}}^{x_{\rho_{ij}}^h}\right) \prod_{r=1}^{j-1} \pi_{\rho_{ir}}^{x_{\rho_{ir}}^h}, \qquad (3.2)$$

$$\text{s.t.:} \quad \sum_{j=1}^{m} x_j^h \le Z_h, \quad h = 1, 2, \qquad (3.3)$$

$$\sum_{j=1}^{m} y_j \le D, \qquad (3.4)$$

$$x_j^h \le My_j, \quad h = 1, 2; j = 1, 2, \ldots, m, \qquad (3.5)$$

$$x_j^h \ge y_j, \quad h = 1, 2; j = 1, 2, \ldots, m, \qquad (3.6)$$

$$x_j^h \in \mathbb{Z}^{0+}, \quad h = 1, 2; j = 1, 2, \ldots, m, \qquad (3.7)$$

$$y_j \in \{0, 1\}, \quad j = 1, 2, \ldots, m. \qquad (3.8)$$

The first objective function includes two components: the construction cost of EMS centers and the total cost of buying ambulances. Contrary to model (1), the second objective function (3.2) is generalized to two types of ambulances to maximize the overall expected survival probability of patients, which is crucial for designing an EMS system. Note that similar to model (1), ρ_{ij} denotes the jth preferred station for the ith demand point and $x_{\rho_{ij}^h}$ is the number of allocated ambulance type h to the preferred station. Constraint (3.3) ensure that the total number of allocated type h ambulances is at most Z_h. Constraint (3.4) guarantees that the total number of constructed stations is at most D. Constraints (3.5) and (3.6) help define the zero-one variables.

The demand distribution of each demand point is estimated using historical call data collected by the emergency system on a daily basis.

Notably, the above model cannot be easily solved using the method proposed by Knight *et al.* (2012) since solving two conflicting objective functions on the basis of queuing theory is quite complex and adds to the computational time. Since real-world EMS problems are dynamic, it is also highly desirable to analyze the problem over time, for which the existing optimization tools are not sufficiently appropriate. Therefore, we present a structured framework in this chapter to achieve our objectives.

To find an initial solution for model (3), the stochastic variables are first assumed to be represented by their expected values. Therefore, the stochastic model is transformed into a deterministic model. The solution methodology is based on a series of steps, as follows:

(1) Determine the potential locations for an emergency station based on various factors, such as population density and accessibility.
(2) Define the demand points. All people living in a city are potential demand points. For simplicity, the gravity centers of areas in the city are considered the demand points.
(3) Estimate traveling time between each station and the demand points in light of various important factors, such as traffic and weather conditions.
(4) Estimate the costs of buying ambulances and constructing emergency stations.
(5) Solve the proposed model (3) under the static condition to reach the primary solution. This solution is used as a primary input for the computer simulation model in Step 6.
(6) Apply the simulation to analyze the different scenarios, which can help consider the real-world conditions in the mathematical model. A range of scenarios can be defined by changing (i) the number of emergency stations, (ii) the allocated ambulances, and (iii) the probability of ambulances' availability.
(7) Use DEA to assess and rank the results of the scenarios obtained from the simulation in Step 6. The scenario with the superior score is the best arrangement (design) for the EMS system.

4. Case Study

In this section, we present an application involving the province of Isfahan in Iran. Isfahan is located in southern central Iran, 1,430 m above sea level, with a population of about 5 million and an area of nearly 107,027 km^2. The current state of the Isfahan emergency system shows that the costs associated with the emergency service time for serving patients (transport, immediate treatment, and medical transmission) are relatively high. This issue has been discussed with experts and managers in several lengthy meetings through brainstorming sessions and the Delphi method. Isfahan health officials have concluded that the root cause of the problem is the improper locations of emergency centers and ineffective ambulance allocation. Moreover, the Isfahan health authority has decided to rise the survival rate by increasing the number of ambulances.

The city of Isfahan is the provincial capital, consisting of 15 areas. The most crowded areas of Isfahan are areas 1, 3, 5, and 6. Because of its aging population, the average age of the population in these areas is higher than that in the other areas. Historically, more than 41% of the total emergency calls have been received from these four areas of Isfahan. Therefore, their EMS coverage plays a vital role for the healthcare managers of the city. For this reason, we apply the method proposed in this chapter to areas 1, 3, 5, and 6 of Isfahan. Each area is divided into three smaller sections, resulting in 12 urban (demand) points. In addition, patients are classified into three types, as shown in Table 2.

We point out that the standard response times in Table 2 are determined by the Ministry of Health, Treatment and Medical

Table 2. Patient types and their standard response times.

Type	**Description**	**Standard response time (minutes)**
I	Urgent and emergency help	9
II	Serious but not urgent	15
III	Non-life-threatening	20

Note: This table shows how patients are classified into three types and their standard response times for Isfahan, Iran.

Education of Iran. The different response times reflect the heterogenous groups of the population that the EMS system serves. As the demand rate and traffic conditions affect response time, we partition the hours of the day into four smaller intervals: dawn (0:00–5:59), morning (6:00–11:59), afternoon (12:00–17:59), and evening (18:00–23:59). The demand rate of the different demand points in terms of each patient type was collected for a two-month period of time, as summarized in Table 3.

Table 3. Demand rate of each patient type.

Time interval	Demand points	Patient types			Time interval	Demand points	Patient types		
		I	II	III			I	II	III
Dawn	1	52	61	60	Dawn	7	32	38	41
Morning		81	83	81	Morning		36	40	39
Afternoon		111	116	95	Afternoon		68	70	64
Evening		163	158	102	Evening		67	84	72
Dawn	2	42	46	40	Dawn	8	26	34	29
Morning		63	64	62	Morning		45	48	37
Afternoon		104	106	114	Afternoon		48	52	48
Evening		125	169	171	Evening		53	68	55
Dawn	3	76	82	62	Dawn	9	21	29	24
Morning		77	86	52	Morning		30	33	28
Afternoon		131	142	130	Afternoon		46	48	41
Evening		122	130	141	Evening		48	50	49
Dawn	4	65	66	68	Dawn	10	12	16	18
Morning		83	85	75	Morning		16	18	22
Afternoon		92	94	94	Afternoon		25	29	35
Evening		121	127	103	Evening		62	65	58
Dawn	5	33	36	28	Dawn	11	18	27	23
Morning		89	92	74	Morning		19	21	20
Afternoon		127	140	135	Afternoon		26	29	34
Evening		156	161	124	Evening		38	42	36
Dawn	6	47	54	46	Dawn	12	15	18	18
Morning		48	54	38	Morning		18	23	17
Afternoon		82	92	84	Afternoon		48	52	41
Evening		51	92	75	Evening		54	61	57

Notes: This table provides information on the demand rates for each patient type. The time intervals are partitioned as: dawn (0:00–5:59), morning (6:00–11:59), afternoon (12:00–17:59), and evening (18:00–23:59). The demand rate of the different demand points in terms of each patient type was collected for a two-month period of time.

The healthcare management team decided to establish 12 emergency stations at most to cover all sections in the four areas. In addition, in collaboration with several practitioners, a potential location for an emergency center in every section is determined, and the population gravity centers are used to form the coordinates of all the demand points. Table 4 displays the travel times between emergency station j, $j \in \{A, B, C, D, E, F, G, H, I, J, K, L\}$ and demand point i, $i \in \{1, 2, 3, 4, 5, 6, 7, 8, 9, 10, 11, 12\}$ in four predetermined time intervals, which are obtained using a GIS software and Google Maps.

A wide variety of survival functions can be used to show the probability of survival. There are two general survival functions, consisting of exponential decay functions and step functions. In the exponential decay function, the survival probability decreases over a period of time, while the step function is a piecewise constant function for the survival probability that jumps by a given amount at a finite number of points. In this case study, a standard response time is considered to define the step survival functions. Table 5 shows the survival functions and relative importance associated with each patient type.

After consultations with the healthcare management team of Isfahan, the maximum number of ambulances that the EMS procurement team can provide includes 37 general ambulances and 12 advanced ambulances. The CPLEX solver in the general algebraic modeling system (GAMS) is used to solve the proposed model (3) as a mixed-integer linear programming (MILP) problem and obtain a primary solution. The weighted sum method is used to solve the two-objective optimization problem. The number of ambulances allocated to the emergency stations is reported in Table 6, and the respective total cost is \$2,152,037. In addition, the survival rates of patient Types I, II, and III are 0.125, 0.638, and 0.803, respectively. It should be noted that we do not consider the dynamic conditions and divided time when running the model. It means that an equal amount is considered for demand rate and travel time at all times.

It is of interest to study the design problem over time. The complexity of observing the dynamic conditions in the proposed model leads us to apply the simulation model and derive insight based on the results. Simulation is an advantageous method to assess and analyze the situation of current and newly designed systems (Robinson, 2005; McCormack & Coates, 2015).

Table 4. Travel time between demand points and service nodes.

Time interval	Demand points	Service nodes (minutes)											
		A	**B**	**C**	**D**	**E**	**F**	**G**	**H**	**I**	**J**	**K**	**L**
Dawn	1	3.1	12.4	13.3	4.1	3.6	14.5	15.0	3.2	2.0	14.0	15.1	13.8
Morning		3.7	12.6	13.3	4.1	3.7	14.8	15.1	3.4	2.4	14.2	15.4	14.0
Afternoon		3.9	13.5	13.8	4.9	4.0	15.4	15.6	3.7	2.5	14.5	16.0	14.6
Evening		4.2	14.0	14.1	5.3	5.1	15.8	16.0	14.1	3.0	15.3	16.2	15.0
Dawn	2	12.5	14.1	13.6	15.0	2.4	13.1	13.0	14.8	16.5	17.3	14.1	2.0
Morning		12.7	14.2	13.7	15.0	2.5	13.3	13.2	15.0	16.8	17.5	14.4	2.2
Afternoon		13.0	14.5	14.0	15.4	2.9	13.9	13.4	15.2	17.2	17.7	14.6	2.4
Evening		13.5	15.0	14.8	16.1	3.5	14.7	14.0	15.8	17.6	18.2	15.4	3.3
Dawn	3	17.5	15.0	3.7	16.8	4.1	12.2	13.8	14.9	6.4	8.2	3.4	2.7
Morning		17.8	15.4	4.0	17.0	4.5	12.5	14.2	15.4	7.0	8.5	3.8	13.4
Afternoon		18.2	16.0	4.5	17.5	5.0	13.0	14.9	16.0	7.7	9.1	4.2	14.0
Evening		18.5	16.6	5.2	18.1	5.7	13.6	15.4	16.7	8.2	9.8	4.9	14.6
Dawn	4	19.5	5.2	8.2	14.2	21.1	17.5	25.1	9.3	8.8	22.4	25.4	3.5
Morning		19.7	5.4	8.5	14.6	21.3	17.8	25.6	9.5	9.0	22.6	25.7	3.7
Afternoon		20.0	6.1	8.9	15.0	21.6	18.2	25.9	10.1	9.2	23.1	26.3	4.2
Evening		20.3	6.1	9.2	15.3	22.2	18.6	26.0	10.6	9.6	23.4	26.7	4.8
Dawn	5	9.9	28.1	18.1	16.8	26.2	8.5	4.2	16.2	18.3	21.1	18.4	3.2
Morning		10.2	28.4	18.3	16.9	26.2	8.6	4.3	16.2	18.5	21.3	18.6	3.5
Afternoon		10.6	28.5	18.8	17.5	26.5	9.0	4.6	16.5	18.9	21.6	19.0	3.7
Evening		11.0	28.7	19.2	17.9	26.9	9.4	4.6	17.0	19.2	21.8	19.4	4.0
Dawn	6	14.2	8.2	6.4	12.2	18.1	5.4	4.3	3.1	17.9	4.5	5.0	6.8
Morning		14.5	8.4	6.4	12.4	18.1	5.7	4.5	3.4	18.2	4.8	5.3	7.0
Afternoon		14.7	8.7	6.6	12.8	18.5	6.0	4.8	3.6	18.3	5.0	5.5	7.2
Evening		15.1	9.0	6.9	13.0	18.7	6.2	5.0	3.8	18.7	5.5	6.0	7.8

(*Continued*)

Table 4. (*Continued*)

Time interval	Demand points	Service nodes (minutes)											
		A	**B**	**C**	**D**	**E**	**F**	**G**	**H**	**I**	**J**	**K**	**L**
Dawn	7	3.2	4.1	5.6	10.3	11.4	5.6	17.8	14.3	3.1	2.2	5.7	16.3
Morning		3.3	4.3	5.9	10.5	11.6	5.8	18.0	14.5	3.3	2.5	6.0	16.5
Afternoon		3.5	4.5	6.2	10.7	11.8	6.0	18.2	14.7	3.5	2.7	6.3	16.8
Evening		3.7	4.8	6.5	11.0	12.0	6.3	18.5	15.0	3.8	3.0	6.7	17.2
Dawn	8	5.4	10.0	9.8	8.7	13.4	15.4	6.1	17.4	14.2	16.3	11.2	13.0
Morning		5.6	10.5	10.2	9.0	13.5	15.5	6.3	17.5	14.4	16.5	11.5	13.3
Afternoon		6.0	10.8	10.5	9.8	13.8	15.8	6.5	17.7	14.7	16.8	11.8	13.5
Evening		6.6	11.3	11.0	10.2	14.2	16.2	6.8	18.2	15.0	17.5	12.3	13.8
Dawn	9	15.8	14.0	10.2	8.9	13.5	16.8	3.2	4.1	5.3	4.8	18.3	6.8
Morning		16.0	14.0	10.5	9.1	13.7	17.0	3.5	4.3	5.5	5.0	18.5	7.0
Afternoon		16.2	14.5	10.8	9.4	14.0	17.3	3.8	4.5	5.8	5.3	18.7	7.2
Evening		16.5	14.7	11.2	9.6	14.2	17.5	4.2	5.0	6.3	5.8	19.0	7.8
Dawn	10	3.4	5.5	4.2	12.8	8.5	6.7	7.8	10.2	11.3	15.4	17.2	9.1
Morning		3.6	5.7	4.5	13.0	8.7	7.0	8.0	10.5	11.5	15.8	17.5	9.3
Afternoon		4.0	6.0	4.8	13.3	9.0	7.2	8.3	10.7	11.7	16.0	17.9	9.5
Evening		4.5	6.4	5.3	13.5	9.5	7.5	8.8	11.1	12.0	16.5	18.3	10.0
Dawn	11	12.2	10.1	6.8	4.2	3.3	14.8	15.0	12.1	3.4	2.1	5.8	14.8
Morning		12.5	10.3	7.0	4.5	3.5	15.0	15.2	12.3	3.6	2.2	6.0	15.0
Afternoon		12.8	10.5	7.3	4.8	6.0	15.2	15.5	12.5	4.0	2.5	6.3	15.3
Evening		13.2	10.8	7.5	5.2	6.5	15.5	16.0	12.8	4.5	3.0	6.8	15.8
Dawn	12	3.2	2.0	4.6	8.8	9.8	12.3	13.5	16.8	7.8	9.2	13.7	5.5
Morning		3.5	2.2	5.0	9.0	10.0	12.5	13.8	17.0	9.0	9.6	14.0	5.9
Afternoon		3.9	2.8	5.4	9.5	10.4	12.8	14.2	17.3	9.4	10.0	14.3	6.4
Evening		4.4	3.3	6.0	9.8	11.0	13.0	14.5	17.6	9.9	10.6	14.7	7.0

Note: This table provides information on the travel time between emergency station j, $j \in \{A, B, C, D, E, F, G, H, I, J, K, L\}$ and demand point i, $i \in \{1, 2, 3, 4, 5, 6, 7, 8, 9, 10, 11, 12\}$ in four predetermined time intervals, which are obtained using GIS software and Google Maps. The time intervals are partitioned as: dawn (0:00–5:59), morning (6:00–11:59), afternoon (12:00–17:59), and evening (18:00–23:59).

Table 5. Proposed survival functions and relative importance of each patient type.

Patient type	**Step survival functions**	**Weights**
I	$s_I(t) = \begin{Bmatrix} 1 \text{ for } 0 \leq t \leq 9 \\ 0 \text{ for } t > 9 \end{Bmatrix}$	4
II	$s_{II}(t) = \begin{Bmatrix} 1 \text{ for } 0 \leq t \leq 15 \\ 0 \text{ for } t > 15 \end{Bmatrix}$	2
III	$s_{III}(t) = \begin{Bmatrix} 1 \text{ for } 0 \leq t \leq 20 \\ 0 \text{ for } t > 20 \end{Bmatrix}$	1

Note: This table provides information on the proposed survival functions for each patient type and the relative importance assigned to them. Patient types are I: Urgent and emergency help, II: Serious but not urgent, and III: Non-life threatening.

Table 6. Number of ambulances (by type) allocated to the emergency stations.

Service nodes	**A**	**B**	**C**	**D**	**E**	**F**	**G**	**H**	**I**	**J**	**K**	**L**
General type	4	5	3	5	4	3	3	2	2	2	2	2
Advanced type	1	2	1	1	1	0	2	1	1	0	1	1

Note: This table provides information on the number of ambulances (by type) allocated to emergency station j, $j \in \{\text{A, B, C, D, E, F, G, H, I, J, K, L}\}$.

While an optimization model aims to calculate the optimal values of decision-making variables, solving the model with multiple dynamic and uncertain parameters is often intricate and uneconomical. To solve and analyze these types of problems, simulation models can be combined with optimization models, which is referred to as *simulation-based optimization*. This is a structural method for determining the optimal parameters in the system, and, in turn, the objective function is measured based on the simulation model (Aiassi *et al.*, 2020). The simulation-optimization in a constant time enables us to evaluate the system's behavior at specific values of input variables. A *simulation experiment* is defined as a test or a series of tests in which significant variations are made to the input variables of the simulation optimization. Thereby, the results obtained from simulation experiments can help the decision-maker analyze

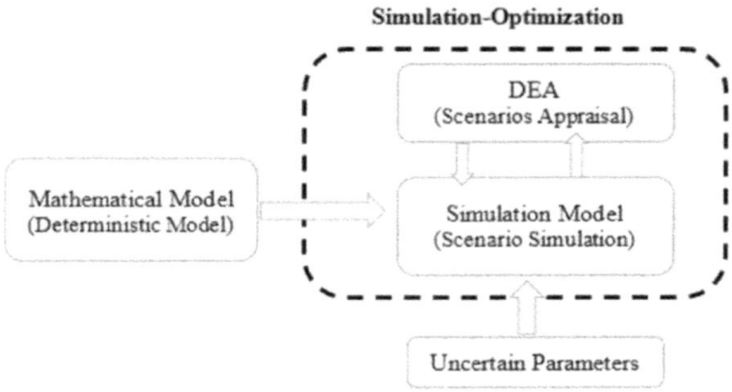

Figure 3. A simulation-based optimization model.

Note: The output of a simulation model is used by an optimization strategy to find the optimal values for output variables, along with feedback on how improvements can be made based on the (nearly) optimal solution.

changes in input variables (parameters) and select the best values for them (Carson & Maria, 1997). Generally, a simulation optimization model includes n input variables $(x_1, x_2, \ldots, x_n)$ and m output variables $(y_1, y_2, \ldots, y_m)$, with the aim of defining a set of optimal input variables which optimize the output variables. The simulation-optimization model is a suitable method for solving complicated and dynamic problems. As shown in Figure 3, the output of a simulation model is used by an optimization strategy to find the optimal values for output variables along with getting feedback on how improvements can be made based on the (nearly) optimal solution. Obviously, defining an appropriate optimization strategy is of great importance in this approach.

Let us construct the following six steps with the aim of analyzing the problem dynamically via simulation and reaching a more realistic and reliable solution:

(1) Calculate the primary solution using model (3).
(2) Assign an emergency station to each demand point with a maximum of three general ambulances and one advanced ambulance.
(3) Assign an emergency station to every two neighboring demand points with a maximum of six general ambulances and two advanced ambulances.
(4) Assign a fixed emergency station to every two neighboring demand points with a maximum of six general and two

advanced ambulances, and allocate a temporary emergency center (a mobile general ambulance located in crowded points temporarily) to each pair of demand points.

(5) Allocate a fixed emergency station to every two neighboring demand points with a maximum of three general ambulances and one advanced ambulance, and allocate a temporary emergency center (a mobile general ambulance located in crowded points temporarily) to each pair of demand points.

(6) Allocate an emergency center to demand points that have the highest call rate (80%) for patient Type I with a maximum of three general ambulances and one advanced ambulance. The remaining demand points are treated in Scenario 3.

We run the simulations of the model corresponding to Figure 1 for the defined scenarios in the Arena software. Notably, historical data collected from Isfahan Health Organization is fitted to a theoretical distribution of the proposed framework. The validation of the proposed model for the case study is carried out by tracking the entities, i.e., ambulances, patients, and resources, of the simulation model, and their adaptability to the expected cases is inspected by Isfahan Health Organization experts. The subject matter experts, therefore, follow the entities to make sure that all the constraints in the mathematical model have been covered by the simulation model. Furthermore, the outputs of the simulation model for certain situations and well-known conditions have been considered (Hatami-Marbini *et al.*, 2020). The simulation model includes probabilistic parameters. The historical data are used to estimate the probability distribution for each parameter, as shown in Table 7.

The demand rate for each patient type and ambulance response time are reported in Tables 3 and 4, respectively. Furthermore, the constant parameters are shown in Table 8.

As it is not straightforward to define the start and end points of the simulation model, a steady-state model is presumed. Therefore, the number of replications needs to be first determined. First, the model is run 30 times, and the statistical description of the outputs is shown in Table 9.

The equation $n \cong z^2_{1-\frac{\alpha}{2}} \frac{s^2}{h^2}$ is used to identify the total number of replications, in which s is the standard deviation of the initial replications and h is a fixed number denoting the half domain of the

Table 7. Parameters' distributions for the simulation model determined by historical data.

Receiving phone call	**Time for getting initial information**	**Probability of physician requirement**	**Time for identifying the level of emergency care**	**Probability of ambulance requirement**	**Ambulance allocation time**	**Service time on the site**
Poisson distribution with a mean of 860 calls per day	Triangular distribution with (1, 2, 3) minutes	Bernoulli distribution probability 75%	Exponential distribution with an average of 1 minute	55% for the general type, 35% for the advanced type, and 10% does not need an ambulance	Exponential distribution with a mean of 1 minute	Exponential distribution with a mean of 4 minutes

Table 8. Constant parameters used in simulation.

Cost of the EMS construction ($)	Purchasing cost of ambulance vehicles (by type; $)		Maximum ambulances (by type)		Maximum EMS
	General	Advanced	General	Advanced	
275,000	62,500	87,500	37	12	12

confidence interval. Table 10 consequently shows the total number of replications for each scenario.

Having gone through Table 10, the results allow us to think of 110 imitations for the simulation model. The replication length in this model is 24 hours for each repetition. A warm-up period, T, is then applied to help determine the running time of a simulation and ensure that the objective function values are measured in a steady state. Let us focus on Scenario 1, for example. As our model seeks to obtain the values of three survival rates and total cost, four Ts are considered for Scenario 1. Comparing the four trends in Figure 4 shows that the longest convergence time is related to the survival rate (c). In other words, the survival rate for Scenario 1 reaches steady-state status when $T = 8{,}000$ h. Likewise, T can be estimated for the remaining scenarios.

Table 11 lists the survival rates and total cost of each scenario after simulating the model 110 times.

Generally, results show that additional service stations and ambulances lead to an increase in survival rates and total cost. Scenario 2, as the simplest scenario, is not acceptable to some extent because it has the maximum cost compared with other scenarios and the survival rates of A and B are minimal. It means that increasing cost in Scenario 2 does not improve all survival rates considerably. In addition, it might be difficult to make decisions on the basis of Scenarios 3–6 due to the conflicting results for the four decision variables, as summarized in Figure 5. That is, though the survival rate plays a key role in an EMS system, resource limitations do not allow decision-makers to improve this factor through unreasonable and infeasible decisions.

Table 9. Statistical output of initial replications.

	Average				Standard deviation				Minimum output				Maximum output			
	Survival rate				Survival rate				Survival rate				Survival rate			
Scenario	I	II	III	Cost	I	II	III	Cost	I	II	III	Cost	I	II	III	Cost
1	0.123	0.640	0.8000	2,000,540	0.011	0.059	0.080	224,338	0.117	0.570	0.790	2,000,003	0.127	0.642	0.806	2,107,854
2	0.121	0.624	0.8490	2,320,001	0.012	0.061	0.081	231,109	0.119	0.618	0.848	2,478,945	0.125	0.627	0.855	2,647,542
3	0.130	0.697	0.0607	1,628,798	0.017	0.068	0.059	163,594	0.129	0.694	0.605	1,435,379	0.137	0.701	0.702	1,725,386
4	0.131	0.685	0.7400	1,943,762	0.028	0.059	0.070	189,341	0.127	0.679	0.738	1,799,856	0.135	0.687	0.747	2,001,360
5	0.135	0.692	0.7120	1,822,395	0.011	0.057	0.070	179,835	0.134	0.690	0.711	1,796,853	0.136	0.697	0.714	1,856,347
6	0.124	0.643	0.8180	1,998,879	0.013	0.059	0.078	188,576	0.123	0.641	0.817	1,987,874	0.127	0.645	0.825	2,201,347

Notes: This table provides information on the mean, standard deviation, and minimum and maximum values of our input and output variables for the six scenarios and three survival rates, including the total cost. All units are in US$.

Table 10. Required number of replications for each scenario.

Scenario	Total required replications
1	106
2	112
3	101
4	107
5	103
6	110

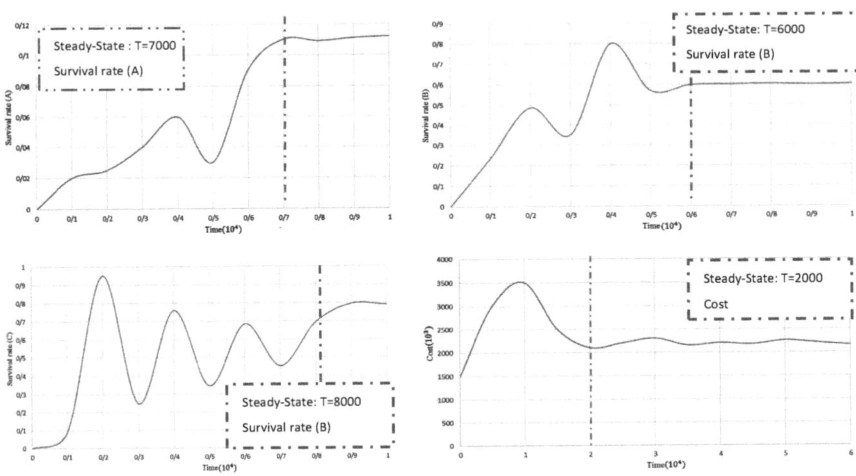

Figure 4. Steady state (warm-up period).

Note: This figure shows the survival rate plotted against convergence time.

Let us use the DEA model to compare these six scenarios and choose the best scenario. To do so, there is a need to define the inputs and outputs for each scenario. The inputs are:

(1) the number of selected EMS stations,
(2) the number of general ambulances allocated to stations, and
(3) the number of advanced ambulances allocated to stations.

Table 11. Results of simulation.

Scenario	Survival rate I	Survival rate II	Survival rate III	Cost ($)	Estimation interval for cost ($)
1	0.125	0.638	0.803	2,031,665	2,031,665 ± 41,923.22
2	0.122	0.623	0.852	2,587,221	2,587,221 ± 43,188.55
3	0.132	0.698	0.609	1,639,999	1,639,999 ± 30,571.67
4	0.130	0.683	0.743	1,953,359	1,953,359 ± 35,383.14
5	0.133	0.694	0.710	1,839,025	1,839,025 ± 33,606.71
6	0.122	0.642	0.821	2,011,628	2,011,628 ± 35,240.18

Note: This table lists the survival rates and total cost (in US$) of each scenario after simulating the model 110 times.

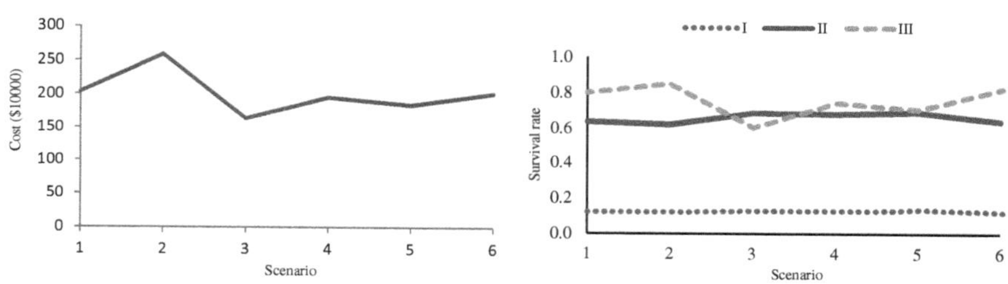

Figure 5. Survival rates and total cost of each scenario.

Table 12. DEA results for the six scenarios.

Scenario	DEA score
1	0.72
2	0.94
3	0.86
4	1.00
5	1.00
6	0.88

The outputs are:

(1) the total cost and
(2) the survival rates of Types I, II, and III.

The results obtained from the DEA model are presented in Table 12.

On the basis of the DEA analysis, Scenarios 4 and 5 are superior followed by Scenario 2. Although the survival rates of Scenarios 4 and 5 are not the greatest, DEA compares the scenarios based on both inputs and outputs to find the best choices. Comparing Scenario 4 with Scenarios 1 and 2 shows that adding a portable station can help increase the survival rate while incurring less cost. Looking at the results in Table 12, it can be inferred that patient Types I and II have a significant effect on choosing Scenarios 4 and 5. Thus, building portable stations in the regions that have the most number of calls from patient Type I can lead to an increase in the survival rate. Scenario 3 is convincingly ranked penultimate, while it has the highest survival rate for Types I and II and a lower cost compared with Scenarios 4 and 5.

According to the developed approach in this case study, Scenarios 4 and 5 are the best configurations for the EMS system for areas 1, 3, 5, and 6 in Isfahan. Let us compare these findings with the existing survival rates in other areas (see Table 13). Clearly, the survival rates of Types I and II for areas 1, 3, 5, and 6 are higher than those in other areas at the minimum cost, while the survival rate of Types III is acceptable.

Table 13. Survival rate of Isfahan municipality regions.

	Survival rate		
Areas in Isfahan	**I**	**II**	**III**
1, 3, 5, and 6 (Scenario 4)	0.130	0.683	0.743
1, 3, 5, and 6 (Scenario 5)	0.133	0.694	0.710
2	0.121	0.639	0.802
4	0.116	0.583	0.765
7	0.110	0.551	0.553
8	0.102	0.621	0.664
9	0.113	0.589	0.816
10	0.117	0.579	0.746
11	0.103	0.563	0.589
12	0.100	0.521	0.486
13	0.109	0.549	0.568
14	0.110	0.572	0.621
15	0.100	0.512	0.439

5. Conclusion

The model described in this chapter integrates simulation and optimization methods to find the best EMS configuration and ambulance allocation with the aim of increasing survival rate and decreasing cost across the EMS system. It is difficult to embed dynamic situations, such as traffic conditions, in mathematical models when analyzing EMS systems. Hence, this model leverages the simulation method to solve the problem across different scenarios.

The simulation-based optimization model was implemented in four selected municipal regions of Isfahan to obtain a proper design for emergency center locations and ambulance allocation. In this regard, six scenarios were defined to simulate the model in a dynamic environment and measure the survival rate and total cost of each scenario. In view of the survival rate and costs, DEA was then used to rank the scenarios and select the best ones (scenarios 4 and 5). The chosen scenarios show that patient Types I and II play a crucial role in increasing the survival rate and are essential to be regarded in designing EMS facilities, which can help to improve the survival rate.

For future research, the proposed approach can be extended by considering some other key factors, such as seasonal variations and weather, that affect the travel time of ambulances. With the growing applications of machine learning techniques, this method could also be studied in combination with ongoing research on identifying the locations of expected emergency events through techniques such as artificial neural networks and decision models for location–allocation. Additionally, it could be a worthwhile study to apply the method proposed in this chapter to similar problems in other emergency response systems, such as urban firefighting facilities.

References

Aboueljinane, L., Sahin, E., & Jemai, Z. (2013, December). A review on simulation models applied to emergency medical service operations. *Computers & Industrial Engineering,* 66(4), 734–750. https://doi.org/10.1016/j.cie.2013.09.017.

Ahmadi-Javida, A., Jalali, Z., & Klassen, K.J. (2017, April). Outpatient appointment systems in healthcare: A review of optimization studies.

European Journal of Operational Research, 258(1), 3–34. https://doi.org/10.1016/j.ejor.2016.06.064.

Aiassi, R., Sajadi, S.M., Hadji-Molana, S.M., & Zamani-Babgohari, A. (2020, November). Designing a stochastic multi-objective simulation-based optimization model for sales and operations planning in built-to-order environment with uncertain distant outsourcing. *Simulation Modelling Practice and Theory,* 104, 102103. https://doi.org/10.1016/j.simpat.2020.102103.

Andersson, H., Granberg, T.A., Christiansen, M., Aartuna, E.S., & Leknesa, H. (2020, January). Using optimization to provide decision support for strategic emergency medical service planning — Three case studies. *International Journal of Medical Informatics,* 133, 103975. https://doi.org/10.1016/j.ijmedinf.2019.103975.

Aringhieri, R., Knight, V., & Smith, H. (2016). OR applied to health in a modern world. *Health Systems,* 5(3), 163–165. https://doi.org/10.1057/s41306-016-0012-5.

Athanassopoulos, A. & Gounaris, C. (2001, January). Assessing the technical and allocative efficiency of hospital operations in Greece and its resource allocation implications. *European Journal of Operational Research,* 133(2), 416–431. https://doi.org/10.1016/S0377-2217(00)00180-6.

Banker, R., Charnes, A., & Cooper, W. (1984). Some models for estimating technical and scale inefficiencies in data envelopment analysis. *Management Science,* 30(9), 1078–1092. https://doi.org/10.1287/mnsc.30.9.1078.

Başar, A., Bülent, Ç., & Ünlüyurt, T. (2011, December). A multi-period double coverage approach for locating the emergency medical service stations in Istanbul. *Journal of the Operational Research Society,* 62(4), 627–637. https://doi.org/10.1057/jors.2010.5.

Bélanger, A. & Edmonston, B. (2018). Implications of world peak population for Canada. *Canadian Studies in Population,* 45(1–2), 1–10. https://doi.org/10.25336/csp29372.

Beraldi, P., Bruni, M., & Conforti, D. (2004, October). Designing robust emergency medical service via stochastic programming. *European Journal of Operational Research,* 158(1), 183–193. https://doi.org/10.1016/S0377-2217(03)00351-5.

Boubaker, S., Le, T.D.Q., & Ngo, T. (2022). Managing bank performance under COVID-19: A novel inverse DEA efficiency approach. *International Transactions in Operations Research*, 30(2023), 2436–2452. https://doi.org/10.1111/itor.13132.

Boubaker, S., Manita, R., & Rouatbi, W. (2021). Large shareholders, control contestability and firm productive efficiency. *Annals of*

Operations Research, 296, 591–614. https://doi.org/10.1007/s10479-019-03402-z.

Boujemâa, H. & Halima, N.B. (2019). Throughput of cooperative HARQ protocols for underlay cognitive radio networks using adaptive and fixed transmit power. *International Journal of Ad Hoc and Ubiquitous Computing,* 32(3). https://doi.org/10.1504/IJAHUC.2019.103129.

Boujemaa, R., Jebali, A., Hammami, S., & Ruiz, A. (2020, November). Multi-period stochastic programming models for two-tiered emergency medical service system. *Computers & Operations Research,* 123, 104974. https://doi.org/10.1016/j.cor.2020.104974.

Bricon-Souf, N., Anceaux, F., Bennani, N., Dufresne, E., & Watbled, L. (2005, October). A distributed coordination platform for home care: Analysis, framework and prototype. *International Journal of Medical Informatics,* 74(10), 809–825. https://doi.org/10.1016/j.ijmedinf.2005.03.020.

Brooks, S.C., Hsu, J.H., Tang, S.K., Jeyakumar, R., & Chan, T.C. (2013, May). Determining Risk for Out-of-Hospital Cardiac Arrest by Location Type in a Canadian Urban Setting to Guide Future Public Access Defibrillator Placement. *Annals of Emergency Medicine,* 61(5), 530–538.e2. https://doi.org/10.1016/j.annemergmed.2012.10.037.

Brotcorne, L., Laport, G., & Semet, F. (2003, June 16). Ambulance location and relocation models. *European Journal of Operational Research,* 147(3), 451–463. https://doi.org/10.1016/S0377-2217(02)00364-8.

Burgess Jr., J.F. & Wilson, P.W. (1995). Decomposing hospital productivity changes, 1985–1988: A nonparametric Malmquist approach. *Journal of Productivity Analysis,* 6, 343–363. https://doi.org/10.1007/BF01073525.

Carson, Y. & Maria, A. (1997). Simulation optimization: Methods and applications. In *Proceedings of the 29th conference on Winter Simulation* (pp. 18–126). https://doi.org/10.1145/268437.268460.

Chanta, S., Mayorga, M.E., Kurz, M.E., & McLay, L.A. (2011). The minimum p-envy location problem: A new model for equitable distribution of emergency resources. *IIE Transactions on Healthcare Systems Engineering,* 1(2), 101–115. https://doi.org/10.1080/19488300.2011.609522.

Charnes, A., Cooper, W., & Rhodes, E. (1978, November). Measuring the efficiency of decision making units. *European Journal of Operational Research,* 2(6), 429–444. https://doi.org/10.1016/0377-2217(78)90138-8.

Chilingerian, J.A. (1995, February). Evaluating physician efficiency in hospitals: A multivariate analysis of best practices. *European Journal of*

Operational Research, 80(3), 548–574. https://doi.org/10.1016/0377-2217(94)00137-2.

Chilingerian, J.A. & Sherman, H.D. (2004). Health care applications. In: W.W. Cooper, L.M. Seiford & J. Zhu (Eds.), *Handbook on Data Envelopment Analysis* (pp. 481–537). International Series in Operations Research & Management Science (Vol. 71). Boston, MA: Springer. https://doi.org/ 10.1007/1-4020-7798-X_17.

Church, R. & ReVelle, C. (1974, December). The maximal covering location problem. In *Papers of the Regional Science Association* (Vol. 32, pp. 101–118). http://yalma.fime.uanl.mx/~roger/work/teaching/class_tso/docs_project/problems/MCLP/1974-prs-Church%20ReVelle-maximal%20covering%20location.pdf.

Cook, W.D. & Seiford, L.M. (2009, January). Data envelopment analysis (DEA) — Thirty years on. *European Journal of Operational Research,* 192(1), 1–17. https://doi.org/10.1016/j.ejor.2008.01.032.

Coppola, N.M., Ozcan, Y.A., & Bogacki, R. (2003). Evaluation of performance of dental providers on posterior restorations: Does experience matter? A Data Envelopment Analysis (DEA) approach. *Journal of Medical Systems Volume,* 27, 445–456. https://doi.org/10.1023/A: 1025659822427.

Daskin, M.S. (1983). A maximum expected covering location model: Formulation, properties and heuristic solution. *Transportation Science,* 17, 48–70. https://doi.org/10.1287/trsc.17.1.48.

Department of Health and Children (DOHC). (1997). The emergency ambulance services: Comptroller and Auditor General report on value for money examination. Stationery Office. http://hdl.handle.net/10147/81266.

Dieleman J.L., Squires E., Bui A.L., Campbell M., Chapin A., Hamavid H., Horst C., Li Z., Matyasz T., Reynolds A., & Sadat N. (2017). Factors associated with increases in US health care spending, 1996–2013. *JAMA*, 318(17), 1668–1678.

Eilat, H., Golany, B., & Shtub, A. (2008, October). R&D project evaluation: An integrated DEA and balanced scorecard approach. *Omega,* 36(5), 895–912. https://doi.org/10.1016/j.omega.2006.05.002.

Emrouznejad, A., Parker, B.R., & Tavares, G. (2008, September). Evaluation of research in efficiency and productivity: A survey and analysis of the first 30 years of scholarly literature in DEA. *Socio-Economic Planning Sciences,* 42(3), 151–157. https://doi.org/10.1016/j.seps.2007.07.002.

Erkut, E., Ingolfsson, A., & Erdoğan, G. (2008, February). Ambulance location for maximum survival. *Naval Research Logistics,* 55(1), 42–58.

Färe, R., Grosskopf, S., Lindgren, B., & Roos, P. (1992). Productivity changes in Swedish pharmacies 1980–1989: A non-parametric Malmquist approach. *Journal of Productivity Analysis,* 3, 85–101. https://doi.org/10.1007/BF00158770.

Farrell, M.J. (1957). The measurement of productive efficiency. *Journal of the Royal Statistical Society. Series A (General),* 120(3), 253–290. https://doi.org/10.2307/2343100.

Ferrier, G.D., Rosko, M.D., & Vald, V.G. (2006). Analysis of uncompensated hospital care using a DEA model of output congestion. *Health Care Management Science,* 9, 181–188. https://doi.org/10.1007/s10729-006-7665-8.

Fritze, R., Graser, A., & Sinn, M. (2018, March). Combining spatial information and optimization for locating emergency medical service stations: A case study for Lower Austria. *International Journal of Medical Informatics,* 111, 24–36. https://doi.org/10.1016/j.ijmedinf.2017.12.008.

Galvão, R.D. & Morabito, R. (2008). Emergency service systems: The use of the hypercube queueing model in the solution of probabilistic location problems. *International Transactions in Operations Research,* 15(5), 525–549. https://doi.org/10.1111/j.1475-3995.2008.00654.x.

Galvãoa, R.D. & ReVelle, C. (1996, January). A Lagrangean heuristic for the maximal covering location problem. *European Journal of Operational Research,* 88(1), 114–123. https://doi.org/10.1016/0377-2217(94)00159-6.

Gendreau, M., Guertin, F., & Potvi, J.-Y. (1999, November). Parallel Tabu search for real-time vehicle routing and dispatching. *Transportation Science,* 33(4), 341–428. https://doi.org/10.1287/trsc.33.4.381.

Gendreau, M., Laporte, G., & Semet, F. (1997, August). Solving an ambulance location model by tabu search. *Location Science,* 5(2), 75–88. https://doi.org/10.1016/S0966-8349(97)00015-6.

Gendreau, M., Laporte, G., & Semet, F. (2001). A dynamic model and parallel tabu search heuristic for real-time ambulance relocation. *Parallel Computing,* 27(12), 1641–1653. https://doi.org/10.1016/S0167-8191(01)00103-X.

Gendreau, M., Laporte, G., & Semet, F. (2006). The maximal expected coverage relocation problem for emergency vehicles. *Journal of the Operational Research Society,* 57(1), 22–28. https://doi.org/10.1057/palgrave.jors.2601991.

Golabian, H., Jamal, A., Tavakkoli-Moghad, R., & Faroughi, H. (2022). A multi-verse optimizer algorithm for ambulance repositioning in emergency medical service systems. *Journal of Ambient Intelligence and Humanized Computing,* 13, 549–570. https://doi.org/10.1007/s12652-021-02918-2.

Goldberg, J.B. (2004). Operations research models for the deployment of emergency services vehicles. *EMS Management Journal,* 1(1), 20–39.

Goldberg, J.B., Dietrich, R., Chen, J., Mitwasi, M., Valenzuela, T., & Criss, E. (1990). A simulation model for evaluating a set of emergency vehicle base locations: Development, validation, and usage. *Socio-Economic Planning Sciences,* 24(2), 125–141. https://doi.org/10.1016/0038-0121(90)90017-2.

Grekousis, G. & Liu, Y. (2019, July). Where will the next emergency event occur? Predicting ambulance demand in emergency medical services using artificial intelligence. *Computers, Environment and Urban Systems,* 76, 110–122. https://doi.org/10.1016/j.compenvurbsys.2019.04.006.

Günal, M.M. & Pidd, M. (2010). Discrete event simulation for performance modelling in health care: A review of the literature. *Journal of Simulation,* 4(1: Special Issue: Simulation in Healthcare: Part 2), 42–51. https://doi.org/10.1057/jos.2009.25.

Hammami, H., Ngo, T., Tripe, D., & Vo, D.-T. (2022). Ranking with a Euclidean common set of weights in data envelopment analysis: With application to the Eurozone banking sector. *Annals of Operations Research,* 311, 675–694. https://doi.org/10.1007/s10479-020-03759-6.

Harewood, S.I. (2002). Emergency ambulance deployment in Barbados: A multi-objective approach. *Journal of Operational Research Society,* 53, 185–192. https://doi.org/10.1057/palgrave.jors.2601250.

Hatami-Marbini, A., Emrouznejad, A., & Tavana, M. (2011, November). A taxonomy and review of the fuzzy data envelopment analysis literature: Two decades in the making. *European Journal of Operational Research,* 214(3), 457–472. https://doi.org/10.1016/j.ejor.2011.02.001.

Hatami-Marbini, A., Sajadi, S.M., & Malekpour, H. (2020, August). Optimal control and simulation for production planning of network failure-prone manufacturing systems with perishable goods. *Computers & Industrial Engineering,* 146, 106614. https://doi.org/10.1016/j.cie.2020.106614.

Hatami-Marbini, A., Varzgani, N., Sajadi, S.M., & Kamali, A. (2022). An emergency medical services system design using mathematical modeling. *Decision Analytics Journal,* 3, 100059. https://doi.org/10.1016/j.dajour.2022.100059.

Hollingsworth, B. (2003). Non-parametric and parametric applications measuring efficiency in health care. *Health Care Management Science,* 6, 203–218. https://doi.org/10.1023/A:1026255523228.

Hollingsworth, B. (2008). The measurement of efficiency and productivity of health care delivery. *Health Economics,* 17, 1107–1128. https://doi.org/10.1002/hec.1391.

Hollingsworth, B., Dawson, P., & Maniadakis, N. (1999). Efficiency measurement of health care: A review of non-parametric methods and applications. *Health Care Management Science,* 2, 161–172. https://doi.org/10.1023/A:1019087828488.

Iannoni, A.P., Morabito, R., & Saydam, C. (2009, June). An optimization approach for ambulance location and the districting of the response segments on highways. *European Journal of Operational Research,* 195(2), 528–542. https://doi.org/10.1016/j.ejor.2008.02.003.

Ingolfsson, A., Budge, S., & Erkut, E. (2008). Optimal ambulance location with random delays and travel times. *Health Care Management Science,* 11, 262–274.

Karatas, M. & Yakıcı, E. (2021, November). A multi-objective location analytics model for temporary emergency service center location decisions in disasters. *Decision Analytics Journal,* 1, 100004. https://doi.org/10.1016/j.dajour.2021.100004.

Knight, V., Harper, P., & Smith, L. (2012, December). Ambulance allocation for maximal survival with heterogeneous outcome measures. *Omega,* 40(6), 918–926. https://doi.org/10.1016/j.omega.2012.02.003.

Leknes, H., Aartun, E.S., Henrik, A., Christiansen, M., & Granberg, T.A. (2017, July). Strategic ambulance location for heterogeneous regions. *European Journal of Operational Research,* 260(1), 122–133. https://doi.org/10.1016/j.ejor.2016.12.020.

Li, X., Zhao, Z., Zhu, X., & Wyatt, T. (2011). Covering models and optimization techniques for emergency response facility location and planning: A review. *Mathematical Methods of Operations Research,* 74, 281–310. https://doi.org/10.1007/s00186-011-0363-4.

Liu, K., Li, Q., & Zhang, Z.-H. (2019). Distributionally robust optimization of an emergency medical service station location and sizing problem with joint chance constraints. *Transportation Research Part B: Methodological,* 119, 79–101. https://doi.org/10.1016/j.trb.2018.11.012.

Liu, J.S., Lu, L.Y., Lu, W.-M., & Lin, B.J. (2013). A survey of DEA applications. *Omega,* 41(5), 893–902. https://doi.org/10.1016/j.omega.2012.11.004.

Liu, Y., Roshandeh, A.M., Li, Z., Kepaptsoglou, K., Patel, H., & Lu, X. (2014, September). Heuristic Approach for optimizing emergency medical services in road safety within large urban networks. *Journal of Transportation Engineering,* 140(9), 04014043.

McCormack, R. & Coates, G. (2015, November). A simulation model to enable the optimization of ambulance fleet allocation and base station location for increased patient survival. *European Journal of Operational Research,* 247(1), 294–309. https://doi.org/10.1016/j.ejor.2015.05.040.

McLay, L.A. (2009). A maximum expected covering location model with two types of servers. *IIE Transactions*, 41(8), 730–741. https://doi.org/10.1080/07408170802702138.

McLay, L.A. & Mayorga, M.E. (2010). Evaluating emergency medical service performance measures. *Health Care Management Science,* 13, 124–136. https://doi.org/10.1007/s10729-009-9115-x.

Miller, J.L. & Adam, E.E. (1996, November). Lack and performance in health care delivery. *International Journal of Quality & Reliability Management,* 13, 63–74. https://doi.org/10.1108/02656719610128501.

National Ambulance Service (2015). *National Ambulance Service Plan.* Ireland: Health Service Executive (HSE).

Nayar, P. & Ozcan, Y.A. (2008). Data envelopment analysis comparison of hospital efficiency and quality. *Journal of Medical Systems,* 32, 193–199. https://doi.org/10.1007/s10916-007-9122-8.

Nickel, S., Reuter-Oppermann, M., & Saldanha-da-Gama, F. (2016, March). Ambulance location under stochastic demand: A sampling approach. *Operations Research for Health Care,* 8, 24–32. https://doi.org/10.1016/j.orhc.2015.06.006.

Nogueira, L., Pinto, L., & Silva, P. (2016). Reducing emergency medical service response time via the reallocation of ambulance bases. *Health Care Management Science,* 19, 31–42. https://doi.org/10.1007/s10729-014-9280-4.

Oliveira, M.D. & Bevan, G. (2006, February). Modelling the redistribution of hospital supply to achieve equity taking account of patient's behaviour. *Health Care Management Science,* 9(1), 19–30. https://doi.org/10.1007/s10729-006-6277-7.

Peleg, K. & Pliskin, J.S. (2004). A geographic information system simulation model of EMS: Reducing ambulance response time. *The American Journal of Emergency Medicine,* 22(3), 164–170. https://doi.org/10.1016/j.ajem.2004.02.003.

Rajagopalan, H.K. & Saydam, C. (2009, December). A minimum expected response model: Formulation, heuristic solution, and application. *Socio-Economic Planning Sciences,* 43(4), 253–262. https://doi.org/10.1016/j.seps.2008.12.003.

Rajagopalan, H.K., Saydam, C., & Xiao, J. (2008, March). A multiperiod set covering location model for dynamic redeployment of ambulances.

Computers & Operations Research, 35(3), 814–826. https://doi.org/10.1016/j.cor.2006.04.003.

ReVelle, C. & Hogan, K. (1989, August). The maximum availability location problem. *Transportation Science,* 23, 151–229. https://doi.org/10.1287/trsc.23.3.192.

Robinson, S. (2005). Discrete-event simulation: From the pioneers to the present, what next? *Journal of the Operational Research Society,* 56(6), 619–629. https://doi.org/10.1057/palgrave.jors.2601864.

Sah, B., Titiyal, R., & Sonia (2017). A goal programming and simulation based study for overall process improvement in an Indian hospital. *International Journal of Services and Operations Management,* 27(4), 439–456. https://doi.org/10.1504/IJSOM.2017.085211.

Saydam, C., Repede, J., & Burwell, T. (1994). Accurate estimation of expected coverage: A comparative study. *Socio-Economic Planning Sciences,* 28(2), 113–120. https://doi.org/10.1016/0038-0121(94)90010-8.

Schmid, V. & Doerner, K.F. (2010, December 16). Ambulance location and relocation problems with time-dependent travel times. *European Journal of Operational Research,* 207(3), 1293–1303. https://doi.org/10.1016/j.ejor.2010.06.033.

Simar, L. & Wilson, P.W. (2007). Estimation and inference in two-stage, semi-parametric models of production processes. *Journal of econometrics,* 136(1), 31–64. https://doi.org/10.1016/j.jeconom.2005.07.009.

Syam, S.S. & Côté, M.J. (2010). A location–allocation model for service providers with application to not-for-profit health care organizations. *Omega,* 38(3–4), 157–166. https://doi.org/10.1016/j.omega.08.001.

Toregas, C., Swain, R., ReVelle, C., & Bergman, L. (1971). The location of emergency service facilities. *Operations Research,* 19, 1363–1373. https://doi.org/10.1287/opre.19.6.1363.

Toro-Díaz, H., Mayorga, M.E., Chanta, S., & McLay, L.A. (2013, April). Joint location and dispatching decisions for emergency medical services. *Computers & Industrial Engineering,* 64(4), 917–928. https://doi.org/10.1016/j.cie.2013.01.002.

Toro-Diaz, H., Mayorga, M.E., McLay, L.A., Rajagopalan, H.K., & Saydam, C. (2015). Reducing disparities in large-scale emergency medical service systems. *Journal of the Operational Research Society,* 66(7), 1169–1181. https://doi.org/10.1057/jors.2014.83.

Tu, L.D., Ho, T.H., Nguyen, D.T., & Ngo, T. (2021). Fintech credit and bank efficiency: International evidence. *International Journal of Financial Studies,* 9(3). https://doi.org/10.3390/ijfs9030044.

Ünlüyurt, T. & Tunçer, Y. (2016, December). Estimating the performance of emergency medical service location models via discrete event simulation. *Computers & Industrial Engineering,* 102, 467–475. https://doi.org/10.1016/j.cie.2016.03.029.

Van Barneveld, T., Bhulai, S., & van der Mei, R. (2016, July). The effect of ambulance relocations on the performance of ambulance service providers. *European Journal of Operational Research,* 252(1), 257–269. https://doi.org/10.1016/j.ejor.2015.12.022.

Wang, Y., Luangkesorn, L.K., & Shuman, L. (2012, December). Modeling emergency medical response to a mass casualty incident using agent based simulation. *Socio-Economic Planning Sciences,* 46(4), 281–290. https://doi.org/10.1016/j.seps.2012.07.002.

White, C. (2007). Health care spending growth: How different is the United States from the rest of the OECD? *Health Affairs,* 26(1), 154–161. https://doi.org/10.1377/hlthaff.26.1.154.

Zaffar, M.A., Rajagopalan, H.K., & Saydam, C. (2016, December). Coverage, survivability or response time: A comparative study of performance statistics used in ambulance location models via simulation–optimization. *Operations Research for Health Care,* 11, 1–12. https://doi.org/10.1016/j.orhc.2016.08.001.

https://doi.org/10.1142/9781800615786_0002

Chapter 2

Technological Gaps in the Performance of Victorian Local Government: An Application of the Meta-Frontier Network Data Envelopment Analysis Approach

Carolyn-Thi Thanh Dung Tran[*,†,‡,§,¶,**,††] **and Brian Dollery**[†,‖,‡‡]

[*]*Business Analytics, Australian Institute of Higher Education, Sydney, Australia*

[†]*Business School, University of New England, Armidale NSW, Australia*

[‡]*School of Education and Social Work, University of Sydney, Sydney, Australia*

[§]*School of Business, University of Technology, Sydney, Australia*

[¶]*Institute for Culture and Society, Western Sydney University, Sydney, Australia*

[‖]*College of Economics, Faculty of International Social Sciences, Yokohama National University, Yokohama, Japan*

[**]*c.tran@aih.nsw.edu.au*

[††]*ttran43@une.edu.au*

[‡‡]*bdollery@une.edu.au*

Abstract

The empirical analysis of local government efficiency has spawned a substantial literature. However, to date, few studies have investigated municipal performance by council type under both a meta-frontier framework and a network structure approach. This chapter addresses this gap in the literature by examining the performance of 41 urban and 38 regional Victorian councils under a common framework and in a network structure of financial activity and service provision. Using a meta-frontier network data envelopment analysis for the period of 2014/15–2018/19, we find that regional councils have divisional and network efficiency scores slightly higher than those of urban councils. However, when all councils are placed in a common framework, urban councils are more efficient than their regional counterparts using their own production technology, with meta-technological gap ratios of 0.880 and 0.826, respectively. A sensitivity analysis of divisional weights in a network operational structure indicates that a higher weight placed on service provision could improve the technological gaps of councils and bring them closer to the full meta-frontier production technology of the whole sector. This underscores the importance of public service provision in the performance of Victorian local government. Various policy implications are thus indicated.

Keywords: Financial activities, financial efficiency, local government, meta-frontier, network data envelopment analysis, production technology, regional service provision, urban service provision, regional councils, urban councils, Victorian local government system.

1. Introduction

National and regional governments worldwide remain concerned over the performance of their respective local government systems, especially in terms of their financial and operational problems (see, for instance, Hood & Dixon, 2015; Kuhlmann & Bouckaert, 2016; Naschold, 2017). These concerns are reflected in the scholarly literature on local government. For example, a voluminous empirical effort has been invested in the analysis of municipal performance (see, for instance, Narbón-Perpiñá & De Witte, 2018a, 2018b, for a comprehensive survey of the empirical literature).

Empirical work on the operational efficiency of local governments has pursued two broad lines of inquiry. In the first place, substantial empirical literature has examined the operational efficiency of specific local services, especially urban transport, waste management, and water services (Rogge *et al.*, 2017). By contrast, a second line of inquiry has focused on system-wide operational efficiency in local government jurisdictions across the world (Dollery *et al.*, 2020).

However, to date, little effort has been devoted to investigating the performance of local governments by local council type under both a meta-frontier framework and a network structure approach.

The current chapter thus seeks to address this gap in the empirical literature on local governments by examining the performance of Victorian local government, divided into two categories of 41 urban and 38 regional local councils, under a common framework as well as in a network structure of financial activity and service provision. Using a meta-frontier network data envelopment analysis (NDEA) for the period of 2014/15–2018/19, we demonstrate that regional councils have divisional and network efficiency scores slightly higher than those of urban councils with respect to their own group frontier.

Following this approach, the chapter contributes to the literature by shedding light on the importance of local councils improving their performance across all their divisional activities, such as service provision and financial efficiency, in order to maximize their overall performance. From a public policy perspective, our findings also point to systematic differences between different kinds of local government and thus a need to tailor policies to meet the needs of the various types of local government.

The chapter consists of six main parts. Section 2 provides a synoptic account of municipal efficiency analysis, whereas Section 3 briefly describes the chief characteristics of the Victorian state local government system by way of institutional background. Section 4 presents the integrated meta-frontier and NDEA models used to estimate the overall network, financial and service efficiencies under the common framework for the entire system and the group frontier for each category of council (regional and urban). Section 5 presents the empirical results of the NDEA model estimations, which are then discussed in Section 6. The chapter ends with a brief consideration of the policy implications of our results.

2. Analysis of Local Government Performance

Over several decades, substantial research has focused on the empirical analysis of local government efficiency. A number of exhaustive surveys of the empirical work on local government efficiency have been published. For instance, Worthington & Dollery (2000) examined the empirical literature on municipal efficiency from its origins until 1999. Along similar lines, Afonso & Fernandes (2008) and Benito *et al.* (2014) considered non-parametric empirical research on local government efficiency. In addition, Da Cruz & Marques (2014) created a generic taxonomic system of the determinants of local government performance. However, to date, the most extensive synoptic account of the empirical literature was given by Narbón-Perpiñá & De Witte (2018a, 2018b), who examined all relevant articles published over the period 1990–2016. Moreover, in their survey, they provided a detailed assessment of variable selection as well as the empirical approaches adopted.

In their review of 84 studies, Narbón-Perpiñá & De Witte (2018a) presented a comprehensive assessment of municipal efficiency analysis at the national level, the empirical methodology employed and the variables most frequently used. Narbón-Perpiñá & De Witte (2018a, pp. 451–452) offered four main conclusions from their survey. First, significant variation existed in research efforts at the national level, with Spain, Belgium and Germany being the most scrutinized nations. Second, most researchers used cross-sectional data, with only "a minority of papers" employing a panel structure and a few studies addressing dynamic efficiency analysis of efficiency trends over time. Third, whereas "there is a wide variety of input and output variables to measure local government efficiency," variable selection was contingent to a significant degree on "the availability of data and the specific services and facilities that local government must provide in each country." Finally, substantial variety "in the number of output variables included" was noted.

Narbón-Perpiñá & De Witte (2018b) focused largely on empirical studies of efficiency that employed non-discretionary environmental variables. Narbón-Perpiñá & De Witte (2018b, p. 1126) drew four main conclusions for future empirical work in the area. First, they contended that the inclusion of the impact of environmental variables on local government efficiency was vital since all local authorities are "affected by factors beyond their control." Second, while "there is a wide variety of determinants of local government efficiency," the empirical literature thus lacked "a clear and standard classification for the inclusion of environmental variables in empirical efficiency analysis." Third, empirical research to date has mostly "neglected endogeneity issues in the data" and "in general, more research on the issue of causality and on how endogeneity biases the efficiency results is needed." Finally, in line with Simar & Wilson (2007), Narbón-Perpiñá & De Witte (2018b, p. 1126) argued that "non-parametric methods are based on the two-stage analysis, mainly using Tobit or OLS," and are unsuitable for tackling environmental factors. Future research should thus employ other approaches.

Aiello & Bonanno (2019) conducted a meta-study of 54 studies published between 1993 and 2016 on the efficiency of local government. They found that the efficiency score of studies using the free disposal hull (FDH) approach was higher than that of studies utilizing the DEA approach and that the efficiency of local authorities is positively associated with the national integrity of the legal system. However, none of these studies examined the network structure of local councils in terms of divisional activities. Most recently, Mergoni & De Witte (2022) conducted a systematic review of policy evaluation and efficiency for various sectors. Thirteen publications on local governments were considered in this systematic review. However, none of these studies investigated local government performance in a structural network or examined whether there was a gap between the efficiency of different types of local council under the same operational framework. In this chapter, we attempt to fill this gap in the empirical literature on local governments.

3. The Victorian State Local Government System

Victoria is the second-biggest Australian state in terms of population, with some 5.5 million people in 1.9 million households (ABS, 2016, 2019). The Victorian local government system comprises 79 "general-purpose" local councils with an aggregate income of $9.9 billion per year, generated largely from property taxes, fees and charges, together with around $700 million annually in intergovernmental grants. In 2016, the Victorian state government imposed property tax increase limitations in the form of an annual rate cap that has constrained the capacity of Victorian councils to raise revenue (Drew & Dollery, 2016). With 43,469 employees, Victorian local government administers $91.2 billion in bridges and local roads, libraries, municipal buildings, parks, recreational facilities and other local infrastructure (Municipal Association of Victoria, MAV, 2019).

In line with other Australian state and territory local government systems, the Victorian local government displays substantial differences in the characteristics of its local authorities (MAV, 2019). Its 79 councils can be divided into 31 metropolitan municipalities, together with 48 regional councils, including 10 regional cities. These 79 local councils are classified into four groups under Local Government Victoria's official council typology: cities, shires, regional cities and boroughs (VAGO, 2018).

Considerable diversity is evident among these councils. For example, the geographical size of local councils ranges from 10.8 km^2 to 22,000 km^2, with the population size extending from 2,900 to 313,500 local residents. Moreover, significant differences are evident between Victorian councils in their fiscal capacity, with annual budgets running from $69 million for regional councils to $201 million for metropolitan councils. Since Victorian councils must provide roughly similar services to their local residents, substantial differences in the quality and mix of these local services exist. To a significant degree, there is little that local authorities can do to address these differences due to environmental factors that fall outside their control, such as population density, population size and remoteness.

In common with other Australian state and territory local government systems, the Victorian local government concentrates on the provision of "services to property," such as domestic waste disposal, local bridges and roads, parks and gardens, as well as wastewater (Dollery *et al.*, 2008). Furthermore, in contrast to most similar nations, local authorities in all Australian municipal systems deliver only a limited range of "services to people." While there has recently been steady growth in the provision of services to people, particularly in aged care, it has been far from uniform across and within Australian local government jurisdictions (Dollery *et al.*, 2006a).

Like all Australian state and territory municipal systems, the Victorian local government system operates under a Local Government Act that prescribes the functions, powers and responsibilities of the local councils (Grant & Drew, 2017). Moreover, in common with all other Australian local government jurisdictions, Victorian local councils fall under the regulatory direction of its Department of Local Government (Dollery *et al.*, 2006a). In Victoria, local authorities have a stylized Westminster political structure in which elected representatives (known colloquially as "councilors"), typically led by an indirectly elected mayor, decide municipal policy, which is then implemented by a local public bureaucracy under a general manager.

4. Method of Analysis

In this section, we present the empirical integrated meta-frontier and NDEA model used to estimate the overall network, financial, and service efficiencies under the common framework for the entire sector and the group frontier for each type of council, both regional and urban. The meta-frontier model is an integrated NDEA model, which is then placed in a common framework to estimate the technological gap ratio of the regional and urban groups to the meta-frontier production technology, holding other external factors constant. Finally, a sensitivity analysis is simulated in terms of different weights of financial and service nodes to examine the importance of financial activities and service provision in the operational efficiency of regional and urban councils. We also outline the data and data sources employed in the empirical analysis.

4.1. *Two-division network data envelopment analysis (NDEA)*

As we have seen, DEA is a non-parametric approach which is widely used in operations research to measure the efficiency of organizations. First proposed by Farrell (1957), then Charnes *et al.* (1978) and Banker *et al.* (1984), DEA has been continuously improved, developed, and applied in various sectors, recently including banking (Boubaker *et al.*, 2021, 2022a, 2022c), financial technology (Boubaker *et al.*, 2022b; Le *et al.*, 2021), and airlines (Ngo *et al.*, 2021). Although the main advantage of DEA is its ability to account for multiple inputs and multiple outputs without making assumptions about the relationships between variables, this method does not consider the internal structure of organizations, which has been addressed through NDEA. In essence, NDEA is an extension of the DEA approach, developed by Färe (1991) and

Färe & Grosskopf (1996a, 1996b, 2000) and then extended by Lewis & Sexton (2004) by adding a complex internal structure in which some divisions use resources produced by other divisions while some divisions produce resources used by other divisions.

Tone & Tsutsui (2009) adapted the framework to construct a network of divisions that were considered together in the operational process of a decision-making unit (DMU). NDEA models have been applied in different sectors to investigate the (in)efficiency of intermediate steps to determine more specific and better solutions for improving the efficiency of organizations (Villano *et al.*, 2020). In higher education, some studies, such as Tran & Villano (2018, 2021), Lee & Worthington (2016), Ding *et al.* (2020), Rayeni & Saljooghi (2010), and Monfared & Safi (2013), have employed NDEA in temporal periods with different divisions/nodes in their research contexts. However, to date, no effort has been made to examine the performance of Victorian local government in a network structure to account for the effects of contextual factors on network and structural efficiencies.

In this chapter, we employ the NDEA slacks-based measurement methodology proposed by Tone & Tsutsui (2009) to assess the performance of Victorian local councils in a network structure. This model possesses advantages for assessing the performance of the financial operations and service provision of each Victorian local council. This approach is useful for determining standard practices in municipal performance in the presence of contextual factors that affect performance. Aside from administrative operations, financial management and service provision play a crucial role in the operation of municipalities and thereby contribute to sustainable organizational development. Accordingly, these two divisions are chosen for this study in our NDEA examination of Victorian municipal performance.

As noted earlier, we have adapted the network slacks-based approach proposed by Tone & Tsutsui (2009) to investigate the performance of municipalities. The theoretical model is briefly outlined as follows.

Assume that n local councils (LCs) $(j = 1, \ldots, n)$ consist of K divisions $(k = 1, \ldots, K)$. Let m_k and r_k be the number of inputs to and outputs from division k, respectively. The link from division k

to division h is denoted by $(kh)_i$ and the set of links by L_{kh}. The inputs, outputs, and linking variables are described as follows:

$x_{ijk} \in R_+$ $(i = 1, \ldots, m_k; j = 1, \ldots, K)$ is input resource i to LC_j for division k;
$y_{rjk} \in R_+$ $(r = 1, \ldots, r_k; j = 1, \ldots, K)$ is output product r from LC_j for division k;
$z_{j(kh)l} \in R_+$ $(j = 1, \ldots, m_k; l = 1, \ldots, L_{kh})$ are the links from the intermediate products of UNI_j from division k to division h in period t, where L_{kh} is the number of items in the link from k to h.

Regarding the objective function, input-output constraints and linking constraints, we define the production possibility as set $P = \{x_k, y_k, z_{kh}\}$, as defined by

$$x_k \geq \sum_{j=1}^{n} x_j^k \lambda_j^k (\forall k), \tag{1}$$

$$y_k \leq \sum_{j=1}^{n} y_j^k \lambda_j^k (\forall k), \tag{2}$$

$$z^{kh} = \sum_{j=1}^{n} z_j^{kh} \lambda_j^k (\forall (kh)_l)(\text{as outputs from}\, k), \tag{3}$$

$$z^{kh} = \sum_{j=1}^{n} z_j^{kh} \lambda_j^h (\forall (kh)_l)(\text{as inputs from}\, h), \tag{4}$$

$$\sum_{j=1}^{n} \lambda_j^k = 1(\forall k), \lambda_j^k \geq 0(\forall j, \forall k), \tag{5}$$

where $\lambda^k \in R_+^n$ is the intensity vector corresponding to division $k, \forall k$. In this model, we only focus on two divisions: financial and service nodes. We thus assume that there are no linking inputs to starting divisions and no linking outputs from terminal divisions. In the New South Wales (NSW) local government context, the variable return-to-scale approach is used in this scenario to take into account the effects of contextual factors on municipal performance with respect to finance and service provision capacities.

4.1.1. *Input output constraints*

For $\mathrm{UNI}_o(o = 1, \ldots, n \in \mathrm{P})$, the input and output constraints can be expressed by

$$x_0^k = X^k\lambda^k + s_0^{k-}; \quad y_0^k = Y^k\lambda^k - s_0^{k+}; \quad \text{and} \quad \lambda^k = 1(\forall k), \tag{6}$$

where

$$X^k = (x_1^k, K, x_n^k) \in R^{m_k \times n} \quad \text{and} \quad Y^k = (y_1^k, K, y_n^k) \in R^{r_k \times n} \tag{7}$$

are input and output matrices and s_0^{k-} and s_0^{k+} are, respectively, input and output slacks.

4.1.2. *The linking constraints*

Regarding the linking constraints, a fixed link value is used in this study. This implies that the linking activities remain unchanged (i.e., nondiscretionary) between the two capacities. The linking constraint is given by

$$z^{kh} = Z^{kh}\lambda^h(\forall(k,h)), \tag{8}$$

$$z^{kh} = Z^{kh}\lambda^k(\forall(k,h)). \tag{9}$$

4.1.3. *The objective function*

The input-oriented approach is used to measure the efficiency score of municipalities. This is because the objective of local councils is to minimize inputs in generating their existing financial and service outputs. Accordingly, the overall and divisional efficiencies are depicted by the following formulae:

Input-oriented overall efficiency

$$\theta_0^* = \min_{\lambda^k s^{k-}} \sum_{k=1}^{K} w^k \left[1 - \frac{1}{m_k}\left(\sum_{i=1}^{m_k} \frac{s_{io}^{k-}}{x_{iok}^t}\right)\right], \tag{10}$$

where $\sum_{k=1}^{K} w^k = 1, w^k \geq 0(\forall k)$, w^k is the relative weight of division k, which is determined with respect to its importance and subject to input-output and linking constraints, as mentioned above.

Input-oriented divisional efficiency

$$\theta_k = 1 - \frac{1}{m_k}\left(\sum_{i=1}^{m_k} \frac{s_{io}^{k-*}}{x_{iok}^{t}}\right) (\forall k), \tag{11}$$

where s_{io}^{k-*} are optimal input slacks for the input-oriented overall efficiency. θ_k is the divisional efficiency index which optimizes the overall efficiency θ_0^*. If $\theta_k = 1$, then LC_0 is called input-efficient for division k. The overall input-oriented efficiency score is the weighted arithmetic mean of the divisional scores:

$$\theta_0^* = \sum_{k=1}^{K} w^k \theta_k. \tag{12}$$

This model is applied to two groups of local councils — regional and urban councils — representing the individual group frontier. The concept illustrated in Figure 1 is used to define the appropriate output and input variables. As noted earlier, we focus on the stages in the network process when local councils are maximizing their ability to undertake financial operations and service provision activities.

A major objective in municipal financial management is for local councils to use total revenue efficiently through expenditure outlays on different operations to meet the demands of the local community. This stage provides financial resources for service divisions to provide services to local residents. By contrast, the main objective of the service division is to provide local public services to the local community, such as environmental services, infrastructure maintenance,

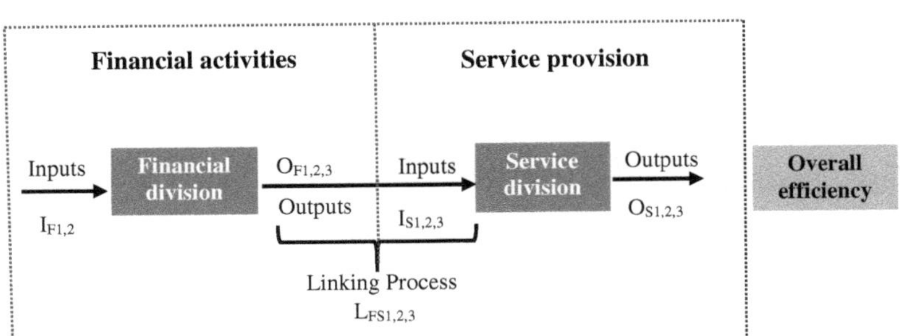

Figure 1. Network of the local council performance in Victoria, Australia.

and community services. Both divisional capacities are crucial to the overall operational processes of local councils. The network model enables us to examine the activities of both divisions independently and observe how they contribute to the overall performance of local councils.

The inputs, outputs and linking variables used in this study are presented in Table 1. The inputs for the financial division include the total own-source revenue (I_{F1}), comprising all revenue obtained by local councils in the financial years in question and used for funding continuing council operations, and the total grants from government (I_{F2}). These types of capital sources were used as inputs in previous studies (Cordero *et al.*, 2017; Tran & Villano, 2018, 2021).

Table 1. Description of variables used in the empirical NDEA model.

Nodes/Variables	Category	Description
Financial division		
I_{F1}	Input	Total own-source revenue of each council.
I_{F2}	Input	Total grants from government of each council.
O_{F1}	Output	Employee costs refer to expenses for full-time equivalent staff working in an individual local council.
O_{F2}	Output	Materials costs are expenses for materials for service provision.
O_{F3}	Output	Other costs necessary for operations of each local council.
Linking process		
L_{FS1}	Link	Employee costs (being an input in the service division).
L_{FS2}	Link	Materials (being an input in the service division).
L_{FS3}	Link	Other costs (being an input in the service division).
Service division		
O_{S1}	Output	The number of households in each council.
O_{S2}	Output	The number of businesses established in each council.
O_{S3}	Output	Total road length (km) in each council to be maintained.

The output variables for the financial division are employee costs (O_{F1}), materials costs (O_{F2}), and other costs (O_{F3}). These variables do not necessarily equally affect the financial division. The output variables for the financial division are linked through the provision of financial resources to the municipal service division as linking variables. These outputs are used for the financial division in a NDEA structure, as shown by Tran & Villano (2018, 2021).

For the service division, the input variables are the three output variables of the financial division with respect to the three types of costs described above. These inputs constitute the resources available to meet the service demands of local residents in individual councils. These cost inputs are widely used in the empirical literature on local governments (Drew *et al.*, 2015, 2017; Cordero *et al.*, 2017; Tran & Dollery, 2020). The key outputs of the service division include the total number of households, the number of established businesses in each local council, and the total length of roads that local councils must maintain (Drew *et al.*, 2017; Tran & Dollery, 2020).

The linking variable acts as an intermediate product, which is a major difference between the NDEA model and the traditional DEA model. This variable is treated as an output in the financial division and becomes an input in the service division. Using an intermediate product, the multiple production stages are estimated among divisions in a local council. In this chapter, three types of expenses are used as the linking variables from the financial division to the municipal services division. These expenditures for different operations represent mediating variables that facilitate the municipal service provision process.

For the NDEA approach, weights for divisions/capacities need to be predetermined based on the practical circumstances. We assume that local councils' ability to implement financial and service provision activities is equally important. Accordingly, the importance of financial operations and service provision tasks is assumed to be initially equal, that is, 0.5 for each division. A sensitivity analysis will then be presented in the following section, which allows us to change different values of the weights for each division to determine the real importance of division in a network organizational structure.

4.2. NDEA meta-frontier production technology

The meta-frontier method was first developed in the late 1960s by Hayami (1969) and Hayami & Ruttan (1970) in agriculture and then expanded to stochastic frontier analysis (Battese & Rao, 2002; Battese *et al.*, 2004) and a DEA framework (O'Donnell *et al.*, 2008). The meta-frontier DEA approach has been applied in various sectors, including hospitality, industry, environment, higher education, and local government (Beltrán-Esteve *et al.*, 2014; Chiu *et al.*, 2013; Hsiao, *et al.*, 2012; Yu & Choi, 2015; Tran & Villano, 2018, 2021; Tran & Dollery, 2020), given its usefulness in capturing the difference between the group frontier and meta-frontier efficiencies (Chao *et al.*, 2018; Chen *et al.*, 2022). Integrating the meta-frontier approach into a NDEA model was first conducted for a local government to measure the gap between the individual group frontier (regional and urban) and the meta-frontier local government system in Victoria when all local councils are placed in a common framework.

According to O'Donnell *et al.* (2008), the meta-frontier framework refers to an overarching boundary encompassing the entire group frontiers. It is assumed that there exists no substantial change in the production technology of each council in the surveyed report, given that local councils are public-owned organizations, and thus any change in technology might take time to get government budget approval. The linear programming of the meta-frontier framework is similar to that of individual group frontiers. However, inputs and outputs in this framework are combined with those of each DMU for the whole system (Tran & Villano, 2018; Tran & Dollery, 2020). The meta-frontier linear programming solved in a network structure in terms of two divisions — financial activities and service provision — produces efficiency scores for each division and the overall network structure for single councils under the meta-frontier framework for the whole local government system. The gap of each council in its own group frontier relative to the whole system meta-frontier is measured as follows:

$$\mathrm{MTGR} = \frac{\mathrm{NE}}{\mathrm{NE}_j}, \tag{13}$$

where MTGR is the meta-frontier technological gap ratio, NE_j is network efficiency for individual group frontier j, and NE is network efficiency for the meta-frontier framework. MTGR ranges within 0–1. If MTGR for any single DMU in each group is less than 1, then this DMU needs to improve its performance to reach full meta-frontier efficiency when we place it together with all the others in a common framework.

4.3. *Sensitivity analysis*

A sensitivity analysis has been proposed to assess the preferred model or the importance of operational divisions in a network structure (Carrington *et al.*, 2005; Tran, 2021). By simulating the divisional weights of each division, the divisional efficiencies would change, thus affecting the overall network efficiency and MTGR of a single council to the meta-frontier framework. Given that local councils provide public services, the importance of financial activities and service provision is relatively comparable, and thus the divisional weights for these divisions are expected to be varied within an interval of 0.4–0.6, so that the sum of weights for two divisions is always equal to one (Tone & Tsutsui, 2014; Tran, 2021). An increasing or decreasing trend in the efficiency of group frontier and meta-frontier for the whole system will identify the importance of divisions with respect to their different weights.

A summary of the variables used in the NDEA analysis is presented in Table 2. On average, the total revenue of municipalities

Table 2. Summary statistics of variables used in NDEA, 2014/15–2018/19 ($n = 79$).

Variables	Type/Node	Mean	Standard deviation
Total own-source revenue ($ million)	Input/Financial	93.46	84.44
Total grants from government ($ million)	Input/Financial	17.29	11.23
Employee costs ($ million)	Output/Financial link to service as input	42.24	35.39
Material costs ($ million)	Output/Financial link to service as input	37.38	33.45
Other costs ($ million)	Output/Financial link to service as input	24.32	18.92
Households	Output/Service	7,242	6,948
Businesses	Output/Service	26,742	23,327
Total roads (km)	Output/Service	2,215	3,590

in Victoria is \$110.75 million, including \$93.46 million from their total own-source revenue and \$17.29 million from total government grants. The total expenditure of local councils covers employee costs (\$42.24 million), material costs (\$37.38 million), and other related costs (\$24.32 million). These expenses are used to provide local public services to households, businesses, and road maintenance in the regions. Overall, Victoria municipalities secured a net operating surplus (\$6.81 million on average) in the reported period. However, the empirical literature on local government (see, for instance, Marquardt & Wiedman, 2004; Stalebrink, 2007; Pilcher & Van der Zahn, 2010; Drew, 2018) reveals that residents are assumed not to support local councils that accrue more revenues (via taxation, fees, and charges collected from residents) than required to cover expenditure on local public services. By contrast, residents also dislike ongoing negative net operating results by local councils. This argument received recent support from Tran & Dollery (2021) on the negative relationship between financial performance and resident satisfaction in Victorian local government. Thus, financial efficiency is of interest for improving the performance of local councils, which can contribute to the effectiveness of local councils, as proxied by resident satisfaction.

5. Empirical Results

This section presents the estimated results of the main research objective by determining the technological gap in the efficiency of municipalities between the regional and urban groups. In so doing, we estimate the group frontier efficiency of councils in separate regional and urban groups in a network structure of financial and service divisions. All councils were then placed into an overarching framework to estimate the meta-frontier network efficiency. Finally, the meta-frontier technological gap (MTGR) of councils is measured by the ratio of their own group frontier efficiency to the meta-frontier efficiency. The sensitivity analysis is simulated by using different divisional weights to inform the importance of divisions in a network structure through variation in the group, meta-frontier efficiencies and MTGRs.

Table 3 presents the estimated results of the group frontier and meta-frontier efficiencies in a network structure for Victorian local government for the period of 2014/15–2018/19. For the urban group, the average efficiency of financial activities is 0.638, which implies that urban councils could potentially ameliorate their efficiency by 0.362 to reach full efficiency in this division. This efficiency score is less than the average allocative efficiency of 0.813 found by Worthington (2000) for 177 NSW councils and the cost efficiency of German municipalities on average at 0.88 (Kalb *et al.*, 2012). By contrast, the efficiency of service provision is higher at 0.861, showing improved practice in providing local services to local residents. The difference between the financial and service efficiencies is significant at the 1% level. However, urban councils still need to improve the efficiency of their service provision by 0.139 to obtain

Table 3. Group frontier and meta-frontier network efficiency of the Victorian local government, 2014/15–2018/19.

	Mean	Standard deviation	Min	Max	Hotelling test
Urban group frontier network performance ($n = 41$)					
Financial efficiency	0.638	0.156	0.289	1.000	
Service efficiency	0.861	0.119	0.595	1.000	
Urban group frontier network efficiency	0.750	0.096	0.550	1.000	255.77***
Rural group frontier network performance ($n = 38$)					
Financial efficiency	0.757	0.126	0.499	1.000	
Service efficiency	0.865	0.118	0.508	1.000	
Rural group frontier network efficiency	0.811	0.085	0.626	1.000	71.36***
Meta-frontier network performance for the whole sample ($n = 79$)					
Financial efficiency	0.546	0.140	0.250	1.000	
Service efficiency	0.777	0.151	0.451	1.000	
Meta-frontier network efficiency	0.661	0.101	0.477	1.000	473.18***
Technological gap ratios to the meta-frontier technology					
Urban councils	0.880	0.100	0.579	1.000	
Rural councils	0.826	0.119	0.552	1.000	20.41***
Unity MTGR		Urban councils	05		
		Rural councils	03		

Note: *** denotes the 1% level of significance.

unity efficiency. Overall, using their own frontier, urban municipalities obtain an average efficiency score of 0.75 in a network framework, meaning that urban councils can improve their performance by reducing input resources in producing the existing outputs. The network efficiency of urban councils under a network structure is less than the average efficiency score of 0.912 found in metropolitan councils in South Australia (Tran & Dollery, 2020).

As can be seen from Table 3, the regional group frontier network efficiency is 0.811, higher than the network efficiency of urban municipalities, using their own technology. However, to reach their full frontier efficiency, regional councils must improve their performance by 0.191. This finding is marginally higher than the average efficiency of medium and small agriculture at 0.785 and slightly less than the average efficiency of the very large and large agriculture group at 0.868 in South Australia (Tran & Dollery, 2020). In common with the urban group, regional councils obtain an efficiency in financial activities that is less than what they achieve in service provision (0.757 vs. 0.865). It is noted that the financial efficiency in the regional group is higher than the cost efficiency found by Worthington (2000) at 0.7 for 177 NSW councils. However, it is less than the cost efficiency of German municipalities on average at 0.88 (Kalb *et al.*, 2012). The Hotelling test indicates that the difference between the service and financial efficiencies for the regional group is statistically significant at the 1% level. This implies that using their own frontier, both the regional and urban groups have demonstrated improved practice in service provision to residents, given federal and state government financial support through annual grants.

The meta-frontier network efficiency for the whole system is estimated for financial activities and service provision when all municipalities are placed in a common framework. The estimated results of these divisions are relatively consistent with the findings in the separate regional and urban groups. Service efficiency (0.777) for the whole system is higher than financial efficiency (0.546) at the 1% level of significance. Overall meta-frontier network efficiency is 0.661, less than the overall group frontier network efficiency of the regional and urban groups. This result is also less than the overall meta-frontier efficiency of local councils in South Australia at 0.84 (Tran & Dollery, 2020) and the standard efficiency of local government in NSW at 0.90 (Drew *et al.*, 2017), but greater than the variable return-to-scale efficiency score of local councils in Western Australia at 0.616 (Fogarty & Mugera, 2013). This network efficiency score is nonetheless relatively consistent with the overall efficiency of Victorian local councils at 0.660, obtained using the standard DEA model for the same reported period (Tran & Dollery, 2023b).

In order to understand the effect of financial activities and service provision on the overall and group efficiencies of local councils in a network structure, a Pearson correlation test was conducted, which is presented in Table 4. We observe that in the urban group, both correlation coefficients of financial activities and service production with group frontier network efficiency are positive and significant at the 1% level and that the correlation coefficient of financial activities is higher than the correlation coefficient of service provision (0.783 vs. 0.568). This implies that financial efficiency has a stronger influence on the overall group network efficiency in providing public services to residents. The substantial empirical literature on local governments confirms a significant relationship between financial performance and the effectiveness of local councils proxied by resident satisfaction (Drew, 2018; Stalebrink, 2007; Pilcher & Van der Zahn, 2010; Tran & Dollery, 2023a). Thus, financial efficiency is a key enhancer of the performance of local councils through service efficiency in public service provision.

Table 4. Pairwise Pearson correlation between financial, service, and overall efficiency for group frontier and meta-frontier NDEA models.

	Financial	**Service**	**Overall efficiency**
Urban group frontier			
Financial	1		
Service	−0.0456	1	
Group network efficiency	0.783***	0.586***	1
Rural group frontier			
Financial	1		
Service	−0.022	1	
Group network efficiency	0.725***	0.672***	1
Meta-frontier for the whole system			
Financial	1		
Service	−0.0498	1	
Overall network efficiency	0.659***	0.718***	1

Note: *** denotes the 1% level of significance.

For the regional group, financial efficiency is more strongly correlated to the group network frontier efficiency than to service efficiency (0.725 vs. 0.672 at the 1% level of significance). This result is consistent with the findings for the urban group. However, the difference between these two correlation coefficients of the urban group is marginal in absolute value. A closer look at Table 4 reveals that the correlation coefficient of service provision in regional councils is higher than that of service provision in urban councils, whereas the correlation coefficient of financial activities in regional councils is less than that of their urban counterpart. This implies that financial efficiency is more important for the urban group, while the importance of finance and service efficiencies is relatively equal for the regional group. Tran & Dollery (2022) found that the influence of administrative intensity differs between regional and urban groups in relation to the financial performance of Australian local councils. The empirical literature generally finds that contextual factors affect the performance of local councils, and their impact depends on the local circumstances (Da Cruz & Marques, 2014; Drew *et al.*, 2016).

Under the meta-frontier framework, the overall network efficiency of councils is more strongly correlated to service efficiency than to financial efficiency (0.718 vs. 0.659 at the 1% level). This result is opposite to the individual group frontiers, where the impact of financial efficiency on the overall network group frontier efficiency is stronger than the effect of service provision, although the difference in the absolute values of these two correlation coefficients is minimal. This underlines the importance of service provision for the whole local government system, although financial operations are recognized as a key enhancer of the efficiencies of both the regional and urban groups in a network structure using their own technology. Tran & Villano (2021) arrived at the same conclusion in their study of the network efficiency of higher education in Vietnam. Specifically, the correlation coefficient of the financial division is less than that of the academic division in relation to the overall network efficiency of universities, underlining the importance of the academic division in the performance of Vietnamese universities.

The meta-frontier technological gap ratios in Table 5 show that urban councils are likely to be more efficient than their regional counterparts in using inputs to produce outputs under unrestricted meta-technology. Specifically, the average MTGRs of urban and

Table 5. Sensitivity analysis for different divisional weights.

	Divisional weight		**Group frontier efficiency**		**Meta-frontier efficiency**		**MTGR**	
Cases	**Financial**	**Service**	**Urban**	**Rural**	**Urban**	**Rural**	**Urban**	**Rural**
1	0.60	0.40	0.727	0.800	0.622	0.656	0.862	0.822
2	0.55	0.45	0.738	0.806	0.638	0.662	0.871	0.824
3	0.50	0.50	0.750	0.811	0.655	0.668	0.880	0.826
4	0.45	0.55	0.761	0.816	0.672	0.674	0.888	0.828
5	0.40	0.60	0.772	0.822	0.688	0.680	0.896	0.830
t-test between two groups			-6.932^{***}		-1.042		8.640^{***}	

Note: *** denotes the 1% level of significance.

regional councils are 0.880 and 0.826, respectively, implying that the associated provision technology of urban councils represents 88% of the existing outputs, while the associated provision technology of regional councils represents 82.6% of the existing outputs that can be supplied to local residents using the same inputs and technology represented by the meta-frontier. In other words, urban councils could potentially ameliorate their performance by 12% providing services to residents, whereas regional councils could increase their efficiency by 18.4% under unrestricted meta-technology. The divergence in MTGRs between regional and urban municipalities is significant at the 1% level. The MTGRs of both the urban and regional groups are greater than the meta-technological gap of the metropolitan and two rural groups in South Australia at 0.843, 0.786, and 0.781 respectively (Tran & Dollery, 2020). For the whole survey period, five urban councils obtained an MTGR of unity, while three rural councils reached the full MTGRs under the same meta-technological framework.

Figure 2 demonstrates the variation in the MTGRs of regional and urban councils over time. While the MTGRs of urban councils exhibit a rising trend, their regional counterparts display an opposite tendency. This indicates that under the same meta-technology, urban councils are closer to their meta-frontier framework, whereas regional councils face a larger gap between their own frontier technology and the meta-technology. This has policy implications for the regional and urban groups with respect to improving their performance in the whole system.

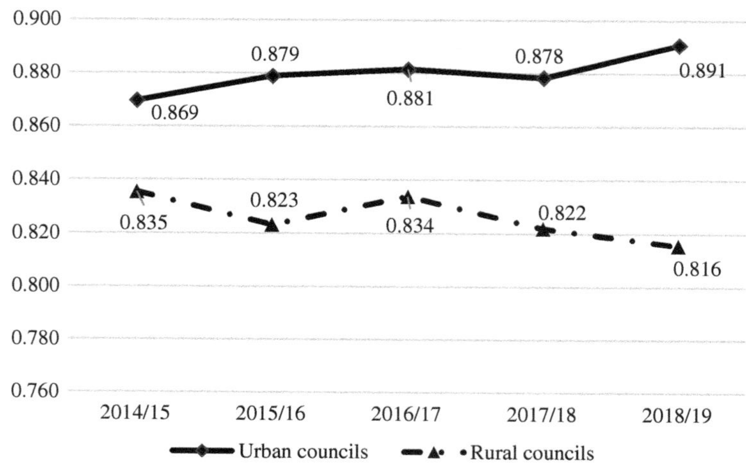

Figure 2. Meta-frontier technological gap ratio of Victorian local councils.

As noted earlier, a sensitivity analysis was conducted by simulating divisional weights to evaluate changes in group frontier, meta-frontier efficiency and MTGR and thereby determine which division — financial or service provision — in a network structure plays an important role in the performance of regional and urban councils under the meta-technology. Table 5 presents the results of efficiency and MTGRs for five cases of different divisional weights for the two divisions (financial and services) under the regional and urban groups. It can be seen that when the weights of service divisions increase (implying that the weights of financial divisions decrease so that the total sum of weights equals one in a network structure), the group frontier efficiencies, meta-frontier efficiencies and MTGRs all increase in the same direction for both the regional and urban groups. Put differently, higher weights of services generate higher MTGRs, which enable local councils to get closer to the meta-frontier technology. The t-test shows that there is a significant difference in the group frontier efficiency with respect to the five cases of different divisional weights between the regional and urban groups at the 1% level. However, the meta-frontier efficiency does not respond to changes in divisional weights. There is thus no difference in the meta-frontier efficiency between the regional and urban groups.

As indicated earlier, the overall standard DEA efficiency of Victorian local councils reported by Tran & Dollery (2022a) is similar to our result obtained using NDEA efficiency in a common framework. This implies that NDEA is effective for the group frontiers classified by council type rather than the whole frontier framework. Future studies may want to add divisions to a network model to investigate their impact on the meta-frontier network efficiency. We find that the MTGRs of urban councils are significantly higher than those of regional councils at the 1% level when the divisional weights for service provision increase from 0.4 to 0.6. This underlines the importance of the service division in improving the performance of local councils in providing public services under unrestricted meta-technology using the same input resources.

6. Discussion of Empirical Findings

The empirical literature has demonstrated that a significant benefit of the NDEA method is that it allows investigators to monitor efficiency variation separately for various subdivisions in a network structure (Tone & Tsutsui, 2009; Tran & Villano, 2021). Although each municipality may have a different organizational structure to provide public services to local residents, financial activities and service provision are considered the two crucial divisions determining the performance of a local council. In essence, the financial division provides financial resources to the service division, which is the main division responsible for organizing and delivering public services to users (Dollery & Grant, 2011; Tran & Dollery, 2021). In the context of local governments, while the efficiency of financial activities and service provision mainly determines overall efficiency, local councils can split these divisions into subdivisions to further explore the operational efficiency of these subdivisions. More effective policies can thus be developed for specific divisions that meet their unique characteristics and responsibilities (Tran & Villano, 2018, 2021). This financial–service network efficiency model has been applied to Victorian regional and urban groups to estimate the overall efficiency of local councils within their own group. In addition, a meta-frontier framework has been deployed to determine whether meta-frontier technological gaps exist in Victorian local councils when we place them all into an unrestricted meta-technological framework.

Our findings reveal that the efficiency of service provision is superior to the efficiency of financial divisions for both the regional (0.865 vs. 0.757) and urban groups (0.861 vs. 0.638). Similarly, when we place all councils into a meta-technology framework, service efficiency is greater than financial efficiency, at 0.777 and 0.546, respectively, at the 1% level of significance. The meta-frontier network efficiency of councils in Victoria at 0.661 is less than the standard DEA efficiency of New South Wales councils at 0.9 (Drew *et al.*, 2017) and of South Australian councils at 0.84 (Tran & Dollery, 2020), but slightly greater than the efficiency of Western Australian councils at 0.616 (Fogarty & Mugera, 2013). This implies that the network efficiency analysis of an entity technically provides a stricter assessment than standard efficiency methods because the former is estimated through the efficiency of each division based on divisional weighting (Tran & Villano, 2021) and that local councils achieve full efficiency only when all their departments are operating at full efficiency level. From a managerial perspective, local councils should have an appropriate strategy for each department to improve their operating efficiency and thereby the overall performance of local councils and the entire local government system.

Using the correlation analysis, we find that under meta-frontier framework, the overall network efficiency is strongly correlated to both financial and service efficiencies. Moreover, the impact of service provision is slightly stronger than the influence of financial activities (0.718 vs. 0.659). By contrast, the group network efficiency of both urban and regional groups is more strongly related to financial activities than to service provision. This reflects the important role of financial operations in ameliorating the performance of local councils (Jurewicz, 2018; Tran & Dollery, 2021).

The meta-frontier technological gap ratios reveal that urban councils are more efficient than regional councils (0.88 vs. 0.826), implying that urban councils can provide 88% of existing outputs, whereas regional councils provide 82.6% of existing outputs, under the same unrestricted meta-technology structure. Given the differences between regional and urban categories of council, our findings highlight the need for public policy intervention to be tailored to meet the specific circumstances of different kinds of local council in any given local government system, such as Victorian local government (Drew, 2018; Tran & Dollery, 2020).

The sensitivity analysis shows that the higher divisional weights placed on service provision could potentially improve the performance of both regional and urban councils in providing existing outputs using the same inputs under an unrestricted meta-technology framework. This underlines the importance of service provision to local residents to the performance of councils during the surveyed period, despite the strong influence of financial resources on the group network efficiency of the regional and urban groups. Put differently, municipalities should be aware that they only obtain full overall efficiency when they are efficient in all internal operations, including financial operations, service provision and community engagement. This implies that local councils should have a specific strategy for each division to reduce technical inefficiency in their divisional operations (Tran & Villano, 2021).

7. Concluding Remarks

This chapter has examined the overall performance of Victorian municipalities and the difference in efficiency between regional and urban groups of councils in service provision to their local residents over the period of 2014/15–2018/19. The NDEA model was used to analyze the impact of the internal structure of each local authority through the two key divisions of financial operations and service provision operations to determine the network efficiency of councils relative to their own group frontiers. The meta-frontier network framework was then employed to estimate the overall meta-frontier network efficiency for the whole system and the MTGRs of urban and regional councils in an unrestricted meta-technology structure.

Our approach generated several illuminating findings. First, under the group frontier, service efficiency was higher than financial efficiency, thereby illustrating that Victorian municipalities were efficiently providing local public services to their residents during the surveyed period, although they could still improve their performance to obtain full frontier efficiency. Second, under the unrestricted meta-frontier framework, urban councils exhibited superior practice in using their associated technology to provide a higher proportion of existing outputs to the whole local government system than their regional counterparts using the same input resources. The MTGRs of urban councils displayed an increasing trend, whereas those of regional councils indicated a declining trend over the reported period. Five urban councils displayed the ability to provide 100% of existing outputs to the whole local government system (MTGR = 1) compared to only three regional councils. These differences between the regional and urban groups of councils were significant at the 1% level. Finally, although the overall meta-frontier network efficiency of Victorian municipalities in an unrestricted meta-technology is more strongly linked to service efficiency than to financial efficiency, the overall frontier network efficiency for the separate regional and urban groups is associated more strongly with financial efficiency than with service efficiency. This indicates that the efficiency of their own group frontiers depends on financial efficiency to boost the overall performance of councils through increased efficiency in service provision. In other words, financial performance plays an important role in boosting the efficiency of both service provision and the whole system.

These findings suggest the following policy implications. First, from a managerial perspective, the full efficiency of municipalities can be achieved if and only if each division in the network organizational structure of local councils is operating at full efficiency, regardless of variations in divisional weightings. This implies that local councils should have specific strategies for each division to improve their operational efficiency and thereby improve their performance using associated production technology. Second, financial efficiency is pivotal in improving the network efficiency of both urban and regional councils, given that the financial division provides the financial resources for service provision. Local councils should be aware of these factors to tailor an appropriate strategy to improve their financial performance. This would also increase their effectiveness through enhanced resident satisfaction. Finally, from a public policy perspective, "one-size-fits-all" policy intervention would be counterproductive, given the differences between urban and regional councils in their ability to provide public services to residents in an unrestricted common framework using the same input resources. Our findings thus highlight the need for public policy intervention to be tailored to meet the specific circumstances of different kinds of local council in any given local government system, such as the Victorian state local government system.

Although our study has provided some informative results using the new network meta-frontier framework in the context of local governments, more empirical work should be undertaken to explore the question further. At least three potentially fruitful avenues for future research are apparent: (1) contextual variables (e.g., demographics, socioeconomics) should be tested in a second-stage regression model to yield an in-depth analysis of the impact of exogenous variables on municipal performance; (2) the meta-frontier dynamic network efficiency model could be developed in future studies to capture the changes in efficiency of each division, overall performance and MTGRs over time to compare with our findings; and (3) additional municipal divisions should be considered, including council administration, which could shed light on important policy questions, such as the impact of administrative intensity on overall municipal performance.

References

Afonso, A. & Fernandes, S. (2008). Assessing and explaining the relative efficiency of local government. *Journal of Socioeconomics*, 37(5), 1946–1979.

Aiello, F. & Bonanno, G. (2019). Explaining differences in efficiency: A meta-study on local government literature. *Journal of Economic Surveys*, 33(3), 999–1027.

Australian Bureau of Statistics (ABS) (2016). *Census of Population and Housing: Socio-Economic Indexes for Areas (SEIFA), Australia, 2016.* Canberra: Australian Bureau of Statistics.

Australian Bureau of Statistics (ABS) (2019). *Population and People, LGA, 2011 to 2018.* Canberra: Australian Bureau of Statistics.

Banker, R.D., Charnes, A., & Cooper, W.W. (1984). Some models for estimating technical and scale inefficiencies in data envelopment analysis. *Management Science*, 30(9), 1078–1092.

Battese, G.E. & Rao, D.S.P. (2002). Technology gap, efficiency, and a stochastic metafrontier function. *International Journal of Business and Economics*, 1, 87–93.

Battese, G.E., Rao, D.S.P., & O'Donnell, C.J. (2004). A metafrontier production function for estimation of technical efficiencies and technology gaps for firms operating under different technologies. *Journal of Productivity Analysis*, 21, 91–103.

Beltrán-Esteve, M., Gómez-Limón, J.A., Picazo-Tadeo, A.J., & Reig-Martínez, E. (2014). A metafrontier directional distance function approach to assessing eco-efficiency. *Journal of Productivity Analysis*, 41, 69–83.

Benito, B., Bastida, F., & García, J.A. (2010). The determinants of efficiency in municipal governments. *Applied Economics*, 42(4), 515–528.

Boubaker, S., Le, T.D.Q., & Ngo, T. (2022a). Managing bank performance under COVID-19: A novel inverse DEA efficiency approach. *International Transactions in Operational Research*, 30(2023), 2436–2452. https://doi.org/10.1111/itor.13132.

Boubaker, S., Manita, R., & Mefteh-Wali, S. (2022b). Foreign currency hedging and firm productive efficiency. *Annals of Operations Research*, 313, 833–854.

Boubaker, S., Manita, R., & Rouatbi, W. (2021). Large shareholders, control contestability and firm productive efficiency. *Annals of Operations Research*, 296(1), 591–614.

Boubaker, S., Do, D.T., Hammami, H., & Ly, K.C. (2022c). The role of bank affiliation in bank efficiency: A fuzzy multi-objective data envelopment analysis approach. *Annals of Operations Research*, 311, 611–639. https://doi.org/10.1007/s10479-020-03817-z.

Carrington, R., Coelli, T., & Rao, P. (2005). The performance of Australian Universities: Conceptual issues and preliminary results. *Economic Papers*, 24, 145–163.

Chao, C.M., Yu, M.M., Hsiung, N.H., & Chen, L.H. (2018). Profitability efficiency, marketability efficiency and technology gaps in Taiwan's banking industry: Meta-frontier network data envelopment analysis. *Applied Economics*, 50(3), 233–250.

Charnes, A., Cooper, W.W., & Rhodes, E. (1978). Measuring the efficiency of decision-making units. *European Journal of Operational Research*, 2(6), 429–444.

Chen, Y., Ni, L., & Liu, K. (2022). Innovation efficiency and technology heterogeneity within China's new energy vehicle industry: A two-stage NSBM approach embedded in a three-hierarchy meta-frontier framework. *Energy Policy*, 161, 112708. https://doi.org/10.1016/j.enpol.2021.112708.

Chiu, C.-R., Lu, K.-H., Tsang, S.-S., & Chen, Y.-F. (2013). Decomposition of metafrontier inefficiency in the two-stage network directional distance function with quasi-fixed inputs. *International Transactions in Operational Research*, 20, 595–611.

Cordero, J.M., Pedraja-Chaparro, F., Pisaflores, E.C., & Polo, C. (2017). Efficiency assessment of Portuguese municipalities using a conditional nonparametric approach. *Journal of Productivity Analysis*, 48(1), 1–24.

Da Cruz, N.F. & Marques, R.C. (2014). Revisiting the determinants of local government performance. *Omega*, 44, 91–103.

Ding, T., Yang, J., Wu, H., Wen, Y., Tan, C., & Liang, L. (2020) (in press). Research performance evaluation of Chinese university: A non-homogeneous network DEA approach. *Journal of Management Science and Engineering*. https://doi.org/10.1016/j.jmse.2020.10.003.

Dollery, B.E. & Grant, B. (2011). Financial sustainability and financial viability in Australian local government. *Public Finance and Management*, 11(1), 28–47.

Dollery, B.E., Byrnes, J.D., & Crase, L. (2008). A note on structural reform in Australian local government. *Australian Journal of Political Science*, 43(2), 333–339.

Dollery, B.E., Crase, L., & Johnson, A. (2006a). *Australian Local Government Economics*. Sydney: University of New South Wales Press.

Dollery, B.E., Wallis, J.L., & Allan, P. (2006b, November). The debate that had to happen but never did: The changing role of Australian local government. *Australian Journal of Political Science*, 41(4), 553–567.

Dollery, B.E., Kitchen, H., McMillan, M., & Shah, A. (2020). *Local Public, Fiscal and Financial Governance: An International Perspective*. New York: Palgrave McMillan.

Drew, J. (2018). Playing for keeps: Local government distortion of accruals in response to high-stakes public policy-making. *Public Money and Management*, 38(1), 57–64.

Drew, J. & Dollery, B.E. (2016). Does size still matter? An empirical analysis of the effectiveness of Victorian local authorities. *Local Government Studies*, 42, 15–28.

Drew, J., Kortt, M.A., & Dollery, B.E. (2015). What determines efficiency in local government? A DEA analysis of NSW local government. *Economic Papers: A Journal of Applied Economics and Policy*, 34(4), 243–256.

Drew, J., Kortt, M.A., & Dollery, B.E. (2016). Can't get no satisfaction: The association between community satisfaction and population size for Victoria. *Australian Journal of Public Administration*, 75(1), 65–77.

Drew, J., Kortt, M.A., & Dollery, B.E. (2017). No Aladdin's cave in New South Wales? Local government amalgamation, scale economies, and data envelopment analysis specification. *Administration & Society*, 49(6), 1450–1470.

Färe, R. (1991). Measuring Farrell efficiency for a firm with intermediate inputs. *Academia Economic Papers*, 19, 329–340.

Färe, R. & Grosskopf, S. (1996a). Productivity and intermediate products: A frontier approach. *Economics Letters*, 50, 65–70.

Färe, R. & Grosskopf, S. (1996b). *Intertemporal Production Frontiers: With Dynamic DEA.* Amsterdam: Kluwer Academic.

Färe, R. & Grosskopf, S. (2000). Network DEA. *Socio-Economic Planning Sciences*, 34, 35–49.

Farrell, M.J. (1957). The measurement of productive efficiency. *Journal of the Royal Statistical Society: Series A (General)*, 120(3), 253–281.

Fogarty, J. & Mugera, A. (2013). Local government efficiency: Evidence from Western Australia. *Australian Economic Review*, 46(3), 300–311.

Grant, B. & Drew, J.D. (2017). *Local Government in Australia.* Amsterdam: Springer.

Hayami, Y. (1969). Sources of agricultural productivity gap among selected countries. *American Journal of Agricultural Economics*, 51, 564–575.

Hayami, Y. & Ruttan, V.W. (1970). Agricultural productivity differences among countries. *The American Economic Review*, 60: 895–911.

Hood, C. & Dixon, R. (2015). *A Government that Worked Better and Cost Less? Evaluating Three Decades of Reform and Change in UK Central Government.* Oxford: OUP.

Hsiao, B., Chern, C.-C., & Yu, M.-M. (2012). Measuring the relative efficiency of IC design firms using the directional distance function and a metafrontier approach. *Decision Support System*, 53: 881–891.

Jurewicz, D. (2018). Financial sustainability of local government. *Stabilność podatkowa i finansowa w zmieniających się warunkach makro-i mikroekonomicznych*, 78(1), 103–116.

Kalb, A., Geys, B., & Heinemann, F. (2012). Value for money? German local government efficiency in a comparative perspective. *Applied economics*, 44(2), 201–218.

Kuhlmann, S. & Bouckaert, G. (2016). *Local Public Sector Reforms in Times of Crisis.* London: Palgrave Macmillan.

Le, T.D.Q., Ho, T.H., Nguyen, D.T., & Ngo, T. (2021). Fintech credit and bank efficiency: International evidence. *International Journal of Financial Studies*, *9*(3), 44. https://www.mdpi.com/2227-7072/9/3/44.

Lee, B.L. & Worthington, A.C. (2016). A network DEA quantity and quality-orientated production model: An application to Australian university research services. *Omega*, 60, 26–33.

Lewis, H.F. & Sexton, T.R. (2004). Network DEA: Efficiency analysis of organisations with complex internal structure. *Computers and Operations Research*, 31, 1365–1410.

Marquardt, C.A. & Wiedman, C.I. (2004). How are earnings managed? An examination of specific accruals. *Contemporary Accounting Research*, 21(2), 461–491.

Mergoni, A. & De Witte, K. (2022). Policy evaluation and efficiency: A systematic literature review. *International Transactions in Operational Research*, 29(3), 1337–1359.

Monfared, M.A.S. & Safi, M. (2013). Network DEA: An application to analysis of academic performance. *Journal of Industrial Engineering International*, 9(1), 1–10.

Municipal Association of Victoria (MAV) (2019). *Annual Report 2017–18*. Melbourne: MAV.

Narbón-Perpiñá, I. & De Witte, K. (2018a). Local governments' efficiency: A systematic literature review — Part I. *International Transactions in Operational Research*, 25(2), 431–468.

Narbón-Perpiñá, I. & De Witte, K. (2018b). Local governments' efficiency: A systematic literature review — Part II. *International Transactions in Operational Research*, 25(4), 1107–1136.

Naschold, F. (2017). *New Frontiers in the Public Sector Management: Trends and Issues in State and Local Government in Europe* (Vol. 69). Amsterdam: Walter de Gruyter and Co.

Ngo, T., & Tsui, K.W.H. (2022). Estimating the confidence intervals for DEA efficiency scores of Asia-Pacific airlines. *Operational Research*, 22(4), 3411–3434.

O'Donnell, C.J., Rao, D.S.P., & Battese, G.E. (2008). Metafrontier frameworks for the study of firm-level efficiencies and technology ratios, *Empirical Economics*, 34, 231–255.

Pilcher, R. & Van der Zahn, M. (2010). Local governments, unexpected depreciation and financial performance adjustment. *Financial Accountability and Management*, 26(3), 299–323.

Rayeni, M.M. & Saljooghi, F.H. (2010). Network data envelopment analysis model for estimating efficiency and productivity in universities. *Journal of Computer Science*, 6(11), 1252–1257.

Rogge, N., Agasisti, T., & De Witte, K. (2017). Big data and the measurement of public organizations' performance and efficiency: The state-of-the-art. *Public Policy and Administration*, 32(4), 263–281.

Simar, L. & Wilson, P.W. (2007). Estimation and inference in two-stage, semi-parametric models of production processes. *Journal of Econometrics*, 136(1), 31–64.

Stalebrink, O.J. (2007). An investigation of discretionary accruals and surplus-deficit management: Evidence from Swedish municipalities. *Financial Accountability and Management*, 23(4), 441–458.

Tone, K. & Tsutsui, M. (2009). Network DEA: A slacks-based measure approach. *European Journal of Operational Research*, 197, 243–252.

Tone, K. & Tsutsui, M. (2014). Dynamic DEA with network structure: A slacks-based measure approach. *Omega*, 42(1), 124–131.

Tran, C.-D.T.T. (2021). Efficiency of the teaching-industry linkage in the Australian vocational education and training. *Empirical Research in Vocational Education and Training*, 13(1), 1–28.

Tran, C.-D.T.T. & Dollery, B.E. (2020). Examining efficiency differentials by council type in South Australian local government: An environmentally adjusted meta-frontier data envelopment approach. *Public Administration Quarterly*, 44(2), 201–235.

Tran, C.-D.T.T. & Dollery, B.E. (2021). All in the mind: Citizen satisfaction and financial performance in the Victorian local government system. *Australian Accounting Review*, 31(1), 51–64.

Tran, C.-D.T.T. & Dollery, B.E. (2022). Administrative intensity and financial sustainability: An empirical analysis of the Victorian local government system. *Politics & Policy*, 50(3), 540–561.

Tran, C.-D.T.T., & Dollery, B.E. (2023a). Does financial sustainability affect local resident satisfaction? The case of the Victorian local government system. *Public money & management*, 43(4), 302–310.

Tran, C.T.T.D., & Dollery, B.E. (2023b). An empirical analysis of the impact of administrative intensity on the operational efficiency of the Victoria state local government system in Australia. *Local Government Studies*, 49(3), 544–567.

Tran, C.-D.T.T. & Villano, R.A. (2018). Measuring efficiency of Vietnamese public colleges: An application of the DEA-based dynamic network approach. *International Transactions in Operational Research*, 25(2), 683–703.

Tran, C.-D.T.T. & Villano, R.A. (2021). Financial efficiencies of Vietnamese public universities: A second-stage dynamic network data envelopment analysis approach. *Singapore Economic Review*, 1–22. DOI: 10.1142/S0217590818500133.

Victoria Auditor-General's Office (VAGO). (2018). *Reports*. Melbourne: VAGO.

Villano, R.A., Magcale-Macandog, D.B., Acosta, L.A., Tran, C.-D.T.T., Eugenio, E.A., & Macandog, P.B.M. (2020). Measuring disaster resilience in the Philippines: Evidence using network data envelopment analysis. *Climate and Development*, 12(1), 67–79.

Worthington, A.C. (2000). Cost efficiency in Australian local government: A comparative analysis of mathematical programming and econometrical approaches. *Financial Accountability & Management*, 16(3), 201–223.

Worthington, A.C. & Dollery, B.E. (2000). An empirical survey of frontier efficiency measurement techniques in local government. *Local Government Studies*, 26(2), 23–52.

Yu, Y. & Choi, Y. (2015). Measuring environmental performance under regional heterogeneity in China: A metafrontier efficiency analysis. *Computational Economics*, 46(3), 375–388.

https://doi.org/10.1142/9781800615786_0003

Chapter 3

Data Envelopment Analysis for Supply Chain Management: An Overview on Applications, Major Findings, and Future Directions

Hakan Yildiz

Department of Global Supply Chain Management,
Mike Ilitch School of Business, Wayne State University,
Detroit, MI, USA
hakan@wayne.edu

Abstract

Performance measurement of supply chains and supply chain processes is an important area of research that allows for benchmarking and improvement. Among the many methods, data envelopment analysis (DEA) is an interesting and useful quantitative method that is capable of considering multiple key performance indicators simultaneously. DEA has been widely used in the area of supply chain management, which spans sourcing decisions, logistics management, supply chain performance evaluation, and supply chain risk management. In this chapter, we provide an overview of these application areas, highlight major findings, and point to future research directions.

Keywords: Supply chain management, supply chain performance, supplier evaluation, supplier selection, logistics performance, supply chain risk management, risk monitoring, advantages of DEA for supply chain management, challenges of DEA for supply chain management, trends and developments in DEA for supply chain management.

1. Introduction

Supply chain management includes the handling of all activities related to the flow of goods or services from the raw material stage to the final product, including the procurement of raw materials, parts, and components, production, and storage and transportation. Its goal is to perform these activities efficiently and effectively to meet and exceed the needs of the customers. Performance measurement of supply chains, processes, and outputs is essential for companies to fulfill their goals and to remain competitive in a marketplace, where the amount of competition has increased significantly due to increased globalization. Higher level performance measures for supply chains can guide managers in determining where to conduct a more in-depth analysis, which can allow them to identify areas for improvement and coordinate the activities of various entities in a supply chain.

In addition to performance measurement, benchmarking a system against other similar systems is also very important, and it is often done to measure effectiveness and identify areas for improvement. Among the many benchmarking methods, data envelopment analysis (DEA) is a quantitative one that can consider multiple key performance indicators (KPIs) simultaneously. DEA measures the efficiency and productivity of decision-making units (DMUs), such as production locations or supply chain entities. DEA uses mathematical programming techniques to compare the performance of multiple DMUs based on various inputs and outputs, where inputs are the resources a system uses and outputs are the outcomes that are measured by performance and activity metrics. DEA works by constructing a linear programming model that represents the inputs and outputs of each DMU as a linear combination of variables, and the model is then used to determine each DMU's relative efficiency. The method produces an efficiency score between 0 and 1, with the latter being the most efficient. Efficiency is calculated by dividing the weighted outputs by weighted inputs. The first DEA model, also known as the CCR model, was first introduced by Charnes *et al.* (1978) and is provided in the following for reference. In this model, x_{ij} is the ith input of the jth DMU. Similarly, y_{kj} is the kth output of the jth DMU; p is the index corresponding to the DMU that is being evaluated; v_i and u_k are the weights assigned to ith input and kth output, respectively. Given these, the CCR model is as follows:

$$\max_{v,u} \ \theta_p = \frac{\sum_{k=1}^{n} u_k y_{kp}}{\sum_{i=1}^{m} v_i x_{ip}},$$

such that

$$\frac{\sum_{k=1}^{n} u_k y_{kj}}{\sum_{i=1}^{m} v_i x_{ij}} \leq 1, \quad \forall j, \quad u_k, v_i \geq 0, \quad \forall k, i.$$

While this model is nonlinear, it is easily converted to a linear model. Moreover, the dual equivalent of the DEA linear programming model allows the inclusion of a convexity constraint, which is known as the BCC model (Banker *et al.*, 1984), and facilitates the assumption of variable returns to scale instead of the constant returns to scale assumption required by the CCR model.

DEA efficiency scores can indicate whether a DMU is using its inputs effectively or whether it is producing enough outputs. While "efficiency" is the standard term for the optimal value of the linear program, this value does not necessarily correspond to "production efficiency" in many contexts. For example, in some contexts, the optimum DEA score can be a quality index or a risk index (Zhu, 2022).

DEA has also been widely used in supply chain management. Major application areas span supplier relationships, logistics management, supply chain performance, and supply chain risk management. The motivation for this chapter is to provide a comprehensive analysis of the use of DEA in supply chain management, given DEA's increased popularity in recent years, and discuss the major application areas, advantages, challenges, trends, and developments. The contribution of this chapter is to synthesize the existing literature on DEA and supply chain management, identify knowledge gaps, propose new research directions, and provide insights for practitioners and decision-makers to enhance the efficiency and effectiveness of their supply chains. While this chapter provides a good overview of the topic, it does not intend to provide a comprehensive review of the literature on DEA in supply chain management research. For that, readers can refer to Ertek *et al.* (2012), Soheilirad *et al.* (2018), and Dutta *et al.* (2022), who provide comprehensive reviews of the applications of DEA in supply chain management and supplier selection.

The rest of the chapter is organized as follows: First, the advantages of using DEA for supply chain management are discussed, followed by the challenges. Then, some important methodological developments are presented. Subsequently, the major application areas are analyzed in sections dedicated to supply chain management, supplier evaluation and selection, logistics, and supply chain risk mitigation and monitoring. The final section provides the conclusion and future research directions.

2. Advantages of Using DEA for Supply Chain Management

DEA has several useful characteristics, such as its relative mathematical simplicity, the ease of understanding the efficiency scores, and its ability to provide results that are relatively robust. In the context of supply chain management, the following are some other key advantages, which are also summarized in Table 1.

First is DEA's nonparametric nature, which does not require any assumptions about the functional form of the production process. This is well suited for supply chain management, as it involves complex and dynamic processes with many factors that can influence efficiency.

Second, the multi-dimensionality of DEA also makes it advantageous for supply chain applications. In supply chain management, decisions often involve balancing multiple factors, such as cost, quality, and speed. DEA can incorporate these multiple inputs and outputs, allowing for the evaluation of efficiency using several factors simultaneously. This also allows for the consideration of trade-offs

Table 1. Advantages and challenges of DEA for SCM.

Advantages	Challenges
Non-parametric	Data requirements
Can incorporate multiple factors	Efficiency scores can be difficult to interpret
Applicable to different contexts	Traditional DEA models are also not suitable for problems with uncertain variables

between different inputs and outputs. For example, a production line may increase its output by using more labor; however, this may come at the cost of increased energy consumption. DEA can be used to evaluate the efficiency of a production line in the presence of such trade-offs. Thus, DEA can be used to analyze the impact of different inputs and outputs on supply chain efficiency, which can help managers make better decisions about resource allocation.

In addition to the traditional use of DEA in evaluating entities, it can evaluate the efficiency of different supply chain operational or risk mitigation strategies, such as outsourcing, offshoring, dual sourcing, and utilizing safety stocks (e.g., Talluri *et al.*, 2013). Thus, DEA's applicability to different contexts makes it very advantageous for supply chain management.

3. Challenges of Using DEA for Supply Chain Management

While DEA can be a valuable tool for evaluating the efficiency of DMUs in a supply chain, there are also some limitations and disadvantages to consider, as summarized in Table 1.

First, in the context of supply chain management, data related to inputs and outputs may be difficult to measure, and necessary data may only sometimes be available. Moreover, DEA assumes that all DMUs have the same technology, which may not be the case in practice. Another key challenge of using DEA is the selection of appropriate inputs and outputs. Many potential inputs and outputs could be considered, and selecting the most relevant ones can be difficult.

Furthermore, the interpretation of the efficiency scores can be difficult, as they do not necessarily reflect the absolute level of efficiency of a DMU. For example, a score of 0.8 may be considered good in some cases but not in others. It is essential to consider the specific goals and objectives of the system when interpreting DEA efficiency scores.

Traditional DEA models are also not suitable for problems with uncertain variables, for which Chen *et al.* (2006) have extended those models. Specifically, traditional DEA models assume that production at different time periods is independent of each other, which does not apply to the supply chain context. Moreover, traditional DEA models do not consider the structure of DMUs and typically treat them as a "black box." While supply chains can also be treated as a "black box," their internal structure-related information can also provide more information, for which network DEA models have been developed (e.g., Fare & Grosskopf, 2000; Castelli *et al.*, 2001, 2004).

4. Trends and Developments

We would like to highlight a few key methodological developments in recent years. First is the contribution of Simar & Wilson (2007), who described a sensible data-generating process (DGP) for popular two-stage approaches that first calculate efficiency scores using DEA and then use regression to identify environmental variables whose impact is statistically significant. They also proposed single and double bootstrap procedures that permit valid inference and showed that statistical efficiency is improved in the second-stage regression when the double bootstrap procedure is used.

Another interesting and useful development is the introduction of an inverse DEA model by Wei *et al.* (2000), which is used to estimate inputs (or outputs) for a DMU when some outputs (or inputs) are changed while assuming that the technical efficiency score of that DMU is maintained and unchanged. Various versions of inverse DEA have been developed and subsequently applied to various domains (e.g., Emrouznejad *et al.*, 2019; Boubaker *et al.*, 2022), including supply chain management (e.g., Moghaddas *et al.*, 2022).

Finally, while DEA was originally developed for comparing homogenous DMUs and homogeneity is a fundamental assumption of many DEA models, new DEA models have also been developed, which relax this homogeneity assumption and can handle non-homogenous DMUs with different production technologies or environmental conditions. Such models can also be used to evaluate DMUs with different numbers of inputs or outputs (e.g., Li *et al.*, 2016). For instance, a manufacturer may have multiple production lines, each with its unique set of inputs, such as supplies, and outputs, such as final products, and DEA can be used to compare the efficiency of these production lines.

5. Supply Chain Management

Measuring the performance of a supply chain can be difficult, as a supply chain consists of several companies that work independently and yet rely on each other to serve their respective customers. Moreover, various metrics can be of interest, including cost, lead times, customer service levels, and asset utilization. In addition, gathering comprehensive data when there is a lack of standardization across various parties can be quite challenging. Despite these challenges, performance measurement of an entire supply chain can help in decision-making about resources, supply chain partners, and processes. Moreover, it can promote better accountability.

DEA has been used to assess the overall performance of a supply chain as a unit of analysis (e.g., Liang *et al.*, 2006; Chen & Yan, 2011; Lau, 2013), where the supply chain is seen as a single entity rather than a combination of separate parts. This approach is useful in optimizing the overall supply chain performance, increasing efficiency, and enhancing customer satisfaction.

A DEA-based supply chain model was introduced by Zhu (2002) to evaluate the supply chain's overall efficiency and the efficiency of its members. From a game theoretic standpoint, Liang *et al.* (2006) and Chen *et al.* (2006) suggest efficiency evaluation methods for a two-tier supply chain model. Yang *et al.* (2011) presented a DEA model for a central decision-maker making decisions for both the supplier and the manufacturer. Dev *et al.* (2014) combined simulation with the DEA approach for comparing several performance metrics in a supply chain network. Narasimhan *et al.* (2004) used DEA to evaluate the different competencies of a supply chain.

More recently, network DEA and dynamic DEA models have been used in the context of big data (Zhu, 2022). For example, Kahi *et al.* (2017) and Badiezadeh *et al.* (2018) used network DEA to study the sustainability of supply chains.

6. Supplier Evaluation and Selection

Sourcing goods and services has been an important area for supply chain managers due to the large share of procurement costs within the total cost structure of a company. Moreover, the quality of goods and services sourced from suppliers has a significant impact on the quality of the finished goods and customer satisfaction. More recently, environmental and social sustainability concerns have turned the spotlight on companies' sourcing practices. Because of these, supplier evaluation and selection have been important areas of research for both practitioners and academics. To determine which partners are contributing to inefficiencies and where changes can be made, a company may utilize DEA to assess the effectiveness of its supply chain partners, such as suppliers and distributors.

Measuring supplier performance, choosing the best suppliers, and evaluating the success of programs involving suppliers are all common themes in supplier relationships. Such decisions are quite complex, as various criteria, some of which are qualitative, are of importance. The effectiveness of suppliers can be compared using DEA based on their costs, levels of quality, and other relevant aspects. Supply chain managers can use this information to choose which suppliers to work with and find the most effective suppliers. DEA has been shown to be an effective tool for supplier evaluation (e.g., Garfamy, 2006; Ross *et al.*, 2006; Narasimhan *et al.*, 2001; Wu & Olson, 2008; Tavana *et al.*, 2016; Goswami & Ghadge, 2020) and supplier selection (e.g., Talluri & Baker, 2002; Wu & Barnes, 2011). In addition, DEA can also be used to evaluate the efficiency of suppliers over time, allowing firms to track supplier performance and identify performance changes.

In addition, DEA models are used to compare the procurement efficiency of companies (Easton *et al.*, 2002), evaluate supplier development programs' effectiveness (Su *et al.*, 2018), and evaluate the efficiency of different sourcing strategies, such as single-sourcing, multiple-sourcing, or strategic partnerships.

7. Logistics

Efficient and effective logistics management is very important for companies to ensure timely delivery of products between different parties in a supply chain. In addition, logistics costs are a major component of the total cost of a product. Thus, to be competitive in the marketplace, companies need to be efficient in their logistics activities, which include transportation, warehousing, inventory management, and facility network design.

DEA can be applied to evaluate transportation alternatives, facility performance, and logistics procedures. DEA can be used to evaluate the efficiency of different inventory management strategies, such as just-in-time (JIT) inventory and economic order quantity (EOQ) models. This can help supply chain managers identify potential improvements in inventory management and reduce costs.

DEA has been used in logistics network design to evaluate the performance of different logistics providers and identify the most efficient logistics network configuration. This is important as it allows organizations to optimize their logistics operations and reduce costs by selecting the best logistics providers and designing the most efficient logistics network.

Chen *et al.* (2010) used DEA to evaluate how well order-picking systems performed in warehouses. Reiner *et al.* (2013) used DEA to evaluate logistics procedures in stores. De Koster & Balk (2008) used DEA to compare distribution facilities.

8. Supply Chain Risk Mitigation and Monitoring

The effectiveness of risk mitigation techniques against supply chain risks has also been evaluated using DEA. Talluri *et al.* (2013) used DEA to determine the effectiveness of seven risk mitigation strategies. They conducted a discrete-event simulation of a supply chain and measured the efficiency of the strategies by applying DEA to the outputs of the simulation model.

Firms must evaluate their performance effectively over time. Charnes *et al.* (1985) created the idea of window analysis, which can offer an analysis of the relative efficacy of DMUs over time. Sengupta (1995) and Färe & Grosskopf (1996) later provided a more thorough analysis of the inter-temporal elements of DEA. Cooper *et al.* (1995) and Dinc *et al.* (1997) both employed a particular DMU's evaluation against its historical performance over time.

Similar temporal methods were used by Parker & Wu (1998) to examine the British steel industry across time. An integrated benchmarking system was presented by Ross & Droge (2002) using DEA and window analysis to assess the productivity of distribution centers and identify trends. To find trends over time, they solved the DEA model for each time period and rank-ordered the distribution centers' efficiencies. Talluri & Sarkis (2002) created a framework for DEA score-based manufacturing system monitoring and improvement.

Yildiz *et al.* (2022) used window analysis to track network performance over time by considering each period as a separate DMU. With a window width of n periods, they solved a DEA model for each new period and the prior $n - 1$ periods. Comparing performance in the most recent period to its performance in prior periods provides a relative efficiency score for the most recent period. In this way, Yildiz *et al.* (2022) provided a method for evaluating and monitoring performance by considering various performance measures using DEA in conjunction with a multivariate statistical process control (SPC). Their findings show a considerable improvement in the performance of the SPC chart as a result of including the DEA scores.

9. Conclusion and Future Research

DEA, a multi-criteria decision-making method, has been increasingly used as a powerful tool for evaluating and improving supply chain performance and supporting strategic decision-making. As organizations continue to face challenges in managing their supply chains, DEA is likely to play an increasingly important role in helping companies evaluate and improve their supply chain operations. Major application areas include supplier evaluation and selection, logistics, and supply chain risk mitigation and monitoring. While there are several key challenges to using DEA for supply chain management, such challenges can also provide many opportunities for future research.

While the existing network DEA models try to capture the relationship between supply chain partners while evaluating supply chain performance, typically these models consider only two-stage supply chains. Thus, expanding these models to address supply networks that consist of a greater number of stages is another potential area of research. Future research in developing network DEA models for more general network structures may necessitate the development and utilization of nonlinear optimization methods (Zhu, 2022).

Another key challenge is the selection of appropriate performance measures. DEA is a multi-criteria decision-making method, and selecting the appropriate performance measures is crucial for the validity of the results. Therefore, novel approaches for choosing the most appropriate performance measures while considering the specific context of the supply chain can be useful.

Dealing with uncertain and incomplete data is also another challenge. DEA requires accurate and complete data to be effective. However, in many cases, the data are uncertain or incomplete, leading to inaccurate or biased results. Future research can focus on developing methods to deal with uncertain and incomplete data, such as robust DEA methods or methods that incorporate uncertainty.

In terms of opportunities, there is a growing need for integrating DEA with other analytical tools, such as machine learning and optimization methods. Integrating DEA with these tools can enhance the ability to analyze large and complex datasets and improve decision-making. Additionally, there is a potential for DEA to be applied in new areas of supply chain management, such as digitization and sustainability. Technologies such as blockchain and the internet of things (IoT) result in large amounts of data that can be used to measure supply chain performance more accurately. Therefore, future research could focus on the use of DEA in analyzing and optimizing digital supply chains.

Another opportunity is to use DEA for real-time or near real-time monitoring and assessment of supply chain performance. This would enable organizations to quickly identify and respond to performance issues and optimize their operations in real time.

References

Badiezadeh, T., Saen, R.F., & Samavati, T. (2018). Assessing sustainability of supply chains by double frontier network DEA: A big data approach. *Computers & Operations Research*, 98, 284–290.

Banker, R.D., Charnes, A., & Cooper, W.W. (1984). Some models for estimating technical and scale inefficiencies in data envelopment analysis. *Management Science*, 30(9), 1078–1092.

Boubaker, S., Le, T.D.Q., & Ngo, T. (2022). Managing bank performance under COVID-19: A novel inverse DEA efficiency approach. *International Transactions in Operational Research*, 30(2023), 2436–2452. https://doi.org/10.1111/itor.13132.

Charnes, A., Cooper, W.W., & Rhodes, E. (1978). Measuring the efficiency of decision making units. *European Journal of Operational Research*, 2(6), 429–444.

Charnes, A., Clark, T., Cooper, W.W., & Golany, B. (1985). A developmental study of data envelopment analysis in measuring the efficiency of maintenance units in the US Air Forces. *Annals of Operations Research*, 2(1), 95–112.

Chen, C. & Yan, H. (2011). Network DEA model for supply chain performance evaluation. *European Journal of Operational Research*, 213(1), 147–155.

Chen, Y., Liang, L., & Yang, F. (2006). A DEA game model approach to supply chain efficiency. *Annals of Operations Research*, 145(1), 5.

Chen, C.M., Gong, Y., De Koster, R.B., & Van Nunen, J.A. (2010). A flexible evaluative framework for order picking systems. *Production and Operations Management*, 19(1), 70–82.

Cooper, W.W., Kumbhakar, S., Thrall, R.M., & Yu, X. (1995). DEA and stochastic frontier analyses of the 1978 Chinese economic reforms. *Socio-Economic Planning Sciences*, 29(2), 85–112.

De Koster, M.B.M. & Balk, B.M. (2008). Benchmarking and monitoring international warehouse operations in Europe. *Production and Operations Management*, 17(2), 175–183.

Dev, N.K., Shankar, R., & Debnath, R.M. (2014). Supply chain efficiency: A simulation cum DEA approach. *The International Journal of Advanced Manufacturing Technology*, 72(9–12), 1537–1549.

Dinc, M., Gulledge, T.R., & Stough, R.R. (1997). Analysis of regional economic performance: An extended DEA application. In *WRSA 36th Annual Meeting in Hawaii*, 23–27 February.

Dutta, P., Jaikumar, B., & Arora, M.S. (2022). Applications of data envelopment analysis in supplier selection between 2000 and 2020: A literature review. *Annals of Operations Research*, 315, 1–56.

Easton, L., Murphy, D.J., & Pearson, J.N. (2002). Purchasing performance evaluation: With data envelopment analysis. *European Journal of Purchasing & Supply Management*, 8(3), 123–134.

Emrouznejad, A., Yang, G.L., & Amin, G.R. (2019). A novel inverse DEA model with application to allocate the CO_2 emissions quota to different regions in Chinese manufacturing industries. *Journal of the Operational Research Society*, 70(7), 1079–1090.

Ertek, G., Akyurt, N., & Tillem, G. (2012). *DEA-based Benchmarking Models in Supply Chain Management: An Application-oriented Literature Review* (pp. 200–218). Istanbul, Turkey: Istanbul Kemerburgaz University & Logistics Association (LODER) & Ideal Kultur Publishing.

Färe, R. & Grosskopf, S. (1996). Productivity and intermediate products: A frontier approach. *Economics Letters*, 50(1), 65–70.

Garfamy, M.R. (2006). A data envelopment analysis approach based on total cost of ownership for supplier selection. *Journal of Enterprise Information Management*, 19(6), 662–678.

Goswami, M. & Ghadge, A. (2020). A supplier performance evaluation framework using single and bi-objective DEA efficiency modelling approach: Individual and cross-efficiency perspective. *International Journal of Production Research*, 58(10), 3066–3089.

Kahi, V.S., Yousefi, S., Shabanpour, H., & Farzipoor Saen, R. (2017). How to evaluate sustainability of supply chains? A dynamic network DEA approach. *Industrial Management & Data Systems*, 117(9), 1866–1889.

Lau, K.H. (2013). Measuring distribution efficiency of a retail network through data envelopment analysis. *International Journal of Production Economics*, 146(2), 598–611.

Li, W., Liang, L., Cook, W.D., & Zhu, J. (2016). DEA models for non-homogeneous DMUs with different input configurations. *European Journal of Operational Research*, 254(3), 946–956.

Liang, L., Yang, F., Cook, W., & Zhu, J. (2006). DEA models for supply chain efficiency evaluation. *Annals of Operations Research*, 145(1), 35–49.

Moghaddas, Z., Tosarkani, B.M., & Yousefi, S. (2022). Resource reallocation for improving sustainable supply chain performance: An inverse data envelopment analysis. *International Journal of Production Economics*, 252, 108560.

Narasimhan, R., Talluri, S., & Das, A. (2004). Exploring flexibility and execution competencies of manufacturing firms. *Journal of Operations Management*, 22(1), 91–106.

Narasimhan, R., Talluri, S., & Mendez, D. (2001). Supplier evaluation and rationalization via data envelopment analysis: An empirical examination. *Journal of Supply Chain Management*, 37(2), 28–37.

Parker, D. & Wu, H.S. (1998). Privatization and performance: A study of the British Steel industry under public and private ownership. *Economic Issues*, 3, 31–50.

Reiner, G., Teller, C., & Kotzab, H. (2013). Analyzing the efficient execution of in-store logistics processes in grocery retailing — The case of dairy products. *Production and Operations Management*, 22(4), 924–939.

Ross, A. & Droge, C. (2002). An integrated benchmarking approach to distribution center performance using DEA modeling. *Journal of Operations Management*, 20(1), 19–32.

Ross, A., Buffa, F.P., Dröge, C., & Carrington, D. (2006). Supplier evaluation in a dyadic relationship: An action research approach. *Journal of Business Logistics*, 27(2), 75–101.

Sengupta, J. (1995). *Dynamics of Data Envelopment Analysis: Theory of Systems Efficiency*. Boston, MA: Springer Science & Business Media.

Simar, L. & Wilson, P.W. (2007). Estimation and inference in two-stage, semi-parametric models of production processes. *Journal of Econometrics*, 136, 31–64.

Soheilirad, S., Govindan, K., Mardani, A., Zavadskas, E.K., Nilashi, M., & Zakuan, N. (2018). Application of data envelopment analysis models in supply chain management: A systematic review and meta-analysis. *Annals of Operations Research*, 271, 915–969.

Su, H.C., Chen, Y.S., & Kao, T.W. (2018). Enhancing supplier development: An efficiency perspective. *Journal of Business Logistics*, 39(4), 248–266.

Talluri, S. & Baker, R. (2002). A multi-phase mathematical programming approach for effective supply chain design. *European Journal of Operational Research*, 141(3), 544–558.

Talluri, S. & Sarkis, J. (2002). A model for performance monitoring of suppliers. *International Journal of Production Research*, 40(16), 4257–4269.

Talluri, S., Kull, T.J., Yildiz, H., & Yoon, J. (2013). Assessing the efficiency of risk mitigation strategies in supply chains. *Journal of Business Logistics*, 34(4), 253–269.

Tavana, M., Li, Z., Mobin, M., Komaki, M., & Teymourian, E. (2016). Multi-objective control chart design optimization using NSGA-III and MOPSO enhanced with DEA and TOPSIS. *Expert Systems with Applications*, 50, 17–39.

Wei, Q., Zhang, J., & Zhang, X. (2000). An inverse DEA model for inputs/outputs estimate. *European Journal of Operational Research*, 121(1), 151–163.

Wu, C. & Barnes, D. (2011). A literature review of decision-making models and approaches for partner selection in agile supply chains. *Journal of Purchasing and Supply Management*, 17(4), 256–274.

Wu, D. & Olson, D.L. (2008). Supply chain risk, simulation, and vendor selection. *International Journal of Production Economics*, 114(2), 646–655.

Yang, F., Wu, D., Liang, L., Bi, G., & Wu, D. (2011). Supply chain DEA: Production possibility set and performance evaluation model. *Annals of Operations Research*, 185(1), 195–211.

Yildiz, H., Talluri, S., Xie, X., Yoon, J., Qiu, P., & Wassick, J.M. (2022). Evaluating and monitoring distribution network efficiency with multivariate process control methods. *International Journal of Production Research*, 60(2), 517–533.

Zhu, J. (2002). *Quantitative Models for Performance Evaluation and Benchmarking: Data Envelopment Analysis with Spreadsheets and DEA Excel Solver*. Boston: Kluwer Academic Publishers.

Zhu, J. (2022). DEA under big data: Data enabled analytics and network data envelopment analysis. *Annals of Operations Research*, 309(2), 761–783.

https://doi.org/10.1142/9781800615786_0004

Chapter 4

Efficiency of Water Provision Service: A Visual Study of Data Envelopment Analysis

Ane Elixabete Ripoll-Zarraga

Business Department,
Universitat Autònoma de Barcelona (UAB),
Cerdanyola del Vallès (Barcelona), Spain
ane.ripoll-zarraga@uab.cat

Abstract

Efficiency analyses of public services, such as water resource management, is in vogue owing to investments financed by taxpayers. However, in regulated sectors, i.e., without competition, data envelopment analysis (DEA) results for decision-making are questionable. Indeed, DEA is sensitive to the data, inputs, and outputs chosen by the researcher and can be highly influenced by outliers. There is no space for specialization: a decision-making unit (DMU) may be inefficient generally but efficient in one particular activity (output). Previous studies on DEA of water efficiency use second-stage analysis with exogenous factors, i.e., not controlled by management, making empirical applications unfeasible. In addition, models do not reflect changes in the specialization of DMUs through time caused by financial crises or lack of infrastructure. DEA visualization combines the standard DEA analysis with multivariate statistical methods. In this chapter, water supply and quality in 31 provinces of China are benchmarked for 2020. The results show nine efficient provinces for the traditional DEA model (all the variables). Other provinces become efficient when considering infrastructure for water supply versus water quality (few pollutants). Two provinces are efficient due to a surplus of natural water resources (overcapacity) in relation to their population, allowing for high-quality water provision, which may be transferred to regions with high water scarcity.

Keywords: Data envelopment analysis (DEA), multidimensional scaling (MDS), DEA visualization, factor analysis, ProFit, cluster analysis, water efficiency measurement, water provision, water quality, water scarcity.

1. Introduction

The study of water provision efficiency is an extensive area of interest. Water provision is a public service that is financed by the taxpayers and regulated. However, investment decisions usually rely on politicians who may choose to make decisions that increase their popularity rather than making appropriate decisions based on facts, as needed by the society and industry. Indeed, there is generally an interest in knowing the relationship between public infrastructure investment and the provision, particularly quality, of the service, which includes water collection, treatment, distribution, drainage, and waste treatment.

A popular technique in estimating efficiency is data envelopment analysis (DEA), a non-parametric method in which we do not have to define a specific production function. DEA studies find extensive applications in diverse fields, such as banking and finance (Tripe, 2010; Le, 2018; Boubaker *et al.*, 2022b) and manufacturing (Becchetti & Sierra, 2003; Wang & Chin, 2009; Vu, 2016; Boubaker *et al.*, 2023). However, for benchmarking purposes, regulators question the use of DEA or other techniques in industries where there are no competitors (Pinto *et al.*, 2017). There are alternatives for assessing the water providers' performance according to, for example, performance indicators (Alegre *et al.*, 2003), where yardstick competition methodology could be more convenient (Bogetoft, 1997). Coelli *et al.* (2003) suggest different approaches to measuring performance. To assess the degree of achieving a specific key performance indicator (KPI), which is known as effectiveness, does not imply doing it in the "shortest" way, i.e., using fewer resources than another water firm provider or maximizing water provision with limited resources compared to other suppliers, which is known as efficiency. Even if the industry is regulated, utility firms must provide citizens with water while minimizing production costs. Water provision is a public service. Hence, it is essential to assess to what extent the service is delivered in the most efficient way to avoid increasing taxes due to inefficiency procedures.

Previous studies in water provision have used conventional DEA models, i.e., assessing the efficiency (or inefficiency) overall, according to the variables used. Second-stage analysis has been used to account for external factors (non-controllable by the management), such as ownership and privatization process (e.g., Le Lannier & Porcher, 2014), distribution networks management (Choi *et al.*, 2015), or location (Romano & Guerrini, 2011; Long & Pijanowski, 2017), potentially concerning scarcity. However, these studies do not seem to provide the relevant background for managerial and policymaker recommendations. There is no robust evidence to conclude that water providers' environment is the main reason for inefficiency. Water service provision consists of a production process. Hence, it must be assessed from water collection until waste treatment operations, i.e., in terms of operational efficiencies. Water is a public utility and essential for living. Therefore, water provision is a requirement. Any inefficiency is likely to be corrected to the extent of being controllable by the management. Exogenous factors are, by definition, non-controllable and not changeable in the short term.

DEA estimates efficiency according to the resources (inputs) and products and services generated (outputs) by a decision-making unit (DMU). The efficiency score is a relative measure that questions if any other competitor has performed better than the DMU analyzed. However, DEA has certain well-known pitfalls. Researchers have shown an interest in improving the standard models, e.g., Charnes *et al.* (1978) and Banker *et al.* (1984), and adding extensions to overcome their main problems, including searching for sources of inefficiency by applying different techniques, for example, bootstrapping (Simar & Wilson, 2007). Previous literature on water provision in China concludes that there is a need to focus on water use management and efficiency improvements (Long & Pijanowski, 2017). Evidence shows the interest of researchers in using DEA for analysis, but most importantly, to prove DEA as a tool for managerial and policymaking decisions.

In this chapter, different models are implemented (DEA specifications), and the DEA efficiency score obtained in each model for each DMU, in this case, provinces, are analyzed using multivariate statistical methods. The aim is to learn from the intrinsic data characteristics, explain sources of inefficiency, and identify extreme cases without losing data. One important matter is to assume that the DMUs are either efficient or inefficient concerning the inputs and outputs chosen. However, a DMU could be inefficient overall but efficient in part of the production process. Indeed, the DMU could become more efficient in specific operations but not others. DEA visualization provides insights regarding similar production practices and specialization. It has been applied to several sectors, for example, higher education institutions (Sagarra *et al.*, 2017), microfinance institutions (Gutierrez-Nieto *et al.*, 2007), and airports (Ripoll-Zarraga & Mar-Molinero, 2020), as well as to panel data to assess the time trend or change in production specialization (Ripoll-Zarraga *et al.*, 2023). To my knowledge, this study is the first to use DEA visualization of public water services in China. The scope of this chapter is to provide insights into the DEA visualization technique for management decisions and policymaking recommendations.

The results show that most provinces increase their inefficiency significantly when not accounting for the quality of the water but the provision of water. Therefore, with few exceptions, most provinces are committed to a high-quality water provision service. The wealthiest provinces are not necessarily the most efficient in water resource utilization and water provision but in accessibility to natural water resources. Population does not seem to influence water resource use, raising questions about whether or not all citizens have access to water. Some provinces have an overcapacity of water resources, which can be used for water provision in other areas with more population or less water supply. However, this may require infrastructure investment and regional agreements.

The remaining chapter is organized as follows: Section 2 provides the Chinese background in water provision. Section 3 reviews the relevant literature. Section 4 introduces the methodology. Section 5 describes the data used. Section 6 discusses the main results, and Section 7 concludes.

2. China Water Supply Background

In the early 1960s, the Red Flag Canal was constructed with the potential of plenty of access to public water through an irrigation system for water provision using surface water (Tang *et al.*, 2019). The increase in consumption and water pollution reduced the available water resources from 41.90 billion cubic meters per year in 1956 to 32.50 in 2000, with an increase in household consumption by 150% from 1980 to 2015 (Pietz, 2015), leaving the North China Plain in a worse situation for accessing natural water resources due to receiving less rainfall than the south. The infrastructure depends on the financing of the state government. Hence, the situation is unlikely to change, at least in the short term. Rural areas have suffered from a lack of investment dating to the 1940s, resulting in poor-quality water access. In 1991, the Chinese government stopped treating water as a public good and implemented water rates and meters, forcing state companies to operate for profits. The provision of public water was transferred to municipal governments, facilitating regional mergers, which were profit-focused. These companies remain state-owned, i.e., with a majority owned by the government, but operate as private companies. The Chinese government allowed the investment of national and foreign capital, as it was unable to provide the necessary infrastructure. Although between 1990 and 2012, the access to improved water supply increased, the lack of infrastructure remained significant. For example, in 2012, more than 290 million rural people lacked access to improved water supply and sanitation services (The World Bank, 2020). China has 31 regions classified at the provincial level as municipalities and autonomous regions. For a review of their water characteristics, see, for example, Hu *et al.* (2006).

The research related to the urban water industry dates back to the 1970s. However, remote areas are in a worse position in terms of access to water and infrastructure for extensive distribution and alternative applications, apart from consumption, to support regional development. Water is a fundamental right and resource for human and animal life. Water shortage becomes a barrier to economic and regional development, influencing life expectancy worldwide. Yet, China is one of the 13 countries with severe water shortage due to the low water availability per capita and having the highest water consumption worldwide. A significant use is in agriculture, which is reflected in the interest of recent studies in assessing the water utilization efficiency in agriculture (e.g., Geng *et al.*, 2019; Huang *et al.*, 2021). There is a need to produce relevant research for public decision-making to ensure citizens have access to water provision for subsistence and for regional areas to develop locally. Water resource efficiency is required to solve water-related problems, such as water shortage and pollution.

3. Literature Review

There is a relationship between external factors and water access (Deng *et al.*, 2016). For example, areas with a higher gross domestic product (GDP) are expected to have better infrastructure and hence water quality. However, the available resources may be underused or wasted. Indeed, before deciding the investment in pipes, sewerage capacity, etc., usually funded by taxpayers, it is required to assess the extent to which a particular city or region is inefficient in using the current water resources and the reasons behind that underutilization, including water supply. For example, the wasting of relevant water resources and a lack of proper sewage treatment show the inefficiency of water management in Guangdong, the wealthiest province in 2020 (11 million yuan), followed by Jiangsu (10.3 million). Inefficiency is exacerbated by increasing water scarcity due to urbanization or industrialization (Tang *et al.*, 2015). In regard to empirical efficiency approaches, we refer to alternative applications of the standard DEA models to evaluate water utilization for decision-making (e.g., Hu *et al.*, 2019), second-stage analysis searching for factors affecting inefficiency, and stochastic frontier analysis (SFA) and its extensions (e.g., Le Lannier & Porcher, 2014; Mu *et al.*, 2016; Laureti *et al.*, 2021).

Exogenous factors are relevant to the extent that they constitute partially or fully the water providers' environment. Hence, these influence — somehow — the decisions regarding inputs or outputs. However, these factors are external and, therefore, non-controllable by managers. Even if the results identify an external factor as a source of inefficiency by using, for example, SFA or a second-stage procedure applied to the efficiencies obtained through DEA, decisions are unlikely to be made to address it. Traditional DEA applications assume that all the DMUs behave similarly according to inputs and outputs. Therefore, a DMU may be inefficient when compared to others overall, i.e., accounting for all the variables chosen. However, a DMU may specialize in part of the production process or specific output, not in all of them, and be (partially) efficient rather than overall (specialization). Indeed, the need for solid results for decision-making is still pending. Previous research assesses efficiency overall without concluding which factors affect water use (or treatment) efficiency per area. The findings show overall that water use efficiency in China has improved but remains at a low level (Mu *et al.*, 2016; Geng *et al.*, 2019).

In terms of methodological improvements in DEA, the previous findings highlight several issues, with the most relevant being zero weights associated with a DMU, which is removed from the assessment set (Dyson *et al.*, 2001); the number of efficient units depends on the number of variables (Pedraja Chaparro *et al.*, 1999); or extreme cases affecting the other units' efficiencies. However, these may be a consequence of the variables chosen, i.e., inputs and outputs (Serrano-Cinca *et al.*, 2016). Another problem is that frontier units (efficient) cannot be ranked against each other. Efficient units can reach the frontier in different ways, i.e., depending on the usage of inputs to produce one or more outputs. However, alternative production procedures are not analyzed. Network DEA allows for identifying different activities developed by a DMU within a "black box." Yet, inefficiency is an overall measure, i.e., it is not possible to differentiate between inefficiencies due to one activity or another. Further, the model forces all the DMUs to specialize in all the activities, i.e., to produce all the outputs. Some recent empirical developments have attempted to confront these and other issues, such as challenging the assumption of prices and quantities of inputs and outputs being deterministic using stochastic DEA to explain deviations in frontiers by differentiating inefficiencies and the errors, for example, caused exogenously, with applications in airlines (Ngo & Tsui, 2021); inverse DEA to estimate output or input quantities with a fixed efficiency level (Zhang & Zhang, 2000), which is used to generate an empirical resilience response to the COVID-19 crisis (Boubaker *et al.*, 2022a); or a common set of weights determined endogenously, not allowing a different set of weights to classify a DMU as efficient (or inefficient), with applications in banking (Hammami *et al.*, 2022). The assumption that all the units operate similarly is not close to reality, i.e., not all DMUs perform the same as they are all not specialized in all the water resource utilization and water provision operations, but partially. In the search for a more robust approach to decision-making, this chapter is an application of a model first proposed by Serrano-Cinca & Mar-Molinero (2004) that combines DEA and multivariate statistical techniques. The main advantage is that we avoid removing units from the sample by estimating a variety of model specifications and performing a two-stage analysis to explain the reasons behind inefficiency. The zero-weight issue is overcome. Frontier units can be ranked against each other since the visualization helps us determine sources of inefficiency (or efficiency) by reflecting on (diverse) production functions, despite the method being non-parametric.

4. Methodology

The methodology involves estimating the efficiencies for each DMU, in this case, provinces, according to all the possible DEA specifications in the first stage (see Appendix A). The two-way table of efficiencies — DEA specifications by DMUs — is analyzed using multivariate statistical analysis, i.e., factor analysis, property fitting, and cluster analysis. The multivariate analysis allows for visualizing the results graphically (data visualization), which helps in interpretation without requiring the addition of external information or further stages. Following Banker *et al.* (1984), The DEA model (1) is output-oriented and exhibits variable returns to scale (Banker *et al.* 1984). The orientation choice is followed to control the production, making decisions among the outputs, not the inputs. The mathematical formulation for s outputs, m inputs, and n DMUs, with "o" denoting the unit of interest, is as follows:

$$\text{Max } \Phi$$

$$\text{s.t.:} \quad \sum_j Y_{rj}\lambda_j \geq \Phi\, Y_{rO}, \quad r = 1, \ldots, s \tag{1}$$

$$X_{iO} - \sum_j X_{rj}\lambda_j \geq 0, \quad i = 1, \ldots, m$$

$$\sum_j \lambda_j = 1, \quad j = 1, \ldots, n.$$

A specification consists of a combination of inputs and outputs. The inputs are identified by letters and the outputs by numbers. The traditional DEA model accounts for all the inputs and outputs chosen *a priori* by the researcher, in this case, ABCD12 (see Table 1). As is well known, DEA is sensitive to the variables used (Dyson *et al.*, 2001), meaning that it can result in very different efficiencies if the inputs and outputs are changed. In this regard, all the possible combinations are estimated, including one input and one output. Some specifications would not make sense, for example, specification D2, which accounts for the population as an input and the quality of the water produced. However, DEA visualization will be able to disregard any nonsensical (non-feasible) combination in the factorial analysis and map outliers (Ashkiani, 2019) without the requirement of removing them from the sample, as is done traditionally. The total number of DEA specifications accounting for four inputs and two outputs is 51.

Table 1. Descriptive statistics inputs and outputs.

Variable	**Type**	**Label**	**Obs.**	**Mean**	**Std. dev.**	**Min.**	**Max.**
Water resources (billion m^3)	Input	A	31	1,019.53	1,020.98	11.00	4,597.30
Drainage pipe length (km)	Input	B	31	23,999.42	22,829.64	831.00	98,633.00
Sewage capacity (10,000 m^3/day)	Input	C	31	618.43	531.77	29.90	2,453.10
Population	Input	D	31	1,670.26	1,226.17	100.40	6,146.90
Water supply (billion m^3)	Output	1	31	187.51	141.81	24.30	572.00
Quality (percentage)	Output	2	31	0.79	0.27	0.00	0.99

Notes: The table shows the four inputs (2019) and the two outputs (2020) used to estimate efficiencies. The inputs are the natural water resources obtained in billion cubic meters per year, distribution infrastructure in kilometers, infrastructure for the residual water treatment in 10,000 cubic meters per day, and the population. The outputs are water supply in billion cubic meters, and the quality of the water, the percentage of purified water contained in the water resources naturally obtained.

The calibration of the model consists of evaluating superefficiency units as per influencing the rest. The Efficiency Measurement System program (Scheel, 2000) shows Beijing, Hainan, Ningxia, Qinghai, Xinjiang, and Tibet as very efficient (see Appendix B). Ningxia and Tibet are the most efficient. Tibet is not a reference unit for any other province. Rather than removing these units from the sample and, therefore, losing data, as is performed traditionally, the methodology will identify peculiarities of these provinces in the provision of water supply, i.e., extreme cases as outliers or mavericks. For example, Ripoll-Zarraga & Mar-Molinero (2020) identified two airports behaving differently from the others. However, these were not superefficient units, not even efficient in all the aeronautical operations. These airports were finally labeled "cargo-oriented" rather than "passenger-oriented," i.e., they were efficient for one output (cargo) but inefficient when accounting for all the outputs (i.e., passengers, movement, and cargo).

5. Data

The choice of inputs and outputs for efficiency estimation is based on literature and data availability. The sample consisted of 31 Chinese provinces (see Figure 1). Hong Kong SAR (Special Administrative Region), Macau, and Taiwan PRC (People's Republic of China) were not analyzed. The water resource system is divided into water usage and waste treatment (e.g., Zhao *et al.*, 2017). The variables used in water efficiency are infrastructure (capital), labor, water flow, pollutants (Hu *et al.*, 2006; Zhou *et al.*, 2008), and fresh and reused water (Bian *et al.*, 2014). The efficiency of water provision represents the relationship between the water produced given limited resources, such as the length of the pipes, drains, and sewerage capacities, which require engines and pumps to increase the water capacity treatment, and the geographical location that decides access to natural resources. Accordingly, there are four inputs: water resources (resources) as naturally obtained through the rain, rivers, etc., surface and groundwater, drainage pipe length (drain), and sewerage capacity (capacity), which are proxies of infrastructure and population. Two outputs represent the water production process: water without pollutants (supply), equivalent to the water consumed, and

Figure 1. Chinese provinces (excluding Hong Kong, Macau, and Taiwan).
Note: The figure shows the 31 Chinese provinces analyzed in the study.

pollutants as an undesirable output. Rather than using an undesirable output, we convert the chemical pollutants into the percentage of purified water contained in the water resources naturally obtained — without pollutants — as a proxy of water quality (quality). The higher the value, i.e., closer to the unity value, the lower the pollutants, i.e., the purer the water provision, potentially avoiding water treatment processes. When a province has a higher level of chemical pollutants versus water resources, the quality indicator is zero, which means that the water contains 100% pollutants. For example, in the eastern region, the provinces with fewer pollutants are Zhejiang (0.95), Jiangxi (0.94), Fujian (0.92), and Anhui (0.91). Shandong is the province with more pollutants (0.59). Except for Henan (0.65), the southcentral and southwest regions have an overall high quality of water produced (0.90–0.99). The north of China, on the other hand, has less quality water, with extreme cases such as

Hebei (0.13) or Tianjin (0.00) and Ningxia in the northwest (0.00). Only Neimenggu (Inner Mongolia) shows a relatively high level of purity (0.86), followed by Beijing (0.79).

All outputs are data from 2020, and the inputs are from 2019 (lagged variables). The assumption is that the investment in the infrastructure by the end of one year and the population changes will impact demand in the following year.

Access to water resources is different across the provinces. Tibet (southwest) is the region with the highest water obtained naturally, and Ningxia (northwest) is the lowest. On the other hand, Tibet has the lowest value of the remaining variables, i.e., drainage pipe length, sewage capacity, and population level. The region with the highest investment in infrastructure and population is Guangdong. The provinces with worse quality water are Ningxia, Tianjin, and Hebei, as previously discussed.

According to the four inputs and two outputs, 15 possible combinations of inputs and three of outputs exist. Since it is required to combine inputs with outputs, there are 45 DEA model specifications, i.e., one input must be combined at least with one output, and vice versa, while one output without input is not feasible. For each specification and province, the efficiency scores according to the BCC model are estimated (see Appendix C).

6. Results and Discussion

The traditional DEA model would account for all the inputs and outputs (ABCD12). There are nine provinces that are technically efficient, i.e., whose efficiency score is unity. Tianjin is highly inefficient with a 260% level, i.e., an inefficiency of 160%, followed by Shandong (147.30%), Henan (137.60%), and Liaoning (135.47%). The rest of the provinces are overall highly efficient. By removing output two from the sample (quality) — specification ABCD1— most provinces increase their inefficiencies significantly, including previous frontier units, e.g., Beijing (207.26%), with an average of 266%. Jiangsu, Ningxia, Qinghai, Xinjiang, and Tibet remain frontiers. Consequently, part of the efficiency is explained by the quality of the water supply and not by the quantity of water provision.

By removing output one, water resources that include pollutants, and accounting for the water quality (ABCD2), most provinces have lower inefficiencies, with an average of 137%, except Hebei (698%) and Tianjin (extreme case). Hebei is one of the provinces with worse water quality, along with Ningxia and Tianjin. However, Ningxia is efficient (100%) for this specification. Therefore, the DEA results are not straightforward. Not accounting for population (D) seems to have an insignificant impact (ABC12), owing to the fact that the efficiencies are very similar to the traditional DEA model and the two DEA specifications (ABC1 and ABC2). As seen, we could treat some DMUs as inefficient unfairly when they are not, but the efficiency scores depend on the researchers' personal choice of variables.

At a glance, Tianjin seems like an extreme case, maverick or outlier, as previously identified in the superefficiency analysis. This analysis showed Ningxia, Qinghai, and Tibet as potential extreme cases as well. However, these analyses accounted for all the inputs and outputs jointly (traditional DEA model), which may not be the best representation of the true production function of the DMUs. Indeed, Ningxia is inefficient when considering either drainage pipe length or sewage capacity and water resources (B1 and C1, respectively) and highly inefficient regarding the natural water quality (B2 and C2) but is efficient (frontier) when accounting for input A. Traditionally, outliers were removed from the sample according to their significant influence on the efficiency scores of the other DMUs. However, DEA visualization avoids having to remove the DMUs. First, data are lost; second, the data (sample) are not the same as the original one. Hence, the increase in efficiency may not be due to removing a competitor. DEA visualization will be able to identify outliers and mavericks without missing information about their location on the maps.

6.1. *DEA visualization: Factor analysis*

The first stage consists of reducing the dimensionality of the data. The factor analysis reveals five extracted factors following the Kaiser criterion (eigenvalue > 0.7), although we account for six factors overall. Table 2 shows the variance explained by the first six factors. Most of the variability of the data is explained by the first two factors (accumulated 83.88).

The factor analysis also provides us with the loading component matrix, in which for each specification, there is a significant load in one or more factors (see Table 3). To understand the meaning of the

Table 2. Factor analysis: Variance explained (unrotated).

Component	Eigenvalue	% Variance	% Cumulative
1	29.98	66.63	66.63
2	7.76	17.25	83.88
3	3.13	6.96	90.84
4	2.05	4.55	95.38
5	1.21	2.69	98.07
6	0.59	1.31	99.39

Note: The table shows the variance explained by each factor and cumulative. Factors with eigenvalues higher than 0.7 are extracted. Three factors explain up to 91% of the variability. However, the remaining two may contain special features not explicitly revealed.

extracted components, in this case, five factors are required to refer to the factor scores shown in the component matrix, with a significant value of, for example, 0.40 (Ripoll-Zarraga & Mar-Molinero, 2020). ABCD12 has, e.g., a positive weight in the first factor (0.93) compared to ABCD1, which mainly contributes to the second factor (0.95). Indeed, most specifications for input A have a higher load and positive loading weight in the first factor when accounting for outputs one and two together, or output two. Whereas, when accounting for output one, most specifications have the highest loadings in Factor 2 (e.g., AB2 is better explained by Factor 1, i.e., 0.98; AB1 is better explained by Factor 2, i.e., 0.83, and Factor 5, i.e., 0.45). The correlations between most DEA specifications and Factor 1 are positive and high. Since we are estimating provinces' efficiency in providing public water services, Factor 1 could be interpreted *a priori* as the overall efficiency under different ways of defining the threshold "efficiency" in this type of service, such as the water provision quality. From the component matrix, we can gain some insights. However, it becomes more complicated with a higher number of inputs and outputs used, i.e., more DEA specifications.

Only analyzing the factor loading matrix makes it challenging to understand the meaning behind the extracted factors. What do these mean? How can we use this information to understand and interpret the intrinsic characteristics of the data? What does this represent in term of efficiency, and even more, how can we differentiate between provinces specialized in different productions, which may not

Table 3. Factor loadings: DEA specifications.

	Component matrix				
DEA specifications	1	2	3	4	5
ABCD12	0.93	−0.04	−0.19	0.24	−0.01
ABCD1	0.06	0.95	0.22	0.21	−0.06
ABCD2	0.98	−0.03	−0.09	0.07	0.06
ABC12	0.93	−0.04	−0.19	0.24	−0.01
ABC1	0.06	0.95	0.22	0.21	−0.06
ABC2	0.98	−0.03	−0.09	0.07	0.06
ABD12	0.94	−0.04	−0.20	0.22	−0.02
ABD1	0.05	0.95	0.21	0.21	−0.04
ABD2	0.98	−0.03	−0.09	0.07	0.06
ACD12	0.93	−0.05	−0.19	0.24	−0.02
ACD1	0.06	0.95	0.21	0.22	−0.06
ACD2	0.98	−0.03	−0.09	0.07	0.06
BCD12	0.98	−0.06	−0.08	0.06	0.04
BCD1	0.80	0.39	−0.12	−0.41	−0.11
BCD2	0.76	−0.28	0.58	−0.05	0.00
AB12	0.94	−0.04	−0.20	0.22	−0.02
AB1	0.01	0.83	0.14	0.00	0.45
AB2	0.98	−0.03	−0.09	0.07	0.06
AC12	0.93	−0.05	−0.19	0.24	−0.02
AC1	0.06	0.95	0.21	0.22	−0.06
AC2	0.98	−0.03	−0.09	0.07	0.06
AD12	0.94	−0.04	−0.20	0.22	−0.03
AD1	0.05	0.95	0.20	0.22	−0.05
AD2	0.98	−0.03	−0.09	0.07	0.06
BC12	0.98	−0.06	−0.08	0.06	0.04
BC1	0.80	0.39	−0.12	−0.41	−0.11
BC2	0.76	−0.28	0.58	−0.05	0.00
BD12	0.98	−0.06	−0.08	0.06	0.04
BD1	0.80	0.40	−0.12	−0.42	−0.09
BD2	0.76	−0.28	0.58	−0.05	0.00
CD12	0.99	−0.07	−0.05	0.05	0.03
CD1	0.81	0.38	−0.09	−0.42	−0.11
CD2	0.76	−0.28	0.58	−0.05	0.00
A12	0.91	−0.04	−0.22	0.15	−0.07
A1	−0.08	0.19	−0.04	−0.18	0.94

(*Continued*)

Table 3. (*Continued*)

	Component matrix				
DEA specifications	1	2	3	4	5
A2	0.98	−0.03	−0.09	0.07	0.06
B12	0.98	−0.06	−0.08	0.06	0.04
B1	0.77	0.39	−0.13	−0.47	0.05
B2	0.76	−0.28	0.58	−0.05	0.00
C12	0.99	−0.08	−0.04	0.05	0.03
C1	0.81	0.38	−0.08	−0.42	−0.11
C2	0.76	−0.28	0.58	−0.05	0.00
D12	0.99	−0.07	−0.05	0.05	0.03
D1	0.81	0.39	−0.10	−0.43	−0.09
D2	0.76	−0.28	0.58	−0.05	0.00

Notes: The table shows the relevance of each extracted factor on each DEA specification. The highest the value (i.e., close to the unitary value), the more relevant the factor related to the specific DEA model. These loadings help to identify the meaning of each factor.

be efficient overall, but efficient in a specific type of water provision? We overcome these issues with the help of visualization, i.e., mapping the DMUs (in our case, provinces) in a two-dimensional space at a time (common map). To visualize the DMUs analyzed, we generate the component matrix. This technique allows for mapping the DMUs across different dimensions (extracted factors). See Figures 2–4.

The component matrix shows us the highest and lowest values for each province and extracted component without knowing what these factors represent. For example, Tianjin is likely an extreme value when considering Factor 1 (5.28), with the rest of the provinces having values lower than unity. Other relevant, although not necessarily extreme, values are: for Factor 2, Chonqing (3.50) and Ningxia (−1.93); for Factor 3, Ningxia (4.81); for Factor 4, Beijing (−3.99); and for Factor 5, Qinghai (4.19) and Tibet (2.40).

Figure 2 shows that Tianjin is located on the right-hand side of Factor 1. This technique allows us to see the meaning of Factor 1 in relation to another dimension, such as Factor 2, which is located near zero. Similarly, Chongqing and Guizhou are located close to zero values for Factor 1 but high values for Factor 2. In other words, accounting for the specific efficiency to which Factor 1 and Factor 2

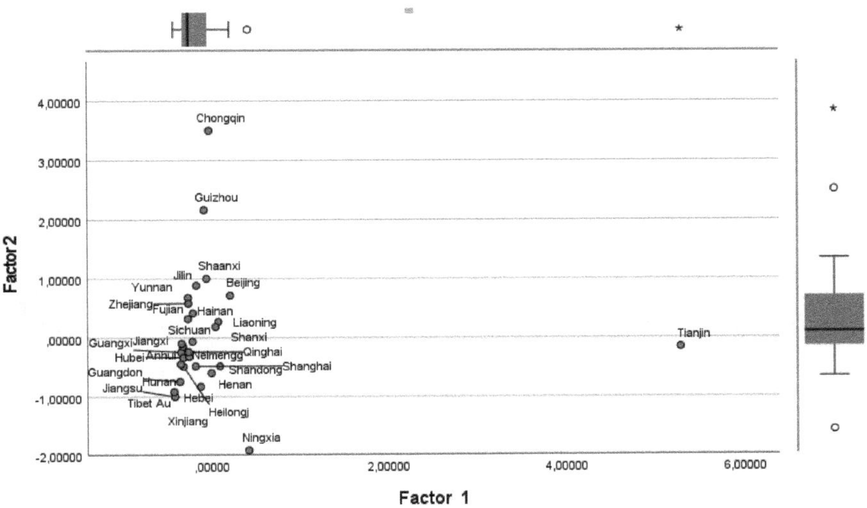

Figure 2. Dimension 1 and Dimension 2.

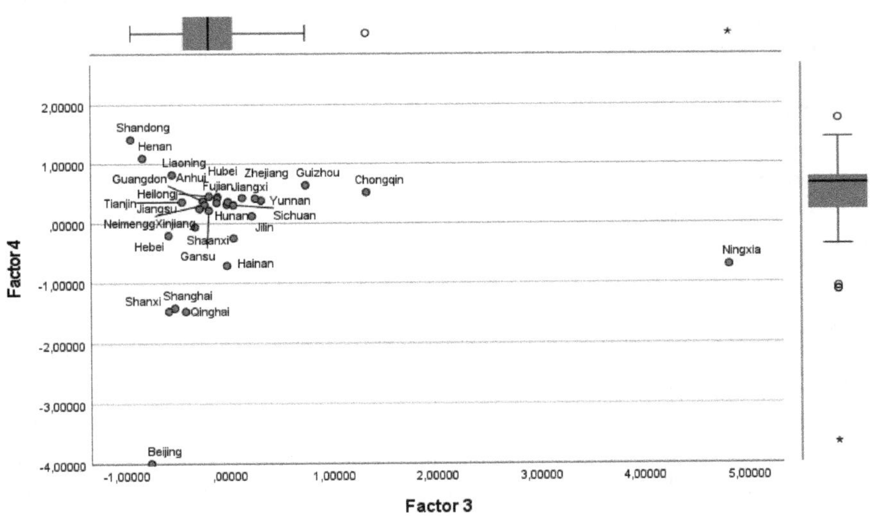

Figure 3. Dimension 3 and Dimension 4.

refer — intrinsic characteristics of the data — these imply different provinces' behavior regarding water provision. In this regard, we cannot assume that the DMUs are either efficient or inefficient generally, i.e., according to all the inputs and outputs used (ABCD12).

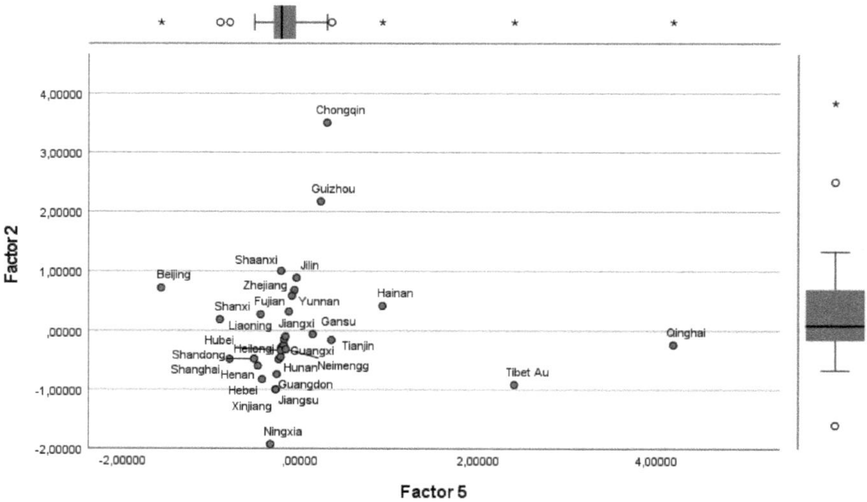

Figure 4. Dimension 2 and Dimension 5.

Notes: Figures 2–4 show the locations of the DMUs (provinces) according to the intrinsic meaning of each extracted factor, which help identify what explains each factor and how the DMUs behave in terms of efficiency concerning them, including extreme cases (outliers and mavericks), specialization, and similar production processes.

Indeed, they could be efficient according to different combinations of the decisional variables (DEA specifications). One relevant stage behind naming the factors is establishing sources of inefficiency.

6.2. *DEA visualization: Property fitting*

The DEA specifications are plotted on a common map using the property fitting approach (Schiffman *et al.*, 1981; Gower & Hand, 1996) to understand the meaning behind the extracted factors. Mar-Molinero & Mingers (2007) demonstrated a mathematical method to represent variables and specifications in the same space. ProFit can be used to draw each DEA specification as a vector, which has its starting point at the origin of the coordinates and shows the direction toward a particular model specification change, i.e., a specific data feature. To draw the vectors, we need to perform a regression. The independent variables are each DEA specification for each province, i.e., the efficiencies estimated at the first stage for each combination of inputs and outputs. The dependent variables are the extracted factors. Finally, ProFit vectors are normalized to unit length as follows:

$$\beta_i^* = \frac{\beta_i}{\sqrt[2]{\sum_{i=1}^{5} \beta_i^2}}, \tag{2}$$

where β_i is the ith regression coefficient (Factor i) associated with each DEA specification and β_i^* is the normalized vector to unit length. The origin of all ProFit vectors is the center of the coordinates. When a vector has a unit length regarding a particular coordinate axis, the factor is relevant for interpretation. On the other hand, if it is close to the center of the coordinates, then it is not relevant. The normalized coefficients are shown in Table 5.

By combining both matrices (Tables 4 and 5), the meaning of a factor which contains the data becomes more accurate.

Figure 5 shows the location of DEA specifications according to Factors 1 and 2 using the property fitting vectors. The traditional DEA model values, i.e., including all the inputs and outputs ABCD12 and any other specification with input A and output two, are located on the right-hand side, whose value is unity or close to it (x-axis). Other relevant specifications exclude input A but include other inputs with outputs one or two. Factor 1 is interpreted as inefficiency in using the water resources and infrastructure for quality water supply (purified water). Since water resources are obtained naturally, this factor may represent a proxy for the scarcity of natural water resources. Tianjin is one province, along with Ningxia, with a significantly low quantity of natural water resources, and Tianjin has a significant amount of investment in infrastructure (underused) compared to Ningxia. Indeed, Tianjin is highly inefficient for all the specifications and extremely inefficient when accounting for output two and inputs B and C. Ningxia is efficient when accounting for input A. The order of provinces concerning Factor 1 provides an idea of how relevant provinces are to producing purified water compared to each other (see Table 4).

Turning our attention to Factor 2, most of the highest values correspond to the DEA specifications accounting for input A and output one, which ignores output two. The ProFit normalized coefficients show, for example, that ABCD1, ABC1, ABD1, ACD1, AC1, and

Table 4. Factor loadings: Provinces.

Provinces	Component matrix				
	1	2	3	4	5
Anhui	−0.30	−0.29	−0.11	0.45	−0.22
Beijing	0.22	0.72	−0.76	−3.99	−1.57
Chongqing	−0.02	3.50	1.32	0.51	0.29
Fujian	−0.26	0.32	0.13	0.43	−0.13
Gansu	−0.21	−0.06	−0.20	0.22	0.13
Guangdong	−0.35	−0.74	−0.25	0.38	−0.27
Guangxi	−0.34	−0.24	−0.02	0.32	−0.19
Guizhou	−0.08	2.17	0.74	0.64	0.22
Hainan	−0.21	0.42	−0.02	−0.71	0.92
Hebei	−0.12	−0.83	−0.59	−0.20	−0.43
Heilongjiang	−0.31	−0.49	−0.20	0.45	−0.25
Henan	0.00	−0.60	−0.83	1.10	−0.48
Hubei	−0.31	−0.34	−0.12	0.42	−0.22
Hunan	−0.34	−0.44	−0.12	0.35	−0.22
Jiangsu	−0.41	−1.00	−0.29	0.25	−0.28
Jiangxi	−0.32	−0.15	−0.01	0.36	−0.19
Jilin	−0.17	0.89	0.22	0.12	−0.05
Liaoning	0.08	0.28	−0.55	0.82	−0.45
Neimenggu (Inner Mongolia)	−0.25	−0.32	−0.24	0.31	−0.17
Ningxia Hui Autonomous Region	0.42	−1.93	4.81	−0.70	−0.34
Qinghai	−0.26	−0.24	−0.42	−1.47	4.19
Shaanxi	−0.05	1.01	0.04	−0.24	−0.22
Shandong	0.10	−0.48	−0.95	1.41	−0.52
Shanghai	−0.18	−0.48	−0.53	−1.41	−0.80
Shanxi	0.05	0.19	−0.59	−1.47	−0.91
Sichuan	−0.33	−0.10	0.04	0.31	−0.17
Tianjin	5.28	−0.16	−0.46	0.36	0.34
Xinjiang Uygur Autonomous Region	−0.41	−1.00	−0.29	0.25	−0.28
Tibet Autonomous Region (Xizang)	−0.42	−0.91	−0.33	−0.06	2.40
Yunnan	−0.26	0.68	0.31	0.38	−0.07
Zhejiang	−0.26	0.59	0.25	0.42	−0.10
Max.	5.28	3.50	4.81	1.41	4.19
Min.	−0.42	−1.93	−0.95	−3.99	−1.57

Notes: The table shows the relative relevance of the extracted factors concerning the provinces. The higher the value (positive or negative), the more relevant the factor is according to the intrinsic characteristics of the DMUs, the production process, and the differences between them.

Table 5. ProFit normalized regression coefficients.

DEA specifications	β_1^*	β_2^*	β_3^*	β_4^*	β_5^*	R^2
ABCD12	0.95	−0.04	−0.19	0.24	−0.01	0.96
ABCD1	0.06	0.95	0.22	0.21	−0.06	1.00
ABCD2	0.99	−0.03	−0.09	0.07	0.06	0.98
ABC12	0.95	−0.04	−0.19	0.24	−0.01	0.96
ABC1	0.06	0.95	0.22	0.21	−0.06	1.00
ABC2	0.99	−0.03	−0.09	0.07	0.06	0.98
ABD12	0.95	−0.04	−0.20	0.23	−0.02	0.96
ABD1	0.05	0.95	0.21	0.21	−0.04	0.99
ABD2	0.99	−0.03	−0.09	0.07	0.06	1.00
ACD12	0.95	−0.05	−0.19	0.24	−0.02	0.97
ACD1	0.06	0.95	0.21	0.22	−0.06	1.00
ACD2	0.99	−0.03	−0.09	0.07	0.06	0.98
BCD12	0.99	−0.06	−0.08	0.06	0.04	0.98
BCD1	0.81	0.39	−0.12	−0.41	−0.11	1.00
BCD2	0.76	−0.28	0.58	−0.05	0.00	1.00
AB12	0.95	−0.04	−0.20	0.23	−0.02	0.96
AB1	0.01	0.87	0.15	0.00	0.47	0.91
AB2	0.99	−0.03	−0.09	0.07	0.06	0.98
AC12	0.95	−0.05	−0.19	0.24	−0.02	0.97
AC1	0.06	0.95	0.21	0.22	−0.06	1.00
AC2	0.99	−0.03	−0.09	0.07	0.06	0.98
AD12	0.95	−0.04	−0.20	0.22	−0.03	0.96
AD1	0.05	0.95	0.20	0.22	−0.05	0.99
AD2	0.99	−0.03	−0.09	0.07	0.06	0.98
BC12	0.99	−0.06	−0.08	0.06	0.04	0.98
BC1	0.81	0.39	−0.12	−0.41	−0.11	1.00
BC2	0.76	−0.28	0.58	−0.05	0.00	1.00
BD12	0.99	−0.06	−0.08	0.06	0.04	0.98
BD1	0.80	0.40	−0.12	−0.42	−0.09	1.00
BD2	0.76	−0.28	0.58	−0.05	0.00	1.00
CD12	0.99	−0.07	−0.05	0.05	0.03	0.98
CD1	0.81	0.38	−0.09	−0.42	−0.11	1.00
CD2	0.76	−0.28	0.58	−0.05	0.00	1.00
A12	0.96	−0.04	−0.23	0.16	−0.07	0.90
A1	−0.09	0.19	−0.04	−0.19	0.96	0.95
A2	0.99	−0.03	−0.09	0.07	0.06	0.98
B12	0.99	−0.06	−0.08	0.06	0.04	0.98
B1	0.78	0.40	−0.13	−0.47	0.05	0.98
B2	0.76	−0.28	0.58	−0.05	0.00	1.00

(*Continued*)

Table 5. (*Continued*)

DEA specifications	β_1^*	β_2^*	β_3^*	β_4^*	β_5^*	R^2
C12	0.99	−0.08	−0.04	0.05	0.03	0.99
C1	0.81	0.38	−0.08	−0.42	−0.11	1.00
C2	0.76	−0.28	0.58	−0.05	0.00	1.00
D12	0.99	−0.07	−0.05	0.05	0.03	0.98
D1	0.81	0.39	−0.10	−0.43	−0.09	1.00
D2	0.76	−0.28	0.58	−0.05	0.00	1.00
Max.	0.99	0.95	0.58	0.24	0.96	—
Min.	−0.09	−0.28	−0.23	−0.47	−0.11	—

Notes: The table shows the ProFit vectors' normalized coefficients, which are used to draw the DEA specifications in the same space. The value of a coefficient close or equal to the unity reflects the relevance of the factor concerning the specific DEA model, i.e., revealing the inputs used and the outputs produced which underlay the meaning of the factor.

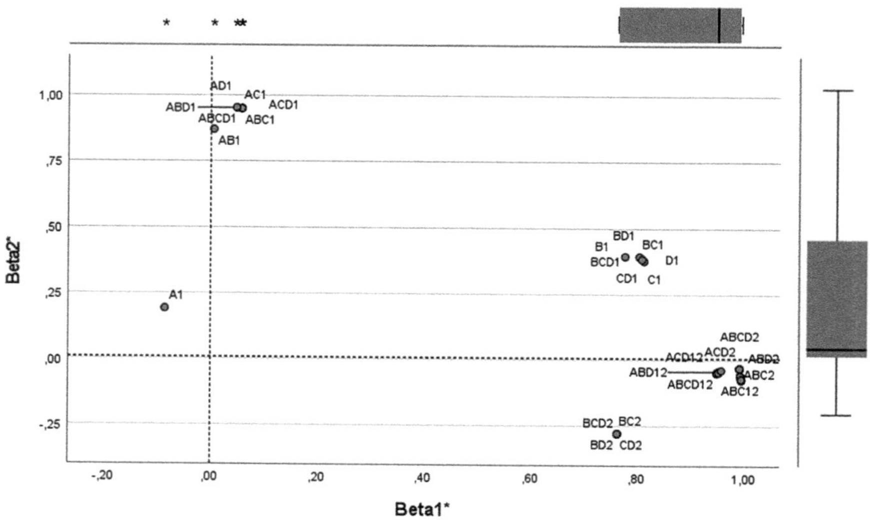

Figure 5. DEA specifications projection (Dimension 1 and Dimension 2).

Notes: The figure shows the ProFit vectors' normalized coefficients for Factors 1 and 2, which are used to draw the DEA specifications. Factor 1 contains the traditional DEA model with all the inputs and outputs analyzed (ABCD12). Factor 2 relates to the DEA specifications concerning output one but not output two.

AD1 have 0.95 and AB1 has 0.83, i.e., almost close to unity, indicating the relevance of Factor 2 in these specifications. Factor 2 seems to relate to the use of natural water resources in supplying water overall. Indeed, Figure 5 shows most of the DEA specifications containing input A and output one in the end-vector points. However, Figure 2 shows provinces with a higher quality of water, such as Chongqing and Guizhou, with high values on the y-axis and lower-quality water toward negative values. These two provinces produce water with very few pollutants. Therefore, Factor 2 is interpreted as efficiency in producing high-quality water. The other extreme is Ningxia since it supplies water with a concentration of 100% pollutants. And between zero and Ningxia, other provinces with quality water range within 0.58–0.69.

Factor 3 seems to have higher values when combining inputs BCD, altogether or separately, and output two (0.58). These inputs refer to infrastructure, i.e., pipe length and sewerage capacity, concerning the population to produce purified water (without pollutants). Therefore, Factor 3 represents the inefficiency in using water infrastructure regarding the population to supply quality water. Factor 4 shows negative loads and relatively low values (−0.42) when accounting for input B, C, or D and output one, i.e., input A and output two. Similarly to Factor 3, these inputs refer to infrastructure for water provision (with pollutants) to the population. Factor 5 shows a significant load for A1 (0.94), the overall natural water resources for water provision, i.e., excluding infrastructure (B, C), population (D), and the quality of the water (2). Indeed, Qinghai is a province with a high level of water resources. However, the water supply is significantly lower compared to provinces with similar resources (e.g., Guizhou, Zhejiang). Hence, this factor represents overcapacity in terms of natural water resources for water provision before being treated. Beijing, Shanghai, and Shanxi would be the most efficient provinces since they have water resources based on the water produced, i.e., these provinces are opposite to Qinghai, meaning that, due to having a low level of natural resources, all of these are used for water supply. The lower the value of Factor 5, the better the use of natural resources for producing water, i.e., fewer natural resources are underused. Tibet is the province with the highest cubic meters of natural resources, yet the supply of water provision is significantly low, despite having the lowest population. However, Tibet is not located at the extreme part of the x-axis for Factor 5. Further research is required to learn from the exogenous province environment, potentially making it less inefficient (with fewer resources underused) than it is. For example, Tibet and Qinghai have the highest water quality (0.99).

6.3. *DEA visualization: Cluster analysis*

Finally, to assess similarities in operations within the sample, we use a combination of ProFit and cluster analyses. Each province is a point in a 51-dimension space, i.e., five factors enclose all the data dimensionality of the DEA specifications. The projections of the 51 models have projected accounting for two dimensions at once, i.e., to be able to visualize them. However, the mapping is in a five-dimensional space. Provinces may appear close to each other in a two-dimension space without necessarily implying that they perform similarly concerning one factor. The cluster analysis will help identify the provinces behaving similarly, i.e., according to the extracted factors. We standardized the data since the efficiency scores range between one and infinite.

For input-oriented models, there is no need for standardization as efficiencies range between 0 and 100.

For input-oriented models, there is no need for efficiencies to be standardized between 0 and 100. There are different methods for cluster analysis. Ward's approach is based on variance analysis to maximize homogeneity within a cluster and heterogeneity between clusters simultaneously. The relative position between two provinces is measured using the squared Euclidean distance. With the help of the dendrogram in Ward's method, seven clusters are identified (see Appendix F). The cluster analysis shows most provinces are classified into clusters 1 or 4 (see Figures 6 and 7).

Tianjin belongs to a cluster located on the right-hand side of Figure 6. No other provinces belong to this cluster. Thus, Tianjin is an extreme value — a maverick — i.e., not affecting the efficiency of other provinces. Tianjin has few natural water resources and significant investment in infrastructure, i.e., drainage pipe length, but produces very little water supply with poor quality, as previously discussed. Therefore, Tianjin is the most inefficient province for high-quality water provision, followed by Ningxia, Beijing, and Shandong (Factor 1). These results contrast previous findings that

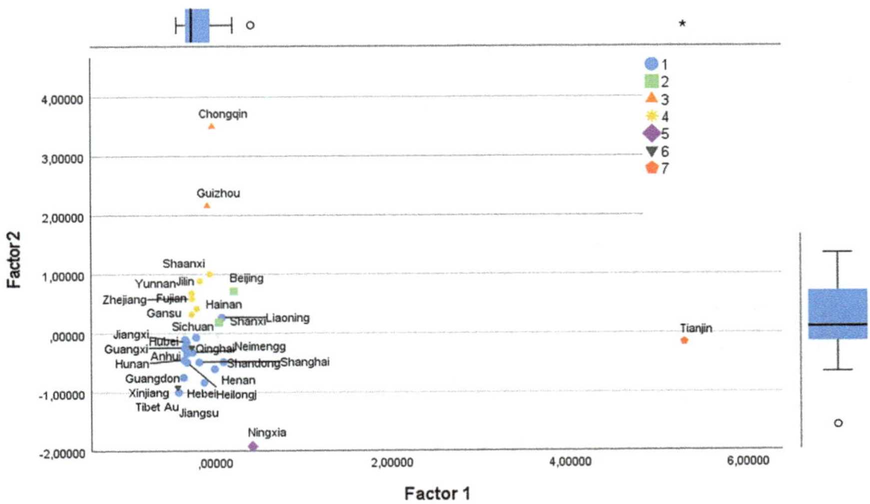

Figure 6. Provinces projection per clusters (Dimension 1 and Dimension 2).

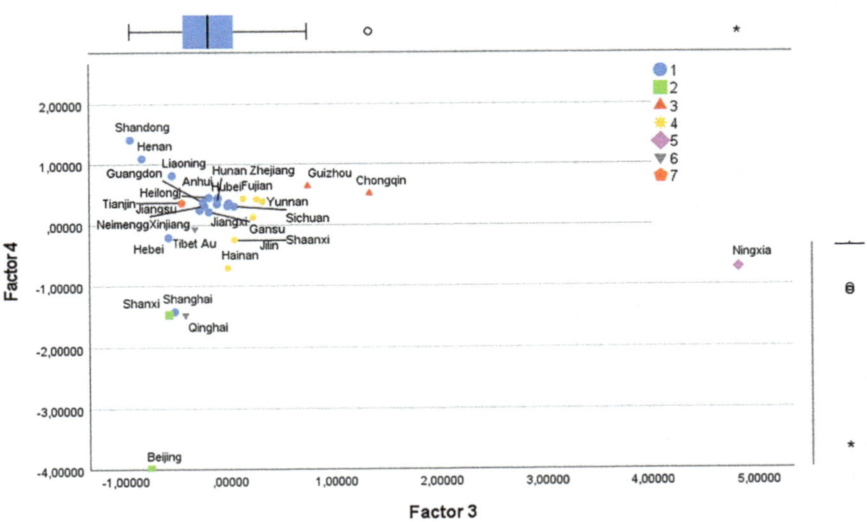

Figure 7. Provinces projection per clusters (Dimension 3 and Dimension 4).

Notes: Figures 6 and 7 are similar to Figures 2 and 3, showing the location of provinces according to the extracted factors, i.e., in these cases, Factors 1–4, with the added value of clusters identified across the DMUs which reveal their similar behavior regarding the intrinsic information contained in the factors (within) and different from the other clusters (between).

the economically developed provinces, such as Beijing, Shanghai, and Tianjin, are highly efficient (Deng *et al.*, 2016). The most efficient provinces are Xinjiang and Tibet owing to their locations on the left-hand side of the x-axis (Figure 6), as confirmed by the efficiency scores, where all the DEA specifications (except A1 for Tibet, 1,776.40%) become frontier or highly efficient. Ningxia, a province with a similar level of water resources and also with the water provision being poor quality, has 10 times less pipe length and a third of Tianjin's sewage capacity. Therefore, the level of inefficiency is significantly lower, located almost at value zero. Tianjin was identified as a potential outlier or maverick and is confirmed to be the most inefficient province producing water without pollutants (purified water). However, this is in terms of not providing purified water (Factor 1). For the other factors, Tianjin may not be an extreme case. Indeed, Ningxia, the only province in cluster 5, is also an extreme case with its value located on the right-hand side of Figure 7. The rest of the clusters containing a few provinces share similar characteristics concerning the water provision production process, i.e., Beijing and Shanxi (cluster 2), Chongqing and Guizhou (cluster 3), or Qinghai and Tibet (cluster 6). Chongqing and Guizhou are the most efficient provinces in water provision, i.e., accounting for the quality (Factor 2) and using the available infrastructure to supply high-quality water (Factor 3), same as Beijing. Even though Ningxia was the frontier for most of the specifications combining input A, it is the most inefficient province using the infrastructure to supply high-quality water compared to the other provinces. Beijing has made a significant investment in water infrastructure (pipe length and sewerage capacity) for the water supply (Factor 4). In other words, Beijing is inefficient in using the infrastructure for water provision. Therefore, Beijing has overcapacity compared to, for example, Guangxi, which produces almost seven times more than Beijing or eight times more than Hunan. Finally, Qinghai and Tibet are the provinces with less scarcity of natural water resources for water provision (Factor 5). Another way to read these results is that these provinces have an excess of natural water resources (A) based on the amount of water produced (1). On the other hand, Beijing has few natural resources for the water supplied.

Combining Factors 1 and 5, we can observe that although Tibet, for example, is efficient in using the water resources for high-quality water provision (Factor 1), it suffers from overcapacity of these resources due to the current level of water produced (Factor 5). Tibet was the poorest province in 2020 and shows high efficiency in using infrastructure for water provision (Factor 4), as seen by the adequate investment in pipes and sewerage capacity. Consequently, these results confront previous findings in terms of GDP, wealth development, and water efficiency utilization. This excess of water resources could be transferred to water supply in other provinces suffering from scarcity and increase the province's income. Qinghai is a province relatively close to Tibet, which also suffers from overcapacity (Factor 5). In this regard, the geographical location of the provinces according to the water natural resources rather than wealth specifies efficiency in water provision according to different output characteristics, i.e., water and high-quality water. Apart from the geographical location with access to mountains during the heavy winter season, another possible explanation could be the population. Regardless of having a low number of inhabitants, the natural resources remain significantly higher compared to other provinces. At the same time, both provinces provide the highest-quality water. Both are classified in the same cluster (cluster 6), with no other province sharing similar characteristics in terms of the production process.

7. Conclusions

This chapter estimates the efficiencies of 31 Chinese provinces for water provision in 2020 according to different combinations of inputs and outputs (DEA specifications). The efficiency level depends on the inputs and outputs chosen by the researcher. Thus, it is a relative measure which cannot be straightforwardly used, as it refers to the efficiency of a DMU benchmarked against the rest and according to the variable used. Nevertheless, DMUs may be more efficient in parts or in certain operations of the production process rather than in all the activities. The DEA results show that some provinces are more efficient in water provision (quantity) or quality, and vice versa. At the same time, some units are efficient in using water infrastructure for water supply. Therefore, the first stage analysis determines that not all the provinces have a similar production function, use the inputs in the same way, or produce the same type of outputs, and are equally efficient in water supply and high-quality water provision. DEA visualization helps identify which DMUs are efficient according to the intrinsic characteristics of the data, determining sources of inefficiency for managerial and policymakers' recommendations.

The findings show that geographical location, according to accessibility to more natural water resources, determines the efficiency of water provision. Indeed, provinces with more GDP (Statista, 2022) are not the most efficient in water provision, confronting most studies in water management efficiency, which either use GDP as a desirable output of water provision or as an explanatory value. Traditionally, Chinese east provinces are more efficient, followed by central areas, the provinces from the west being the most inefficient in water resource efficiency. Coastal territories such as Fujian, Guangdong, and Hainan are economically developed regions. Hence, water resource efficiency is better than in arid areas with less precipitation, for example, Anhui, Gansu, Jilin, Shanxi, and Xinjiang. However, the findings show that wealthy territories, Fujian and Hainan, and arid areas, Anhui, Jilin, and Shanxi, are highly inefficient when not accounting for the water quality provision, i.e., only the water supply. Beyond the natural water resources, these provinces become efficient due to being able to provide the population with high-quality water. Deng *et al.* (2016) find that a low scarcity of water resources may influence water resource use efficiency as per individual awareness of the lack of water. These provinces are located near each other, concerning the inefficiency of using water resources to produce high-quality water (Factor 1). Tibet is the poorest province in 2020 and highly efficient in using naturally obtained water resources; the current infrastructure is significantly lower, and it produces a low level of water because of its low population but high quality, similar to Qinghai. Ningxia, also one of the poorest provinces in 2020, is highly efficient (an extreme case) in the infrastructure used to produce purified water (with fewer pollutants) (Factor 3). The rest of the provinces are not efficient or very inefficient.

The results also show that the provinces in the east, i.e., Qinghai, Tibet, and Xinjiang, are highly efficient for most specifications. Indeed, the first two appear as extreme values for Factor 5, i.e., natural water resources for water provision, with low scarcity and the highest quality of water. Tibet and Qinghai have an excess of natural resources per geographical location for water provision and population, which could provide water for other regions nearby. Large cities, such as Beijing or Shanxi, have a high degree of scarcity of natural water resources compared to the level of water provision, resulting in low-quality water.

Tianjin is an extreme value, i.e., highly inefficient when accounting for all the inputs and outputs, i.e., the water resources and infrastructure to produce high-quality water (Factor 1). Tianjin has a high scarcity of resources but overinvestment in pipes, yet the water quality is unacceptable. On the other hand, Ningxia, with its high scarcity, has adequate infrastructure, even though the quality is also low. Chongqing, although not efficient overall, is the province with better water resources naturally obtained for water provision, followed by Guizhou. Qinghai becomes efficient for most of the DEA specifications. However, it shows an overcapacity of resources and infrastructure (pipe length) for the water supply (Factor 5). Tibet (with an overcapacity of water resources) and Hainan (pipe length) behave similarly. Ningxia is the province with a higher water supply ratio versus available water resources (Factor 2).

The north and east of China have low water quality overall. Ningxia is the worst province in terms of using the infrastructure to produce high-quality water. On the other hand, this could be related to the poor access to natural water resources (high scarcity) in the Northwest. Except for Beijing, there is a correlation between the high scarcity of natural resources and the provision of low water quality overall. For example, Ningxia, Tianjin, has less than 15 cubic meters of resources and provides the worst water quality. Ditto Hebei, Shanxi, Shandong, Henan, and Liaoning. These provinces are located geographically beside each other in the east of China. Also, these provinces have a similar location in the maps regarding Factor 2 and Factor 3, with values lower or equal to zero. However, for provinces with a scarcity of 580-800 billion cubic meters, the water has few pollutants, for example, Jilin, Fujian, Chongqing, and Xinjiang with ranges between 0.90 and 0.96, or Hainan has 270 billion cubic meters and 0.93. These provinces are mapped similarly regarding Factor 3 as per the efficient use of infrastructure to supply high-quality water.

For illustration purposes, two outputs and four inputs are used. However, more inputs are likely required according to the water providers' environment, i.e., the external characteristics of the territory, including the orography. DEA visualization is also feasible for panel data (Ripoll-Zarraga *et al.*, 2023). More years are required for effective policy decisions, i.e., public investment, avoiding over (or under) investments according to persistent water provision needs and, thus, assuring access to water for all citizens, including those in remote areas. Further analysis can remove the identified extreme cases from the sample to confirm these results and account for the geographical and spatial distribution of the water resources and altitude. It has been shown that for some models, the population is not relevant. Hence, it is essential to analyze the reasons behind it. One explanation may be that not all citizens are in the census nor have access to water provision, or, for example, have a tap (meter) in their households. If there is, for example, a public fountain in some communities to obtain water from, travel-model connectivity must be accounted for. Residents in remote areas may not have the resources to collect water for personal use. Further research will analyze these matters to determine the relevance of water as a human right.

Appendix A

Table A1. DEA specifications.

Inputs/Outputs	A	B	C	D	CD
AB (Resource-drain)	n/a	n/a	Resource-drain-capacity	Resource-drain-population	Resource-drain-capacity-population
AC (Resource-capacity)	n/a	Equivalent AB-C	n/a	Resource-capacity-population	n/a
AD (Resource-population)	n/a	Resource-population-drain	Equivalent AC-D	n/a	n/a
BC (Drain-capacity)	Equivalent AC-B	n/a	n/a	Drain-capacity-population	n/a
BD (Drain-population)	Equivalent AD-B	n/a	Equivalent BC-D	n/a	n/a
1 (Supply)					
2 (Quality)					
12 (Supply-quality)					

Appendix B

Table B1. Superefficiency analysis: Returns to scale.

Output-oriented models				
Province	**CRS (%)**	**Benchmark (Provinces)**	**VRS (%)**	**Benchmark (Provinces)**
Anhui	273.40	n/a	105.21	n/a
Beijing	48.66	3	34.36	0
Chongqing	270.22	n/a	101.05	n/a
Fujian	206.74	n/a	103.92	n/a
Gansu	122.87	n/a	109.62	n/a
Guangdong	409.73	n/a	103.78	n/a
Guangxi	248.80	n/a	100.65	n/a
Guizhou	211.69	n/a	107.57	n/a
Hainan	53.22	22	36.79	7
Hebei	228.86	n/a	104.20	n/a
Heilongjiang	190.44	n/a	105.87	n/a
Henan	198.75	n/a	137.60	n/a
Hubei	363.47	n/a	104.44	n/a
Hunan	291.45	n/a	102.16	n/a
Jiangsu	151.84	n/a	73.06	4
Jiangxi	204.36	n/a	102.08	n/a
Jilin	188.23	n/a	104.97	n/a
Liaoning	233.36	n/a	135.47	n/a
Neimenggu (Inner Mongolia)	131.24	n/a	108.56	n/a
Ningxia Hui Autonomous Region	21.41	16	big	2
Qinghai	63.39	8	big	17
Shaanxi	166.16	n/a	110.23	n/a
Shandong	249.21	n/a	147.30	n/a
Shanghai	104.39	n/a	80.96	3
Shanxi	99.09	4	96.90	0
Sichuan	445.79	n/a	100.14	n/a
Tianjin	305.32	n/a	260.32	n/a
Xinjiang Uygur Autonomous Region	34.74	18	33.56	21
Tibet Autonomous Region (Xizang)	24.83	0	big	0
Yunnan	229.07	n/a	101.10	n/a
Zhejiang	335.82	n/a	102.41	n/a

Appendix C

Table C1. Results DEA specifications VRS-output oriented.

Provinces	ABCD12 (%)	ABCD1 (%)	ABCD2 (%)	ABC12 (%)	ABC1 (%)
Anhui	105.21	212.74	109.27	105.21	212.74
Beijing	100.00	207.26	100.00	100.00	207.26
Chongqing	101.05	796.74	101.53	101.05	796.74
Fujian	103.92	305.44	105.91	103.92	305.44
Gansu	109.62	196.71	110.70	109.62	196.71
Guangdong	103.78	141.20	110.07	103.78	141.20
Guangxi	100.65	218.53	104.23	100.65	218.53
Guizhou	107.57	584.80	108.71	107.57	584.80
Hainan	100.00	242.79	100.00	100.00	242.79
Hebei	104.20	104.20	698.00	104.20	104.20
Heilongjiang	105.87	181.62	110.79	105.87	181.62
Henan	137.60	155.84	146.32	137.60	155.84
Hubei	104.44	204.68	108.63	104.44	204.68
Hunan	102.16	187.03	106.58	102.16	187.03
Jiangsu	100.00	100.00	122.89	100.00	100.00
Jiangxi	102.08	233.69	105.50	102.08	233.69
Jilin	104.97	378.44	106.09	104.97	378.44
Liaoning	135.47	271.74	137.73	135.47	271.74
Neimenggu (Inner Mongolia)	108.56	197.37	110.87	108.56	197.37
Ningxia Hui Autonomous Region	100.00	100.00	100.00	100.00	100.00
Qinghai	100.00	100.00	100.00	100.00	100.00
Shaanxi	110.23	375.42	111.04	110.23	375.42
Shandong	147.30	182.25	159.54	147.30	182.25
Shanghai	100.00	118.01	100.00	100.00	118.01
Shanxi	100.00	209.26	107.71	100.00	209.26
Sichuan	100.14	240.96	103.32	100.14	240.96
Tianjin	260.32	260.32	123,119,716.51	260.32	260.32
Xinjiang Uygur Autonomous Region	100.00	100.00	106.79	100.00	100.00
Tibet Autonomous Region (Xizang)	100.00	100.00	100.00	100.00	100.00
Yunnan	101.10	361.71	103.08	101.10	361.71
Zhejiang	102.41	348.54	104.57	102.41	348.54

ABC2 (%)	ABD12 (%)	ABD1 (%)	ABD2 (%)	ACD12 (%)	ACD1 (%)	ACD2 (%)
109.27	105.21	212.79	109.27	105.21	212.74	109.27
100.00	100.00	207.26	100.00	100.00	207.26	100.00
101.53	101.05	814.05	101.53	101.05	796.74	101.53
105.91	103.92	308.56	105.91	103.92	305.44	105.91
110.70	109.62	295.74	110.70	109.62	196.71	110.70
110.07	103.78	141.20	110.07	103.78	141.20	110.07
104.23	100.65	218.53	104.23	100.65	218.55	104.23
108.71	107.57	609.62	108.71	107.57	584.80	108.71
100.00	100.00	288.38	100.00	100.00	242.79	100.00
698.00	104.20	104.20	698.00	108.16	108.16	698.00
110.79	105.87	181.62	110.79	105.87	181.64	110.79
146.32	137.60	155.84	146.32	137.60	162.02	146.32
108.63	104.44	204.68	108.63	104.44	204.70	108.63
106.58	102.16	187.03	106.58	102.16	187.08	106.58
122.89	100.00	100.00	122.89	100.00	100.00	122.89
105.50	102.08	233.75	105.50	102.08	233.69	105.50
106.09	104.97	378.44	106.09	104.97	383.05	106.09
137.73	135.47	271.74	137.73	135.47	298.57	137.73
110.87	108.56	204.70	110.87	108.56	197.37	110.87
100.00	100.00	100.00	100.00	100.00	100.00	100.00
100.00	100.00	100.00	100.00	100.00	100.00	100.00
111.04	110.23	375.42	111.04	110.23	383.16	111.04
159.54	147.30	185.87	159.54	147.30	182.25	159.54
100.00	100.00	118.01	100.00	100.00	118.01	100.00
107.71	110.03	210.26	127.72	100.00	209.26	107.71
103.32	100.14	241.04	103.32	100.14	240.96	103.32
123,119,717.45	260.32	260.32	123,119,715.73	260.32	260.32	123,119,715.79
106.79	100.00	100.00	106.79	100.00	100.00	106.79
100.00	100.00	100.00	100.00	100.00	100.00	100.00
103.08	101.10	365.70	103.08	101.10	361.71	103.08
104.57	102.41	348.57	104.57	102.41	348.54	104.57

(Continued)

Table C1. (*Continued*)

Provinces	BCD12 (%)	BCD1 (%)	BCD2 (%)	AB12 (%)	AB1 (%)
Anhui	105.21	212.74	109.27	105.21	212.79
Beijing	124.68	1,405.40	125.15	100.00	207.26
Chongqing	102.77	813.82	103.48	101.05	814.05
Fujian	105.38	311.77	108.00	103.92	308.56
Gansu	114.40	264.08	116.09	109.62	295.74
Guangdong	103.78	141.20	110.07	103.78	141.20
Guangxi	100.65	218.53	104.23	100.65	218.53
Guizhou	107.57	584.80	108.71	107.57	633.10
Hainan	105.77	465.93	106.11	100.00	529.13
Hebei	312.16	312.16	768.33	104.20	104.20
Heilongjiang	105.87	181.62	110.79	105.87	181.62
Henan	146.31	240.74	153.44	137.60	155.84
Hubei	104.44	204.68	108.63	104.44	204.68
Hunan	102.16	187.03	106.58	102.16	187.03
Jiangsu	100.00	100.00	127.49	100.00	100.00
Jiangxi	102.08	233.69	105.50	102.08	233.75
Jilin	108.05	484.68	109.68	104.97	378.44
Liaoning	141.27	441.37	144.57	135.47	271.74
Neimenggu (Inner Mongolia)	112.16	234.22	115.38	108.56	204.70
Ningxia Hui Autonomous Region	173.69	173.69	990,193,410.64	100.00	100.00
Qinghai	100.00	383.12	100.00	100.00	654.10
Shaanxi	115.00	629.63	116.34	110.23	375.42
Shandong	159.75	256.79	167.76	147.30	185.87
Shanghai	111.86	585.30	113.24	100.00	118.01
Shanxi	169.77	781.34	172.23	110.03	210.26
Sichuan	100.14	240.96	103.32	100.14	241.04
Tianjin	2,051.86	2,051.86	991,530,784.10	260.32	260.32
Xinjiang Uygur Autonomous Region	100.00	100.00	108.55	100.00	100.00
Tibet Autonomous Region (Xizang)	100.00	100.00	100.00	100.00	100.00
Yunnan	101.10	361.71	103.08	101.10	365.73
Zhejiang	102.41	348.54	104.57	102.41	348.57

AB2 (%)	AC12 (%)	AC1 (%)	AC2 (%)	AD12 (%)	AD1 (%)	AD2 (%)
109.27	105.21	212.74	109.27	105.21	212.79	109.27
100.00	100.00	207.26	100.00	100.00	207.26	100.00
101.53	101.05	796.74	101.53	101.05	814.29	101.53
105.91	103.92	305.44	105.91	103.92	311.86	105.91
110.70	109.62	196.71	110.70	109.62	296.59	110.70
110.07	103.78	141.20	110.07	103.78	141.20	110.07
104.23	100.65	218.62	104.23	100.65	218.55	104.23
108.71	107.57	584.80	108.71	107.57	609.62	108.71
100.00	100.00	242.79	100.00	100.00	288.38	100.00
698.00	108.16	108.16	698.00	108.16	108.16	698.00
110.79	105.87	181.64	110.79	105.87	181.70	110.79
146.32	137.60	162.02	146.32	137.60	184.78	146.32
108.63	104.44	204.70	108.63	104.44	204.84	108.63
106.58	102.16	187.08	106.58	102.16	187.12	106.58
122.89	100.00	100.00	122.89	100.00	100.00	122.89
105.50	102.08	233.69	105.50	102.08	233.77	105.50
106.09	104.97	383.05	106.09	104.97	392.94	106.09
137.73	135.47	298.57	137.73	135.47	323.41	137.73
110.87	108.56	197.37	110.87	108.56	205.58	110.87
100.00	100.00	100.00	100.00	100.00	100.00	100.00
100.00	100.00	100.00	100.00	100.00	100.00	100.00
111.04	110.23	383.16	111.04	110.23	404.87	111.04
159.54	147.30	182.25	159.54	147.30	185.87	159.54
100.00	100.00	118.01	100.00	100.00	118.01	100.00
127.72	100.00	209.26	107.71	114.22	231.33	139.79
103.32	100.14	240.96	103.32	100.14	241.21	103.32
123,119,716.12	260.32	260.32	123,119,715.73	260.32	260.32	123,119,717.21
106.79	100.00	100.00	106.79	100.00	100.00	106.79
100.00	100.00	100.00	100.00	100.00	100.00	100.00
103.08	101.10	361.71	103.08	101.10	365.70	103.08
104.57	102.41	348.54	104.57	102.41	348.66	104.57

(*Continued*)

Table C1. (*Continued*)

Provinces	BC12 (%)	BC1 (%)	BC2 (%)	BD12 (%)	BD1 (%)
Anhui	105.21	212.74	109.27	105.21	212.79
Beijing	124.68	1,405.40	125.15	124.68	1,405.40
Chongqing	102.77	813.82	103.48	102.77	814.05
Fujian	105.38	311.77	108.00	105.38	311.80
Gansu	114.40	264.08	116.09	114.40	388.07
Guangdong	103.78	141.20	110.07	103.78	141.20
Guangxi	100.65	218.53	104.23	100.65	218.53
Guizhou	107.57	584.80	108.71	107.57	609.62
Hainan	105.77	465.93	106.11	105.77	481.13
Hebei	312.16	312.16	768.33	312.16	312.16
Heilongjiang	105.87	181.62	110.79	105.87	181.62
Henan	146.31	240.74	153.44	146.31	240.74
Hubei	104.44	204.68	108.63	104.44	204.68
Hunan	102.16	187.03	106.58	102.16	187.03
Jiangsu	100.00	100.00	127.49	100.00	100.00
Jiangxi	102.08	233.69	105.50	102.08	233.75
Jilin	108.05	484.68	109.68	108.05	484.68
Liaoning	141.27	441.37	144.57	141.27	441.37
Neimenggu (Inner Mongolia)	112.16	234.22	115.38	112.16	293.44
Ningxia Hui Autonomous Region	173.69	173.69	990,193,423.17	173.69	173.69
Qinghai	100.00	383.12	100.00	100.00	449.52
Shaanxi	115.00	629.63	116.34	115.00	629.63
Shandong	159.75	256.79	167.76	159.75	256.92
Shanghai	111.86	585.30	113.24	111.86	585.30
Shanxi	169.77	781.34	172.23	169.77	783.57
Sichuan	100.14	240.96	103.32	100.14	241.04
Tianjin	2,051.86	2,051.86	991,530,792.62	2,052.80	2,052.80
Xinjiang Uygur Autonomous Region	100.00	100.00	108.55	100.00	100.00
Tibet Autonomous Region (Xizang)	100.00	100.00	100.00	100.00	100.00
Yunnan	101.10	361.71	103.08	101.10	365.70
Zhejiang	102.41	348.54	104.57	102.41	348.57

BD2 (%)	CD12 (%)	CD1 (%)	CD2 (%)	A12 (%)	A1 (%)	A2 (%)
109.27	105.21	212.74	109.27	105.21	213.19	109.27
125.15	124.68	1,405.89	125.15	100.00	207.26	100.00
103.48	102.77	813.82	103.48	101.05	815.98	101.53
108.00	105.38	311.77	108.00	103.92	312.57	105.91
116.09	114.40	264.08	116.09	109.62	404.35	110.70
110.07	103.78	141.20	110.07	103.78	141.20	110.07
104.23	100.65	218.55	104.23	100.65	219.07	104.23
108.71	107.57	584.80	108.71	107.57	634.85	108.71
106.11	105.77	465.93	106.11	100.00	700.64	100.00
768.33	312.23	312.23	768.33	108.16	108.16	698.00
110.79	105.87	181.64	110.79	105.87	182.11	110.79
153.44	146.31	240.80	153.44	137.60	187.66	146.32
108.63	104.44	204.70	108.63	104.44	205.09	108.63
106.58	102.16	187.08	106.58	102.16	187.48	106.58
127.49	100.00	100.00	127.49	100.00	100.00	122.89
105.50	102.08	233.69	105.50	102.08	234.33	105.50
109.68	108.05	484.72	109.68	104.97	485.98	106.09
144.57	141.27	441.63	144.57	135.47	335.74	137.73
115.38	112.16	234.22	115.38	108.56	275.09	110.87
990,193,413.79	242.31	242.31	991,530,783.72	100.00	100.00	100.00
100.00	100.00	383.12	100.00	100.00	2353.91	100.00
116.34	115.00	629.68	116.34	110.23	502.56	111.04
167.76	159.75	256.79	167.76	147.30	185.87	159.54
113.24	111.86	585.56	113.24	100.00	118.01	100.00
172.23	169.77	781.34	172.23	153.59	231.33	154.92
103.32	100.14	240.96	103.32	100.14	241.45	103.32
991,530,783.71	2,051.86	2,051.86	991,530,783.71	260.32	260.32	123,119,725.69
108.55	100.00	100.00	108.55	100.00	100.28	106.79
100.00	100.00	100.00	100.00	100.20	1,776.40	100.31
103.08	101.10	361.71	103.08	101.10	366.67	103.08
104.57	102.41	348.54	104.57	102.41	348.99	104.57

(*Continued*)

Table C1. (*Continued*)

Provinces	B12 (%)	B1 (%)	B2 (%)	C12 (%)
Anhui	105.21	212.79	109.27	105.21
Beijing	124.68	1,405.40	125.15	124.68
Chongqing	102.77	814.05	103.48	102.77
Fujian	105.38	311.80	108.00	105.38
Gansu	114.40	411.97	116.09	114.40
Guangdong	103.78	141.20	110.07	103.78
Guangxi	100.65	218.53	104.23	100.65
Guizhou	107.57	633.10	108.71	107.57
Hainan	105.77	788.89	106.11	105.77
Hebei	312.16	312.16	768.33	312.23
Heilongjiang	105.87	181.62	110.79	105.87
Henan	146.31	240.74	153.44	146.31
Hubei	104.44	204.68	108.63	104.44
Hunan	102.16	187.03	106.58	102.16
Jiangsu	100.00	100.00	127.49	100.00
Jiangxi	102.08	233.75	105.50	102.08
Jilin	108.05	484.68	109.68	108.05
Liaoning	141.27	441.37	144.57	141.27
Neimenggu (Inner Mongolia)	112.16	293.47	115.38	112.16
Ningxia Hui Autonomous Region	173.69	173.69	990,193,410.90	267.53
Qinghai	100.00	786.78	100.00	100.00
Shaanxi	115.00	629.63	116.34	115.00
Shandong	159.75	256.92	167.76	159.75
Shanghai	111.86	585.30	113.24	111.86
Shanxi	169.77	783.57	172.23	169.77
Sichuan	100.14	241.04	103.32	100.14
Tianjin	2,052.80	2,052.80	991,530,783.75	2,051.86
Xinjiang Uygur Autonomous Region	100.00	100.00	108.55	100.00
Tibet Autonomous Region (Xizang)	100.00	100.00	100.00	100.00
Yunnan	101.10	365.73	103.08	101.10
Zhejiang	102.41	348.57	104.57	102.41

C1 (%)	C2 (%)	D12 (%)	D1 (%)	D2 (%)
212.74	109.27	105.21	212.79	109.27
1,405.89	125.15	124.68	1,406.46	125.15
813.82	103.48	102.77	814.29	103.48
311.77	108.00	105.38	311.86	108.00
264.08	116.09	114.40	388.07	116.09
141.20	110.07	103.78	141.20	110.07
218.62	104.23	100.65	218.55	104.23
584.80	108.71	107.57	609.62	108.71
465.93	106.11	105.77	481.13	106.11
312.23	768.33	312.40	312.40	768.33
181.64	110.79	105.87	181.70	110.79
240.80	153.44	146.31	241.01	153.44
204.70	108.63	104.44	204.84	108.63
187.08	106.58	102.16	187.12	106.58
100.00	127.49	100.00	100.00	127.49
233.69	105.50	102.08	233.77	105.50
484.72	109.68	108.05	484.79	109.68
441.63	144.57	141.27	441.84	144.57
234.22	115.38	112.16	293.44	115.38
267.53	991,530,784.00	242.31	242.31	991,530,783.99
383.12	100.00	100.00	449.52	100.00
629.68	116.34	115.00	629.84	116.34
256.79	167.76	159.75	257.08	167.76
585.56	113.24	111.86	586.03	113.24
781.34	172.23	169.77	783.83	172.23
240.96	103.32	100.14	241.21	103.32
2,051.86	991,530,806.00	2,052.85	2,052.85	991,530,798.78
100.00	108.55	100.00	100.00	108.55
100.00	100.00	100.00	100.00	100.00
361.71	103.08	101.10	365.70	103.08
348.54	104.57	102.41	348.66	104.57

Appendix D

Table D1. Factor analysis unrotated.

Provinces	Factor 1	Factor 2	Factor 3	Factor 4	Factor 5
Anhui	−0.302	−0.286	−0.112	0.446	−0.218
Beijing	0.217	0.718	−0.758	−3.995	−1.568
Chongqing	−0.021	3.505	1.323	0.515	0.293
Fujian	−0.261	0.324	0.126	0.428	−0.131
Gansu	−0.211	−0.060	−0.197	0.219	0.134
Guangdong	−0.352	−0.740	−0.254	0.377	−0.267
Guangxi	−0.339	−0.241	−0.022	0.317	−0.191
Guizhou	−0.078	2.170	0.735	0.636	0.223
Hainan	−0.205	0.420	−0.022	−0.707	0.919
Hebei	−0.119	−0.825	−0.586	−0.201	−0.432
Heilongjiang	−0.312	−0.486	−0.196	0.455	−0.245
Henan	0.002	−0.596	−0.834	1.098	−0.480
Hubei	−0.314	−0.336	−0.117	0.420	−0.220
Hunan	−0.343	−0.445	−0.121	0.348	−0.223
Jiangsu	−0.408	−0.997	−0.286	0.250	−0.279
Jiangxi	−0.319	−0.147	−0.010	0.364	−0.186
Jilin	−0.166	0.888	0.219	0.123	−0.046
Liaoning	0.082	0.275	−0.551	0.815	−0.448
Neimenggu (Inner Mongolia)	−0.249	−0.315	−0.239	0.313	−0.167
Ningxia Hui Autonomous Region	0.423	−1.929	4.812	−0.705	−0.338

Table D1. (*Continued*)

Provinces	**Factor 1**	**Factor 2**	**Factor 3**	**Factor 4**	**Factor 5**
Qinghai	−0.256	−0.236	−0.422	−1.470	4.188
Shaanxi	−0.052	1.006	0.043	−0.244	−0.219
Shandong	0.100	−0.478	−0.946	1.414	−0.521
Shanghai	−0.175	−0.481	−0.526	−1.414	−0.799
Shanxi	0.048	0.189	−0.585	−1.466	−0.909
Sichuan	−0.333	−0.097	0.041	0.308	−0.170
Tianjin	5.282	−0.157	−0.456	0.361	0.344
Xinjiang Uygur Autonomous Region	−0.408	−0.997	−0.286	0.250	−0.279
Tibet Autonomous Region (Xizang)	−0.418	−0.913	−0.332	−0.057	2.402
Yunnan	−0.259	0.683	0.307	0.383	−0.069
Zhejiang	−0.255	0.586	0.251	0.418	−0.099

Notes:

F1: Inefficiency (scarcity) of natural water resources concerning infrastructure to produce high-quality water.

F2: Efficiency in high-quality water provision.

F3: Inefficiency in using water infrastructure (pipes and sewerage capacity) concerning the population to supply purified water.

F4: Inefficiency in using infrastructure (pipes and sewerage capacity) concerning the population to supply water.

F5: Total natural water resources (scarcity) for water provision.

Appendix E

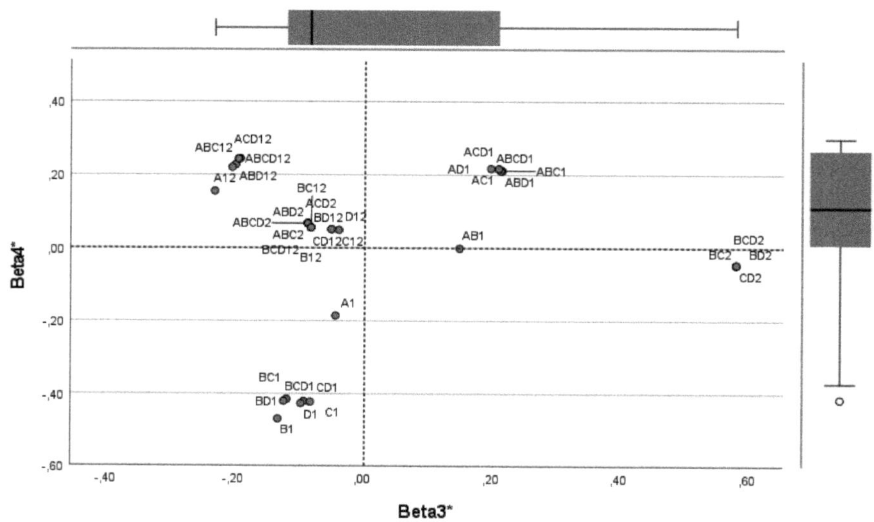

Figure E1. DEA specifications projection (Dimension 3 and Dimension 4).

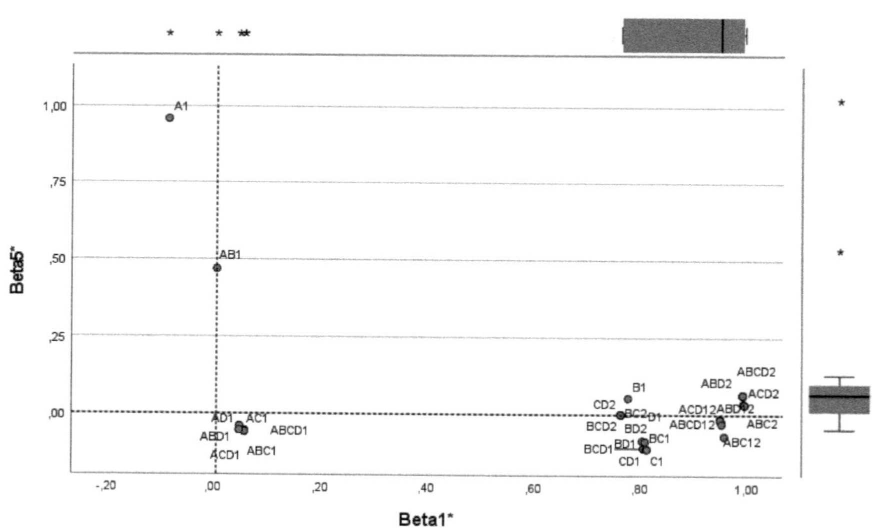

Figure E2. DEA specifications projection (Dimension 1 and Dimension 5).

Appendix F

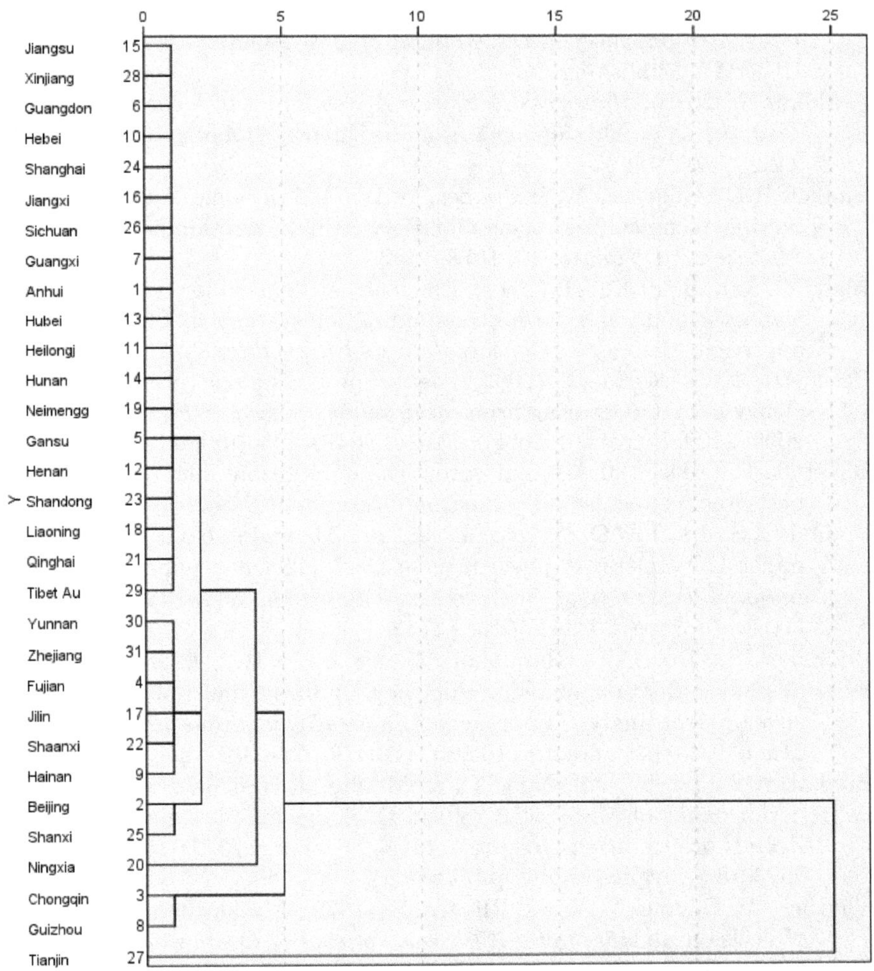

Figure F1. Ward's hierarchical cluster analysis: Dendrogram.

References

Alegre, H., Baptista, J.M., Cabrera Jr., E., Cubillo, F., Duarte, P., Hirner, W., Merkel, W., & Parena, R. (2003). *Performance Indicators for Water Supply Services. Manual of Best Practice Series.* London, UK: IWA Publishing.

Ashkiani, S. (2019). *Mavericks Revisited: A New Index to Identify Maverick Units in Data Envelopment Analysis.* https://doi.org/10.13140/RG.2.2.29147.72482.

Banker, R.D., Charnes, A., & Cooper, W.W. (1984). Some models for estimating technical and scale efficiency in data envelopment analysis. *Management Science,* 30, 1078–1092.

Bian, Y., Yan, S., & Xu, H. (2014). Efficiency evaluation for regional urban water use and wastewater decontamination systems in China: A DEA approach. *Resources, Conservation and Recycling,* 83, 15–23.

Becchetti, L. & Sierra, J. (2003). Bankruptcy risk and productive efficiency in manufacturing firms. *Journal of Banking & Finance,* 27(11), 2099–2120. https://doi.org/10.1016/S0378-4266(02)00319-9.

Bogetoft, P. (1997). DEA-based yardstick competition: The optimality of best practice regulation. *Annals of Operational Research,* 73, 277–298.

Boubaker, S., Le, T.D.Q., & Ngo, T. (2022a). Managing bank performance under COVID-19: A novel inverse DEA efficiency approach. *International Transactions in Operational Research*, 30(2023), 2436–2452. https://doi.org/10. 1111/itor.13132.

Boubaker, S., Do, D.T., Hammami, H., & Ly, K.C. (2022b). The role of bank affiliation in bank efficiency: A fuzzy multi-objective data envelopment analysis approach. *Annals of Operations Research,* 311, 611–639. https://doi.org/10.1007/s10479-020-03817-z.

Boubaker, S., Le, T.D.Q., Ngo, T., & Manita, R. (2023). Predicting the performance of MSMEs: A hybrid DEA-machine learning approach. *Annals of Operations Research.* https://doi.org/10.1007/s10479-023-05230-8 (Advance online publication).

Charnes, A., Cooper, W.W., & Rhodes, E. (1978). Measuring the efficiency of decision making units. *European Journal of Operational Research,* 2, 429–444.

Choi, T., Kang, K., & Koo, J. (2015). Efficiency evaluation of leakage management using data envelopment analysis. *Journal American Water Works Associations,* 107(1), 1–11. https://doi.org/10.5942/jawwa.2015.107.0002.

Coelli, T., Estache, A., Perelman, S., & Trujillo, L. (2003). *A Primer on Efficiency Measurement for Utilities and Transport Regulators.* WBI Development Studies. Washington, DC, USA: The World Bank. https://doi.org/10.1596/0-8213-5379-9.

Deng, G., Li, L., & Song, Y. (2016). Provincial water use efficiency measurement and factor analysis in China: Based on SBM-DEA model. *Ecological Indicators,* 69, 12–18. https://doi.org/10.1016/j.ecolind.2016.03.052.

Dyson R.G., Allen, R., Camanho, A.S., Podinovski, V., Sarrico, C.S., & Shale, E.A. (2001). Pitfalls and protocols in DEA. *European Journal of Operational Research,* 132, 245–259. https://doi.org/10.1016/S0377-2217(00)00149-1.

Geng, Q., Ren, Q., Nolan, R.H., Wu, P., & Yu, Q. (2019). Assessing China's agricultural water use efficiency in a green-blue water perspective: A study based on data envelopment analysis. *Ecological Indicators,* 96, 329–335. https://doi.org/10.1016/j.ecolind.2018.09.011.

Gower, J.C. & Hand, D.J. (1996). *Biplots.* London: Chapman and Hall.

Gutierrez-Nieto, B., Serrano-Cinca, C., & Mar-Molinero, C. (2007). Microfinance institutions and efficiency. *Omega. The International Journal of Management Science,* 35, 131–142.

Hammami, H., Ngo, T., Tripe, D., & Vo, D.-T. (2022). Ranking with a Euclidean common set of weights in data envelopment analysis: With application to the Eurozone banking sector. *Annals of Operations Research,* 311(2), 675–694.

Hu, J.-L., Wang, S.-C., & Yeh, F.-Y. (2006). Total-factor water efficiency of regions in China. *Resource Policy,* 31(4), 217–230.

Hu, M., Hu, Y., & Yuan, J. (2019). Benchmarking on water resource utilization efficiency of prefecture-level cities in Jiangxi, China: A Bootstrap-DEA approach with three-stage DEA models. *Environment and Natural Resources Research,* 9, 14. https://doi.org/10.5539/enrr.v9n3p14.

Huang, Y., Huang, X., Xie, M., Cheng, W., & Shu, Q. (2021). A study on the effects of regional differences on agricultural water resource utilization efficiency using super-efficiency SBM model. *Scientific Reports,* 11, 9953. https://doi.org/10.1038/s41598-021-89293-2.

Laureti, T., Benedetti, I., & Branca, G. (2021). Water use efficiency and public goods conservation: A spatial stochastic frontier model applied to irrigation in southern Italy. *Socio-Economic Planning Sciences,* 73, 100856. https://doi.org/10.1016/j.seps.2020.100856.

Le, T.D.Q. (2018). Bank risk, capitalisation and technical efficiency in the Vietnamese banking system. *Australasian Accounting Business & Finance Journal,* 12(3), 42–61. https://doi.org/10.14453/aabfj.v12i3.4.

Le Lannier, A. & Porcher, S. (2014). Efficiency in the public and private French water utilities: Prospects for benchmarking. *Applied Economics,* 46(5), 556–572.

Long, K. & Pijanowski, B.C. (2017). Is there a relationship between water scarcity and water use efficiency in China? A national decadal assessment across spatial scales. *Land Use Policy,* 69, 502–511. https://doi.org/10.1016/j.landusepol.2017.09.055.

Mar-Molinero, C. & Mingers, J. (2007). Mapping MBA programmes: An alternative analysis. *Journal of the Operational Research Society,* 58(1), 874–886.

Mu, L., Fang, L., Wang, H., Chen, L., Yang, Y., & Qu, X.J. (2016). Exploring Northwest China's agricultural water-saving strategy: Analysis of water use efficiency based on an SE-DEA model conducted in Xian, Shaanxi Province. *Water Science & Technology,* 74(5), 1106–1115. https://doi.org/10.2166/wst.2016.286.

Ngo, T., & Tsui, K.W.H. (2021). Estimating the confidence intervals for DEA efficiency scores of Asia-Pacific airlines. *Operational Research,* 22, 3411–3434.

Pedraja Chaparro, F., Salinas Jimenez, J., & Smith, P. (1999). On the quality of the data envelopment analysis model. *Journal of the Operational Research Society,* 50, 636–645.

Pietz, D.A. (2015). *The Yellow River. The Problem of Water in Modern China.* Cambridge, MA, London: Harvard University Press.

Pinto, F.S., Simoes, P., & Marques, R.C. (2017). Raising the bar: The role of governance in performance assessments. *Utilities Policy,* 49(C), 38–47.

Romano, G. & Guerrini, A. (2011). Measuring and comparing the efficiency of water utility companies: A data envelopment analysis approach. *Utilities Policy,* 19, 202–209.

Ripoll-Zarraga, A.E. & Mar-Molinero, C. (2020). Exploring the reasons for efficiency in Spanish airports. *Transport Policy,* 99, 186–202. https://doi.org/10.1016/j.tranpol.2020.08.021.

Ripoll-Zarraga, A.E., Portillo, F., & Mar-Molinero, C. (2023). The impact of the economic crisis on the efficiency of Spanish airports: A DEA visualisation analysis. *Research in Transportation Business & Management S.I. Transport Efficiency,* 100689. https://doi.org/10.1016/j.rtbm.2021.100689.

Sagarra, M., Mar-Molinero, C., & Agasisti, T. (2017). Exploring the efficiency of Mexican universities: Integrating data envelopment analysis and multidimensional scaling. *Omega, the International Journal of Management Science,* 67, 123–133.

Scheel, H. (2000). Efficiency Measurement System (EMS). http://www.holger-scheel.de/doordea.htm, http://www.holger-scheel.de/ems/.

Schiffman, J.F., Reynolds, M.L., & Young, F.W. (1981). *Introduction to Multidimensional Scaling: Theory, Methods, and Applications.* London: Academic Press.

Serrano-Cinca, C. & Mar-Molinero, C. (2004). Selecting DEA specifications and ranking units via PCA. *Journal of the Operational Research Society,* 55, 521–528.

Serrano-Cinca, C., Fuertes Callén, Y., & Mar-Molinero, C. (2005). Measuring DEA efficiency in internet companies. *Decision Support Systems,* 38(4), 557–573.

Serrano-Cinca, C., Mar-Molinero, C., & Fuertes Callén, Y. (2016). Input and output search in DEA: The case of financial institutions. In: S.N. Hwang, H.S. Lee & J. Zhu (Eds.), *Handbook of Operations Analytics Using Data Envelopment Analysis*, Chapter 3. New York: Springer.

Simar, L. & Wilson, P. (2007). Estimation and inference in two-stage, semiparametric models of production processes. *Journal of Econometrics,* 136(1), 31–64.

Statista. https://www.statista.com/statistics/278557/gdp-of-china-by-region/ (Accessed November 2022).

Tang, J., Folmer, H., & Xue, J. (2015). Technical and allocative efficiency of irrigation water use in the Guanzhong Plain, China. *Food Policy,* 50, 43–52.

Tang, Y., Zhang, F., Wang, S., Zhang, X., Guo, S., & Guo, P. (2019). A distributed interval nonlinear multiobjective programming approach for optimal irrigation water management in an arid area. *Agricultural Water Management*, 220(1), 13–26. https://doi.org/10.1016/j.agwat.2019.03.052.

The World Bank (2020). Access to sanitation services improves the quality of life in rural China. https://www.worldbank.org/en/news/feature/2020/03/10/access-to-sanitation-services-improves-the-quality-of-life-in-rural-china (Accessed January 2023).

Tripe, D. (2010). Using DEA to investigate bank safety and soundness — Which approach works best? *Journal of Financial Economic Policy,* 2(3), 237–250. http://www.emeraldinsight.com/doi/abs/10.1108/17576381011085449.

Vu, H.D. (2016). Technical efficiency of FDI firms in the Vietnamese manufacturing sector. *Review of Economic Perspectives,* 16(3), 205–230. https://doi.org/10.1515/revecp-2016-0013.

Wang, Y.M. & Chin, K.S. (2009). A new approach for the selection of advanced manufacturing technologies: DEA with double frontiers. *International Journal of Production Research,* 47(23), 6663–6679. https://doi.org/10.1080/00207540802314845.

Zhang, Q.W. & Zhang, X. (2000). An inverse DEA model for inputs/outputs estimate. *European Journal of Operational Research,* 121(1), 151–163.

Zhou, P., Ang, B.W., & Poh, K.L. (2008). Measuring environmental performance under different environmental DEA technologies. *Energy Economics,* 30, 1–14.

https://doi.org/10.1142/9781800615786_0005

Chapter 5

Assessing the Efficiency of Contagion Control and Medical Treatment for COVID-19 in OECD Countries Using Data Envelopment Analysis

Ana Martins* and David Sousa†

University of Algarve, Portugal

**aimartin@ualg.pt*

†*dhsousa@ualg.pt*

Abstract

The disease caused by the coronavirus (COVID-19) was one of the most significant health problems worldwide. The World Health Organization declared the event a global public health emergency in late January 2020 and then declared it a global pandemic in March 2020. The impressive and rapid spread of the virus was unprecedented and surpassed all expectations. With nearly every country in the world infected, the challenge of containing the virus became increasingly serious. The main objective of this study is to explore the potential of using data envelopment analysis (DEA) to establish international comparisons on the efficiency of implementing programs to fight the COVID-19 pandemic, thus allowing for the identification of a set of good practices. To do this, data from 18 countries belonging to the Organization for Economic Co-operation and Development were used to assess the efficiency of infection control and medical treatment, models A and B, respectively. The results indicate that there is a notable variation in efficiency, suggesting that a better use of resources could prevent the spread of the virus and high mortality. The most efficient countries for both models are Slovenia and Latvia. Less efficient countries were Denmark, Iceland, and Switzerland in the control of infection and Germany, Austria, and Switzerland in medical treatment. These results also demonstrate the potential strategic role of the DEA methodology for the efficient and effective planning of scarce resources to fight the pandemic.

Keywords: Data envelopment analysis, infection control, COVID-19, efficiency, medical treatment, OECD countries, technical efficiency, coronavirus, pandemic, health system.

1. Introduction

1.1. *Definition of the topic*

The world has witnessed several emerging diseases that have spread to different countries. According to the Portuguese general health direction (*Direção Geral de Saúde* — DGS), severe acute respiratory syndrome coronavirus (SARS-CoV) led to an epidemic in 2002–2003, and Middle East respiratory syndrome coronavirus (MERS-CoV) appeared in 2012, causing sporadic cases of human infection or small clusters of respiratory disease cases. More recently, a new SARS-CoV-2 coronavirus was discovered in China's Wuhan city in December 2019. The disease caused by the new coronavirus is designated COVID-19.

Coronaviruses belong to the family *Coronaviridae*, which includes viruses that can cause infection in humans, other mammals, and birds. Typically, these infections affect the respiratory system and can resemble the common cold or progress to more serious illnesses, such as pneumonia (DGS, 2021). It is assumed that SARS-CoV-2 was introduced to humans by zoonotic transmission and transmitted from person to person through close contact with people who were infected (direct transmission) or through contact with contaminated surfaces and objects (indirect transmission). Initially, the incubation period of the disease was estimated to be 1–14 days, which made it difficult to control the spread (DGS, 2021).

According to the World Health Organization (WHO), a pandemic begins when new diseases appear in the population that spread easily and infect humans, causing serious illness (WHO, 2022).

The outbreak of COVID-19 represented one of the most critical events in the world. WHO declared the event a global public health emergency in late January 2020 and then declared it a global pandemic in March 2020. The impressive and rapid spread of the virus was unprecedented and has exceeded all expectations. With almost every country in the world infected, the challenge of containing the virus became increasingly serious. Within a few months, COVID-19 emerged on every continent. The global situation on February 15, 2021, was 108,484,802 confirmed cases and 2,394,323 deaths (WHO, 2021).

Viruses do not respect borders. Even the Spanish flu pandemic of 1918 reached remote islands within months, long before the globalization of air travel. Humanity has previously eradicated infectious diseases, but only twice: rinderpest in 2011, a disease that mainly infected cattle, and smallpox in 1980. Smallpox was eradicated through global vaccination campaigns, becoming one of humanity's greatest success stories, where more than 100 million lives were saved (WHO, 2021).

COVID-19 has put the world to the test in terms of political leadership, national health systems, social welfare services, and solidarity. In the face of this challenge, there is the possibility of increasing social and health inequalities. The key strategies to combat a pandemic are mitigation and suppression. Mitigation aims to slow but not entirely stop its spread, i.e., to reduce the demand for healthcare by protecting those at greatest risk of serious illness. Suppression aims to stop the growth of the pandemic by reducing the number of positive cases to low levels for an indefinite period (Kavaliunas *et al.*, 2020).

While some countries failed in their response to the pandemic, others met the challenge much more successfully, limiting the direct and indirect impacts of the pandemic. The countries that responded most successfully were able to avoid the choice between a high mortality rate and a high socioeconomic impact of the pandemic. At this point, the most important thing to retain is that it is possible to fight the pandemic (Ritchie *et al.*, 2020a).

However, it is important to note that governments had to act without prior knowledge during the rapid spread of the disease. Primarily, with the appearance of a new pandemic, the knowledge base needed to design interventions was uncertain. After several months of intensive study and experience fighting COVID-19, knowledge about the key parameters of the virus was known with some degree of certainty. These parameters include transmission, symptoms, the incubation period, the ability to recover, and the segments of the population that are most vulnerable. Despite the findings to date, there are still many questions that need to be answered (Capano *et al.*, 2020).

The response to the pandemic did not occur at the same time, in the same way, or with the same accuracy. Some countries implemented mandatory regulations, while others only issued recommendations. The pandemic crisis involved many types of uncertainty: issues around immediacy, data limitations, a lack of consensus among experts and variations in expertise, the possibility of over-or under-reactions, and different levels of trust in government and political support, among others.

The perception and decisions of governments influenced each country's performance in fighting COVID-19, and studying different performances will be a good tool for comparison. Countries can learn from each other and avoid repeating the mistakes made by other countries (Ghasemi *et al.*, 2020; Yuen *et al.*, 2022).

Healthcare systems aim to maintain, restore, and improve the health of the population. Therefore, it is crucial to evaluate efficiency in relation to the health resources consumed. Increasing demand for healthcare, growing expenditures, and decreasing budgets force a search for priorities (Pelone *et al.*, 2015; Zakowska & Godycki-Cwirko, 2020).

Health systems are different between countries in terms of infrastructure, patient numbers, financing, and government; however, they currently face common challenges and goals. Assessing and comparing the performance of various health systems provides an opportunity for policymakers to determine how a given health system performs relative to its international peers, understand how it works to identify good and bad practices, and ultimately find more effective approaches to achieve sustainability and better quality (Nolte *et al.*, 2006). The implemented measures and their results are frequently compared.

One question to be answered in the literature is whether the resources used in the epi-curve reduction are used efficiently.

This study seeks to fill this gap in the literature by using the Efficiency Measurement System (EMS) software, which allows the evaluation of the efficiency of various healthcare systems. It can be done by analyzing data related to resources and results achieved in the first 100 days of the pandemic since the first confirmed case. In this study, we apply a non-parametric method called data envelopment analysis (DEA) to analyze the efficiency of contagion control and medical treatment in 18 countries belonging to the Organization for Economic Cooperation and Development (OECD).

While there are several alternative techniques available to perform efficiency analysis in healthcare, a predominant use of DEA methodology focusing on hospitals, nursing homes, and physicians has been highlighted by several reviews in measuring frontier efficiency in healthcare. The DEA methodology is an attractive tool for efficiency measurement because it can handle the existence of multiple resources (inputs) and multiple health outcomes (outputs) in the transformation process. Moreover, it does not require robust assumptions about the underlying technology that links resources to outputs and measures efficiency in relative terms (Pelone *et al.*, 2015).

1.2. *Relevance of the topic*

The pandemic has affected numerous countries, among which we can identify the following three distinct forms of the impact on the population: (i) some countries have not been able to contain the pandemic, and the death toll has increased rapidly; (ii) other countries have had large outbreaks but managed to decrease the number of deaths; (iii) other countries have had rapid outbreaks but were able to reduce the number of deaths quickly, i.e., they were able to prevent major outbreaks. Although some comments on the pandemic reflect that all countries did not respond well to the pandemic, major differences among them are noted. In addition to illness and deaths, COVID-19 also has an indirect impact on countries. When a country has a rapid outbreak of the disease, it can only respond with one of two undesirable options: (1) leave the pandemic unchecked, in which a large portion of the population will become ill and die, or (2) apply containment measures that entail increased unemployment and poverty rates, lower production, and higher mortality from other causes (Ritchie *et al.*, 2020a).

Despite the great disruption that the virus has caused, preventive measures have also caused huge economic disruption. The failure of the initial measures to contain the rapid worldwide spread of the virus prompted mandatory residency, with significant impacts on the global economy (Balmford *et al.*, 2020).

The problem with rapid outbreaks is the ease with which many people become infected in the same region at the same time, exceeding the capacity of the healthcare system. When this happens, many people die without treatment. Another problem is that healthcare workers are also susceptible to infection, which will also contribute to a decrease in the capacity of healthcare systems when they are most needed. During these periods, all patients, including non-COVID-19 patients, may have a much higher risk. Patients with cancer or cardiovascular diseases may not have access to their regular care. Other users may put off appointments with their doctors for fear of being infected (Ritchie *et al.*, 2020a). However, interventions to prevent COVID-19 also suppressed other deaths, such as those caused by influenza (Balmford *et al.*, 2020).

Different countries have achieved vastly different mortality rates. Understanding what drives these differences is crucial, allowing for better responses to future pandemics. This study aims to answer the critical question of why different countries experience different mortality rates and what can be learned for future policies. All the developments and results presented were obtained in a study carried out by Sousa (2022).

1.3. *Objectives*

The main objective of this study is to explore the potential of using the DEA methodology to make international comparisons of the efficiency of program implementation in combating the COVID-19 pandemic, enabling the identification of a set of best practices. The purpose of the study is identical to that of the coalition of experts, funders, and collaborators called Exemplars in Global Health (EGH), which is "to help inform current and future global health policies and programs based on what has worked in the past" (Kennedy *et al.*, 2020). This is possible by analyzing the most successful countries fighting the pandemic.

The application of two models (A and B) allows a better definition of specific objectives: Model A analyzes the response of the countries in controlling the contagion, allowing the identification of which countries had a better performance in controlling the spread of the virus; Model B allows to identify which countries responded efficiently in medical treatment, avoiding a high number of deaths.

1.4. *Research limits*

Identifying the countries with good practices is not a simple task since it requires data, which are difficult to obtain, and the existing data are far from perfect. An example is the total number of cases, of which only a fraction is known, namely those confirmed by a test. The real number is far higher than the reported one. People with symptoms due to infection with COVID-19 were tested and identified. Most asymptomatic people were not detected and did not count as reported cases.

The first issue in country comparisons is the difference in national statistical approaches. Some countries are much more likely than others to attribute a death to COVID-19. As emphasized by Balmford *et al.* (2020), the degree of underreporting is far from constant across countries. While almost all countries counted only deaths that were confirmed as caused by COVID-19, Belgium, for example, included deaths where COVID-19 is merely suspected as a contributing factor. This results in much higher mortality rates than in other countries.

The health datasets from the Center for Systems Science and Engineering (CSSE) at Johns Hopkins University and from the OECD and WHO provide some of the best available sources of cross-country comparative data, however, with some limitations. Variables of interest are not collected regularly, and data collection methods are far from consistent.

The concept of health system efficiency is, in practice, strongly contested and difficult to measure accurately across countries. Few attempts have been made to create unique comparative measures of health system efficiency, and all available studies suffer from significant limitations, including the lack of a clear conceptual framework, limited data availability, and difficulties in selecting appropriate and reliable empirical techniques.

1.5. *Organization of the study*

In this study, efficiency evaluation is performed using the DEA methodology. In the first stage, the countries with the best performance in controlling contagion are identified, and in the second stage, countries are evaluated based on medical treatment.

Besides this introduction, the chapter is divided into four sections. Section 2 gathers some contributions from the literature on the DEA methodology. Section 3 presents the methodology applied in the study, as well as the sample and variables used in the models. In Section 4, the results are presented and discussed. Finally, Section 5 concludes the study, presents its limitations, and makes proposals for future research.

2. Literature Review

2.1. *DEA methodology*

The DEA methodology, developed by Charnes *et al.* (1978), found its roots in Farrell's (1957) seminal work and is a non-parametric linear programming technique that has proven to be one of the most important and widely used approaches for measuring the performance of homogenous decision-making units (DMUs) performing the same task. Each DMU has an inherent process of transforming resources (inputs) into results (outputs).

This first DEA model, known as the CCR model, assumes constant returns to scale (CRS). Later, Banker *et al.* (1984) proposed a DEA model, known as the BCC model, which assumes the existence of variable returns to scale (VRS). Despite the numerous existing extensions to DEA models, the CCR and BCC models are the most widely used.

Based on Farrell's (1957) seminal definitions of efficiency, technical efficiency consists of producing the maximum amount of output from a given amount of input (output-oriented), or, alternatively, producing a given output with minimum amounts of input (input-oriented), so that when a firm is technically efficient, it operates on its production frontier.

Table 1 shows the mathematical formulation for both models and orientations. Here, j is the number of DMUs, s is the number of

Table 1. Mathematical formulation of the CRS and VRS models.

Frontier type	Input-oriented	Output-oriented
CRS	$\min \theta - \varepsilon \left(\sum_{i=1}^{m} s_i^- + \sum_{r=1}^{s} s_r^+ \right)$	$\max \phi + \varepsilon \left(\sum_{i=1}^{m} s_i^- + \sum_{r=1}^{s} s_r^+ \right)$
	s.t.:	s.t.:
	$\sum_{j=1}^{n} \lambda_j x_{ij} + s_i^- = \theta x_{io}$	$\sum_{j=1}^{n} \lambda_j x_{ij} + s_i^- = x_{io}$
	$i = 1, 2, \ldots, m;$	$i = 1, 2, \ldots, m;$
	$\sum_{j=1}^{n} \lambda_j y_{rj} - s_r^+ - y_{ro}$	$\sum_{j=1}^{n} \lambda_j y_{rj} - s_r^+ = \phi y_{ro}$
	$r = 1, 2, \ldots, s;$	$r = 1, 2, \ldots, s;$
	$\lambda_j \geq 0, \quad j = 1, 2, \ldots, n$	$\lambda_j \geq 0, \quad j = 1, 2, \ldots, n$
VRS	Add $\sum_{j=1}^{n} \lambda_j = 1$	

Source: Zhu (2014).

outputs, y_{rj} is the amount of output r produced by DMU j, m is the number of inputs, x_{ij} is the amount of input i used by DMU j, λ_j is the coefficient of benchmark j, s_i^- is the slack associated with input i in DMU, s_r^+ is the slack associated with output r in the DMU, ε is a small positive number, θ is the radial reduction to be applied to all inputs of the DMU, and ϕ is the radial increase to be applied to all outputs of the DMU (Zhu, 2014).

The DEA frontier is formed by a linear combination of the points representing the set of best practices, forming a convex set of production possibilities. Because it is a deterministic method, it assumes that all deviations between observed spending and the minimum spending frontier curve are explained by inefficiency. One of the characteristics of the DEA methodology is that it allows each unit to identify a benchmarking group, i.e., a group of units that follow the same goals and priorities but perform better. The goal is to rank each unit as highly as possible compared to the other units. The method then defines benchmarking units for each observation, which allows for calculating the increases in results or decreases in resources necessary for the performance to be optimized (Amado & Dyson, 2009).

There are at least four distinct reasons that make DEA models suitable for use in health system performance evaluation. First, since most DEA models can be conceptualized as linear programs, a wide range of useful information becomes available for performance improvement. Second, the DEA methodology can handle the existence of inputs and outputs without requiring unit prices for each. Third, it does not require any empirical assumptions about the functional form linking inputs to outputs, and this reduces exposure to the theoretical assumptions of its empirical specification compared to studies that adopt parametric approaches. Fourth, the DEA methodology uses all available data to construct the best practice frontier to which each non-optimal output point is compared (Amado & Dyson, 2009).

Despite these important advantages, this technique has some limitations. First, the non-parametric approach is quite sensitive to outliers, extreme values, and measurement errors. An outlier can make other observations appear relatively less efficient if it is used to compare the remaining observations. In addition, units without peers are sometimes rated with full efficiency, even though they are significantly less efficient (Varabyova & Müller, 2016). Therefore, some studies, such as those by Retzlaff-Roberts *et al.* (2004), explicitly excluded countries at different stages of development from the analysis. Second, studies based on the DEA methodology have a higher degree of uncertainty than those based on statistics that use uncertainty models about the robustness of the results. The DEA methodology does not provide any diagnostics to assess the adequacy of the model with respect to several methodological issues, such as discrepant observations, measurement precision, random errors, and the application of sensitivity analysis. Third, since no DEA model can include all potential variables, a DEA analysis can lead to partial and potentially misleading results (Amado & Dyson, 2009). The DEA methodology assumes that it is possible to completely characterize healthcare output by identifying a set of inputs, outputs, and outcomes. However, some of the outputs and outcomes of healthcare systems are not measurable. In addition, there are many factors operating at different levels that impact this transformation, which can lead to a violation of the linear relationship between inputs and outputs. Awareness of these limitations and their potential impact on outcomes is necessary.

As mentioned earlier, the DEA methodology also offers input- or output-oriented models. The input-oriented model should be used when it is needed to control resources since it minimizes the resources (inputs) needed to produce the same amount of result (output). On the other hand, output-oriented models maximize results (output) while using the same amount of resources (input). Ozbugday *et al.* (2020) argue that one should select the orientation according to the highest level of control over resources versus results.

The results of the technical efficiency calculations generate efficiency scores between 0 and 1, with a straightforward interpretation: A score of 1 (100%) indicates that this DMU produces efficiently compared to the other DMUs in the sample, i.e., it demonstrates best practices that other DMUs can learn from. A score below 1 indicates inefficiency, and the score indicates the degree of inefficiency relative to DMUs considered efficient. The DEA methodology involves the mediation of a relative level of efficiency (Althin *et al.*, 2019; Boubaker *et al.*, 2022b, 2023).

For a better evaluation of the results obtained by the DEA methodology, some complementary analyses have emerged, such as superefficiency levels, inverted frontier, and composite efficiencies.

When BCC models are used in DEA methodology, DMUs that present the lowest value for one of the inputs or the highest value for one of the outputs are considered efficient. According to Ali (1993), such DMUs can be considered false-efficient or efficient by default. In this case, an analysis of the results must be done to understand the real performance of each DMU.

According to Martins (2012), the combination of standard efficiency with inverted DEA (IDEA) inefficiency allows for the identification of false-efficient DMUs through the level of composite efficiency. This efficiency indicator requires an efficient DMU to have a high level of standard efficiency and a low level of inverted efficiency. These results can be obtained using the Integrated System for Decision Support (ISYDS) software.

Recently, Boubaker *et al.* (2022a) evaluated the performance and efficiency of 49 Islamic banks in 10 countries during 2019–2020 using the IDEA approach. Their analysis demonstrates the differences between the conventional IDEA and the proposed IDEA, where they indicate that the proposed IDEA promotes efficiency in terms of cost savings.

As mentioned by Boubaker *et al.* (2022a), the setting of IDEA is different from that of conventional DEA, as the problem is solved for a target efficiency score, where the DMUs should optimize the use of their inputs (or outputs), given a certain change in the outputs (or inputs). This inversion of the optimization problem was carried out by Zhang & Cui (1999), Wei *et al.* (2000), and Gattoufi *et al.* (2014) (Table 2).

Here, θ_{j0} is the targeted efficiency score of DMU_{j0} ($j = 1, 2, \ldots, n$), Δy_{rj0} is the (given) change in its output y_r ($r = 1, 2, \ldots, m$), Δx_{ij0} is the estimated change in its input x_i ($r = 1, 2, \ldots, m$), and w^i is the weight assigned to the inputs x_{ij0} of the DMU j_0.

The concept of the inverted frontier applied by Yamada *et al.* (1994), Entani *et al.* (2002), and Mello *et al.* (2005) consists in examining outputs as inputs and inputs as outputs. In other words, while

Table 2. Mathematical formulation of the input-oriented inverted DEA.

Input-oriented

$$\min \Delta x_{j0} = \sum_{i=1}^{k} w^i \Delta x_{ij0}$$

s.t.:

$$\sum_{j}^{n} \lambda_j y_{rj} \geq (y_{rj0} + \Delta y_{rj0}), \quad r = 1, 2, \ldots, m$$

$$\sum_{j}^{n} \lambda_j x_{ij} \leq \theta_{j0}(x_{ij0} + \Delta x_{ij0}), \quad i = 1, 2, \ldots, k$$

$$\lambda_j \geq 0, \quad j = 1, 2, \ldots, n$$

Source: Boubaker *et al.* (2022a).

the classical DEA models allow obtaining the best classification of the DMUs evaluated, the IDEA models represent a pessimistic evaluation of the DMUs since the inverted frontier is composed of the DMUs with the worst practices and is usually called the inefficient frontier. Therefore, the higher the level of inverted efficiency, the lower the efficiency of the DMU (Martins, 2012).

According to Meza *et al.* (2007), the composite efficiency level is defined as the arithmetic mean between the efficiency with respect to the conventional (standard) DEA frontier and the complement of the efficiency with respect to the inverted frontier (see Equation (1)). To obtain an index in which the efficient units have a value of 1, the normalized composite efficiency level is usually used, which is obtained by dividing the composite efficiency values of each DMU by the highest composite efficiency value obtained in the sample.

$$\text{Compound efficiency} = \frac{[\text{Standard efficiency} + (1 - \text{Inverted efficiency})]}{2}. \quad (1)$$

Martins (2012) argues that the normalized composite efficiency level requires an efficient DMU to perform well in areas where it is better (high standard efficiency level) and to perform acceptably well in areas where it is worse (low inverse efficiency level).

Other, more recent, DEA approaches are being developed, such as bootstrapping DEA, common weight DEA, and network DEA.

By applying the DEA methodology, it is possible to evaluate the impact of certain environmental variables on the performance of different DMUs. Simar & Wilson (2007) demonstrate that the application of double bootstrap procedures improves statistical efficiency.

Moreover, assigning weights in the DEA methodology is also a common step when we want to indicate the weight of each variable in the production model. Over the years, new approaches have been suggested to improve the assignment of weights to variables. Hammami *et al.* (2022) recently presented a new method to define a set of common Euclidean weights, where the DEA methodology benefits from better ranking and more realistic weight assignment.

Reference technologies are treated as black boxes in most traditional DEA models, where all relevant inputs are transformed into outputs that characterize the production process. However, Färe & Grosskopf (2000) address a network DEA that allows us to analyze intermediate processes and evaluate organizational and component performances.

2.2. *DEA methodology applied to the health sector*

According to Kohl *et al.* (2019), the DEA methodology is widely applied in the agriculture, banking, education, health, and transportation industries. Regarding its application in healthcare studies, we highlight the following: Retzlaff-Roberts *et al.* (2004), Spinks & Hollingsworth (2007), Cordero-Ferrera *et al.* (2011), Cylus *et al.* (2015), and, more recently, Breitenbach *et al.* (2020), with the application of this methodology in the evaluation of the efficiency of health systems in the fight against COVID-19.

Breitenbach *et al.* (2020) applied the DEA methodology to analyze the efficiency of containing the spread of COVID-19, identifying 12 efficient and 19 less efficient countries in the use of available resources. Among the worst performers were Germany, Austria, Canada, and the USA, with efficiency values of approximately 55%, i.e., less efficient than Belgium, France, and Italy, which were among the countries most affected by the rampant spread of the virus.

The DEA methodology has been widely used to analyze efficiency in the health industry worldwide; however, since the COVID-19 pandemic is a relatively recent event, few papers have used this methodology to analyze the efficiency of health systems across countries fighting against it.

The CCR model has been the most widely used model in studies on the healthcare sector. However, the past decade shows a trend toward the assumption of VRS (the BCC model). However, the application and interpretation of the BCC model should be done with caution, as the measured economies of scale of DMUs not located at the frontier may be incorrect (Kohl *et al.*, 2019).

Few efficiency indicators are available for most countries, which limits the number of countries, service domains, and production processes that can be compared. Using data from the CSSE, OECD, and WHO, only a part of the efficiency indicators can be selected, keeping the set of countries constant across various model specifications. This causes a general problem in comparing the efficiency of health systems and highlights the urgent need to develop a wider range of efficiency indicators that are comparable across countries.

Less efficient use of resources in the healthcare setting means loss of money and lives, which reveals the need for improving efficiency and effectiveness with a new balance between prevention and treatment, particularly significant in pandemics.

It is also important to keep in mind that there are sociocultural, political, and economic factors that are not directly controllable by prevention authorities but that can explain differences in efficiency and help understand the development of the pandemic in each country. Consequently, it is important to include non-discretionary variables in the analysis that will ensure comparable technologies (Santos *et al.*, 2012).

Literature pertaining to DEA methodology for addressing pandemic COVID-19 can be identified in the articles portrayed in the following.

Aydin & Yurdakul (2020) use a new three-stage framework to assess the performances of countries against the COVID-19 outbreak. Breitenbach *et al.* (2021) evaluate resource efficiency in the management of COVID-19 but do not mention some important resources used to fight the pandemic. Hospital beds are a variable that would affect efficiency results because they are intertwined with recovered cases. Cinaroglu (2021a) reviews the preparedness of Turkish intensive care unit (ICU) services in fighting the pandemic. Cinaroglu (2021b) also evaluates Turkish oncology services during pandemic times. Ferraz *et al.* (2021) assess the hospital health resources directly linked to the pandemic but do not include other equally relevant factors such as socioeconomics, infrastructure, and income distribution. Ibrahim *et al.* (2020) evaluate the management of the pandemic in two stages: control and treatment. Otherwise, Klumpp *et al.* (2022) seek to evaluate the impact of interventions in combating the pandemic. Mariano *et al.* (2021) intend to demonstrate how each region is prepared to care for patients infected with COVID-19 compared to the other regions. Md Hamzah *et al.* (2021) capture the surveillance and means necessary to fight the pandemic but do not add the main characters in the community surveillance and medical care models. Healthcare professionals should be implicit in all the processes described. The same is true for the Giménez *et al.* (2021) study, which does not incorporate health professionals into its model, but they are the ones who can detect and treat positive cases.

To capture the efficiency of health systems with populations above 50 million, it would have been more appropriate for Mourad *et al.* (2021) to apply an economic and political stability index because the selected countries have very different characteristics. Nepomuceno *et al.* (2020) focus on the complexity of needs for the evacuation of hospital beds during the COVID-19 outbreak. For a better assessment, the severity of the diseases and the cost of each allocation could be studied.

Su *et al.* (2021) evaluate efficiency in a distinct group of countries, so they should consider economic factors, such as GDP, to compensate for differences in wealth between countries. They could also apply the resources used for pandemic mitigation, such as the number of tests performed. Taherinezhad & Alinezhad (2022) illustrate what an ideal model might look like by identifying the inputs and outputs inherent in each system. However, they left out important variables for evaluating the efficiency of the 50 nations; examples are human and physical resources. Xu *et al.* (2021) and Lupu & Tiganasu (2022) could have included demographic variables to assess their impact, as well as the number of ICU beds to assess their allocation in times of pandemics. Zhang *et al.* (2020) evaluate the community in a very narrow way; they should specify in more detail to build a complex DEA model, such as indicating the stringency index for each community.

To complement Martínez-Córdoba *et al.*'s (2021) study, a demographic variable, such as the population over 65 years old, should be introduced, in addition to the prevalence of non-communicable diseases, for a better understanding of the pandemic. Studies that could also apply the same demographic variable are Mohanta *et al.* (2021) and Tavakoli *et al.* (2022).

Kerstens & Shen (2021) do not mention economic factors, although they mention beds as capital stock, like Di Giorgio *et al.* (2016); however, different countries may be allocating their funds to other health sectors. Since they intend to use COVID-19 mortality to select among hospital plant capacity models, they should include demographic factors, as they allow for distinction between cities.

3. Methodology

3.1. *Characterization of the sample*

This study applied a non-parametric methodology (DEA and IDEA) to identify best practices among some OECD countries during the first 100 days after the first confirmed case of COVID-19. The data collection began during the early stages of the pandemic. Capturing the first 100 days after the first confirmed case in each country allows for a fair comparative analysis across countries during the first and worst wave of the pandemic. A period of 100 days was applied in many technical and specialized mass media analyses, since it is considered sufficient time to capture the major impact and spread of the virus in its early stages.

DEA input-oriented models, Model A: Contagion Control and Model B: Medical Treatment, are used to estimate technical efficiency. Model A uses five inputs and one output, and Model B uses six inputs and one output, as described in this section. In both models, the sample size exceeds the minimum number of inputs and outputs considered necessary.

To ensure fair comparability of data and countries, only relative measures, such as percentages, rates, or indices, are used. The data used were collected from international public sources, including the CSSE, OECD, and WHO databases. While these are reliable sources, some of the information available is provided by each country. Considering that some of these countries have different criteria for collecting statistics, the following analysis should be interpreted with caution.

Initially, it was intended to use data from the 37 OECD countries. However, some countries were excluded due to missing values or significantly different accounting approaches. Only 18 countries were included in the study (Table 3).

Table 4 presents the descriptive statistics of the variables used in the study, which allows for the perception of how different countries had different responses and therefore different results. For a better comparison between countries, the descriptive statistics of the variables are calculated per million inhabitants, except for GDP and population density.

Table 3. Final sample with indication of countries and exclusion criteria.

Initial countries number	Exclusion criteria	Excluded countries	Final countries number
37	Unavailable or unmatched number of tests	Canada, Colombia, France, Hungary, Japan, Mexico, the Netherlands, Norway, Poland, South Korea, and Sweden	26
26	Number of nurses and doctors in practice unavailable	Chile, Slovakia, Greece, Ireland, Portugal, Turkey, and USA	19
19	Unmatched number of deaths	Belgium	18

Table 4. Descriptive statistics of the variables used in DEA models.

Variables	Mean	Std. dev.	Minimum	Maximum
Tests	64.386	43.781	22.576	185.115
Cases	2.259	1.859	272	6.557
Deaths	138	182	4	570
Doctors	3.777	632	2.796	5.174
Nurses	9.537	3.743	4.401	17.436
Hospital beds	4.379	1.713	2.439	7.920
GDP	8.74	1.81	5.40	12.14
Pop. density	127	113	3	403
Pop. $\geq$65 yo	190.615	28.203	126.134	232.680

Notes: Tests, cases, deaths, doctors, nurses, hospital beds, and population $\geq$ 65 years old have million per inhabitant as a unit of measurement; GDP (gross domestic product) is presented as a percentage; Population density is measured by inhabitants per km^2 of land area.

The country with the highest testing is Iceland, and the one with the lowest is Finland. However, the standard deviation is quite high, indicating a large variation in testing between countries.

Australia has the lowest number of cases and deaths among the OECD countries. Luxembourg has the highest number of cases, while Spain has the highest number of deaths per million inhabitants. Austria has the highest number of doctors per million inhabitants, and Switzerland has a higher number of nurses per million inhabitants. Finally, Germany registers the highest measurement regarding hospital beds.

Switzerland has the highest percentage of GDP available for health spending. On average, OECD countries allocate 8.74%. Israel has the highest number of inhabitants per km^2, and Italy registers the highest measurement for population $\geq$ 65 years old.

3.2. *Models and variables included in the study*

Table 5 presents the input and output variables included in models A and B, along with their units of measurement.

Model A evaluates the efficiency of contagion control. Therefore, the selected inputs are the number of tests for COVID-19, percentage of GDP in health, number of doctors, number of nurses, and population spacing.

The number of tests allows us to understand how countries tracked the evolution of the pandemic. Without data about how many are tested for the virus, there is no way to understand the spread of

Table 5. Variables selected to integrate the DEA models.

<table>
<tr><th>Model</th><th>Inputs</th><th>Measurement unit</th><th>Output</th><th>Measurement unit</th></tr>
<tr><td rowspan="5">A</td><td>Number of tests</td><td rowspan="3">Per million inhabitants</td><td rowspan="5">Number of confirmed cases</td><td rowspan="5">Inverse of the number of cases per million population</td></tr>
<tr><td>Number of physicians</td></tr>
<tr><td>Number of nurses</td></tr>
<tr><td>Population spacing</td><td>Inverse of number of inhabitants per km^2 of land area</td></tr>
<tr><td>GDP in health</td><td>%</td></tr>
<tr><td rowspan="6">B</td><td>Number of confirmed cases</td><td rowspan="5">Per million inhabitants</td><td rowspan="6">Number of confirmed deaths</td><td rowspan="6">Inverse of the number of deaths per million inhabitants</td></tr>
<tr><td>Number of physicians</td></tr>
<tr><td>Number of nurses</td></tr>
<tr><td>Number of hospital beds</td></tr>
<tr><td>Population $\geq$65 years old</td></tr>
<tr><td>GDP in health</td><td>%</td></tr>
</table>

Note: GDP – Gross domestic product.

the pandemic. Other authors, such as Aydin & Yurdakul (2020) and Breitenbach *et al.* (2020), have also applied this variable.

Health expenditures measure the final consumption of health goods and services (i.e., current health expenditures), including personal health care (curative care, rehabilitative care, long-term care, ancillary services, and medical goods) and collective services (prevention and public health services, as well as health administration), but excluding investment expenditures (OECD, 2021a). Richer nations are probably in a better position to limit the spread of pandemics, so GDP was used to compensate for differences between countries due to wealth. This variable is often used in health studies, such as those by Aydin & Yurdakul (2020), Breitenbach *et al.* (2020), Retzlaff-Roberts *et al.* (2004), Santos *et al.* (2012), Spinks & Hollingsworth (2007), and Zanakis *et al.* (2007).

The variable number of physicians is defined by practicing physicians who provide direct care to patients. Physicians are usually generalists who take responsibility for the provision of long-term care to individuals and families, or specialists, such as pediatricians, obstetricians/gynecologists, psychiatrists, medical specialists, and surgical specialists (OECD, 2021b). This variable was also used in the studies by Afonso & Aubyn (2011), Amado & Santos (2009), Breitenbach *et al.* (2020), Retzlaff-Roberts *et al.* (2004), and Zanakis *et al.* (2007).

Nurses are defined as the number of all practicing nurses who provide direct healthcare services to patients, including self-employed nurses (OECD, 2021c). Similar studies in healthcare efficiency also applied this variable, such as those by Afonso & Aubyn (2011), Amado & Santos (2009), Nepomuceno *et al.* (2020), and Zanakis *et al.* (2007).

Population density represents the population at mid-year divided by land area in square kilometers (OWD, 2017). Because the virus can be transmitted from person to person through close contact with people infected with SARS-CoV-2 or through contact with contaminated surfaces and objects, proximity between people increases the likelihood of transmitting and contracting the disease. Thus, the variation in population density between different countries may be a determining factor. Aydin & Yurdakul (2020), Ghasemi *et al.* (2020), Shirouyehzad *et al.* (2020), and Zanakis *et al.* (2007), who applied their studies in the context of pandemics, also considered this variable important. However, the input population density in this study is an undesirable input, and to capture the desired effect, it enters the model with the designation of population spacing, which is the mathematical inverse of population density. Furthermore, due to its inability to change, the input is classified as non-discretionary.

The output chosen for Model A is the number of confirmed cases. Cases considered positive meet the following criteria: persons with a positive nucleic acid amplification test (NAAT), persons with a positive SARS-CoV-2 rapid antigen test and meeting probable case definition or suspect criteria A or B, and asymptomatic persons with a positive SARS-CoV-2 rapid antigen test who are a contact of a probable or confirmed case. This variable was also included in the following studies: Aydin & Yurdakul (2020), Ghasemi *et al.* (2020), and Shirouyehzad *et al.* (2020).

Model B evaluates efficiency in medical treatment. The selected inputs are the number of confirmed cases, percentage of GDP in health, number of hospital beds, number of physicians, number of nurses, and percentage of the population over 65 years old (see Table 5).

Hospital beds provide a measure of the resources available for providing services to patients admitted to hospitals. The total number of hospital beds includes intensive care beds, rehabilitation beds, long-stay beds, and other hospital beds (OECD, 2021d). This variable is also present in the following studies: Afonso & Aubyn (2011), Aydin & Yurdakul (2020), Nepomuceno *et al.* (2020), Ozbugday *et al.* (2020), Retzlaff-Roberts *et al.* (2004), and Zanakis *et al.* (2007).

The percentage of the population over the age of 65 represents the annual population over that age with both sexes combined (OWD, 2020). One of the most clearly established mortality risk factors is a positive association with age (Balmford *et al.*, 2020). Older people are more likely to die from contracting COVID-19 than younger people. Therefore, countries that include a higher proportion of elderly people tend to report a higher number of deaths. Ghasemi *et al.* (2020) and Zanakis *et al.* (2007) also used this variable.

The output of Model B is the number of confirmed deaths. A death due to COVID-19 is defined for surveillance purposes, as a death resulting from a clinically compatible disease in a probable or confirmed COVID-19 case. A death due to COVID-19 may not be attributable to another disease (e.g., cancer) and should be counted independently of preexisting diseases that are suspected of triggering a severe course of COVID-19 (WHO, 2020a). This variable was used by Aydin & Yurdakul (2020), Ghasemi *et al.* (2020), and Shirouyehzad *et al.* (2020).

Undesirable outputs are present in both models, namely the number of confirmed cases and the number of confirmed deaths. The undesirable outputs should be reduced to improve performance (Zhu, 2014).

In standard DEA models, such as envelope models with VRS, it is assumed that outputs should be increased and resources should be decreased to improve performance or reach the best practice frontier. If we treat undesirable outputs as inputs so that undesirable outputs can be reduced, the resulting DEA model does not reflect the true production process (Zhu, 2014).

Let x and y be the desirable and undesirable outputs, respectively, which should increase (x) and decrease (y) to improve performance. However, in the output-oriented VRS model, both x and y must increase to improve performance. The solution to increase desirable outputs and decrease undesirable outputs would require multiplying each undesirable output by -1 and then finding an appropriate value to allow all negative undesirable outputs to be positive. Another solution is to apply a monotone decreasing transformation (e.g., $1/x$) to the undesirable outputs and then use the adapted variables as outputs. Since using a linear transformation preserves convexity, it is a good choice for a DEA model (Zhu, 2014). The monotone linear decreasing transformation is applied to handle the undesirable outputs in this study. As a result, convexity and linearity are preserved.

Since governments prefer to provide high-quality healthcare in the long run using the least number of resources and policymakers have more control over inputs, the input-oriented DEA model was considered to be the best option in order to minimize the resources required for contagion control and medical treatment.

VRS was assumed because, by doubling the resources, it was not realistic to expect that the results would also double. Besides this, the BCC model is more appropriate to deal with ratio variables (Hollingsworth & Smith, 2003).

The software used to run the DEA models is Efficiency Measurement System (EMS) 1.3 and Integrated System for Decision Support (ISYDS) 3, developed by Meza *et al.* (2005). Both pieces of software were needed because EMS presents far more outputs to analyze and ISYDS provides inverted DEA calculations.

4. Results and Discussion

4.1. *Model A: Contagion control*

Results analyses should be done with caution due to the exploratory nature of this research and the limitations of the data. Nevertheless, the results presented in Table 6 highlight some relevant aspects of each country's effectiveness in controlling the contagion of COVID-19.

Table 6 shows the pure technical efficiency scores estimated for each country. The average efficiency score of the 18 countries is 92.88%, with a standard deviation of 9.29%, which indicates that

Table 6. Standard efficiency levels of Model A: Contagion control.

DMU	Pure technical efficiency (%)
Australia	98.53
Austria	82.96
Czechia	100.00
Denmark	76.41
Estonia	93.57
Finland	100.00
Germany	85.14
Iceland	76.38
Israel	100.00
Italy	100.00
Latvia	100.00
Lithuania	88.69
Luxembourg	100.00
New Zealand	93.36
Slovenia	100.00
Spain	100.00
Switzerland	76.82
United Kingdom	100.00
Mean	92.88
Standard Deviation	9.29
Max	100.00
Min	76.38

Notes: This table presents the standard efficiency levels for Model A: Contagion Control and their mean, standard deviation, minimum, and maximum values. The inputs for Model A are the number of tests, number of doctors, number of nurses, population spacing, and gross domestic product in health. The output of Model A is the number of confirmed cases.

Table 7. Pairs and lambdas from some countries (Model A).

	Czechia	Israel	Latvia	Luxembourg	Slovenia	UK
Estonia	—	—	81%	14%	—	5%
Germany	1%	11%	—	—	—	88%
Iceland	—	—	20%	40%	40%	—
New Zealand	—	—	7%	6%	87%	—
Switzerland	5%	28%	—	—	1%	66%
No of times country is used as a peer	2	4	7	5	4	5

there are differences among the countries and potential for improvement in some of them. In this study, nine countries are rated as efficient (Slovenia, Spain, Finland, Israel, Italy, Latvia, Luxembourg, the United Kingdom, and the Czech Republic) in controlling the contagion of SARS-CoV-2 transmission. The country with the lowest efficiency score is Iceland, with 76.38%.

DEA also identifies benchmarks (efficient DMUs) for each of the inefficient countries. For example, as illustrated in Table 7, Germany and Switzerland, which share several sociocultural and economic similarities, have similar sets of benchmarks for learning good practices. These two countries register below-average efficiency rates, suggesting great potential for improvement in this context. Switzerland achieved 76.82% pure technical efficiency, and its benchmarks are the United Kingdom, Israel, the Czech Republic, and Slovenia. On the other hand, Germany has achieved 85.14% pure technical efficiency, and its benchmarks are also the United Kingdom, Israel, and the Czech Republic. Thus, in developing their strategies, policies, and programs for contagion control, Germany and Switzerland can draw inspiration from the policies, and strategies developed by these countries. These countries were identified as benchmarks for Germany and Switzerland because they share similarities in their social and economic structures, suggesting that similar policies may be effective in combating COVID-19.

In addition to information about the benchmarks from which less efficient countries can learn, Table 7 also provides useful information for defining the goals to be achieved by these countries. This information is contained in the lambda values displayed in each of the rows of Table 7. For example, the lambda values associated with

Iceland (i.e., 0.20, 0.40, and 0.40) represent the proportion of the Latvian, Slovenian, and Luxembourgian service levels that Iceland must achieve to become efficient.

The lambda values allow us to define the targets to be achieved by each country for the different contagion control variables, which can be seen in Table 8. The population spacing variable was not included in the table because countries do not have the ability to change population density. The targets set for each country represent the values that would allow each country to be considered efficient. In the case where the numbers for each variable decrease, it indicates a less efficient use of resources, suggesting the possibility of an improvement in the allocation of resources being used to fight the pandemic.

As presented in Table 7, there are three countries that serve as a benchmark for five or more countries, showing robust efficiency status. These countries are Latvia, Luxembourg, and the United Kingdom. These results indicate that the policies and strategies in

Table 8. Goals for achieving contagion control efficiency.

DMU	Tests	GDP	Doctors	Nurses	Cases
Australia	25,420	9.20	3,215	11,643	272
Austria	42,889	8.47	3,917	5,623	1,873
Czechia	44,840	7.76	4,015	8,022	907
Denmark	62,690	7.66	3,217	7,745	1,935
Estonia	71,718	6.35	3,250	5,512	1,440
Finland	22,576	9.09	3,165	14,076	1,025
Germany	28,273	9.92	2,873	7,407	2,000
Iceland	85,805	6.71	3,094	9,436	5,327
Israel	66,841	7.45	3,359	5,247	1,995
Italy	41,523	8.66	3,969	5,730	3,605
Latvia	62,953	6.26	3,339	4,401	571
Lithuania	92,539	6.03	3,176	7,003	618
Luxembourg	143,174	5.40	2,890	11,351	6,557
New Zealand	47,974	7.95	3,167	9,771	314
Slovenia	40,454	8.26	3,170	10,116	717
Spain	34,774	9.00	4,026	5,876	4,800
Switzerland	36,749	9.33	3,018	7,032	3,596
United Kingdom	23,379	10.25	2,796	7,656	3,104

Notes: Tests, cases, doctors, and nurses have million/inhabitant as the unit of measure; GDP (gross domestic product) is presented as a percentage.

contagion control used by these countries should be identified to provide guidance for other countries.

To determine if these results could register false-efficiency, as stated by Ali (1993), the standard efficiency frontier and the inverted efficiency frontier were used to obtain the composite efficiency levels presented in Table 9.

No country reaches the maximum composite efficiency. Slovenia and Latvia reach values of 1 (or very close) for normalized composite efficiency, thus representing the most consistent countries in efficiency levels. Finland and the Czech Republic also have standardized composite efficiency values above 85%, while Estonia and New Zealand also register quite high values. Spain, Italy, Luxembourg, and the United Kingdom, considered efficient in the standard frontier, are classified as false-efficient, presenting relatively low values (76% of normalized composite efficiency). These are examples of countries positioned simultaneously on both frontiers.

Table 9. Inverted and composite efficiency levels of Model A: Contagion Control.

DMU	Standard efficiency	Inverted efficiency	Compound efficiency	Normalized compound efficiency
Australia	99	100	49	75
Austria	83	100	41	63
Czechia	100	82	59	90
Denmark	76	100	38	58
Estonia	94	73	60	92
Finland	100	83	59	89
Germany	85	100	43	65
Iceland	76	100	38	58
Israel	100	99	50	77
Italy	100	100	50	76
Latvia	100	69	66	100
Lithuania	89	100	44	68
Luxembourg	100	100	50	76
New Zealand	93	82	56	85
Slovenia	100	71	64	98
Spain	100	100	50	76
Switzerland	77	100	38	59
UK	100	100	50	76

Note: Values are presented as a percentage.

Note that there are more countries on the inverted frontier (11) than on the standard frontier (9). This reveals a greater ease in identifying less efficient countries, relative to efficient countries, which effectively demonstrates the great difficulty in dealing with this pandemic and its consequences.

4.2. *Model B: Medical treatment*

The results presented in Table 10 highlight some relevant aspects of each country's efficiency in the medical treatment of COVID-19.

Table 10. Standard efficiency levels of Model B: Medical treatment.

DMU	**Pure technical efficiency (%)**
Australia	100.00
Austria	78.53
Czechia	90.12
Denmark	100.00
Estonia	98.81
Finland	99.97
Germany	75.27
Iceland	95.83
Israel	100.00
Italy	96.72
Latvia	100.00
Lithuania	97.09
Luxembourg	100.00
New Zealand	100.00
Slovenia	100.00
Spain	99.23
Switzerland	74.05
United Kingdom	100.00
Mean	94.76
Standard deviation	9.03
Max	100.00
Min	74.05

Notes: This table presents the standard efficiency levels for Model B: Medical Treatment and their mean, standard deviation, minimum, and maximum values. The inputs for Model B are the number of confirmed cases, the number of doctors, the number of nurses, the number of hospital beds, the population $\geq$65 years old and gross domestic product of health. The output of Model B is the number of confirmed deaths.

Table 11. Pairs and lambdas from some countries (Model B).

	Australia	Israel	Latvia	NZ	Slovenia	UK
Czechia	16%	21%	63%	—	—	—
Germany	—	32%	—	32%	15%	21%
Italy	—	88%	—	—	—	12%
Lithuania	10%	4%	86%	—	—	—
Spain	—	76%	—	—	—	24%
No of times country is used as a peer	2	9	4	4	2	5

The average efficiency score for the 18 countries is 94.76%, with a standard deviation of 9.03%, indicating that there are differences between countries and potential for improvement in some of them. In this study, eight countries are rated as efficient (Australia, Denmark, Slovenia, Israel, Latvia, Luxembourg, New Zealand, and the United Kingdom) in the medical treatment of COVID-19. Switzerland presents the lowest efficiency value (74.05%).

Table 11 shows the benchmarks for each of the least efficient countries. Germany and the Czech Republic were ranked below average in efficiency rates, suggesting great potential for improvement. Germany achieved a 75.27% efficiency score, and its benchmarks are Israel, New Zealand, the United Kingdom, and Slovenia. On the other hand, the Czech Republic achieved a 90.12% efficiency score, and its benchmarks are Latvia, Israel, and Australia. In developing their strategies, policies, and programs for medical care, Germany and the Czech Republic can draw inspiration from the policies and strategies developed by their benchmarks.

The lambda values associated with Italy (i.e., 0.12 and 0.88) represent the proportion of Israel and the United Kingdom service levels that Italy must achieve to become efficient (Table 11).

The lambda values allow us to define the targets to be achieved by each country for the different medical treatment variables, which can be seen in Table 12. The variable population $\geq$ 65 years old was not included in the table because it is considered a non-discretionary input. The numbers for each variable decrease, which indicates less efficient use of resources, indicating the possibility of an improvement in the allocation of resources being used to fight the pandemic.

Table 12. Goals for achieving medical treatment efficiency.

DMU	Cases	GDP	Doctors	Nurses	Hospital beds	Deaths
Australia	272	9.33	3,714	11,816	3,683	4
Austria	1,471	7.21	3,355	5,323	3,802	19
Czechia	818	7.00	3,403	5,764	4,747	11
Denmark	1,935	10.03	4,210	10,137	2,439	95
Estonia	1,423	6.71	3,327	5,061	4,505	20
Finland	1,025	8.89	3,164	9,780	3,600	14
Germany	1,505	8.77	3,221	8,157	3,017	12
Iceland	1,563	8.42	3,533	7,505	2,866	14
Israel	1,995	7.45	3,359	5,247	3,111	34
Italy	2,130	7.80	3,290	5,542	3,031	45
Latvia	571	6.26	3,339	4,401	5,552	14
Lithuania	600	6.61	3,376	5,161	5,269	11
Luxembourg	6,557	5.40	2,890	11,351	4,448	179
New Zealand	314	9.32	3,392	10,466	2,656	5
Slovenia	717	8.26	3,170	10,116	4,417	52
Spain	2,263	8.13	3,223	5,831	2,952	46
Switzerland	2,663	7.93	3,184	6,360	3,075	41
UK	3,104	10.25	2,796	7,656	2,456	463

Notes: Cases, deaths, doctors, nurses, and hospital beds have as unit of measure million per inhabitant; GDP (gross domestic product) is presented as a percentage.

As presented in Table 11, there are four countries that serve as a benchmark for four or more countries, showing robust efficiency status. These countries are Israel, Latvia, New Zealand, and the United Kingdom. These results indicate that the policies and service delivery strategies used by these countries should be identified to provide guidance to other countries.

To identify potential false-efficient countries in Model B, the standard efficiency frontier and the inverted efficiency frontier were used to obtain the composite efficiency levels presented in Table 13.

Again, no country presents the maximum value of composite efficiency. Israel registers a high (but not maximum) composite efficiency, being the only country to reach the maximum value of normalized composite efficiency.

Australia (90%) and New Zealand (93%) show high values on the standardized composite frontier. Finland, Luxembourg, and

Table 13. Inverted and composite efficiency levels of Model B: Medical treatment.

DMU	Standard efficiency	Inverted efficiency	Compound efficiency	Normalized compound efficiency
Australia	100	83	59	90
Austria	79	100	39	61
Czechia	90	93	49	75
Denmark	100	95	53	81
Estonia	99	90	54	84
Finland	100	100	50	77
Germany	75	100	38	58
Iceland	96	100	48	74
Israel	100	70	65	100
Italy	97	100	48	75
Latvia	100	92	54	84
Lithuania	97	99	49	76
Luxembourg	100	100	50	77
New Zealand	100	79	61	93
Slovenia	100	92	54	83
Spain	99	100	50	76
Switzerland	74	100	37	57
UK	100	100	50	77

Note: Values are presented as a percentage.

the United Kingdom, considered efficient on the standard frontier, present a relatively low normalized composite efficiency value (77%).

These results corroborate some findings reported in the literature and highlight some relevant aspects about each country's efficiency in controlling the contagion and medical treatment of COVID-19, which will be discussed later.

4.3. *Discussion of the main results*

In this study, the results for Australia support the findings by Zanakis *et al.* (2007), in which they found evidence that countries with lower population density can provide better health system performance without necessarily having a higher GDP.

Germany spends approximately 12% of its GDP on healthcare. As a result, the capacity of Germany's healthcare system is considered very high. According to Wieler *et al.* (2020), Germany has the largest number of hospital beds and a robust sector of private and public laboratories, of which approximately 200 indicate capacity for SARS-CoV-2 testing. Switzerland, which spends a higher percentage of its GDP on healthcare, has more doctors and nurses than Germany but not a greater number of beds.

Table 14 presents the number of tests, cases, and deaths per million population that are officially recorded as caused by COVID-19 in all selected countries by the 100th day after the first confirmed case.

There are many technologies used for testing for COVID-19. The different tests can be divided into two types: polymerase chain reaction (PCR) tests and rapid immunochromatographic tests

Table 14. Descriptive statistics for the variables: Number of tests, cases, and deaths.

DMU	**Tests per million**	**Cases per million**	**Deaths per million**
Australia	25,799	272	4
Austria	51,698	1,873	75
Czechia	44,840	907	31
Denmark	82,045	1,935	95
Estonia	76,646	1,440	52
Finland	22,576	1,025	46
Germany	33,208	2,000	84
Iceland	185,115	5,327	29
Israel	66,841	1,995	34
Italy	41,523	3,605	502
Latvia	62,953	571	14
Lithuania	115,720	618	26
Luxembourg	143,174	6,557	179
New Zealand	60,363	314	5
Slovenia	40,454	717	52
Spain	34,774	4,800	570
Switzerland	47,837	3,596	224
United Kingdom	23,379	3,104	463
Mean	64,386	2,259	138
Standard deviation	43,781	1,859	182
Max	185,115	6,557	570
Min	22,576	272	4

Note: Values are presented per million inhabitants.

(Ritchie *et al.*, 2020b). Only PCR and antigen tests were included in this study. Confirmation of cases is usually based on a positive result from a PCR test and/or an antigen test, as recommended by the WHO (2020b).

When analyzing the results, it should be kept in mind that no country knows the real total number of people infected with COVID-19. They only know the number of people officially tested, which indicates that the counts of confirmed cases depend on the testing capacity of each country. Thus, caution is needed when looking at the number of confirmed cases, as low testing also implies a low registered number of people infected. An adequate observation of the virus spread implies that countries with more widespread outbreaks need to do more testing (Hasell, 2020).

Iceland performed far better than any other OECD country, reaching 185,115 tests per million inhabitants. The average testing in the selected countries is 64,386 tests per million inhabitants.

On the other hand, when the number of confirmed cases is high relative to testing, this suggests that there may not be enough testing being done to adequately monitor the outbreak. In these countries, the actual number of infections may be much higher than the number of confirmed cases. Countries with the highest proportion of testing relative to confirmed cases are likely to have more accurate knowledge of total cases, not just confirmed cases (Ritchie *et al.*, 2020b).

At the beginning of an outbreak, the possibility of containment increases if testing is done frequently and contacts are followed up. However, once community transmission is widespread and containment is no longer possible, population-wide measures to reduce transmission are less dependent on testing. If these measures are successful in reducing spread, case finding and contact follow-up eventually become feasible again, and widespread testing again becomes critical (Ritchie *et al.*, 2020a).

Wieler *et al.* (2020) claim that Germany did not prevent the COVID-19 outbreak, but the prevention protocols in place facilitated the country's response. These protocols included the early establishment of testing capabilities, high levels of testing, and an effective containment strategy among the elderly, which may explain why Germany has a much lower mortality rate than comparable countries. This study supports these findings, as Germany's performance is below average (4.19%).

Germany has shown that the country's enabling environment, including a good local public healthcare system and specialized scientific institutions, has largely contributed to this broad progress. Wieler *et al.* (2020) mention that there was an efficient use of the country's hospital capacity. However, the results of our study indicate otherwise, with Germany recording a below-average efficiency score (85.14% in contagion control and 75.27% in medical treatment).

According to Wieler *et al.* (2020), in Germany, the first outbreaks affected tourists, festivals, and workplaces and not primarily nursing homes, which may have helped limit infections and deaths among elderly people. According to national guidelines, nursing homes must have tested negative or been quarantined in an isolation area for 14 days, and the elderly were also tested more frequently than younger age groups.

Ritchie *et al.* (2020b) argue that Australia has a positive case rate of approximately 1% because it performs hundreds of tests for each detected case, which is in line with the results obtained in our study. With a value of less than 1%, we have Latvia, which also belongs to the group of countries that are efficient at controlling the contagion. Other countries with higher positive rates indicate that they do only a few tests per detected case. It is unlikely that countries that do fewer tests per confirmed case will be able to find all the infected.

The WHO has suggested, as a general benchmark for adequate testing and as an indicator, that the pandemic is under control when a country achieves a positive rate of less than 5% (WHO, 2020c). Table 15 presents that most of the countries in the sample have a positive rating lower than recommended, with an average score of 4.38%. Spain and the United Kingdom exceed the advised value in a distinct manner. The positive rate can also help understand the spread of the virus. In countries with a high positive rate, the number of confirmed cases is likely to represent only a small fraction of the actual number of infections. And when the positive rate is increasing in a country, this may suggest that the virus is spreading faster than the observed growth in confirmed cases.

The number of tests done by a country to find a positive COVID-19 case can be calculated by the inverse of the positive rate data. It is unlikely that countries that do a very few number of tests per confirmed case are testing widely enough to find all cases. The WHO (2020d) suggested 10–30 tests per confirmed case as a general

Table 15. Rate and inverse relationship of number of cases and deaths.

DMU	Positive rate	Number of tests/positive	Death rate	Number of positive/death
Australia	1.05	95	1.40	71
Austria	3.62	28	3.99	25
Czech Republic	2.02	49	3.38	30
Denmark	2.36	42	4.91	20
Estonia	1.88	53	3.61	28
Finland	4.54	22	4.49	22
Germany	6.02	17	4.19	24
Iceland	2.88	35	0.55	181
Israel	2.98	34	1.69	59
Italy	8.68	12	13.93	7
Latvia	0.91	110	2.39	42
Lithuania	0.53	187	4.16	24
Luxembourg	4.58	22	2.72	37
New Zealand	0.52	192	1.46	68
Slovenia	1.77	56	7.32	14
Spain	13.80	7	11.87	8
Switzerland	7.52	13	6.22	16
United Kingdom	13.28	8	14.91	7
Mean	4.38	55	5.18	38
Standard deviation	4.05	57	4.25	41
Max	13.80	192	14.91	181
Min	0.52	7	0.55	7

Notes: The rate is presented as a percentage. Inverse ratios are presented as whole numbers.

benchmark for adequate testing. The only countries in this study that do not meet the minimum benchmark are Spain and the United Kingdom. New Zealand presents the highest testing per positive case, reaching a value of 192 tests.

Countries with a lower proportion of deaths per confirmed case have more effective responses during the treatment phase. Iceland has 181 confirmed cases per death, making it one of the countries with the most effective medical treatment.

The mortality rate reported in the United Kingdom is the highest, surpassing even that of Italy, which experienced the first outbreaks before the pandemic was declared. These results are consistent with Balmford *et al.*'s (2020) findings, which indicate that the United Kingdom has the highest mortality rate of the OECD countries.

Because older people are more likely than younger people to die from COVID-19, countries with a higher population ≥65 years old have an increased risk (Singh & Adhikari, 2020). Germany's greatest success has been its relatively limited transmission in long-term care facilities. The relatively low rate of infection among the ≥65-year-old German population is likely a factor behind their relatively low mortality rate. Germany is the third country in this study with the most elderly population and has only 84 deaths per million population. However, in large countries, pandemics may evolve differently across cities or regions, and this study did not include analysis at the regional level.

There are several limitations to the process of cross-country comparison. The dynamics of the pandemic may depended on factors beyond a country's control, such as population density, geographical dispersal patterns, the number of borders or air/land entry points, and the volume of tourists and trade (Hazen, 2020). For example, the challenges of controlling an outbreak in an isolated country, like New Zealand, are very different from those in Switzerland, a country in the geographic heart of Europe that shares land borders with five countries.

Another topic of interest is the effectiveness of mobility blocks, which aim to promote social distancing to reduce population crowding and the spread of disease. According to Balmford *et al.* (2020), a clear separation can be observed between Germany, which went quickly into blockade and controlled the growth of the virus, and countries such as the United Kingdom and Spain, where blockade was delayed, resulting in an increase in cases. Governments had to continually decide about the right time to introduce a mobility block as the pandemic developed. According to Balmford *et al.* (2020), the date chosen for the blockade reveals the government's preferences regarding the trade-off between avoided deaths and GDP losses. In this study, the fastest country to impose a blockade was the Czech Republic, with an efficiency score of 100% in controlling the contagion. However, we cannot agree that a quick blockade implies a lower number of deaths. As examples, we have Australia, which imposed the blockade after 68 days and only recorded four deaths, achieving an efficiency score of 98.53% in controlling contagion.

The most effective countries for medical treatment were Australia, Denmark, Slovenia, Israel, Latvia, and New Zealand.

According to Wieler *et al.* (2020), the German government entered the pandemic with a detailed National Pandemic Plan, which allowed the government to activate it quickly without wasting time on disputes related to governance, accounting, or costs. Despite this, Germany did not achieve maximum efficiency in either contagion control (85.14%) or medical treatment (75.27%).

Capano *et al.* (2020) argue that Italy was the first major epicenter of the COVID-19 pandemic in the Western world and had little precedent to formulate its chaotic and slow response. Italy counted more than 500 deaths per million in its first 100 days after the first confirmed case.

In general, the response to the pandemic occurred in the same way across countries: first with the definition of the most vulnerable people, then with the need for greater protection of health care workers, and finally with the prospects for the general population.

5. Conclusion

5.1. *Summary*

In this study, the main socioeconomic factors that collectively affect the COVID-19 indicators are demonstrated, and each country's efficiency in fighting the pandemic is evaluated. It is concluded through DEA models that the countries with robust efficiency in both contagion control and medical treatment are Slovenia and Latvia. Furthermore, the countries with lower population density that can offer better health system performance and per capita support (doctors, nurses, and hospital beds), with a lower percentage of GDP in health, are the most likely to exhibit lower COVID-19 indicators. These findings may be useful for health policymakers and will open doors for further studies in various countries on the determinants of the COVID-19 pandemic and efficient global strategies to combat it while avoiding wasted resources. It is recognized that this study is preliminary and needs further work, such as using inputs more directly related to COVID-19, as there are several limitations in the data and general health variables used in this study. Crucial to bridging the lack of and gaps in data are the ongoing efforts of the

Center for Disease Control, CSSE, OECD, and WHO in collecting complete and reliable global data. Some countries do not have data, and even when they do, they do not cover all the important variables, do not cover more than one year, or do not guarantee the accuracy of the information.

The results of this study indicate wide variation in efficiency across countries, revealing that some countries have been more efficient in allocating resources to combat COVID-19 than others. The analysis described in this study also allows for the identification of appropriate benchmarking for each country as well as performance improvement targets, enabling the identification of the causes of poor performance and the exploration of the best performing practices. This assessment enables the development of solid strategies for improving the control of contagion and the medical treatment of COVID-19 or similar events.

Studies applying the DEA methodology to pandemics reveal a consistent difference in terms of empirical results. This diversity can be partly explained by different methodological constructs, i.e., in the selection of input and output variables, and other model specifications, such as technology assumptions and weight constraints. In addition, it must be taken into consideration that there is a wide range of evaluation contexts in the composition of study designs, such as the type of healthcare organizations, level of analysis, and period.

While valuable, these approaches are limited in conducting comparative analyses because an intervention may not be efficient in some countries but be efficient in others, and they are also limited in identifying best practices since some countries may be better than others using certain indicators.

Medical research and innovations in diagnosis and treatment create a growing number of potential options for improving outcomes at any given resource level. Real-world efficiency studies will be increasingly important to inform policymakers, patients, and the public about how the healthcare system can produce value for patients.

5.2. *Prospects for future research*

In future research, it would be interesting to include the costs of infection control and medical treatment. Examples are the cost of avoiding infections and the cost of medical treatment. The first efficiency indicator measures the cost per unit of service provided by a given prevention program (for example, the cost of mass testing). The second efficiency indicator measures the cost per hospitalization (e.g., cost per day of hospitalization). Although several input indicators can be used in this context, the costs of delivering COVID-19 prevention interventions are arguably the most relevant.

According to Kavaliunas *et al.* (2020), in Sweden, the main goal in fighting the COVID-19 pandemic was to change social behavior to minimize interpersonal contact and thus reduce the spread of the virus. While changing social behavior in some countries would require coercive and fully enforced rules and regulations, in others this could be accomplished with more subtle signals from the government. Sweden stood out from most countries since its containment strategy was not focused on a complete blockade but included more selective measures, such as minimizing people's visits to areas where a high density of inhabitants could be expected. Government agencies preferred to issue recommendations and advice on appropriate social behavior rather than enact coercive legislation or regulations. While these aspects of the Swedish approach have aroused the greatest international curiosity, their effectiveness remains uncertain to date. Other relevant aspects, such as the unpreparedness of local governments and the coordination problems of a highly decentralized healthcare system, have been less noted but have also played an important role in determining the outcome of Sweden's delayed and weak response to the COVID-19 crisis. In future research, it would be interesting to insert Sweden into DEA models to gain insight into how a response so distinct from other countries would perform in terms of efficiency.

Future research studies should investigate the ways in which the DEA methodology can be combined with other methodologies to better inform decision-making and policymaking in combating the pandemic.

References

Afonso, A. & Aubyn, M. (2011). Assessing health efficiency across countries with a two-step and bootstrap analysis. *Applied Economics Letters*, 18, 1427–1430. https://doi.org/10.1080/13504851.2010.541149.

Ali, A. (1993). Streamlined computation for data envelopment analysis. *European Journal of Operational Research*, 64(1), 61–67. https://doi.org/10.1016/0377-2217(93)90008-B.

Althin, R., Färe, R., Gralén, K., Grosskopf, S., Jönsson, B., & Wilking, N. (2019). Efficiency and productivity of cancer care in Europe. *Journal of Cancer Policy*, 21. https://doi.org/10.1016/j.jcpo.2019.100194.

Amado, C. & Dyson, R. (2009). Exploring the use of DEA for formative evaluation in primary diabetes care: An application to compare English practices. *Journal of the Operational Research Society*, 60(11), 1469–1482. http://www.jstor.org/stable/40295709.

Amado, C. & Santos, S. (2009). Challenges for performance assessment and improvement in primary health care: The case of the Portuguese health centres. *Health Policy*, 91(1), 43–56. https://doi.org/10.1016/j.healthpol.2008.11.008.

Aydin, N. & Yurdakul, G. (2020). Assessing countries' performances against COVID-19 via WSIDEA and machine learning algorithms. *Applied Soft Computing Journal*. https://doi.org/10.1016/j.asoc.2020.106792 (Accessed February 18, 2021).

Balmford, B., Annan, J., Hargreaves, J., Altoè, M., & Bateman, I. (2020). Cross-country comparisons of COVID-19: Policy, politics and the price of life. *Environmental and Resource Economics*, 76, 525–551. https://doi.org/10.1007/s10640-020-00466-5.

Banker, R., Charnes, A., & Cooper, W. (1984). Some models for estimating technical and scale inefficiencies in data envelopment analysis. *Management Science*, 9(1), 1078–1092. https://doi.org/10.1287/mnsc.30.9.1078.

Boubaker, S., Le, T.D.Q., & Ngo, T. (2022a). Managing bank performance under COVID-19: A novel inverse DEA efficiency approach. *International Transactions in Operational Research*, 30(2023), 2436–2452. https://doi.org/ 10.1111/itor.13132.

Boubaker, S., Do, D.T., Hammami, H., & Ly, K.C. (2022b). The role of bank affiliation in bank efficiency: A fuzzy multi-objective data envelopment analysis approach. *Annals of Operations Research*, 311, 611–639. https://doi.org/10.1007/s10479-020-03817-z.

Boubaker, S., Le, T.D.Q., Ngo, T., & Manita, R. (2023). Predicting the performance of MSMEs: A hybrid DEA-machine learning approach. *Annals of Operations Research*. https://doi.org/10.1007/s10479-023-05230-8.

Breitenbach, M.C., Ngobeni, V., & Aye, G.C. (2020). Efficiency of healthcare systems in the first wave of COVID-19 — A technical efficiency analysis. https://mpra.ub.uni-muenchen.de/101440/1/MPRA_paper_8872.pdf (Accessed February 12, 2021).

Breitenbach, M.C., Ngobeni, V., & Aye, G.C. (2021). Global healthcare resource efficiency in the management of COVID-19 death and infection prevalence rates. *Frontiers in Public Health*, 9. https://doi.org/10.3389/fpubh.2021.638481.

Capano, G., Howlett, M., Jarvis, D., Ramesh, M., & Goyal, N. (2020). Mobilizing policy (in)capacity to fight COVID-19: Understanding variations in state responses. *Policy and Society*, 39(3), 285–308. https://doi.org/10.1080/14494035.2020.1787628.

Charnes, A., Cooper, W., & Rhodes, E. (1978). Measuring the efficiency of decision-making units. *European Journal of Operational Research*, 2(6), 429–444. https://doi.org/10.1016/0377-2217(78)90138-8.

Cinaroglu, S. (2021a). Intensive care unit services preparedness for the pandemic: An efficiency analysis. *Health Services Insights*, 14. https://doi.org/10.1177/11786329211037527.

Cinaroglu, S. (2021b). Oncology services efficiency in the age of pandemic: A jackknife and bootstrap sensitivity analysis for robustness check of DEA scores. *Journal of Cancer Policy*, 27. https://doi.org/10.1016/j.jcpo.2020.100262.

Cordero-Ferrera, J., Crespo-Cebada, E., & Murillo-Zamorano, L. (2011). Measuring technical efficiency in primary health care: The effect of exogenous variables on results. *Journal of Medical System*, 35(4), 545–554. https://doi.org/10.1007/s10916-009-9390-6.

Cylus, J., Papanicolas, I., & Smith, P. (2015). Using data envelopment analysis to address the challenges of comparing health system efficiency. *Global Policy*. https://doi.org/10.1111/1758-5899.12212 (Accessed March 8, 2021).

Di Giorgio, L., Moses, M.W., Fullman, N., Wollum, A., Conner, R.O., Achan, J., Achoki, T., *et al.* (2016). The potential to expand antiretroviral therapy by improving health facility efficiency: Evidence from Kenya, Uganda, and Zambia. *BMC Medicine*, 14(1). https://doi.org/10.1186/s12916-016-0653-z.

Direção Geral de Saúde (DGS) (2021). *COVID-19.* Ministério da Saúde. Perguntas Frequentes Categoria — COVID-19 (min-saude.pt). https://www.sns24.gov.pt/tema/doencas-infecciosas/covid-19/ (Accessed February 6, 2021).

Entani, T., Maeda, Y., & Tanaka, H. (2002). Dual models of interval DEA and its extension to interval data. *European Journal of Operational Research*, 136(1), 32–45. https://doi.org/10.1016/S0377-2217(01)00055-8.

Färe, R. & Grosskopf, S. (2000). Network DEA. *Socio-Economic Planning Sciences*, 34, 35–49. https://doi.org/10.1016/S0038-0121(99)00012-9.

Farrell, M. (1957). The measurement of productive efficiency. *Journal of the Royal Statistical Society: Series A (General)*, 120(3), 253–290. https://doi.org/10.2307/2343100.

Ferraz, D., Mariano, E.B., Manzine, P.R., Moralles, H.F., Morceiro, P.C., Torres, B.G., de Almeida, M.R., *et al.* (2021). COVID health structure index: The vulnerability of Brazilian microregions. *Social Indicators Research*, 158(1), 197–215.

Gattoufi, S., Amin, G.R., & Emrouznejad, A. (2014). A new inverse DEA method for merging banks. *IMA Journal of Management Mathematics*, 25(1), 73–87. https://doi.org/10.1093/imaman/dps027.

Ghasemi, A., Boroumand, Y., & Shirazi, M. (2020). How do governments perform in facing COVID-19. How do governments perform in facing COVID-19 (Accessed February 11, 2021).

Giménez, V., Prior, D., Thieme, C., & Tortosa-Ausina, E. (2021). International comparisons of the COVID-19 pandemic management: What can be learned from activity analysis techniques? Working Papers 2021/12, Economics Department, Universitat Jaume I, Castellón (Spain).

Hammami, H., Ngo, T., Tripe, D., & Vo, D. (2022). Ranking with a Euclidean common set of weights in data envelopment analysis: With application to the Eurozone banking sector. *Annals of Operations Research*, 311, 675–694. https://doi.org/10.1007/s10479-020-03759-6.

Hasell, J. (2020). Testing early, testing late: Four countries' approaches to COVID-19 testing compared. https://ourworldindata.org/coronavirus-testing (Accessed February 9, 2021).

Hazen, H. (2020). Teaching COVID-19 topics in a geographic framework. *The Geography Teacher*, 17(2), 33–43. https://doi.org/10.1080/19338341.2020.1764375.

Hollingsworth, B. & Smith, P. (2003). Use of ratios in data envelopment analysis. *Applied Economics Letters*, 10(11), 733–735. https://doi.org/10.1080/1350485032000133381.

Ibrahim, M.D., Binofai, F.A.S. and Mm Alshamsi, R. (2020). Pandemic response management framework based on efficiency of COVID-19 control and treatment. *Future Virology*, 15(12), 801–816.

Kavaliunas, A., Ocaya, P., Mumper, J., Lindfeldt, I., & Kyhlstedt, M. (2020). Swedish policy analysis for COVID-19. *Health Policy and Technology*, 9(4), 598–612. https://doi.org/10.1016/j.hlpt.2020.08.009.

Kennedy, D., Seale, A., Bausch, D., Ritchie, H., & Roser, M. (2020). How experts use data to identify emerging COVID-19 success stories. https://ourworldindata.org/identify-covid-exemplars (Accessed February 8, 2021).

Kerstens, K. & Shen, Z. (2021). Using COVID-19 mortality to select among hospital plant capacity models: An exploratory empirical application to Hubei province. *Technological Forecasting and Social Change*, 166. https://doi.org/10.1016/j.techfore.2020.120535 (Accessed February 3, 2022).

Klumpp, M., Loske, D. and Bicciato, S. (2022). COVID-19 health policy evaluation: Integrating health and economic perspectives with a data envelopment analysis approach. *European Journal of Health Economics*. https://doi.org/10.1007/s10198-021-01425-7 (Accessed February 3, 2022).

Kohl, S., Schoenfelder, J., Fügener, A., & Brunner, J. (2019). The use of Data Envelopment Analysis (DEA) in healthcare with a focus on hospitals. *Health Care Management Science*, 22, 245–286. https://doi.org/10.1007/s10729-018-9436-8.

Lupu, D. & Tiganasu, R. (2022). COVID-19 and the efficiency of health systems in Europe. *Health Economics Review*, 12(1). https://doi.org/10.1186/s13561-022-00358-y.

Mariano, E., Torres, B., Almeida, M., Ferraz, D., Rebelatto, D., & Mello, J. (2021). Brazilian states in the context of COVID-19 pandemic: An index proposition using network data envelopment analysis. *IEEE Latin America Transactions*, 19(6), 917–924.

Martínez-Córdoba, P.J., Benito, B., & García-Sánchez, I.M. (2021). Efficiency in the governance of the COVID-19 pandemic: Political and territorial factors. *Globalization and Health*. https://doi.org/10.1186/s12992-021-00759-4 (Accessed February 3, 2022).

Martins, A. (2012). Avaliação da eficiência e identificação dos fatores determinantes da eficiência do setor bancário em Portugal. Sapientia, Repositório da Universidade do Algarve. http://hdl.handle.net/10400.1/6077 (Accessed July 24, 2021).

Md Hamzah, N., Yu, M.-M., & Fong See, K. (2021). Assessing the efficiency of Malaysia health system in COVID-19 prevention and treatment response. *Health Care Management Science*, 24, 273–285.

Mello, J., Gomes, E., Meza, L., Neto, L., & Anna, A. (2005). Fronteiras DEA difusas. *Investigação Operacional*, 25, 85–103. https://docplayer.com.br/80102784-Fronteiras-dea-difusas.html.

Meza, L., Mello, J., Gomes, E., & Fernandes, A. (2007). Seleção de variáveis em DEA aplicada a uma análise do mercado de energia elétrica, *Investigação Operacional*, 27, 21–36. https://www.researchgate.net/publication/256373181_Seleccao_de_variaveis_em_DEA_aplicada_a_uma_analise_do_mercado_de_energia_eletrica (Accessed July 24, 2021).

Meza, L., Mello, J., Neto, L., & Gomes, E. (2005). ISYDS — Integrated system for decision support (SIAD — Sistema Integrado de Apoio à

Decisão): A software package for data envelopment analysis model. *Investigação Operacional*, 25(3), 493–503. https://www.researchgate.net/publication/250031653_ISYDS-Integrated_System_for_Decision_Support_SIAD-Sistema_Integrado_de_Apoio_a_Decisao_A_software_package_for_data_envelopment_analysis (Accessed July 24, 2021).

Mohanta, K.K., Sharanappa, D.S., & Aggarwal, A. (2021). Efficiency analysis in the management of COVID-19 pandemic in India based on data envelopment analysis. *Current Research in Behavioral Sciences*, 2. https://doi.org/10.1016/j.crbeha.2021.100063 (Accessed February 3, 2022).

Mourad, N., Habib, A.M., & Tharwat, A. (2021). Appraising healthcare systems' efficiency in facing COVID-19 through data envelopment analysis. *Decision Science Letters*, 10(3), 301–310.

Nepomuceno, T., Silva, W., Nepomuceno, K., & Barros, I. (2020). A DEA-based complexity of needs approach for hospital beds evacuation during the COVID-19 outbreak. *Journal of Healthcare Engineering*. https://doi.org/10.1155/2020/8857553 (Accessed February 10, 2021).

Nolte, E., Wait, S., & McKee, M. (2006). Investing in health: Benchmarking health systems. https://www.nuffieldtrust.org.uk/files/2017-01/investing-in-health-benchmarking-health-systems-web-final.pdf (Accessed February 12, 2021).

Organization for Economic Co-operation and Development (OECD) (2021a). Health spending (indicator). https://data.oecd.org/healthres/health-spending.htm (Accessed March 27, 2021).

Organization for Economic Co-operation and Development (OECD) (2021b). Doctors (indicator). https://data.oecd.org/healthres/doctors.htm#indicator-chart (Accessed March 27, 2021).

Organization for Economic Co-operation and Development (OECD) (2021c). Nurses (indicator). https://data.oecd.org/healthres/nurses.htm#indicator-chart (Accessed March 27, 2021).

Organization for Economic Co-operation and Development (OECD) (2021d) Hospital beds (indicator). https://data.oecd.org/healtheqt/hospital-beds.htm (Accessed March 28, 2021).

Our World in Data (OWD) (2017). Population density (indicator). https://ourworldindata.org/grapher/population-density?tab=chart (Accessed March 27, 2021).

Our World in Data (OWD) (2020). Population by broad age group (indicator). https://ourworldindata.org/grapher/population-by-broad-age-group?country=~OWID_WRL (Accessed March 28, 2021).

Ozbugday, F., Tirgil, A., & Kose, E. (2020). Efficiency changes in long-term care in OECD countries: A non-parametric Malmquist Index approach. *Socio-Economic Planning Sciences*. https://doi.org/10.1016/j.seps.2019.100733 (Accessed March 3, 2021).

Pelone, F., Kringos D., Romaniello, A., Archibugi, M., Salsiri C., & Ricciardi, W. (2015). Primary care efficiency measurement using data envelopment analysis: A systematic review. *Journal of Medical Systems*, 39(1), 156. https://doi.org/10.1007/s10916-014-0156-4 (Accessed February 20, 2021).

Retzlaff-Roberts, D., Chang, C., & Rubin, R. (2004). Technical efficiency in the use of health care resources: A comparison of OECD countries. *Health Policy*, 69(1), 55–72. https://doi.org/10.1016/j.healthpol.2003.12.002.

Ritchie, H., Ortiz-Ospina, E., Beltekian, D., Mathieu, E., Hasell, J., Macdonald, B., Giattino, C., Appel, C., Rodés-Guirao, L., & Roser, M. (2020a). Overview of the coronavirus pandemic (COVID-19). https://ourworldindata.org/coronavirus-testing (Accessed February 6, 2021).

Ritchie, H., Ortiz-Ospina, E., Beltekian, D., Mathieu, E., Hasell, J., Macdonald, B., Giattino, C., Appel, C., Rodés-Guirao, L., & Roser, M. (2020b). Coronavirus (COVID-19) testing. Coronavirus (COVID-19) testing — Statistics and research — Our world in data (Accessed March 27, 2021).

Santos, S., Amado, C., & Santos, M. (2012). Assessing the efficiency of mother-to-child HIV prevention in low- and middle-income countries using data envelopment analysis. *Health Care Management Science,* 15, 206–222. https://doi.org/10.1007/s10729-012-9196-9.

Shirouyehzad, H., Jouzdani, J., & Khodadadi-Karimvand, M. (2020). Fight against COVID-19: A global efficiency evaluation based on contagion control and medical treatment. *Journal of Applied Research on Industrial Engineering*, 7(2), 109–120. https://www.researchgate.net/publication/340407134_Fight_Against_COVID-19_-_A_Global_Efficiency_Evaluation_based_on_Contagion_Control_and_Medical_Treatment (Accessed February 4, 2021).

Simar, L. & Wilson, P. (2007). Estimation and inference in two-stage, semi-parametric models of production processes. *Journal of Econometrics*, 136(1), 31–64. https://doi.org/10.1016/j.jeconom.2005.07.009.

Singh, R. & Adhikari, R. (2020). Age-structured impact of social distancing on the COVID-19 epidemic in India. (6). https://www.researchgate.net/publication/340209224_Age-structured_impact_of_social_distancing_on_the_COVID-19_epidemic_in_India_Updates_at_httpsgithubcomrajeshrinetpyross (Accessed February 11, 2021).

Sousa, D. (2022). Avaliação da Eficiência do Controlo do Contágio e do Tratamento Médico à COVID-19 em países da OCDE utilizando a Análise Envoltória de Dados. Master's Thesis in Management, University of Algarve, supervision by Martins, A.

Spinks, J. & Hollingsworth, B. (2007). Cross-country comparisons of technical efficiency of health production: A demonstration of pitfalls. *Applied Economics*, 1–11. https://doi.org/10.1080/00036840701604354.

Su, E.C.Y., Hsiao, C.H., Chen, Y.T., & Yu, S.H. (2021). An examination of COVID-19 mitigation efficiency among 23 countries. *Healthcare (Switzerland)*, 9(6). https://doi.org/10.3390/healthcare9060755.

Taherinezhad, A. & Alinezhad, A. (2022). Nations performance evaluation during SARS-CoV-2 outbreak handling via data envelopment analysis and machine learning methods. *International Journal of Systems Science: Operations and Logistics*. https://doi.org/10.1080/23302674.2021.2022243.

Tavakoli, M., Tavakkoli-Moghaddam, R., Mesbahi, R., Ghanavati-Nejad, M., & Tajally, A. (2022). Simulation of the COVID-19 patient flow and investigation of the future patient arrival using a time-series prediction model: A real-case study. *Medical and Biological Engineering and Computing*, 60(4), 969–990.

Varabyova, Y. & Müller, J. (2016). The efficiency of health care production in OECD countries: A systematic review and meta-analysis of cross-country comparisons. *Health Policy*, 120(3), 252–263. https://doi.org/10.1016/j.healthpol.2015.12.005.

Wei, Q., Zhang, J., & Zhang, X. (2000). An inverse DEA model for inputs/outputs estimate. *European Journal of Operational Research*, 121(1), 151–163. https://doi.org/10.1016/S0377-2217(99)00007-7.

Wieler, L., Rexroth, U., & Gottschalk, R. (2020). Emerging COVID-19 success story: Germany's strong enabling environment. Exemplars in Global Health. https://ourworldindata.org/covid-exemplar-germany-2020 (Accessed June 6, 2021).

World Health Organization (WHO) (2020a). International guidelines for certification and classification (coding) of COVID-19 as cause of death. https://www.who.int/docs/default-source/classification/icd/covid-19/guidelines-cause-of-death-covid-19-20200420-en.pdf (Accessed March 28, 2021).

World Health Organization (WHO) (2020b). COVID-19: Surveillance, case investigation and epidemiological protocols. WHO COVID-19 Case definition. https://www.who.int/publications/i/item/WHO-2019-nCoV-Surveillance_Case_Definition-2022.1 (Accessed March 27, 2021).

World Health Organization (WHO) (2020c). Public health criteria to adjust public health and social measures in the context of COVID-19. https://iris.who.int/bitstream/handle/10665/332073/WHO-2019-nCoV-Adjusting_PH_measures-Criteria-2020.1-eng.pdf?isAllowed=y&sequence=1 (Accessed June 13, 2021).

World Health Organization (WHO) (2020d). COVID-19: Virtual press conference. https://www.who.int/docs/default-source/coronaviruse/

transcripts/who-audio-emergencies-coronavirus-press-conference-full-30mar2020.pdf?sfvrsn=6b68bc4a_2 (Accessed June 13, 2021).

World Health Organization (WHO) (2021). Timeline: WHO's COVID-19 response. https://www.who.int/emergencies/diseases/novel-coronavirus-2019/interactive-timeline/ (Accessed February 6, 2021).

World Health Organization (WHO) (2022). Coronavirus disease (COVID-19) pandemic. https://www.who.int/europe/emergencies/situations/covid-19 (Accessed February 2, 2023).

Xu, Y., Park, Y.S. and Park, J.D. (2021). Measuring the response performance of U.S. states against COVID-19 using an integrated DEA, cart, and logistic regression approach. *Healthcare (Switzerland)*, 9(3). https://doi.org/10.3390/healthcare9030268.

Yamada, Y., Matui, T., & Sugiyama, M. (1994). New analysis of efficiency based on DEA. *Journal of the Operations Research Society of Japan*, 37(2), 158–167.

Yuen, M.K., Ngo, T., Le, T.D.Q., & Ho, T.H. (2022). The environment, social and governance (ESG) activities and profitability under COVID-19: Evidence from the global banking sector. *Journal of Economics and Development*, 24(4), 345–364. https://doi.org/10.1108/JED-08-2022-0136.

Zakowska, I. & Godycki-Cwirko, M. (2020). Data envelopment analysis applications in primary health care: A systematic review. *Family Practice*, 37(2), 147–153. https://doi.org/10.1093/fampra/cmz057.

Zanakis, S., Alvarez, C., & Li, V. (2007). Socio-economic determinants of HIV/AIDS pandemic and nations efficiencies. *European Journal of Operational Research*, 176(3), 1811–1838. https://doi.org/10.1016/j.ejor.2005.10.033.

Zhang, X.-S. & Cui, J.-C. (1999). A project evaluation system in the state economic information system of China an operations research practice in public sectors. *International Transactions in Operational Research*, 6(5), 441–452. https://doi.org/10.1111/j.1475-3995.1999. tb00166.x.

Zhang, Y., Cao, P., Meng, J., Qiu, J., Hu, Q., & Cheng, L. (2020). Exploration of the evaluation and optimization of community epidemic prevention in Wuhan based on a DEA model. *International Journal of Environmental Research and Public Health*, 17(20), 1–14.

Zhu, J. (2014). *Quantitative Models for Performance Evaluation and Benchmarking — Data Envelopment Analysis with Spreadsheets* (3rd edn.). International Series in Operations Research & Management Science (Vol. 213). Springer International Publishing. https://doi.org/10.1007/978-3-319-06647-9.

https://doi.org/10.1142/9781800615786_0006

Chapter 6

Two-Stage Data Envelopment Analysis in Healthcare: Concerns, Controversies, and Future Directions

Mehtap Çakmak Barsbay

Department of Health Management,
Faculty of Health Sciences, Ankara University,
Ankara, Turkey
mehtapcakmak@gmail.com

Abstract

Data envelopment analysis (DEA) is the most widely used non-parametric method in healthcare operation management to measure technical, productive, and allocative efficiency. As healthcare is characterized by complex production processes, we need some other subsequent techniques. Thus, integrated with DEA models, additional estimation procedures have been applied to evaluate efficiency. Although the literature on DEA is prevalent, there exists a lack of evidence in the studies using two-stage DEA in healthcare efficiency analysis. This chapter aims to review publications about two-stage DEA, which is a specific variation of conventional DEA, and to explore how two-stage DEA procedures are prevalent in healthcare. Investigating the state of the art of two-stage DEA models can add value for researchers who plan to conduct research using DEA. This chapter offers a rapid review and bibliometric analysis to explore publications regarding the relevant topic. The number of publications reached a peak in 2021. Review articles focused on various healthcare specialties. Seventeen articles were related to hospitals and healthcare centers, and tobit regression remained the primary choice of analysis for the dependent variable in 11 articles. It was widely used across various units, such as health regions, health systems, and patient-level treatment. Some concerns and controversies were addressed to improve validation and prove practical usefulness.

Keywords: Bibliometric review, healthcare efficiency, data envelopment analysis, literature survey, efficiency measurement, performance evaluation, environmental variables, organizational determinants, intermediate factors.

1. Introduction

Efficiency in healthcare is applied to determine how well a healthcare unit is transforming resources into better health- and non-health-related outcomes. Thus, efficiency measurement is key for value-based healthcare. However, it is not deterministic, simply because healthcare efficiency depends on more than inputs and outputs. Moreover, in order to accurately estimate efficiency scores, exogenous/environmental factors, contextual/organizational determinants, or both, are essential. These variables can potentially provide additional information on healthcare services' performance.

Exogenous or environmental variables, such as demographics, socioeconomic status, and cultural factors, are key elements beyond a production system. They may have an impact on health- and non-health-related outcomes. On the other hand, contextual variables, such as hospital size and location, are determinants within a healthcare system that may affect care delivery. Without considering these variables, healthcare providers could be unfairly compared. By accounting for these determinants, benchmarking can provide a more accurate comparison of healthcare providers' performance and help identify areas for improvement in patient care. Thanks to these exogenous and contextual factors, healthcare efficiency estimation can yield more accurate results, creating value for decision-makers, practitioners, employees, and patients.

This chapter aims to conduct a bibliometric and rapid systematic review to explore how two-stage data envelopment analysis (DEA) models are prevalent in the health and healthcare sectors. In this context, this chapter is motivated in several ways. First, it provides a general overview of efficiency measurement focusing on DEA models, and then, it introduces two-stage DEA estimation procedures applied to healthcare. Building on a rapid review and bibliometric analysis, the final section presents the characteristics of decision-making units (DMUs), inputs, outputs, and environmental and/or contextual factors in the reviewed publications. In the discussion, the concerns and controversies of two-stage DEA in healthcare are discussed, and in conclusion, future directions for researchers are presented.

2. Basics of Efficiency Measurement with DEA

Healthcare institutions' efficiency can be measured using either parametric or non-parametric methods. While stochastic frontier analysis is the most commonly used parametric approach, it is characterized by the main assumption that it is impossible to fully specify a production function (Nepomuceno *et al.*, 2022). As a non-parametric method, DEA is also widely used in healthcare operations management.

Since its inception in 1978 by Charnes *et al.*, DEA has been a trendy tool for efficiency measurement. DEA defines the best practice boundary or data envelope as the most appropriate input/output combination (Charnes *et al.*, 1994). In other words, it assumes that not all organizations are efficient. Depending on multiple inputs and outputs, best-practicing units create a production function frontier that allows other units for further comparisons.

DEA assumes a positive correlation between variables, in which input and output processes are specific in production modeling, which is called *the black box* (Färe & Grosskopf, 2000). It presents two technical efficiency assessment methods: variable returns to scale (VRS) using the Banker, Charnes, and Cooper (BCC) model, and constant returns to scale (CRS) using the Charnes, Cooper, and Rhodes (CCR) model.

Either VRS or CRS can be preferred, depending on the research aim. If a researcher aims to evaluate productive efficiency rather than scale efficiency, CRS can be a good choice. However, if a researcher aims to evaluate scale efficiency rather than productive efficiency, VRS should be preferred. In other words, the VRS model measures the efficiency of DMUs that do not operate at optimal scale, while the CRS model is suitable for optimal scale.

Technical efficiency is a widely used efficiency evaluation approach for healthcare providers (Kohl *et al.*, 2019). In the case where a model is determined to be input-oriented, the aim is to keep the current output level constant and minimize input level. In the case of an output-oriented DEA model, it is aimed at increasing outputs while keeping inputs constant. In both DEA models, efficiency measurement can be based on input minimization or output maximization (Çakmak Barsbay, 2021).

In the BCC model, the productivity frontier does not start from the origin to provide variable returns to scale. This ensures that efficiency limit in the BCC model has a concave form. The model gains convexity by adding the constraint that lambda (λ) equals 1. In the CCR model, the efficiency limit is linear. An efficient DMU according to the BCC model may not be efficient according to the CCR model. However, efficient DMUs in the CCR model will also be efficient in the BCC model. This is an expected result of the models. The input-oriented BCC model's mathematical Equations (1)–(5) are given as follows. These equations signify that s outputs are obtained using m inputs belonging to n DMUs (Çakmak Barsbay, 2021):

$$\min Q - E\left(\sum_{i=1}^{m} s_i^- + \sum_{r=1}^{s} s_r^+\right). \tag{1}$$

$$\sum_{j=1}^{n} x_{ij}\lambda_j - Q_{xi0} + s_i^- = 0. \tag{2}$$

$$\sum_{j=1}^{n} y_{rj}\lambda_j - y_{r0} - s_r^+ = 0. \tag{3}$$

$$\sum_{j=1}^{n} \lambda_j = 1, \ \lambda \text{ ve } s_i^-, \ s_r^+ \geq 0. \tag{4}$$

$$i = 1, \ldots, m, \quad r = 1, 2, \ldots, s, \quad j = 1, 2, \ldots, n. \tag{5}$$

2.1. *Overview of DEA studies and its application in healthcare*

A quick search in the Web of Science database shows that, until the first quarter of 2023, DEA studies had accumulated 7,710 scientific articles across many research domains, from business economics to zoology, compared with 889 stochastic frontier analysis studies. Several authors have surveyed the DEA literature in different periods, research domains, and country-specific studies. In a literature survey, Liu *et al.* (2013) identified the most active DEA subdomain during 1978–2010, and they found that two-stage contextual factor evaluation was prominent. In addition, country-specific healthcare efficiency studies originated mainly in Germany and the United Kingdom (UK). While Vrabkova & Lee (2023) presented an overview of the empirical studies on the efficiency of hospitals in Germany, Thai & Noguchi (2019) conducted research to evaluate efficiency in UK hospitals. Kohl *et al.* (2019) attempted to analyze this topic through a literature study. They found that among 262 papers, the majority of DEA research (100 papers) was motivated by examining the association between efficiency and other salient factors, such as quality, ownership type, specialization, and regulations, while others (99 papers) simply conducted pure DEA efficiency analysis. The remaining 48 papers focused on new advanced DEA models, such as the Malmquist index (MPI), and slack-based super efficiency, while 36 papers were on program evaluation using DEA procedures.

Meanwhile, growing evidence has shown that there may not be a relationship between inputs and outputs. There are complex production processes that can be separated into more than one phase. Moreover, DMUs can reduce undesirable outputs with a minus sign (Kordrostami & Amirteimoori, 2005). Although traditional CCR and BCC are still the most preferred DEA methods, bootstrap DEA, MPI, two-stage DEA, and network models are also common in

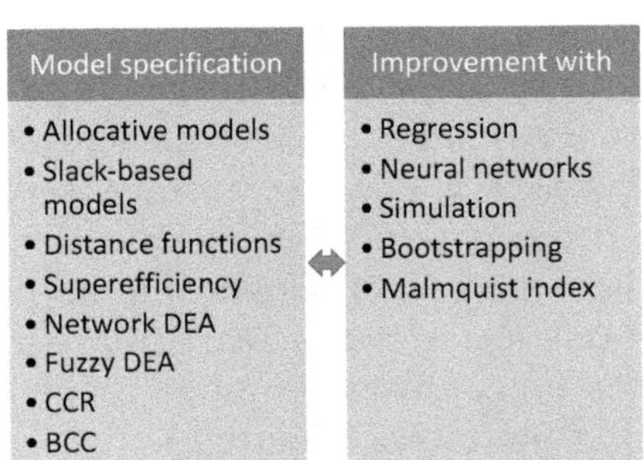

Figure 1. DEA models.

healthcare efficiency measurement (Nepomuceno *et al.*, 2022; Kohl *et al.*, 2019; Ngo & Tsui, 2022; Boubaker *et al.*, 2022). As depicted in Figure 1, DEA models can be integrated with many subsequent techniques.

2.2. *Two-stage data envelopment analysis*

Two-stage DEA is an extension of traditional (radial/basic) DEA methods. Energy, banking, education, transportation, the insurance sector, and healthcare are the most popular domains for efficiency measurement with two-stage DEA (Nepomuceno *et al.*, 2022; Emrouznejad & Yang, 2018; Dobos *et al.*, 2023). Two-stage DEA is often used in process management, where there are multiple stages (subsystems) involved. In other words, there exist more than one separable but networked stages in a process. Halkos *et al.* (2014, p. 1) defined four types of two-stage DEA procedures: "connected, relational, game theoretic and independent." Connected, relational, and game-theoretic two-stage DEA procedures measure the efficiency of a production process involving multiple stages or processes. However, independent two-stage DEA is different. In order to clarify the differences, the former is called "two-stage network DEA," whereas the latter is simply called "two-stage DEA."

In two-stage network DEA, relevant input and output factors are called internal/intermediate variables (x and y) (see Figure 2). Intermediate (by)products (z) are considered outputs from the first stage and inputs to the second stage (Maleki *et al.*, 2019; Liu *et al.*, 2015). In a two-stage network DEA, DMUs have separate but linked subsystems (see Figure 2).

The first stage evaluates the efficiency of subsystem 1 in producing intermediate outputs (z), and the second stage evaluates the efficiency of subsystem 2 assuming the intermediate outputs of the first stage as inputs. In other words, a two-stage network DEA model typically involves two separate DEA models, one for each subsystem, linked together using intermediate inputs or outputs (z). In this regard, in a two-stage network DEA, a process is divided into two stages: The first stage involves input suppliers, and the second stage involves output producers (Figure 3).

The first-stage suppliers provide inputs to the second-stage producers, who then transform these inputs into outputs. Two-stage

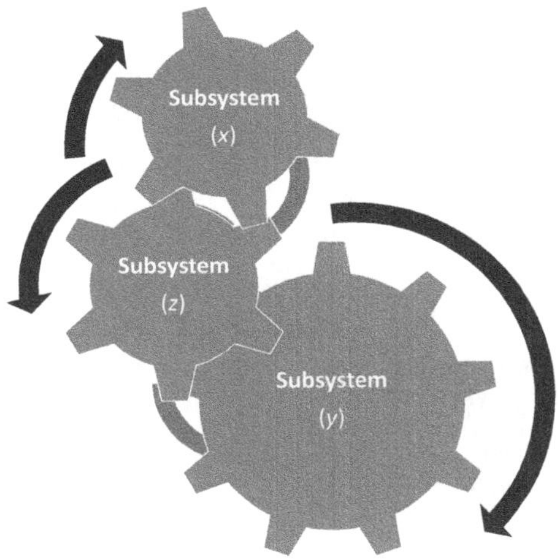

Figure 2. Production phase in two-stage DEA models.

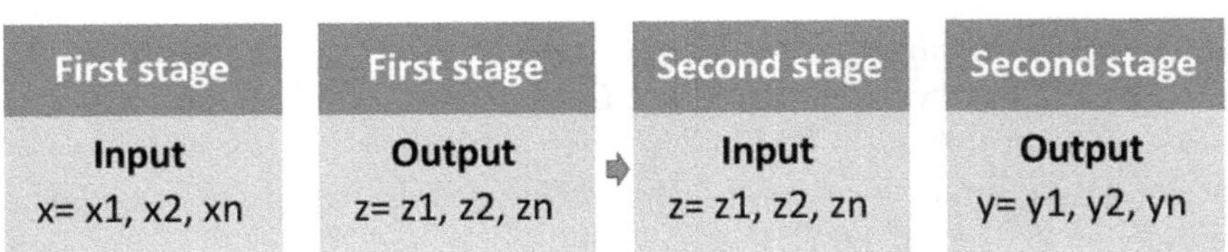

Figure 3. Production phases in a two-stage network DEA model.

network DEA aims to determine the relative efficiency of each stage in a single process as well as the overall efficiency. This is done by comparing the inputs and outputs of each stage to the inputs and outputs of other stages using a mathematical model to calculate efficiency scores. Two-stage network DEA is a useful tool for identifying inefficiencies in a process in terms of the six-sigma approach and for optimizing performance of multi-stage processes. In a study by Cohen-Kadosh & Sinuany-Stern (2020) on 22 orthopedic wards of acute Israeli hospitals, fracture type and Charlson index were used as inputs (x), whereas surgery within 48 h and usage of a drain for 1 day (rate) (z) were the outputs of the treatment process (A). Meanwhile, z was used once again as inputs for process B, and survival rate after surgery and a lower level of infection rate were used as success outputs (Figure 4). Furthermore, the socioeconomic index of patients

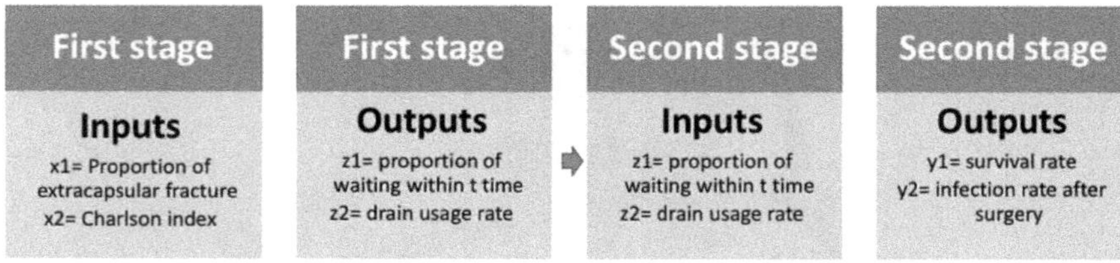

Figure 4. Production phases in a two-stage network model.

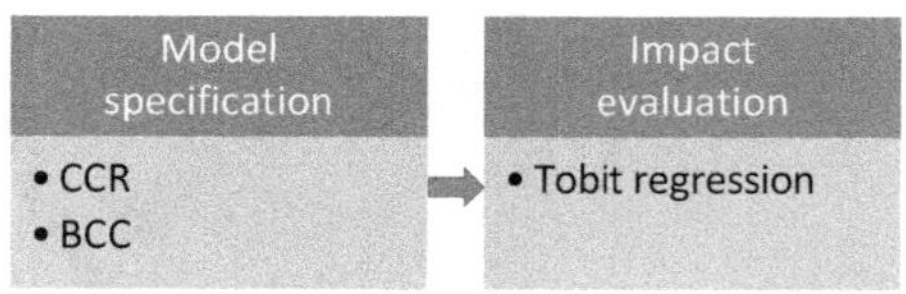

Figure 5. Production phases in a two-stage DEA model.

and the geographical periphery status of a hospital were adopted as environmental variables.

The second type of two-stage DEA is a method to explore possible determinants of DMUs' relative efficiency. A (external) two-stage DEA analysis consists of two stages: In the first stage, the efficiency scores of DMUs are calculated using DEA. This involves identifying the most efficient DMUs and determining the relative efficiency among the DMUs. In two-stage DEA, the second stage is different from that of traditional DEA methods and the two-stage network DEA procedure (Figure 4). Two-stage DEA estimation procedures can involve exogenous variables. They are typically external to the DMUs and may include environmental, economic, or social factors. These factors are called contextual and environmental determinants of efficiency. In the second stage, relevant factors that contribute to DMUs' inefficiencies are identified using regression analysis (Johnson & Kuosmanen, 2012; Vrabkova & Lee, 2023). For instance, Vrabkova & Lee (2023) used a two-step approach in their study. They applied an input-oriented DEA model with CRS and VRS for a six-year period from 2015 to 2020. In the second stage, impact evaluation was conducted using tobit regression analysis (Figure 5). They concluded that the change in efficiency score could be explained by those exogenous factors. As a conclusion, two-stage DEA analysis is a validated approach to providing a more comprehensive evaluation of DMUs by considering both their internal performance (efficiency scores) and external factors that may affect their overall performance. It presents results closer to reality, as a production process is not limited to interactions among inputs and outputs (Henriques *et al.*, 2020).

3. Materials and Methods

We conducted a rapid review in November 2022 to understand how two-stage DEA is performed in healthcare.

3.1. *Search strategy*

A search query (https://www.webofscience.com/wos/woscc/summary/168ae6f8-6a0d-46c3-8dd2-166ce2a7d4b6-7d74c2f4/relevance/1) is built on the bibliometric database Web of Science (WoS). WoS is chosen as a valid database to obtain sufficient information regarding the study's main concepts and keywords, as it is the most widely used and trusted database for quantitative analysis of scientific publications (van Nunen *et al.*, 2018). The WoS core collection includes a high-quality collection of journals, books, and conference proceedings (Clarivate, 2021). As inclusion criteria, the terms "health or hospital" (all fields) and "two-stage data envelopment analysis" (topic) are used to identify relevant publications in the database. No time span is applied. Peer-reviewed journal articles are included. "All languages" criterion is applied. After data extraction, 24 articles are included for analysis.

3.2. *Data analysis*

Bibliometric analysis is used to explore publications regarding a relevant topic. Bibliometric analysis is a technique used to reveal the chronology of scientific literature, publication activity, and trend topics, as well as to create a roadmap for future research, transforming publications into quantitative data (Donthu *et al.*, 2021;

Aria & Cuccurullo, 2017). Microsoft Excel is used to sort and perform classification procedures. Data management is conducted using the R package Bibliometrix, version 4.2.2. Additionally, statistical data in the WoS database are used for descriptive analyses.

4. Results

A total of 35 articles were initially identified, of which 11 were excluded at the first screening step, as they were not related to healthcare. In other words, 24 publications met the inclusion criteria and were selected for a full-text review. As depicted in Figure 6, the number of publications has been increasing since 2006. There was only one publication in 2006. During 2007–2010, no papers on two-stage DEA in healthcare were published. During the 2011–2016 period, publication activity started to increase, and the past five years have been the most productive period. Publication activity is expressed by the number of scientific papers. It is an important indicator to understand the development of a research topic over a time span. From 2018 on, an increasing number of publications can be observed every year. The annual growth rate is 9.05%. A peak of publication activity is observed in 2021.

Recent reviews by Nepomuceno *et al.* (2022) and Emrouznejad & Yang (2018) revealed that three time periods were important for DEA studies. The post-2004 period is coined an *exponential growth stage* of DEA research. During this period, publication activity was about 680 per year, with a peak of 1,000 papers (Emrouznejad & Yang, 2018, p. 5). The coming years will provide evidence on the

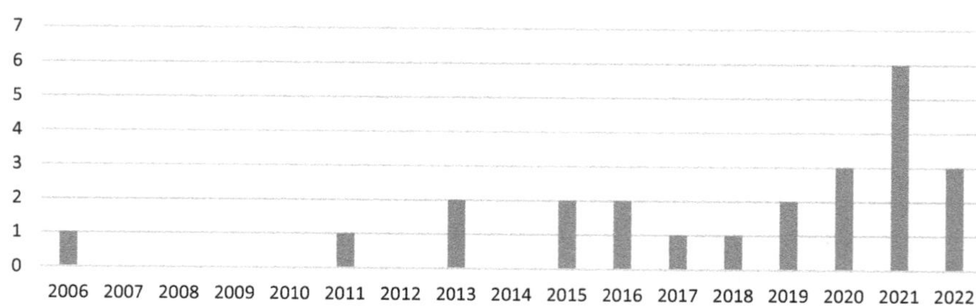

Figure 6. Number of publications in the 2006–2022 period.

prognosis of multi-stage DEA studies; nevertheless, this trend currently seems to be in a *precursor phase*, characterized by a small body of knowledge (van Nunen *et al.*, 2018).

Trend topic analysis is one of the most useful methods for discovering trends and emerging topics. It is based on the keywords, titles, and abstracts in the publications. The prominent trend topic in this research is technical efficiency ($n = 49$, 42%), followed by healthcare ($n = 19, 16\%$) (Figure 7).

Each journal in WoS is assigned to at least one subject category, which is a particular field of research. WoS consists of approximately 250 subject categories. All articles published in a journal inherit the journal's assignment of subject categories (van Nunen *et al.*, 2018). As depicted in Figure 8, 16 research areas were assigned to 28 articles

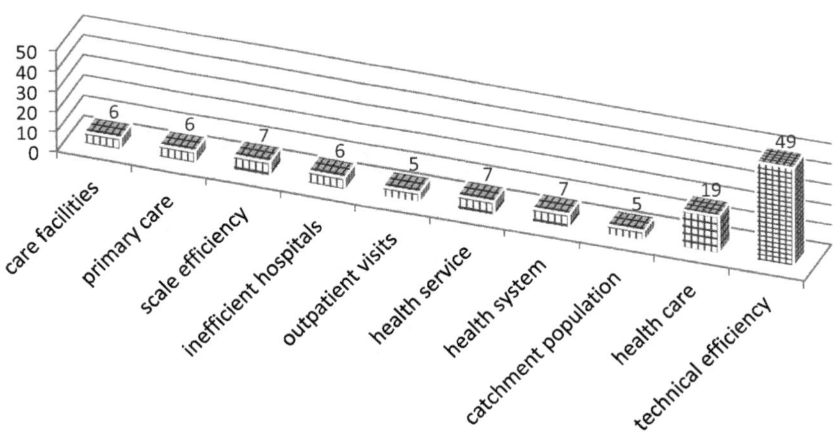

Figure 7. Trend topics.

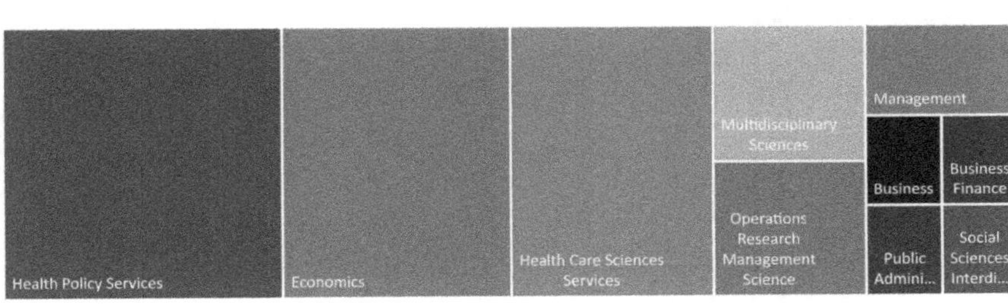

Figure 8. Research areas.

by WoS. The five most commonly assigned research areas are "Health Policy Services" ($n = 11$), "Economics" ($n = 8$), "Health Care Sciences Services" ($n = 8$), "Multidisciplinary Sciences," and "Operations Research Management Science."

The reviewed publications represented six clusters. Similar articles in terms of content were gathered in the same cluster. Given that two articles are cited together by a third article, a relationship emerges between them as they are co-cited. The size of the circles shown in Figure 9 represents the number of citations: the larger a circle, the more citations from a publication. A shorter distance between

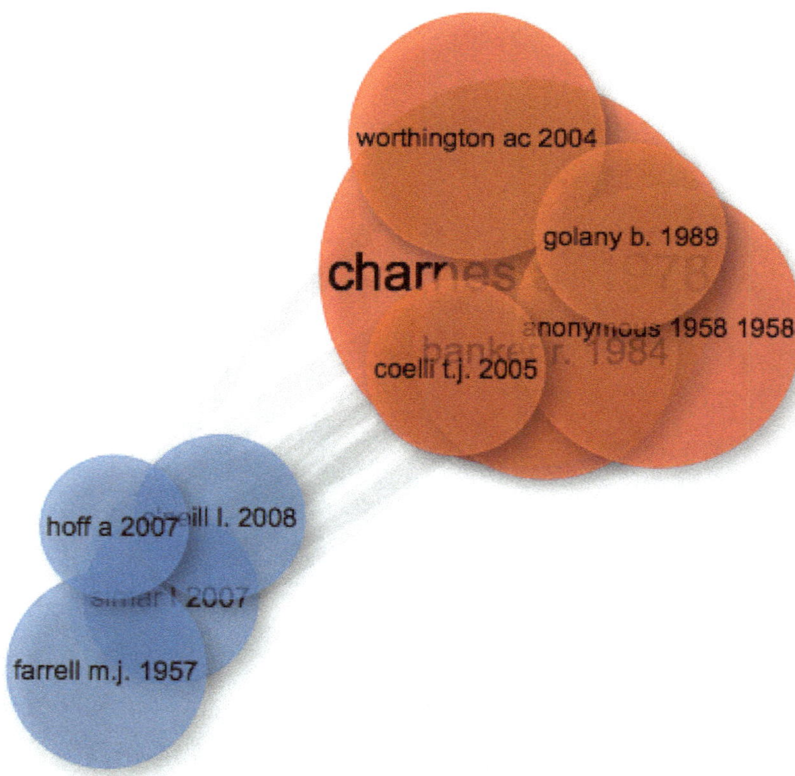

Figure 9. Co-citation network.

the two arcs indicates a stronger relationship and greater similarity between them. The same facet suggests a similar topic between these publications (van Nunen *et al.*, 2018).

The reviewed articles focused on various specialties in healthcare. The maximum number of reports ($n = 17$) were related to hospitals and healthcare centers, while in six articles, district-, region-, or province-level health units were defined as DMUs. The results of the reviewed articles that used two-stage DEA are summarized in Table 1. Ten of the 24 articles used panel data. Tobit regression remains the choice of analysis for the dependent variable in 11 articles.

Most of the DEA researchers measured the pure and total technical or scale efficiency of specific healthcare organizations, or only selected silos, such as clinics and wards. The majority of the researchers used the tobit model in the second stage to estimate explanatory factors (Yitbarek *et al.*, 2019a, 2019b, 2021; Sahin *et al.*, 2021; Ramalho *et al.*, 2022; Lamesgen *et al.*, 2022; Kirigia & Asbu, 2013; Ilgun *et al.*, 2022; Ilgun & Sahin, 2022; Cavalieri *et al.*, 2018; Amare *et al.*, 2020). Furthermore, a variety of regression techniques were used in different research domains, including classic linear regression (Sahin *et al.*, 2021; Allin *et al.*, 2016), multiple regression (Abeney & Yu, 2015), bootstrap truncated regression (Thai & Noguchi, 2020; Heil *et al.*, 2021; Cavalieri, 2017; Barasa *et al.*, 2021), logistic regression (Du *et al.*, 2021), truncated regression (Marschall & Flessa, 2011; Cavalieri *et al.*, 2018; Adang *et al.*, 2016), ordinary least squares (OLS) (Cavalieri *et al.*, 2018; Karagiannis, 2015; Besstremyannaya, 2013), path analysis (Ho & Huang, 2020), and DEA (Yang *et al.*, 2021).

In traditional efficiency measurement methods, exogenous and contextual variables are not included in the set of inputs and outputs. While conventional (basic) DEA models are a valuable tool for evaluating healthcare efficiency, two-stage DEA offers further analysis using environmental and contextual determinants. In particular, two-stage DEA estimation procedures have been widely implemented with bias correction to account for the influence of related factors not directly controlled by healthcare providers (Jung *et al.*, 2023).

Table 1. Results of the reviewed two-stage DEA studies in healthcare.

Author and year	**DMUs**	**Measurement parameters and estimation model in the second stage**
Abeney & Yu (2015)	10 Canadian provinces	**Inputs:** Capital, labor, energy, materials, and services **Outputs:** Waiting time and number of cases treated **Environmental variables:** Incidence of arthritis, high blood pressure, current smoking, healthy eating, post-secondary education, regular access to a doctor, influenza immunization, asthma, physical inactivity, age over 65, obesity **Two-stage:** Multiple regression analysis **Panel data:** 2001–2008
Adang *et al.* (2016)	106 GPs and 1,671 patients in the Netherlands	**Inputs:** Average number of consultations with a GP, average number of consultations with a practice nurse, and average number of consultations with a practice assistant **Outputs:** Statins and anticoagulants prescription according to the relevant guidelines, LDL cholesterol, and systolic blood pressure **Environmental variables:** Practice size, training practice, and task practice nurse **Two-stage:** Truncated regression analysis **Panel data:** Not available
Allin *et al.* (2016)	89 health regions in Canada	**Inputs:** Spending per capita (hospitals, prescription drugs, physicians, residential care facilities, and community nurses) **Outputs:** Age-standardized treatable potential years of life lost, survival rate, and mortality rate **Environmental variables:** Education, non-aboriginal recent immigrants **Two-stage:** Log-linear regression analysis **Panel data:** 2007–2009
Amare *et al.* (2020)	20 public hospitals in Ethiopia	**Inputs:** Salary expenditure, non-salary expenditure, and the number of beds **Outputs:** Antenatal care, skilled delivery, and postnatal care **Environmental and organizational variables:** Total catchment population, availability of nearby health facilities and distance of the facility from the capital city, managerial service years of the chief executive officer (CEO), CEO's educational status, distance of the CEO's residence from the facility, additional benefit packages for the employees, level of facility, service year of facility, and patient waiting time

Table 1. (*Continued*)

Author and year	DMUs	Measurement parameters and estimation model in the second stage
		Two-stage: Tobit regression analysis **Panel data:** Not available
Barasa *et al.* (2021)	47 counties in Kenya	**Inputs:** Public county health expenditure, private county health expenditure, and number of healthcare facilities **Outputs:** Disability-adjusted life years **Environmental variables:** Level of corruption, HIV burden, development budget absorption, recurrent budget absorption, alcohol consumption, level of literacy, quality of care, population density, country economic performance, private health facilities as a proportion of all health facilities in the county, proportion of absenteeism of health workers, absenteeism, proportion of households that with clean water, and proportion of the population that smokes **Two-stage:** Bootstrapped truncated regression analysis **Panel data:** Not available
Besstremyannaya (2013)	832 hospitals in Japan	**Inputs:** Number of doctors, total number of nurses, number of other hospital personnel, number of beds, and labor cost **Outputs:** Discharges and outpatients **Environmental variables:** Government subsidy per bed, affiliation status, emergency beds out of total beds, rural or urban, non-profitable status, accreditation, and reimbursed drug rate **Two-stage:** Least-squares regression analysis with logit transformation **Panel data:** 1999–2009
Cavalieri *et al.* (2018)	492 hospitals in Italy	**Inputs:** Beds, physicians, nurses, and other personnel **Outputs:** Inpatient days, discharged patients, and weighted discharged patients **Environmental variables:** Adoption of financing mechanisms (DRG, global budget, etc.) **Two-stage:** OLS, truncated, and tobit regression analyses **Panel data:** 1999–2010
Cavalieri (2017)	405 hospital infrastructure contracts in Italy	**Inputs:** Actual time of infrastructure completion (days) and actual cost of infrastructure completion **Outputs:** Expected duration (days) and value of the winning bid

(*Continued*)

Table 1. (*Continued*)

Author and year	DMUs	Measurement parameters and estimation model in the second stage
		Environmental variables: Corruption index at provincial level, crimes against public administration per 100,000 inhabitants at provincial level **Two-stage:** Bootstrap truncated regression analysis **Panel data:** 2000–2005
Du *et al.* (2021)	30 pension institutions in China	**Inputs:** Total area, number of beds, medical qualification, total agency services, professional managers, medical staff, other personnel, training, investment in fixed assets, operating costs, government subsidy amount, complete service programming, scientific service process, completeness of service files, and types of services provided **Outputs:** Number of people with long-term medical care insurance, total occupants, operating income, and safety management **Environmental variables:** Number of people with long-term medical care insurance, total occupants, operating income, and safety management **Two-stage:** Binary logistic regression analysis **Panel data:** Not available
Heil *et al.* (2021)	25 Dutch hospitals	**Inputs:** Preoperative involvement of physical therapists, dieticians, and geriatricians **Outputs:** Severe postoperative complications **Environmental variables:** Practice size and hospital type **Two-stage:** Bootstrap truncated regression analysis **Panel data:** Not available
Ho & Huang (2020)	47 hospitals in Washington and California	**Inputs:** Advertisement image influence **Outputs:** Net patient revenue, satisfaction with overall care and service, and referrals (friends/family) **Environmental-intermediate variables:** Total assets, total operating expenses, and total beds in service for the facility **Two-stage:** Path analysis **Panel data:** Not available

Table 1. (*Continued*)

Author and year	**DMUs**	**Measurement parameters and estimation model in the second stage**
Ilgun & Sahin (2022)	81 provinces in Turkey	**Inputs:** Primary healthcare budget allowance per capita, number of physicians, number of nurses and midwives, and number of other health personnel **Outputs:** Number of visits to primary healthcare services per capita, average number of follow-ups per pregnant woman, average number of follow-ups per infant, and infant mortality rate **Environmental-control variables:** Income per capita and university graduates **Two-stage:** Tobit regression analysis **Panel data:** 2012–2014
Ilgun *et al.* (2022)	17 maternity and child hospitals in Turkey	**Inputs:** Number of beds and number of physicians (practitioners + specialists) **Outputs:** Number of outpatient clinics, number of inpatients, number of operations (Type A + Type B + Type C), and number of deliveries (normal deliveries + cesarean deliveries) **Control variables:** Number and status of beds and the provinces where the beds are available **Two-stage:** Tobit regression analysis **Panel data:** 2014–2017
Karagiannis (2015)	53 public hospitals in Greece	**Inputs:** Number of doctors, number of nursing personnel, number of administrative and other personnel, and number of beds **Outputs:** Number of inpatient days and number of laboratory tests **Contextual variables:** Location dummy, size dummy, specialization dummy, average length of stay, number of doctors per patient, number of nurses per patient, and percentage of population up to 65 years old **Two-stage:** Single-equation (OLS) and system-of-equations (SUR) models **Panel data:** Not available
Kirigia & Asbu (2013)	19 public secondary-level community hospitals in Eritrea	**Inputs:** Number of physicians (doctors), number of nurses and midwives, number of laboratory technicians, and number of operational beds and costs

(*Continued*)

Table 1. (*Continued*)

Author and year	**DMUs**	**Measurement parameters and estimation model in the second stage**
		Outputs: Number of outpatient department visits and number of inpatient department discharges **Environmental variables:** Outpatient visits, region population, and hospital location **Two-stage:** Tobit regression analysis **Panel data:** Not available
Lamesgen *et al.* (2022)	15 primary hospitals in Ethiopia	**Inputs:** Human resources (salary of employees), capital (number of beds to offer neonatal health services), and non-salary recurrent expenditures **Outputs:** Number of neonatal outpatient visits, neonatal inpatient admissions, and neonatal referrals **Environmental and organizational variables:** Catchment population, availability of nearby health facilities that offer neonatal health services, distance of the manager's residence from the hospital, service year of the hospital, manager's educational status, manager's service year, availability of incentive packages for the clinical staff, and patient waiting time **Two-stage:** Tobit regression analysis **Panel data:** Not available
Marschall & Flessa (2011)	24 primary care facilities in Northwestern Burkina Faso	**Inputs:** Personnel costs, building area, depreciation of equipment, and vaccination costs **Outputs:** General consultation and nursing care, deliveries, immunization, and special services, e.g., family planning and prenatal and postnatal consultations **Environmental variables:** Value of animal ownership, value of durable household goods, ethnic groups, and religion **Two-stage:** Censored and truncated regression analysis **Panel data:** Not available
Ramalho *et al.* (2022)	55 primary care health centers in Portugal	**Input variables:** Number of doctors, number of nurses, and number of intern doctors **Output variables:** Short-term diabetic complications admission rate, long-term diabetic complications admission rate, uncontrolled diabetes admission rate, rate of lower-extremity amputation among patients with diabetes, and diabetes prevention quality composite

Table 1. (*Continued*)

Author and year	**DMUs**	**Measurement parameters and estimation model in the second stage**
		Environmental variables: Patients per medical doctor, proportion of elderly patients, unemployment rate, prevalence of DM, proportion of patients with DM and record of hemoglobin A1c (HbA1c) in the past six months, proportion of patients with DM and with the last HbA1c ≤8.0%, proportion of patients with DM <65 years old with a HbA1c test ≤6.5%, proportion of patients with DM and with a microalbuminuria test in the past year, proportion of patients with a risk assessment for DM2 (three years), cost of therapy for diabetic patients, and cost of therapy for diabetic patients under control **Two-stage:** Tobit regression analysis **Panel data:** Not available
Sahin *et al.* (2021)	Hospitals affiliated to the Ministry of Health at 81 provinces in Turkey	**Inputs:** Numbers of hospitals, beds, physicians, nurses, midwives, and other health personnel **Outputs:** Number of outpatients, proportion of inpatients, proportion of surgeries, and risk-adjusted crude mortality rate **Environmental-control variables:** Per capita income of provinces, population, proportion of university graduates, and metropolitan status **Two-stage:** Tobit regression analysis and linear regression analysis **Panel data:** 2010–2015
Thai & Noguchi (2020)	84 New Zealand district health boards	**Inputs:** Expenditure **Outputs:** Volume of inpatient activities, volume of outpatient activities, emergency, volume of elective surgeries, the number of patients supported to quit smoking, number of eight-month infants immunized, average life expectancy at birth **Explanatory variables:** Net value of building, net value of information technology, proportions of discharged Maori patients, performance in the target to treat patients within six hours since presentation at the emergency department, variance between actual expenditures and plan, variances between actual and plan volume of outpatients, variances between actual and plan volume of inpatients **Two-stage:** Double bootstrap truncated regression analysis **Panel data:** 2013/2014–2016/2017

(*Continued*)

Table 1. (*Continued*)

Author and year	**DMUs**	**Measurement parameters and estimation model in the second stage**
Yang *et al.* (2021)	18 medical districts in Taiwan	**Inputs:** Total number of beds, physicians, and nursing staff **Outputs:** Average relative value unit of each outpatient case and average relative value unit of each inpatient case **Intermediate variables:** Outpatient visits, emergency room visits, and inpatient days **Two-stage:** DEA **Panel data:** 2003–2009
Yitbarek *et al.* (2019b)	68 health posts and 23 health centers in Southwest Ethiopia	**Inputs:** Non-salary recurrent expenses (expenses for vaccines, drugs, and supplies), administrative staff, clinical and midwife nurses, laboratory technicians and technologists, pharmacy technicians and pharmacists, and health officers **Outputs:** Number of neonatal outpatients, neonatal referrals, and those who received service at maternal and child health units **Environmental and organizational variables:** Service years of the facility, facility head's years of experience, size of the catchment population, availability of nearby health facilities, facility head's level of education, neonates in the catchment population, and waiting time for service **Two-stage:** Tobit regression analysis **Panel data:** Not available
Yitbarek *et al.* (2021)	72 health posts in Southwest Ethiopia	**Inputs:** Number of health extension workers and non-salary recurrent expenses (expenses for supplies, drugs, and vaccines) **Outputs:** Number of health education sessions, antenatal care, family planning services, child immunization, treated diarrhea cases, treated malaria cases, number of households' visits, and number of referrals **Environmental-organizational variables:** Service years, size of the catchment populations, availability of health facilities around, and average waiting time for service **Two-stage:** Tobit regression analysis **Panel data:** Not available

Table 1. (*Continued*)

Author and year	DMUs	Measurement parameters and estimation model in the second stage
Yitbarek *et al.* (2019a)	14 public maternity hospitals in Ethiopia	**Inputs:** Salary for health professionals and other administrative staff **Outputs:** Numbers of antenatal care visits, delivery, postnatal care, family planning, and abortion and post-abortion services **Environmental-organizational variables:** Number of years since the hospital started service delivery, level of the hospital (secondary or primary), qualification of the CEO of the hospital, average waiting time for maternal health services, benefit packages for the employees of the hospital, catchment population, availability of nearby public health facilities that provide maternal and reproductive health services, location of the hospital (urban or rural) **Two-stage:** Tobit regression analysis **Panel data:** Not available

5. Discussion

This chapter presents a macroscopic evaluation of research trends in two-stage DEA models in the healthcare industry up to 2022. Even though two-stage DEA models in healthcare have reached a saturation point over the past five years, publication output remains unstable. The coming years will provide sufficient evidence on publication activity to determine if it will remain at a standstill or be characterized by an increasing or declining trend.

This rapid review of two-stage DEA in healthcare provides evidence that it is a widely used technique in healthcare studies. Two-stage DEA involves a two-step approach for evaluating the efficiency of healthcare systems or providers, hospitals, and patient outcomes. Abeney & Yu (2015) highlighted that there were no differences in the efficiencies of DMUs since they adopted good practices. Therefore, in this case, exogenous factors did not explain differences between efficiency scores. In Adang *et al.* (2016)'s study, as a different type of DMU, the management of cardiovascular risk among patients was used. They found that variation in technical efficiency between primary care practices was associated with training practice status and whether clinical tasks for cardiovascular risk management were performed by a practice nurse or a general practitioner (GP) did not influence technical efficiency in a statistically significant way, nor did practice size. Amare *et al.* (2020) performed a two-stage DEA in which the tobit regression model was used in the second stage to identify potential factors affecting public hospitals' technical efficiency in providing maternal health services.

Two-stage DEA is a valid technique to measure the relative efficiency of DMUs with multiple production or service delivery stages. While it has been widely used in several decision-making units, including health regions, health systems, and patient-level treatment, some concerns and controversies exist surrounding its interpretation that need to be addressed to improve its validity and practical usefulness.

5.1. *Concerns and controversies*

(i) *Choice of variables*: There are controversies over which input and output factors should be included in two-stage DEA models. The choice of relevant variables can be subjective due to data (un)availability. Unless relevant data are accessible to researchers, they may tend to exclude many of them. Thus, the choice of inputs and outputs should be scientifically unbiased and accurate.

(ii) *Environmental variables*: Researchers may use different sets of exogenous variables which produce different results. The inclusion of less relevant exogenous variables may decrease estimations' predictive power. Thus, variables which exhibit cause-and-effect relationships with operational performance should be incorporated into the dataset.

(iii) *Characteristics of DMUs*: Another controversy is the choice of DMUs in two-stage DEA. Researchers may include different DMUs depending on their research question or data availability. The characteristics of DMUs may directly affect the results. An inadequate number of DMUs is another issue that researchers need to deal with.

(iv) *Model choice*: One of the main concerns with two-stage DEA is the model specification in the second stage of the estimation procedure. The choice of estimation model in the second stage may have a significant impact on the final results. Thus, model choice should be evidence-based.

(v) *Data quality*: Since healthcare data can be incomplete, inaccurate, or too complex to (dis)aggregate, they may lead to biased, unvalidated, or unreliable results.
Researchers use different types of datasets and quality assumptions, which may produce some controversial results. Thus, data quality is important, and some quality appraisal methods should be incorporated into DEA studies.

(vi) *Evidence-based decision-making*: Relying on the results of a two-stage DEA can also be challenging, especially for policymakers or managers of healthcare organizations. They may need a deeper understanding of efficiency calculation methodology. Thus, there may be a risk of misinterpreting results, leading to inaccurate evidence for decision-making.

6. Conclusion

This chapter highlights that, to estimate healthcare efficiency, not only inputs and outputs are required, but environmental factors and contextual elements are also important. Thus, future research may focus on developing more comprehensive models that can better capture healthcare delivery systems' complexities, such as hybrid DEA incorporating machine learning and artificial intelligence. This can help improve the accuracy and reliability of results, especially when dealing with complex or big data in healthcare. Using multi-criteria decision-making approaches in two-stage DEA may be another future direction. This may help researchers account for multiple objectives and criteria that healthcare systems or providers need to balance, such as the triple aim of health: cost, quality and value, and access.

References

Abeney, A. & Yu, K. (2015). Measuring the efficiency of the Canadian health care system. *Canadian Public Policy*, 41(4), 320–331.

Adang, E.M.M., Gerritsma, A., Nouwens, E., van Lieshout, J., & Wensing, M. (2015). Efficiency of the implementation of cardiovascular risk management in primary care practices: An observational study. *Implementation Science*, 11, 67. https://doi.org/10.1186/s13012-016-0434-2.

Allin, S., Grignon, M., & Wang, L. (2016). The determinants of efficiency in the Canadian health care system. *Health Economics, Policy, and Law*, 11(1), 39–65.

Amare, T., Yitayal, M., & Amare, G. (2020). Technical efficiency of maternal health services provision in public hospitals of northwest Ethiopia: A two-stage data envelopment analysis. *Risk Management and Healthcare Policy*, 13, 3135–3146.

Aria, M. & Cuccurullo, C. (2017). Bibliometrix: An R-tool for comprehensive science mapping analysis. *Journal of Informetrics*, 11(4), 959–975.

Barasa, E., Musiega, A., Hanson, K., Nyawira, L., Mulwa, A., Molyneux, S., Maina, I., Tsofa, B., Normand, C., & Jemutai, J. (2021). Level and determinants of county health system technical efficiency in Kenya: Two stage data envelopment analysis. *Cost Effectiveness and Resource Allocation*, 19, 78. https://doi.org/10.1186/s12962-021-00332-1.

Besstremyannaya, G. (2013). The impact of Japanese hospital financing reform on hospital efficiency: A difference-in-difference approach. *Japanese Economic Review*, 64, 337–362.

Boubaker, S., Le, T.D.Q., & Ngo, T. (2022). Managing bank performance under COVID-19: A novel inverse DEA efficiency approach. *International Transactions in Operational Research*, 30(2023), 2436–2452. https://doi.org/10.1111/itor.13132.

Çakmak Barsbay, M. (2021). Energy efficiency analysis in public hospitals: An application of radial and non-radial data envelopment analysis models. *Pamukkale University Journal of Engineering Sciences*, 27(1), 35–42.

Cavalieri, M., Guccio, C., & Rizzo, I. (2017). On the role of environmental corruption in healthcare infrastructures: An empirical assessment for Italy using DEA with truncated regression approach. *Health Policy*, 121(5), 515–524.

Cavalieri, M., Guccio, C., Lisi, D., & Pignataro, G. (2018). Does the extent of per case payment system affect hospital efficiency? Evidence from the Italian NHS. *Public Finance Review*, 46(1), 117–149.

Charnes, A., Cooper, W.W., & Rhodes, E. (1978). Measuring the efficiency of decision making. *European Journal of Operational Research*, 2, 429–444.

Charnes, A., Cooper, W.W., Lewin, A.Y., & Seiford, L.M. (1994). Basic DEA models. In: *Data Envelopment Analysis: Theory, Methodology, and Applications*. Dordrecht: Springer.

Clarivate (2021). Data citation index. Retrieved November 25, 2022, from https://clarivate.com/products/scientific-and-academic-research/.

Cohen-Kadosh, S. & Sinuany-Stern, Z. (2020). Hip fracture surgery efficiency in Israeli hospitals via a network data envelopment analysis. *Central European Journal of Operations Research*, 28(1), 251–277.

Dobos, I. & Vörösmarty, G. (2023). Management applications and methodology developments of DEA-an overview of literature reviews. *International Journal of Management and Decision Making.* DOI: 10.1504/IJMDM.2023.10050363.

Donthu, N., Kumar, S., Mukherjee, D., Pandey, N., & Lim, W.M. (2021). How to conduct a bibliometric analysis: An overview and guidelines. *Journal of Business Research*, 133, 285–296.

Du, N., Wu, P., Yuan, M., & Li, Z. (2021). Performance evaluation of combining with medical and old-age care in pension institutions of China: A two-stage data envelopment analysis. *Risk Management and Healthcare Policy,* 14, 4211–4222.

Emrouznejad, A. & Yang, G.L. (2018). A survey and analysis of the first 40 years of scholarly literature in DEA: 1978–2016. *Socio-Economic Planning Sciences*, 61(1), 4–8.

Färe, R. & Grosskopf, S. (2000). Network DEA. *Socio-Economic Planning Science*, 34(1), 35–49.

Halkos, G.E., Tzeremes, N.G., & Kourtzidis, S.A. (2014). A unified classification of two-stage DEA models. *Surveys in Operations Research and Management Science*, 19(1), 1–16.

Heil, T.C., Melis, R.J.F., Maas, H.A.A.M., van Munster, B.C., Olde Rikkert, M.G.M., de Wilt, J.H.W., Adang, E.M.M., & PRECOLO Consortium. (2021). Technical efficiency evaluation of colorectal cancer care for older patients in Dutch hospitals. *PLoS ONE*, 16(12), e0260870. https://doi.org/10.1371/journal.pone.0260870.

Henriques, I.C., Sobreiro, V.A., Kimura, H., & Mariano, E.B. (2020). Two-stage DEA in banks: Terminological controversies and future directions. *Expert Systems with Applications*, 161, 113632.

Ho, F.N. & Huang, C. (2020). The interdependencies of marketing capabilities and operations efficiency in hospitals. *Journal of Business Research*, 113, 337–347. https://doi.org/10.1016/j.jbusres.2019.09.037.

Ilgun, G. & Sahin, B. (2022). Investigation of factors affecting efficiency of primary healthcare in Turkey with two-stage data envelopment analysis. *International Journal of Healthcare Management*, 15(1), 45–51.

Ilgun, G., Sönmez, S., Konca, M., & Yetim, B. (2022). Measuring the efficiency of Turkish maternal and child health hospitals: A two-stage data envelopment analysis. *Evaluation and Program Planning*, 91, 102023. https://doi.org/10.1016/j.evalprogplan.2021.102023.

Johnson, A.L. & Kuosmanen, T. (2012). One-stage and two-stage DEA estimation of the effects of contextual variables. *European Journal of Operational Research,* 220(2), 559–570.

Jung, S., Son, J., Kim, C., & Chung, K. (2023). Efficiency measurement using data envelopment analysis (DEA) in public healthcare: Research trends from 2017 to 2022. *Processes*, 11, 811. https://doi.org/10.3390/pr11030811.

Karagiannis, R. (2015). A system-of-equations two-stage DEA approach for explaining capacity utilization and technical efficiency. *Annals of Operations Research*, 227, 25–43.

Kirigia, J.M. & Asbu, E.Z. (2013). Technical and scale efficiency of public community hospitals in Eritrea: An exploratory study. *Health Economics Review* 3(6), 1–16.

Kohl, S., Schoenfelder, J., Fügener, A., & Brunner, J.O. (2019). The use of data envelopment analysis (DEA) in healthcare with a focus on hospitals. *Health Care Management Science*, 22, 245–286.

Kordrostami, S. & Amirteimoori, A. (2005). Undesirable factors in multi-component performance measurement. *Applied Mathematics and Computation*, 171(2), 721–729.

Lamesgen, A., Miniyihun, A., & Amare, T. (2022). Evaluating the technical efficiency of neonatal health service among primary hospitals of northwest Ethiopia: Using two-stage data envelopment analysis and Tobit regression model. *PLoS ONE*, 17(11), e0277826. https://doi.org/10.1371/journal.pone.0277826.

Liu, J.S., Lu, L.Y.Y., Lu, W., & Bruce, J.Y.L. (2013). Data envelopment analysis 1978–2010: A citation-based literature survey. *Omega*, 41, 3–15.

Liu, W., Zhou, Z., Ma, C., Liu, D., & Shen, W. (2015). Two-stage DEA models with undesirable input-intermediate-outputs. *Omega,* 56, 74–87.

Maleki, S., Ebrahimnejad, A., & Matin, R.K. (2019). Two stage network DEA with convex hull in intermediate products. *Hacettepe Journal of Mathematics and Statistics*, 48(1), 312–331.

Marschall, P. & Flessa, S. (2011). Efficiency of primary care in rural Burkina Faso. A two-stage DEA analysis. *Health Economics Review*, 1, 5. https://doi.org/10.1186/2191-1991-1-5.

Nepomuceno, T.C.C., Piubello Orsini, L., de Carvalho, V.D.H., Poleto, T. & Leardini, C. (2022). The core of healthcare efficiency: A comprehensive bibliometric review on frontier analysis of hospitals. *Healthcare*, 10, 1316.

Ngo, T. & Tsui, K.W.H. (2022). Estimating the confidence intervals for DEA efficiency scores of Asia-Pacific airlines. *Operational Research*, 22, 3411–3434.

Ramalho, A., Souza, J., Castro, P., Lobo, M., Santos, P., & Freitas, A. (2021). Portuguese primary healthcare and prevention quality indicators for Diabetes Mellitus — A data envelopment analysis. *International Journal of Health Policy and Management*, 11(9), 1725–1734.

Şahin, B., İlgün, G., & Sönmez, S. (2021). Determining the factors affecting the technical efficiency scores of public hospitals using different regression methods. *Benchmarking: An International Journal,* 28(7), 2202–2215.

Thai, Q.K. & Noguchi, M. (2019). Evaluating efficiency of English acute foundation trusts under system reform: A two-stage DEA approach. *Health Services and Outcomes Research Methodology*, 19(4), 215–240.

Thai, Q.K. & Noguchi, M. (2020). Measuring efficiency of the New Zealand district health boards: An empirical research using two-stage data envelopment analysis. *The Economic Society of Australia*, 39(1), 58–71.

Van Nunen, K., Li, J., Reniers, G., & Ponnet, K. (2018). Bibliometric analysis of safety culture research. *Safety Science*, 108, 248–258.

Vrabková, I. & Lee, S. (2023). Approximating the influence of external factors on the technical efficiency score of hospital care: Evidence from the federal states of Germany. *Health Economics Review*, 13, 7. https://doi.org/10.1186/s13561-022-00414-7.

Yang, S.W., Chu, K.C., & Kreng, V.B. (2021). The impact of global budgeting on the efficiency of healthcare under a single-payer system in Taiwan. *International Journal of Environmental Research and Public Health*, 18(20), https://doi.org/10.3390/ijerph182010983.

Yitbarek, K., Abraham, G., Adamu, A., Tsega, G., Berhane, M., Hurlburt, S., Mann, C., & Woldie, M. (2019b). Technical efficiency of neonatal health services in primary health care facilities of Southwest Ethiopia: A two-stage data envelopment analysis. *Health Economics Review*, 9, 27. https://doi.org/10.1186/s13561-019-0245-7.

Yitbarek, K., Abraham, G., Berhane, M., Hurlburt, S., Mann, C., Adamu, A., Tsega, G., & Woldie, M. (2021). Significant inefficiency in running community health systems: The case of health posts in Southwest Ethiopia. *PLoS ONE*, 16(2), e0246559. https://doi.org/10.1371/journal.pone.0246559.

Yitbarek, K., Adamu, A., Tsega, G., Siraneh, Y., Erchafo, B., Yewhalaw, D., Tekle, F., & Woldie, M. (2019a). Technical efficiency of maternal and reproductive health services in public hospitals of Oromia Regional State, Ethiopia. *Health Services Insights*, 12, 1178632919837630. https://doi.org/10.1177/1178632919837630.

https://doi.org/10.1142/9781800615786_0007

Chapter 7

Study of the Efficiency and Productivity of the Productive Paths of the Bioeconomy in the CNU Universities

Carlos Alberto Zúniga-González[*,||], **Mario Ramoń López**[†,**], **Jorge Luis Icabalceta-Mairena**[‡,††], **Elgin Antonio Vivas-Viachica**[§,‡‡], and **Napoleon Vicente Blanco-Orozco**[¶,§§]

[*]School of Agricultural and Veterinary Sciences, Department of Agroecology, Bioeconomy and Climate Change Research Center, National Autonomous University of Nicaragua, León, Nicaragua

[†]Department of Agricultural Economics, National Autonomous University of Nicaragua, Managua, Nicaragua

[‡]FAREM Matagalpa, National Autonomous University of Nicaragua, Managua, Nicaragua

[§]Development Department, National Agrarian University of Nicaragua, Nicaragua

[¶]Department of the Electrical Engineering Research Group, Faculty of Electrical Engineering and Computing, National University of Engineering, Nicaragua

[||]czuniga@ct.unanleon.edu.ni

[**]mariolopez2005@gmail.com

[††]jlim@jorgeicabalceta.com

[‡‡]elgin.vivas@gmail.com

[§§]napoleon.blanco@fec.uni.edu.ni

Abstract

This chapter focuses on measuring the impact of the efficiency and productivity of the research units of the member universities of the CNU. The multi-stage DEA methodology was used with the BCC model of variable returns to scale (VRS). The results showed that in competitive conditions, the efficiency indicators of the DMUs for the universities are defined by business borders; however, when efficiency is analyzed in experimental conditions, the border is marked by the Quality Control Laboratory, Entomophagous Laboratory, the Biochemistry and Physiology Laboratory of UNAN León, and the Tissue Culture and Molecular Biology Laboratory. The efficiency indicators were 19% under constant yields, both in competition and experimental conditions, but varied from 27% to 38% under variable yields, surpassing the competition conditions. Finally, in terms of scale production, the efficiency indicators under experimental conditions decreased from 79% to 65%.

Keywords: DEA, BCC, residual biomass, efficiency, CNU, peers, slack, VRSTE, CRSTE, efficiency scale, scale return, bioeconomy, productive path.

1. Introduction

One of the problems in measuring efficiency and productivity has been the conflating of the concept of performance with that of productivity (López-González *et al.*, 2016b). This chapter focuses its attention on the ways to measure efficiency through a frontier function. Frontiers have been studied in many ways in the past 50 years, among which two main ones stand out. One is the enveloping data analysis, and the other is the stochastic frontier. These involve mathematical programming and econometric methods, respectively. This study used the DEAP version 2.1 computer program. The estimates made with this software are based on efficiency, which consists of two components: the technical efficiency, which reflects the ability of a research unit to maximize its output by applying residual biomass, and the allocative efficiency, which reflects the ability of a research unit to use the residual biomass in optimal proportions given their respective market prices (Farrell, 1957).

Another important situation to consider has been the boom that the bioeconomy concept and productive paths have been witnessing (Zúniga-González *et al.*, 2022a, 2022b; López-González *et al.*, 2016a; Colón *et al.*, 2021; Rangel *et al.*, 2015).

The measurement of efficiency and productivity of universities of the National Council of Universities (CNU), like other universities, was a subject little discussed, and even more so if we link it to the productive paths of the bioeconomy (Zúniga-González *et al.*, 2022a; Maza Ávila *et al.*, 2017; Fernández *et al.*, 2013; González-Araya & Vásquez, 2010; Boubaker *et al.*, 2021; Ngo & Tsui, 2021). Universities in the past decade have been creating and developing research units in the productive paths of the bioeconomy using human, animal, and plant residual biomass in such a way that a growth in the number of laboratories and research centers has been observed, where technological and innovative processes to process residual biomass stand out (Kuramoto, 2021; Milan & Zúniga-González, 2021; Nuñez & Montalvo, 2015).

Evaluating and investigating the efficiency and performance of the academic sector by estimating a multidimensional "best practice" frontier for all analyzed universities and comparing these banks to this frontier using a frontier analysis approach is common (Thorndyke *et al.*, 2022; Zhu *et al.*, 2022; Naderi, 2022). Data envelope analysis (DEA) and stochastic frontier analysis (SFA) are the two most popular tools of frontier analysis. DEA is more popular due to its flexibility within the complex environment of the academic sector and its relevance to the academic efficiency literature, even with small sample sizes (Coupet & Ba, 2022; Papadopoulos & Parmeter, 2022). More specifically, a DEA uses a given set of inputs to produce a maximum set of outputs (output-oriented DEA) or uses a minimum set of inputs to produce a given set of outputs (input-oriented DEA), or a combination of both, to assess the efficiency of the university (El Gibari *et al.*, 2022; Olariu & Brad, 2022).

The contribution of this chapter is twofold. First, we examine the impact of bioeconomics on the performance of 46 decision-making units (DMUs) in biomass across four universities by comparing their efficiency during 2022 (Yu *et al.*, 2022; Lubsanova *et al.*, 2022). The efficiency of each DMU is compared, estimating the costs to be reduced to be efficient at the DMU that marks the efficiency frontier (Contreras & Lizano, 2022; Zhu *et al.* 2022a). Universities were compared with private sector companies and laboratories that use biomass. Under conditions of imperfect competition, the efficiencies of the universities were low. The second situation was analyzed only with 36 DMUs of the four universities, of which three stood out for their efficiency. The most important contribution is that it is possible to identify not only the DMU but also the most efficient biomass in the research processes (Yu *et al.*, 2022; Lee, 2022).

Therefore, this chapter is focused on measuring the efficiency of the DMUs in the universities of the (CNU): the National Autonomous University of Nicaragua, León (UNAN-León), the National Autonomous University of Nicaragua, Managua (UNAN-Managua), the National Agrarian University (UNA), and the National Engineering University (UNI). Some private companies and private and public farms were added to the chapter. The purpose of the study was focused on measuring the efficiency of the DMUs under the optimal production frontier of the productive paths identified in the research units under study (Zúniga-González *et al.*, 2014). To offer performance indicators, comparisons of these research units or DMUs is performed in order to provide university decision-makers with indicators of the use of technology information and residual biomass and thus establish synergies between state institutions, universities, and producers. The chapter is organized as follows: In an introductory section, the problem and its importance of this topic are presented. In Section 2, we address the main efficiency concepts, the CCR (Charnes *et al.*, 1978) and BCC (Banker *et al.*, 1984) models, and the DEA methodology. In Section 3, the applied methodology and empirical model are presented. Section 4 presents the results and Section 5 presents the discussion, and finally, the conclusion of the chapter is provided.

2. Literature Review

The purpose of this section was to highlight the importance of measuring efficiency and to discuss how technological efficiency, often represented as a frontier function, has been calculated for academic DMUs where biomass was used (Contreras & Lozano, 2022). The frontiers have been estimated using different methods; however, the following two are the most commonly used: (a) DEA using mathematical programming (Jauhar *et al.*, 2022) and (b) SFA using econometric methods (Coupet & Ba, 2022; Zúniga-González & Jaramillo-Villanueva, 2022). In this chapter, the DEAP 2.1 software relevant to the DEA methods (Coelli, 1996; Zúniga-González & Jaramillo-Villanueva, 2022) was used.

DEA is now widely used to assess the eco-efficiency of local socio-economic development (Zurano-Cervelló *et al.*, 2018; Ratner, 2017).

In Nicaragua, this method is not yet widespread. Few studies have evaluated regional economic systems from the perspective of the comparative sustainability of universities where biomass is considered (Zúniga-González *et al.*, 2014, 2022a; Orozco & Rodriguez, 2022). Georgescu-Roegen (1971) began the approach to a bioeconomy model by asking the following question: Is the availability of natural resources infinite or does it have limitations and what is the capacity of the biosphere to absorb waste (residual biomass)? With this focus on the empirical application of the DEA method, models known as eco-efficiency have been developed (Yu *et al.*, 2022; Lubsanova *et al.*, 2022). This approach has been applied to the efficiency of university management when biomass is used (El Gibari *et al.*, 2022; Contreras & Lozano, 2022; Fernández-Santos *et al.*, 2013). The input-oriented efficiency measure assumes the production function of the fully efficient DMUs is known. Fandel (2007) emphasized an input-oriented approach for students during their initial four semesters. This approach might appear challenging for students, yet the responsibility for its implementation lies with the scientific staff. These staff members have fixed funding allocations tied to seminar and examination outcomes. It is a direct result of the use of input orientation. Agasisti & Salerno (2007) assessed the cost efficiency of Italian universities. Glas *et al.* (2006) examined whether best-practice efficiency measures based on DEA empirically support current policy goals in UK higher education. It is also assessed whether such support depends on the specific types of efficiency measures used in his DEA modeling.

2.1. *Review of DEA studies*

The modern measure of efficiency (Färe *et al.*, 1994) begins with Farrel (1957) improving the work of Debreu (1951) and Koopmans (1951) to define two efficiency components in a DMU. Technical efficiency expresses the ability of a DMU to maximize its revenue given a set of inputs, and allocative efficiency represents the ability of the DMU to optimize its portions of inputs given the variability of market prices. Farrel's initial idea was to determine the input–input space, and from there, the reduced-input approach and the current input-oriented measurement were developed.

The following hypothetical example, shown in Figure 1 and Table 1, explains Farrel's initial idea. Two inputs (x_1 and x_2) are

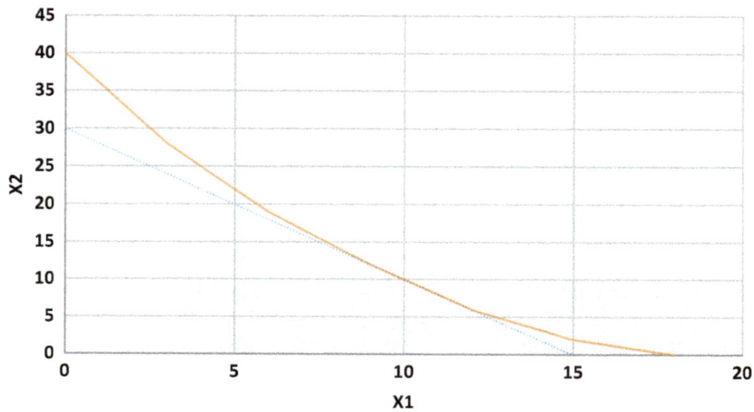

Figure 1. Isoquants and isocosts combinations, $Y = 5$.

Table 1. Combinations of inputs to obtain quantities of the same product. Hypothetical data, $Y = 5$.

x_1	x_2		**Cost**
0	40	—	200
3	28	4.0	170
6	19	3.0	155
9	12	2.3	150
12	6	2.0	150
15	2	1.3	160
18	0	0.7	180
$Y_1 = 5$			
Cost 150			
P_{X1}	10	0	18
P_{X2}	5	40	0

Source: Zúniga-González (2011).

assumed to produce an optimal level of output at $Y = 5$ under the assumption of constant returns to scale. The isoquant indicates the technical efficiency of the DMU given the percentages of x_1/y and x_2/y represented by the solid curve built from the substitution rates indicated in Table 1. The allocative efficiency is indicated by the dotted curve defined by the price ratio P_{X1}/P_{X2}, considering the optimal level of production $Y = 5$. At points $(12, 6)$, this price ratio is equal to the substitution rate of the isoquants, and precisely the point where the isoquant intersects with the isocost is the optimal point of production and therefore of efficiency.

The slack variables in DEA have a direct relationship with the solution to the problem. Thus, for example, if the optimal solution of the problem given by the model turns out to have an efficiency of 1, it means that the DMU evaluated is efficient in relation to the other DMUs, considering that it is not possible to find any other DMU that achieves at least the outputs of the evaluated DMU while handling fewer resources. Thus, the DMU is inefficient when its efficiency is less than 1; then, the slack value is obtained by subtracting from 1 the value of the relative efficiency of the inefficient DMU (Serrano *et al.*, 2000).

Consequently, every DMU with an efficiency of 1 has a slack value equal to zero. That is, any DMU that has a slack value means that it has scale inefficiencies. So, the higher the value of the slack, the lower the efficiency of the DMU evaluated (Oviedo & Rodríguez, 2011).

The analysis of enveloping data, since its appearance in 1978 (Charnes *et al.*, 1978) with the work of Farrell (1957), has been growing in its theoretical (Cooper *et al.*, 2007; Cook & Seiford, 2009) and empirical scope. The objective is to provide input and output data from different DMUs in an arrangement assigning to each of them a quantifying value of relative efficiency in such a way that the most efficient DMUs will be those that obtain the highest output with the lowest possible input, thus determining the efficient frontier.

To estimate the referred efficiency, two strategies have been used depending on whether it is oriented toward inputs or outputs (Charnes *et al.*, 1981). In the first input-oriented model, given the level of outputs, the idea is to obtain the maximum proportional reduction in the vector of inputs, keeping the frontier or level of output constant. The investigation tries to find the efficiency of the DMU. On the other hand, the output-oriented model considers a given input, where the maximum proportional increase in outputs is sought while remaining within the possibilities frontier.

3. Methodology

The variable returns to scale (VRS) model of an input-oriented, multi-stage DEA (Mozaffari *et al.*, 2022; Ashley-Dejo, 2022; Liu *et al.*, 2022) was used for the treatment of the input slack variable, where a sequence of radial linear programming (PLs) is conducted to identify the projected points of efficiency (Banker *et al.*, 1984). The sample involved 46 observations of one output (y) and one input (x_1) (Färe *et al.*, 1994). DEA provides non-parametric mathematical programming estimates focused on calculating the frontier. The research units are decision-making units (DMU's) (Moreira & Bravo, 2009; Vargas-Leiton *et al.*, 2015; Lanteri, 2002; Robles, 2021; Coelli, 1996; O'Neill *et al.*, 2008; Thanker *et al.*, 2021; Tsionas, 2021). Considering that the best way to introduce DEA is via percentages or ratios, for each DMU, it would be necessary to obtain a ratio of all the outputs over all the inputs in such a way that $u'y_i/v'x_i$'X_i could be represented., where u is a vector of $M \times 1$ weights or ratios of outputs and v is a vector of $K \times 1$ weights or ratios of inputs.

To select the optimal weights or ratios, the following mathematical programming models (Equation (1)) are specified:

$$\begin{aligned} &\max_{u,v}(u'y_i/v'x_i),\\ \text{s.t.:}\quad &\frac{u'y_i}{v'x_i} \leq 1,\ \ J = 1, 2, \ldots, N,\\ &u, v \geq 0. \end{aligned} \tag{1}$$

The results of values of u and v represent the efficiency measure for each maximized DMU, subject to the constraint (s.t.) that all measures must be greater than or equal to 1. One of the problems with this estimate is that there are infinitely many solutions, and to avoid this, we have proposed a restriction subject to $v'x_i = 1$ (Equation (2)):

$$\begin{aligned} &\max_{u,v}(u'y_i/v'x_i),\\ \text{s.t.:}\quad &\frac{u'y_i}{v'x_i} = 1,\\ &u'y_i - v'x_i \leq 0,\ \ J = 1, 2, \ldots, N,\\ &u, v \geq 0. \end{aligned} \tag{2}$$

It is observed that the expression of u and v have changed. This form is unknown in the way that the multiplier is developed in the linear programming problem posed. By using the duality in linear programming, an equivalent form (Equation (3)) can be derived:

$$\begin{aligned} & \min_{\theta,\lambda}(\theta), \\ \text{s.t.:} \quad & -y_i + Y\lambda \geq 0, \\ & \theta x_i - X\lambda \geq 0, \\ & \text{NI}'\lambda = 1, \\ & \lambda \geq 0, \end{aligned} \tag{3}$$

where NI is a vector of $N \times 1$. This wrapping or development involves the fewest restrictions that the multiplier form $(K + M < N + 1)$ poses, and for this reason, this is the preferred form to find the solutions. The θ value obtained would be the efficiency measurement score for each DMU. This is satisfied if $\theta \leq 1$, with values of 1 indicating a point on the frontier and hence the measurement of technical efficiency for each DMU, according to Farrel (1957), Lanteri (2002), and Shephard (1970).

The particularity of the model used is that variables are added according to the criteria defined by Georgescu-Roegen (1976), who defines three categories: Service Funds $S(i)$, Flows $F(t)$, and Product Flow (M). In $S(i)$, physical capital (y) and labor force (x_1) are defined, and $F(t)$ refers to the natural resources, equipment, and infrastructure of each research unit. M is defined by the residual biomass (animal or vegetable), or what is identified as waste product flow (x_2) by Zúniga-González *et al.* (2022a).

3.1. *Data*

An information collection instrument was applied to 46 DMUs in universities with companies and the 36 DMUs of only universities under study, including companies and private and public farms. The variables used are the following (Svetlanská *et al.*, 2012; Zúniga-González *et al.*, 2022a; Georgescu-Roegen, 1976; Zúniga-González, 2022):

- y: defined by the average annual revenue generated by the research unit, Product Flow (M).
- x_1: defined by the average annual costs assumed by each research unit; this cost includes the value of residual biomass, labor, supplies, and materials, represented by Funds for Services $S(i)$ and Flows $F(t)$.

Table 2. Descriptive statistics for the inputs and outputs of the sampled DMUs with companies included (2020).

	Output (y_1)	**Input (x_1)**	**Labor (x_2)**
Mean	449.63	312.26	7
Standard deviation	847.36	564.91	7
Minimum	3,000.00	1,999.00	1
Maximum	4,450.80	2,840.00	46

Table 3. Descriptive statistics for the inputs and outputs of the sampled DMUs only with university units (2020).

	Output (y)	**Input (x_1)**	**Labor (x_2)**
Mean	230.03	242.02	6
Standard deviation	364.51	439.93	4
Minimum	3,000.00	1,999.00	2
Maximum	1,906.85	1,940.32	21

x_2: defined by the quantities of labor used in each DMU.
Year: the data were collected in 2022.

The DEAP 2.1 software was used to implement the DEA (Coelli, 1996).

Tables 2 and 3 present the descriptive statistics of our input and output data (in US$, 2020 constant prices), in which one can see a mix of changes happening to the DMU's inputs and outputs with companies and university units. For instance, comparing the companies with the university units, the average expenses (x_1) dropped from $312,000 in 2020, but the labor (x_2) represents seven persons for the same period. From Table 2, considering only university units, the average expenses (x_1) dropped from $242,020, but the labor ($x_2$) represents six persons for the same period. On the other hand, fluctuations in those inputs across the sampled DMU became larger, with the values of the standard deviation being $564,000 and $439,930, respectively.

4. Results

For the analysis of the relative efficiency indices, the DEA methodology was applied to 46 data points that include universities and companies in order to compare the results of the technical efficiency of the productive paths of the bioeconomy. The second part of the analysis focused on 37 academic DMUs (laboratories and research centers). These conditions are linked to the imperfect competitive market. For the academic units, the objective is to inculcate scientific innovation and generate knowledge in the teaching processes. The companies have objectives focused on utility maximization and minimization of costs.

The results under market conditions are presented in Table 4. The company Frozen Fruit S.A. (No. 40) in the business sector and the Center for Research and Biological Controllers of UNAN León with the Laboratory of Entomopathogenic Bacteria *Bacillus subtilis* (No. 9) stand out for their efficiency. Both DMUs mark the production frontier with efficiency and productivity. Overall, on average, all the DMUs present a technical efficiency at constant yields of 19%. This index increases in yields of variable scale, growing by 27%, and if the DMUs decide to increase production at scale, the results show an increase up to 79%. There is a well-marked trend of increasing returns to scale under market conditions when using technologies and innovation with residual biomass in productive paths.

Table 5 analyzes the value projections of the input variables, or what is known as input-oriented analysis, for this study. The lambda ratio (λ_i) can be observed, which indicates the levels of inefficiency (slack) of the DMUs with respect to the two DMUs (Frozen (No. 42) and Entomopathogenic Bacteria Laboratories (No. 9)) that mark the frontier of possibilities and therefore are the most efficient. This defines the radial movement that indicates the value of the inputs to decrease so that the DMU is projected with levels of efficiency. Thus, it can be noted that both companies and academic units need to reduce their costs in the productive path of the bioeconomy. The academic DMUs need to adjust to the indicated cost reduction by improving technology and innovation processes with the residual biomass applied. The DMUs that need immediate adjustment are numbers 7, 16, 25, and 26, as they surpass the required 50%

Table 4. Comparison of relative efficiency indices of productive paths of the bioeconomy.

No.	CRSTE CCR	VRSTE BCC	Efficiency scale	Scale returns
CNU				
1	0.114	0.115	0.995	Increase
2	0.110	0.112	0.977	Increase
3	0.121	0.128	0.950	Increase
4	0.101	0.114	0.889	Increase
5	0.102	0.104	0.982	Increase
6	0.245	0.257	0.953	Increase
7	0.801	0.845	0.948	Increase
8	0.253	0.426	0.594	Increase
9	0.152	1.000	0.152	Increase
10	0.145	0.190	0.764	Increase
11	0.202	0.257	0.788	Increase
12	0.200	0.419	0.476	Increase
13	0.101	0.114	0.885	Increase
14	0.236	0.316	0.747	Increase
15	0.209	0.308	0.678	Increase
16	0.253	0.676	0.374	Increase
17	0.131	0.237	0.555	Growing
18	0.013	0.014	0.942	Growing
19	0.151	0.251	0.603	Growing
20	0.113	0.119	0.945	Growing
21	0.101	0.123	0.820	Growing
22	0.101	0.112	0.902	Growing
23	0.101	0.103	0.983	Growing
24	0.202	0.208	0.973	Growing
25	0.101	0.530	0.191	Increase
26	0.101	0.530	0.191	Increase
27	0.101	0.149	0.678	Increase
28	0.101	0.105	0.966	Increase
29	0.101	0.149	0.678	Increase
30	0.101	0.120	0.840	Increase
31	0.101	0.120	0.840	Increase
32	0.101	0.158	0.638	Increase
33	0.101	0.103	0.980	Increase
34	0.121	0.155	0.784	Increase
35	0.119	0.150	0.797	Increase
36	0.121	0.155	0.784	Increase
37	0.119	0.150	0.797	Increase

Table 4. (*Continued*)

No.	CRSTE CCR	VRSTE BCC	Efficiency scale	Scale returns
Companies				
38	0.157	0.158	0.994	Increase
39	0.101	0.101	0.998	Constant
40	1.000	1.000	1.000	Increase
41	0.101	0.110	0.916	Increase
42	0.539	0.542	0.995	Increase
43	0.202	0.212	0.954	Increase
44	0.446	0.513	0.870	Increase
45	0.233	0.238	0.978	Increase
46	0.149	0.189	0.789	Increase
	0.187	0.265	0.794	

reduction. Under these same conditions, companies must also make adjustments, indicating that they exceed 50% of their costs. These DMUs are numbers 42 and 44.

Table 6 shows that the behavior of the DMUs is different and varied in the productive paths of the bioeconomy studied. The Quality Laboratory of the Faculty of Chemical Sciences (No. 7), the Entomopathogenic Bacteria Laboratory of the Center for Research (No. 9) and Reproduction of Biological Controllers (No. 8), both from UNAN León, the Laboratory of Plant Tissue Cultures and Molecular Biology from UNA (No. 24), and the Biochemistry and Physiology Laboratory of the Faculty of Medical Sciences of UNAN-León (No. 1) mark the efficiency frontier for all DMUs. The performance returns are varied, observing that there are DMUs that grow, while others decrease. It can be noted that the four units that mark efficiency are as follows: the quality laboratory by maintaining constant returns, the entomopathogenic bacteria laboratory by increasing yields, the tissue laboratory and the biochemistry and physiology laboratory by decreasing yields. On overall average as well as in market conditions, the academic DMUs under constant returns have a technical efficiency of 19%, while in variable scale returns, it rises to 39%, higher than in conditions of imperfect competition; however, in scale efficiency, if a decision is made to increase production in the productive paths, the increase would be slightly less than 65%.

Table 5. Comparison of projection in value of input variables by productive path of the bioeconomy (year 2021–2022).

No.	Reference (peers) (weights λi)		Original value	Radial movement	Movement slack	Projected value	Percentage
CNU							
1	40 (0.428)	9 (0.572)	1,685,067.00	−1,491,303.400	0	193,763.633	11.50
2	40 (0.138)	9 (0.862)	567,446.00	−503,690.190	0	63,755.814	11.24
3	40 (0.067)	9 (0.933)	250,000.00	−218,085.920	0	31,914.081	12.77
4	40 (0.029)	9 (0.971)	130,000.00	−115,209.030	0	14,790.971	11.38
5	40 (0.168)	9 (0.839)	742,117.00	−664,688.500	0	77,428.503	10.43
6	40 (0.070)	9 (0.930)	129,566.00	−96,218.312	0	33,347.688	25.74
7	40 (0.064)	9 (0.936)	36,239.98	−5,628.570	0	30,611.410	84.47
8	40 (0.005)	9 (0.995)	9,750.00	−5,598.021	0	4,151.979	42.58
9	1	—	1,999.00	0.000	0	1,999.000	—
10	40 (0.011)	9 (0.989)	37,348.50	−30,257.893	0	7,090.607	18.98
11	40 (0.013)	9 (0.987)	30,800.00	−22,898.563	0	7,901.437	25.65
12	40 (0.003)	9 (0.997)	7,700.00	−4,470.151	0	3,229.849	41.95
13	40 (0.028)	9 (0.972)	126,000.00	−111,611.930	0	14,388.074	11.42
14	40 (0.010)	9 (0.990)	21,000.00	−14,367.688	0	6,632.312	31.58
15	40 (0.007)	9 (0.993)	16,970.00	−11,747.826	0	5,222.174	30.77
16	40 (0.002)	9 (0.998)	3,999.00	−1,294.931	0	2,704.069	67.62
17	40 (0.004)	9 (0.996)	16,029.98	−12,234.711	0	3,795.269	23.68
18	40 (0.057)	9 (0.943)	1,940,322.60	−1,912,629.200	0	27,693.416	1.43
19	40 (0.005)	9 (0.995)	16,901.00	−12,655.750	0	4,245.250	25.12
20	40 (0.061)	9 (0.939)	245,000.00	−215,805.470	0	29,194.528	11.92
21	40 (0.016)	9 (0.984)	75,000.00	−65,748.859	0	9,251.141	12.33
22	40 (0.033)	9 (0.967)	150,000.00	−133,194.550	0	16,805.454	11.20
23	40 (0.182)	9 (0.818)	811,085.40	−727,692.660	0	83,392.738	10.28

24	40 (0.120)	9 (0.880)	269,400.00	−213,432.990	0	55,967.015	20.77
25	40 (0.000)	9 (1.000)	3,956.00	−1,860.708	0	2,095.292	52.96
26	40 (0.000)	9 (1.000)	3,956.00	−1,860.708	0	2,095.292	52.96
27	40 (0.007)	9 (0.993)	35,000.00	−29,777.826	0	5,222.174	14.92
28	40 (0.097)	9 (0.903)	435,110.00	−389,587.080	0	45,522.925	10.46
29	40 (0.064)	9 (0.936)	35,000.00	−29,777.826	0	5,222.174	14.92
30	40 (0.019)	9 (0.981)	86,661.58	−76,235.836	0	10,425.744	12.03
31	40 (0.019)	9 (0.981)	86,661.58	−76,235.836	0	10,425.744	12.03
32	40 (0.006)	9 (0.994)	29,428.00	−24,767.061	0	4,660.939	15.84
33	40 (0.159)	9 (0.841)	709,345.00	−636,199.980	0	73,145.020	10.31
34	40 (0.013)	9 (0.987)	50,000.00	−42,259.722	0	7,740.278	15.48
35	40 (0.014)	9 (0.986)	55,000.00	−46,756.101	0	8,243.899	14.99
36	40 (0.013)	9 (0.987)	50,000.00	−42,259.722	0	7,740.278	15.48
37	40 (0.014)	9 (0.986)	55,000.00	−46,756.101	0	8,243.899	14.99
Companies							
38	40 (0.378)	9 (0.622)	1,087,500.00	−916,082.930	0	171,417.068	15.76
39	40 (0.638)	9 (0.362)	2,840,000.00	−2,552,246.500	0	287,753.494	10.13
40	1		450,000.00	0.000	0	450,000.000	
41	40 (0.039)	9 (0.961)	175,000.00	−155,676.440	0	19,323.559	11.04
42	40 (0.427)	9 (0.573)	356,485.00	−163,412.230	0	193,072.766	54.16
43	40 (0.071)	9 (0.929)	160,000.00	−126,071.440	0	33,928.565	21.21
44	40 (0.024)	9 (0.976)	24,800.00	−12,073.875	0	12,726.125	51.32
45	40 (0.141)	9 (0.859)	273,408.00	−208,270.150	0	65,137.850	23.82
46	40 (0.013)	9 (0.987)	42,000.00	−34,058.273	0	7,941.727	18.91

Table 6. Indices of relative efficiencies of productive paths of the CNU bioeconomy.

	CCR	BCC	Scale	Returns
UNAN-Leon				
1	0.143	1.000	0.143	Decrease
2	0.137	0.616	0.223	Decrease
3	0.151	0.193	0.785	Decrease
4	0.126	0.133	0.948	Increase
5	0.128	0.660	0.194	Decrease
6	0.306	0.474	0.646	Decrease
7	1.000	1.000	1.000	Constant
8	0.316	0.469	0.673	Increase
9	0.189	1.000	0.189	Increase
10	0.181	0.217	0.835	Increase
11	0.252	0.294	0.858	Increase
12	0.250	0.451	0.553	Increase
13	0.126	0.134	0.945	Increase
14	0.295	0.359	0.820	Increase
15	0.260	0.345	0.754	Increase
16	0.316	0.711	0.444	Increase
UNA				
17	0.164	0.259	0.634	Increase
18	0.017	0.017	0.995	Constant
19	0.189	0.277	0.681	Increase
20	0.141	0.141	0.998	Constant
21	0.126	0.142	0.887	Increase
22	0.126	0.131	0.960	Increase
23	0.126	0.680	0.186	Decrease
24	0.252	1.000	0.252	Decrease
UNI				
25	0.126	0.534	0.236	Increase
26	0.126	0.534	0.236	Increase
27	0.126	0.167	0.754	Increase
28	0.126	0.398	0.317	Decrease
29	0.126	0.167	0.754	Increase
30	0.126	0.139	0.905	Increase
31	0.126	0.139	0.905	Increase
32	0.126	0.176	0.717	Increase
33	0.126	0.629	0.201	Decrease
UNAN-Managua				
34	0.151	0.177	0.854	Increase
35	0.149	0.172	0.866	Increase
36	0.151	0.177	0.854	Increase
37	0.149	0.172	0.866	Increase
	0.191	0.386	0.651	

Table 7. Projection in value of input variables by productive path of the CNU bioeconomy (year 2021–2022).

No.	Reference (peers) (weights λi)		Original value	Radial movement	Movement slack	Projected value	Percentage
UNAN-León							
1	1 (1.0) [4]	—	1,685,067.0	0.000	0	1,685,067.00	—
2	24 (0.9439)	1 (0.057)	567,446.0	−218,027.010	0	349,418.98	61.58
3	24 (0.051)	7 (0.949)	250,000.0	−201,781.180	0	48,218.82	19.29
4	7 (0.447)	9 (0.553)	130,000.0	−112,692.620	0	17,307.38	13.31
5	1 (0.156)	24 (0.844)	742,117.0	−252,230.310	0	489,886.69	66.01
6	24 (0.108)	7 (0.108)	129,566.0	−68,164.300	0	61,401.70	47.39
7	7 (1.0) [29]	—	36,240.0	0.000	0	36,240.00	—
8	7 (0.075)	9 (0.925)	9,750.0	−5,174.490	0	4,575.51	46.93
9	9 (1.0) [26]	—	1,999.0	0.000	0	1,999.00	—
10	7 (0.178)	9 (0.822)	37,348.5	−29,256.280	0	8,092.22	21.67
11	7 (0.206)	9 (0.794)	30,800.0	−21,737.450	0	9,062.55	29.42
12	7 (0.043)	9 (0.957)	7,700.0	−4,228.020	0	3,471.98	45.09
13	7 (0.433)	9 (0.567)	126,000.0	−109,174.780	0	16,825.23	13.35
14	7 (0.162)	9 (0.838)	21,000.0	−13,456.230	0	7,543.77	35.92
15	7 (0.113)	9 (0.887)	16,970.0	−11,113.770	0	5,856.23	34.51
16	7 (0.025)	9 (0.975)	3,999.0	−1,156.230	0	2,842.77	71.09
UNA							
17	7 (0.063)	9 (0.937)	16,030.0	−11,881.350	0	4,148.63	25.88
18	7 (0.898)	9 (0.102)	1,940,322.6	−1,907,574.600	0	32,747.96	1.69
19	7 (0.079)	9 (0.921)	16,901.0	−12,213.870	0	4,687.13	27.73
20	7 (0.950)	9 (0.050)	245,000.0	−210,455.630	0	34,544.37	14.10
21	7 (0.253)	9 (0.747)	75,000.0	−64,322.230	0	10,677.77	14.24
22	7 (0.517)	9 (0.483)	150,000.0	−130,281.850	0	19,718.15	13.15
23	24 (0.801)	1 (0.199)	811,085.4	−259,924.560	0	551,160.84	67.95
24	24 (1.0) [7]	—	269,400.0	0.000	0	269,400.00	—
UNI							
25	7 (0.003)	9 (0.997)	3,956.0	−1,841.770	0	2,114.24	53.44
26	7 (0.003)	9 (0.997)	3,956.0	−1,841.770	0	2,114.24	53.44
27	7 (0.113)	9 (0.887)	35,000.0	−29,143.770	0	5,856.23	16.73
28	24 (0.588)	7 (0.412)	435,110.0	−261,749.680	0	173,360.32	39.84
29	7 (0.113)	9 (0.887)	35,000.0	−29,143.770	0	5,856.23	16.73
30	7 (0.295)	9 (0.705)	86,661.6	−74,578.146	0	12,083.43	13.94
31	7 (0.295)	9 (0.705)	86,661.6	−74,578.146	0	12,083.43	13.94
32	7 (0.093)	9 (0.907)	29,428.0	−24,243.410	0	5,184.59	17.62
33	24 (0.875)	1 (0.125)	709,345.0	−263,465.090	0	445,879.91	62.86
UNAN-Managua							
34	7 (0.201)	9 (0.799)	50,000.0	−41,130.310	0	8,869.69	17.74
35	7 (0.218)	9 (0.782)	55,000.0	−45,527.618	0	9,472.38	17.22
36	7 (0.201)	9 (0.799)	50,000.0	−41,130.310	0	8,869.69	17.74
37	7 (0.201)	9 (0.782)	55,000.0	−45,527.618	0	9,472.38	17.22

With these results, the DEA methodology was applied with the CCR and BCC models to study the relative efficiency indices in 36 research units of the member universities of the CNU.

Finally, Table 7 shows the projected values of the cost of the inputs where the residual biomass is included, indicating the necessary values to reach the desired level of efficiency. The quality control laboratory counts 29 times as a pair of DMUs, the entomopathogenic bacteria laboratory 26 times, the tissue laboratory 7 times, and the biochemistry and physiology laboratory 4 times. The DMUs that show a decrease must work on the improvement of residual biomass and technology to reduce their costs to reach the frontier determined by the most efficient DMUs.

5. Discussion

The measurement of the efficiency of laboratories and research centers associated with the CNU has made it possible to identify the average levels of efficiency for each model under conditions of constant, variable, and scale yields (Banker *et al.*, 1984; Charnes *et al.*, 1978; Färe *et al.*, 1994). With this, we identify good practices that allow for improving the current level of efficiency of the DMUs that are not yet fully efficient. This was important for the universities UNAN León, UNA, UNI, and UNAN Managua, where new technologies and innovative processes are being developed in the productive paths of the bioeconomy (Zúniga-González *et al.*, 2014).

The DEA methodology offers a new way of estimating efficiency by adding an alternative enveloping data analysis to the partial approach. With the present charter, the yield measures in residual biomass for crops, milk, meat, wood, etc. are exceeded. The ratios of constant and scale returns are added to these returns in an enveloping manner similar to the projections of the costs to be reduced to achieve optimal technical efficiency (Zúniga-González *et al.*, 2022a; García & Zavala, 2022).

The measurement of the bioeconomy requires methods and techniques such as those used in this chapter that identify deficient DMUs and offer options to improve their optimal level to minimize the costs of the residual biomass and technological processes used (García-Bucio *et al.*, 2022, Catari-Yujra *et al.*, 2022).

On the other hand, it is very important to consider the application of DEA models to eco-intensification approaches, considering the trend of DEA development, such as two-stage DEA (Tran & Ngo, 2014; Wasiaturrahma *et al.*, 2020), bootstrap DEA (Simar & Wilson, 2007), inverse DEA (Boubaker *et al.*, 2022), or common weight DEA (Hammami *et al.*, 2022; Jauhar *et al.*, 2022).

6. Conclusion

Applying the DEA methodology to the research units of the CNU universities was viable because the evaluation of the relative technical efficiency carried out on the 47 DMUs, including the companies and the 36 exclusive DMUs of the research units, allows us to focus mainly on those variables that are controllable as inputs (residual biomass, reagents, materials, etc.). Therefore, the evaluation of the relative technical efficiency was developed to be oriented to these. Thus, it was demonstrated that the DMUs can maintain their current production levels by making significant reductions in their costs, determined by the projection of the value required for the minimization of the costs per productive path of each DMU (see Tables A1 and A2).

Acknowledgments

This research was carried out with CNU Competitive Funds. We thank the authorities at the CNU, the research vice-rectories, for facilitating the research processes. We also thank the authorities at UNAN-León, especially for the coordination of projects.

Appendix

Table A1. DMU, productive path. and residual biomass for universities and companies sample.

No.	Decision-making unit (DMU)	Productive path	Residual biomass
	National Council of Universities (CNU)		
1	Biochemical Lab. and physiol. FCM UNAN-Leon	App. of biotech.	Urine, blood, feces
2	Microbiology Laboratory FF.CC.MM UNAN-León	App. of biotech.	Parasites, infectious agents
3	CEI-Vaccine Center-DEN FF.CC.MM UNAN-León	App. of biotech.	Urine, blood, feces, reagents
4	Laboratory analysis of contam. FCM UNAN-Leon	App. of biotech.	Urine, blood, feces
5	Lab. Mauricio Diaz Muller Farm and prod of Hogar F.CC.QQ UNAN-León	App. of biotech.	Biomass inorganic matter, plant residues
6	Lab. Mauricio Diaz Muller Alimentos F.CC.QQ UNAN-León	App. of biotech.	Vegetable waste (seeds, cereals, fruits, vegetables)
7	Quality Control Laboratory F.C.Q UNAN-León	App. of biotech.	React. Quim. synth. natural
8	Entomop fungi ECAV CIRCB UNAN-León	Ecointensification	*Trichoderma, Beauveria bassiana*
9	Entomopathogenic bacteria CIRCB UNAN-León	Ecointensification	*Bacillus subtilis*
10	NPV ECAV CIRCB UNAN-León	Ecointensification	*Spodop. siuna, Frugiperda exigua*
11	Earthworm fertilizer ECAV CNRA UANA-León	Ecointensification	Cow manure
12	Bioferment compost ECAV CNRA UNAN-León	Ecointensification	Cow manure
13	Mountain microorganisms ECAV CNRA UNAN-León	Ecointensification	Stubble from soil edaphon, MO, semolina, milk, molasses
14	Bokachi ECAV CNRA UNAN-León	Ecointensification	Chicken manure, coffee pulp, charcoal, semolina, molasses
15	Trichogramma ECAV CIRCB UNAN-León	Ecointensification	*Sitotroga cerealella* (hosp., food)
16	Chrysopa ECAV CIRCB UNAN-León	Ecointensification	*Sitotroga cerealella* (hosp., food)
17	FDR-UNA Agroindustry Laboratory	App. of biotech.	Seeds, cereals, fruits, vegetables
18	Lab. entomopathogenic fungi UNA	Ecointensification	Entopathogenic fungi (*Trichoderma, Beauveria bassiana, Bacillus felineus, Bacillus subtilis*)

Table A1. (*Continued*)

No.	Decision-making unit (DMU)	Productive path	Residual biomass
19	Multifunctional Biology Laboratory UNA Camoapa Headquarters	App. of biotech.	Urine, blood, feces
20	Service of Information on Agri Sost (SIMAS) A	Ecointensification	Tree biomass
21	ONE headquarters Juigalpa	Ecointensification	Seeds, cereals, fruits, vegetables
22	Lab. LARENA Fac. of Rec. Nat. and Amb. UNA	Ecointensification	Seeds, cereals, fruits, vegetables
23	Uni. invt. in animal feed strat Fac. Ciencia Animal UNA	Increase efficiency in the value chain	Organic matter, detergent, garlic, chili, natural banana vinegar, manure
24	Lab. Cult of veg tissues and molecular biol UNA	App. of biotech.	Vitroplants, plant tissues
25	PVIT-UNI Biomass Laboratory (Biogas Biofilters)	Ecointensification	Solid and liquid waste
26	Biomass Laboratory PVIT-UNI (Bio-input wastewater)	Bioref.-bioprodtos	Inorganic matter, plant residues
27	FTC-UNI investigation coordination	Bioref.-bioprodtos	Biomass agro-industrial waste
28	FTC-UNI Materials Laboratory	Bioref.-bioprodtos	Biomass agro-industrial waste
29	Research coordination FARQ-UNI (Wood waste)	Bioref.-bioprodtos	Tree biomass
30	Research coordination FARQ-UNI (Math. Const. Waste)	Bioref.-bioprodtos	Biomass agro-industrial waste
31	Research coordination FARQ (Plastics)	Bioref.-bioprodtos	Plastic waste
32	Faculty of Chemical Engineering FIQ-UNI	Biotech app	Organic matter, detergents, garlic, chili, natural banana vinegar, manure
33	FEC-UNI Electrical Engineering research group	Bioref.-bioprodtos	Biomass agro-industrial waste
34	FAREM Matagalpa UNAN-Mga	Ecointensification	Solid and liquid waste
35	FAREM Esteli UNAN-Mga	Ecointensification	Solid and liquid waste
36	FAREM Juigalpa UNAN-Mga	Ecointensification	Solid and liquid waste
37	FAREM Carazo UNAN-Mga	Ecointensification	Solid and liquid waste
	Companies		
38	SOPPEXCCA	Increase efficiency value chain	Chicken manure, brown pulp, charcoal, semolina, molasses

(*Continued*)

Table A1. (*Continued*)

No.	Decision-making unit (DMU)	Productive path	Residual biomass
39	ECOOALPHA	Ecointensification	Coffee pulp
40	Frozen Fruit S.A.	Irrigation energy efficiency	Tree biomass
41	Agricultural Valdivia (Rice)	Ecointensification	Rice husk
42	Esteli Vegetable Nursery	Ecointensification	Substratum
43	Banana producing cooperative	Ecointensification	Plant residual biomass
44	Santa Ana Luisa Estate	Ecointensification	Plant residual biomass
45	Finca Lourdes (The Guardians)	Ecointensification	Plant residual biomass
46	Sacred field farm	Ecointensification	Dehydrated moringa

Table A2. DMU, productive paths, and residual biomass for only universities sample.

No.	Decision-making unit (DMU)	Productive path	Residual biomass
	UNAN-Leon		
1	Biochemical Lab. and physiol. FF.CC.MM	App. of biotech.	Urine, blood, feces
2	Microbiology Laboratory FF.CC.MM	App. of biotech.	Parasites, infectious agents
3	CEI-Vaccine Center-DEN F.C.M	App. of biotech.	Urine, blood, feces
4	Lab. contamination analysis F.CC.MM	App. of biotech.	Urine, blood, feces
5	Lab. Mauricio Diaz Muller Farm and prod of Hogar F.C.Q	App. of biotech.	Biomass inorganic matter
6	Lab. Mauricio Diaz Muller Food	App. of biotech.	Seeds, cereals, fruits
7	F.CC.QQ quality control lab	App. of biotech.	Natural and synthetic chemical reagents
8	Entomop fungi ECAV CIRCB	Ecointensification	*Trichoderma*, *Beauv. bassiana*
9	Entomop bacteria. ECAV CIRCB	Ecointensification	*Bacillus subtilis*
10	NPV ECAV CIRCB	Ecointensification	*Spodop. siuna*, *Frugip. exigua*
11	Earthworm fertilizer ECAV CNRA	Ecointensification	Cow manure
12	Bioferment compost ECAV CNRA	Ecointensification	Cow manure
13	Microorg. mountain ECAV CNRA	Ecointensification	Stubble from soil, semolina, milk, molasses

Table A2. (*Continued*)

No.	Decision-making unit (DMU)	Productive path	Residual biomass
14	Bokachi ECAV CNRA	Ecointensification	Chicken manure, brown pulp, carbon veget, semolina, molasses
15	Trichogramma ECAV CIRCB	Ecointensification	*Sitotroga cerealella*
16	Chrysopa ECAV CIRCB	Ecointensification	*Sitotroga cerealella*
17	FDR-UNA Agroindustry Laboratory	App. of biotech.	Seeds, cereals, fruits
18	Lab. entomopathogenic fungi UNA	Ecointensification	*Trichoderma, Beauveria bassiana, Bacillus felineus, Bacillus subtilis*
19	Lab. multifunction. of biol Camoapa	App. of biotech.	Urine, blood, feces
20	Service of Information on Agri Sost (SIMAS)	Ecointensification	Tree biomass
21	Headquarters Juigalpa	Ecointensification	Seeds, cereals, fruits
22	Lab. LARENA Fac. of Rec. Nat. and Amb.	Ecointensification	Seeds, cereals, fruits, vegetables
23	Uni. invt. in animal feed strat Fac. Ciencia Animal UNA	Increase efficiency value chain	Organic matter, detergents, garlic, chili, banana vinegar, manure
24	Lab. Cult of veg tissues and biol molec. UNI	App. of biotech.	Vitroplants, plant tissues
25	Biomass Lab PVIT-UNI (Biofilt. Biogas)	Ecointensification	Solid and liquid animal waste
26	Biomass Laboratory PVIT-UNI (Bioinput)	Bioref.-biopts	Inorganic matter, plant residues
27	FTC-UNI investigation coordination	Bioref.-biopts	Biomass agro-industrial waste
28	FTC-UNI Materials Laboratory	Bioref.-biopts	Biomass agro-industrial waste
29	FARQ-UNI research coordination	Bioref.-biopts	Tree biomass
30	FARQ-UNI research coordination	Bioref.-biopts	Biomass agro-industrial waste
31	FARQ research coordination	Bioref.-biopts	Plastic waste
32	Faculty of Chemical Engineering FIQ-UNI	Biotechnology applications	Organic matter, detergents, garlic, chili, banana vinegar, manure
33	FEC-UNI Electrical Engineering research group UNAN-Managua	Bioref.-biopts	Biomass agro-industrial waste
34	FAREM Matagalpa	Ecointensification	Solid and liquid waste
35	FAREM Esteli	Ecointensification	Solid and liquid waste
36	FAREM Juigalpa	Ecointensification	Solid and liquid waste
37	FAREM Carazo	Ecointensification	Animal residual biomass (solid and liquid waste)

References

Agasisti, T. & Salerno, C. (2007). Assessing the cost efficiency of Italian universities. *Education Economics*, 15(4), 455–471.

Ashley-Dejo, S.S. (2022). Analysis of technical efficiency of fish hatchery enterprises in Osun State, Nigeria: An application of two-stage DEA approach. *Journal of Bangladesh Agricultural University*, 20(3), 251–258. https://doi.org/10.5455/JBAU.5351.

Banker, R.D., Charnes, A., & Cooper, W.W. (1984). Some models for estimating technical and scale inefficiencies in data envelopment analysis. *Management Science*, 30, 1078–1092. https://doi.org/10.1287/mnsc.30.9.1078.

Boubaker, S., Le, T.D.Q., & Ngo, T. (2022). Managing bank performance under COVID-19: A novel inverse DEA efficiency approach. *International Transactions in Operational Research*, 30(2023), 2436–2452. https://doi.org/10.1111/itor.13132.

Boubaker, S., Manita, R., & Rouatbi, W. (2021). Large shareholders, control contestability and firm productive efficiency. *Annals of Operations Research*, 296(1), 591–614.

Catari-Yujra, G., Guzmán-Moreno, M.A., Hands, M., & Reyes-Osornio, M. (2022). Rendimiento de maíz en sistema de callejones de Inga spp. con diferentes dosis de fertilizantes naturales. *Revista Iberoamericana de Bioeconomia y Cambio Climatico*, 8(15), 1886–1897. https://doi.org/10.5377/ribcc.v8i15.14383.

Charnes, A., Cooper, W.W., & Rhodes, E. (1978). Measuring the efficiency of decision making units. *European Journal of Operational Research*, 2. https://doi.org/10.1016/0377-2217(78)90138-8.

Coelli, T.J. (1996). A guide to DEAP version 2.1: A data envelopment analysis (computer) program. No. 8/96. CEPA Working Papers. Department of Econometrics. Centre for Efficiency and Productivity Analysis, University of New England, Australia, 96(08), 1–49.

Colon-García, A.P., Catari-Yujra, G., & Alvarado, E. (2021). Los senderos productivos de la bioeconomía: El caso Honduras. *Revista Iberoamericana de Bioeconomia y Cambio Climatico*, 7(14), 1713–1726. https://doi.org/10.5377/ribcc.v7i14.12820.

Contreras, I. & Lozano, S. (2022). Size efficiency, splits and merger gains, and centralized resource reallocation of Spanish public universities. *Socio-Economic Planning Sciences*, 81, 101190.

Cook, W. & Seiford, M. (2009). Data envelopment analysis (DEA). Thirty years on. *European Journal of Operational Research*. https://doi.org/10.1016/j.ejor.2008.01.032.

Cooper, W.W., Seiford, M.L., & Karou, T. (2007). Data envelopment analysis. In: *A Comprehensive Text with Models, Applications, References and DEA-solver Software.* New York: Springer. https://doi.org/10.1007/978-0-387-45283-8.

Coupet, J. & Ba, Y. (2022). Benchmarking university technology transfer performance with external research funding: A stochastic frontier analysis. *The Journal of Technology Transfer*, 47(2), 605–620.

Debreu, G. (1951). The coefficient of resource utilization. *Econometrica: Journal of the Econometric Society*, 273–292. https://doi.org/10.2307/1906814.

El Gibari, S., Gómez, T., & Ruiz, F. (2022). Combining reference point based composite indicators with data envelopment analysis: Application to the assessment of universities. *Scientometrics*, 127(8), 4363–4395.

Fandel, G. (2007). On the performance of universities in North Rhine-Westphalia, Germany: Government's redistribution of funds judged using DEA efficiency measures. *European Journal of Operational Research*, 176(1), 521–533.

Färe, R., Grosskopf, S., & Lovell, C.A.K. (1994). *Production Frontiers.* New York: Cambridge University Press. https://doi.org/10.1007/978-1-4612-2626-0_1.

Farrel, M.J. (1957). The measurement of productive efficiency. *Journal of the Royal Statistical Society*, ACXX, Part 3, 253–290. https://doi.org/10.2307/2343100.

Fernández-Santos, Y., Martínez-Campillo, A., & Fernández-Fernández, J.M. (2013). Evaluación de la eficiencia y el cambio de productividad en el sistema universitario público español tras la implantación de la LOU. *Hacienda Pública Española/Review of Public Economics*, 205(2), 71–98.

García, A.R. & Zavala, A.O. (2022). Evaluación de la eficiencia del mezcal en las entidades federativas de México: un análisis de la envolvente de datos (DEA). *Inquietud Empresarial*, 22(1), 83–99.

García-Bucio, P., Sotelo-Navarro, P.X., Poggi-Varaldo, H.M., Cañizares-Villanueva, R.O., & Escamilla-Alvarado, C. (2022). Indicadores de la bioeconomía circular para el aprovechamiento de la fracción orgánica de los residuos sólidos urbanos. *Revista Internacional de Contaminación Ambiental*, 38, 78–92. https://doi.org/10.20937/RICA.54350.

Georgescu-Roegen, N. (1971). *The Entropy Law and the Economic Process.* Cambridge: Harvard University Press.

Georgescu-Roegen, N. (1976). *Energy and Economic Myths: Institutional and Analytical Economic Essays.* New York: Pergamon Press, p. 236.

Glass, J.C., McCallion, G., McKillop, D.G., Rasaratnam, S., & Stringer, K.S. (2006). Implications of variant efficiency measures for policy evaluations in UK higher education. *Socio-economic Planning Sciences*, 40(2), 119–142.

González-Araya, M. & Vásquez, G.V. (2010). Análisis de eficiencia y productividad de las universidades chilenas mediante análisis y encapsulamiento de datos. *Aporte Santiaguino*, 3(2), 245–256. https://doi.org/10.32911/as.2010.v3.n2.446.

Hammami, H., Ngo, T., Tripe, D., & Vo, D.-T. (2022). Ranking with a Euclidean common set of weights in data envelopment analysis: With application to the Eurozone banking sector. *Annals of Operations Research*, 311, 675–694.

Jauhar, S.K., Zolfagharinia, H., & Amin, S.H. (2022). A DEA-ANN-based analytical framework to assess and predict the efficiency of Canadian universities in a service supply chain context. *Benchmarking: An International Journal*. https://doi.org/10.1108/BIJ-08-2021-0458.

Koopmans, T.C. (1951). Efficient allocation of resources. *Econometrica: Journal of the Econometric Society*, 455–465. https://doi.org/10.2307/1907467.

Kuramoto, J. (2021). Ciencia, tecnología e innovación. Balance de investigación en Políticas Públicas 2011–2016. *Agenda de Investigación*, 638–678.

Lanteri, L.N. (2002). Productividad, desarrollo tecnológico y eficiencia. la propuesta de los índices Malmquist. Anales de la Asociación Argentina de Economía Política, XXXVII Reunión Anual, Tucumán, Argentina [en línea]. www.aaep.org.ar.

Lee, D.H. (2022). Production efficiency and economic benefit evaluation of biohydrogen produced using macroalgae as a biomass feedstock in Asian circular economies. *International Journal of Hydrogen Energy*, 47(96), 40532–40551.

Liu, X., Chen, X., Wu, Q., Deveci, M., & Delen, D. (2022). Measuring efficiency of the high-tech industry using uncertain multi-stage nonparametric technologies. *Expert Systems with Applications*, 119490.

López-González, Á.S., Zúniga-González, C.A., Rangel-Cura, R.A., Dios-Palamares, R., Colón-García, A.P., & Quirós-Madrigal, O.J. (2016a). Impacto de la productividad y eficiencia técnica de los granos básicos en Nicaragua, 1961–2013. *Revista Iberoamericana de Bioeconomia y Cambio Climatico*, 2(1), 411–436. https://doi.org/10.5377/ribcc.v2i1.5682.

López-González, Á.S., Zúniga-González, C.A., López, M.R., Quirós-Madrigal, O.J., Colón-García, A.P., Navas-Calderón, J., Martínez-Andrades, E., & Rangel-Cura, R.A. (2016b). Estado del arte de la medición de la productividad y la eficiencia técnica en América Latina: Caso Nicaragua. *Revista Iberoamericana de Bioeconomia y Cambio Climatico*, 1(2), 76–100. https://doi.org/10.5377/ribcc.v1i2.2478.

Lubsanova, N.B., Maksanova, L.B.Z., Eremko, Z.S., Bardakhanova, T.B., & Mikheeva, A.S. (2022). The eco-efficiency of Russian Regions in North Asia: Their green direction of regional development. *Sustainability*, 14(19), 12776.

Maza Ávila, F.J., Vergara Schmalbach, J.C., & Román Romero, R. (2017). Eficiencia y productividad en la cobertura de las universidades públicas colombianas. *Investigación y Desarrollo*, 25(2), 6–33. https://doi.org/10.14482/indes.25.2.10957.

Milán Pérez, J.A. & Zúniga-Gonzalez, C.A. (2021). Necesidades de investigación y transferencia de tecnologías sobre cambio climático en Nicaragua: Una oportunidad en la Bioeconomía. *Revista Iberoamericana de Bioeconomia y Cambio Climatico*, 7(13), 1518–1543. https://doi.org/10.5377/ribcc.v7i13.11270.

Moreira, V. & Bravo, B. (2009). Un estudio de eficiencia técnica en lecherías usando meta regresión: Una perspectiva internacional. *Chilean Journal of Agricultural Research*, (9), 1–2.

Mozaffari, M.R., Ostovan, S., Wanke, P.F., & Tan, Y. (2022). Evaluation of multi-stage fuzzy networks in DEA and DEA-R based on liquidity ratios with undesirable outputs. *International Journal of Fuzzy Systems*, 24(5), 2411–2446.

Naderi, A. (2022). Efficiency measurement of higher education units using multilevel frontier analysis. *Journal of Productivity Analysis*, 57(1), 79–92.

Ngo, T. & Tsui, K.W.H. (2021). Estimating the confidence intervals for DEA efficiency scores of Asia-Pacific airlines. *Operational Research*. https://doi.org/10.1007/s12351-021-00667-w.

Núñez Jover, J. & Montalvo Arriete, L.F. (2015). La política de ciencia, tecnología e innovación en Cuba y el papel de las universidades. *Revista Cubana de Educación Superior*, 34(1), 29–43.

Olariu, G.V. & Brad, S. (2022). Preventive risk management of resource allocation in Romanian higher education by assessing relative performance of study programs with DEA method. *Sustainability*, 14(19), 12527.

O'Neill, L., Rauner, M., Heidenberger, K., & Kraus, M. (2008). A cross-national comparison and taxonomy of DEA-based hospital efficiency studies. *Socio-Economic Planning Sciences*, 42(3), 158–189. https://doi.org/10.1016/j.seps.2007.03.001.

Orozco, N.V.B. & Giraud, L. (2022). Optimización de la gestión ambiental y económica del recurso energético biomasa para la generación de energía eléctrica en Nicaragua. *Fuentes: El reventón Energético*, 20(1), 87–103.

Oviedo, W. & Rodríguez, G. (2011). Medición de la eficiencia técnica relativa de las fincas asociadas a Coounión en Guasca Cundinamarca. *Revista MVZ Córdoba*, 16(2), 2616–2627. https://doi.org/10.21897/rmvz.1036.

Papadopoulos, A. & Parmeter, C.F. (2022). Quantile methods for stochastic frontier analysis. *Foundations and Trends ® in Econometrics*, 12(1), 1–120.

Rangel Cura, R.A., Zúniga González, C.A., Colón García, A.P., Losilla Solano, L., & Berrios-Zepeda, R.A. (2015). Medición de la contribución de la bioeconomía en América Latina: Caso Cuba. *Revista Iberoamericana de Bioeconomia y Cambio Climatico*, 1(1), 223–240. https://doi.org/10.5377/ribcc.v1i1.2150.

Ratner, S.V.E. (2017). Dynamic problems of estimation of ecological-economic efficiency of regions based on basic models of data envelopment analysis. *Upravlenie Bol'shimi Sistemami*, 67, 81–106.

Robles, E.A. (2021). Crecimiento de la productividad total de los factores en Costa Rica e inestabilidad macroeconómica. *Revista de Ciencias Económicas*, 39(1), 1–24. https://doi.org/10.15517/rce.v39i1.47622.

Serrano, V.C. & Blasco, O.M.B. (2000). *Evaluación de la eficiencia mediante el Análisis Envolvente de Datos: Introducción a los modelos básicos*. B-EUMED. Universidad de Valencia.

Shephard, R.W. (1970). *Theory of Cost and Production Functions*. Princeton: Princeton University Press.

Simar, L. & Wilson, P.W. (2007). Estimation and inference in two-stage, semi-parametric models of production processes. *Journal of Econometrics*, 136, 31–64.

Svetlanská, T., Turčeková, N., & Adamičková, I. (2015). Efficiency of Biomass Production–Methodological approaches. *Visegrad Journal on Bioeconomy and Sustainable Development*, 4(1), 2–6.

Thaker, K., Charles, V., Pant, A., & Gherman, T. (2021). A DEA and random forest regression approach to studying bank efficiency and corporate governance. *Journal of the Operational Research Society*, 1–28. https://doi.org/10.1080/01605682.2021.1907239.

Thorndyke, L.E., Milner, R.J., & Jaffe, L.A. (2022). Endowed chairs and professorships: A new frontier in gender equity. *Academic Medicine*, 97(11), 1643–1649.

Tran, D.H. & Ngo, D.T. (2014). Performance of the Vietnamese automobile industry: A measurement using DEA. *Asian Journal of Business and Management*, 2(3), 184–191.

Tsionas, M.G. (2021). Optimal combinations of stochastic frontier and data envelopment analysis models. *European Journal of Operational Research*. https://doi.org/https://doi.org/10.1016/j.ejor.2021.02.003.

Vargas-Leitón, B., Solís-Guzmán, O., Sáenz-Segura, F., & León-Hidalgo, H. (2015). Eficiencia técnica en hatos lecheros de Costa Rica. *Agronomía Mesoamericana*, 26(1), 1–15. https://doi.org/10.15517/am.v26i1.16886.

Wasiaturrahma, Sukmana, R., Ajija, S.R., Salama, S.C.U., & Hudaifah, A. (2020). Financial performance of rural banks in Indonesia: A two-stage DEA approach. *Heliyon*, 6(7), 1–9.

Yu, S., Zheng, Y., Hu, X., & Shu, K. (2022). Spatial impacts of biomass resource endowment on provincial green development efficiency. *Renewable Energy*, 189, 651–662.

Zhu, Q., Aparicio, J., Li, F., Wu, J., & Kou, G. (2022b). Determining closest targets on the extended facet production possibility set in data envelopment analysis: Modeling and computational aspects. *European Journal of Operational Research*, 296(3), 927–939.

Zhu, Z., Jin, Y., Su, Y., Jia, K., Lin, C.L., & Liu, X. (2022a). Bibliometric-based evaluation of the neuromarketing research trend: 2010–2021. *Frontiers in Psychology*, 13, 872468.

Zúniga-González, C.A. (2011). Texto básico de economía agrícola: Su Importancia para el Desarrollo Local Sostenible. Editorial Universitaria, UNAN León. DOI: 10.22004/ag.econ.111604. Registro de la propiedad Intelectual No OL-019-2011. https://doi.org/10.22004/ag.econ.111604.

Zúniga-González, C.A. (2022). Data for: Study of the efficiency and productivity by the productive paths of the Bioeconomy in the CNU Universities. figshare. Book. https://doi.org/10.6084/m9.figshare.21625517.v1.

Zúniga-González, C.A. & Jaramillo-Villanueva, J.L. (2022). Frontier model of the environmental inefficiency effects on livestock bioeconomy [version 1; peer review: Awaiting peer review]. *F1000Research*, 11, 1382. https://doi.org/10.12688/f1000research.128071.1.

Zúniga-González, C.A., Moreno-Mayorga, L.F., & Quiroz-Medina, C.R. (2022a). Estudio de la eficiencia técnica en escuelas de campo de Nicaragua. *Revista Tecnología en Marcha*, 128–140. https://doi.org/10.18845/tm.v35i3.5696.

Zúniga-González, C.A., López, M.R., Icabaceta, J.L., Vivas-Viachica, E.A., & Blanco-Orozco, N. (2022b). Epistemología de la Bioeconomia. *Revista Iberoamericana de Bioeconomia y Cambio Climatico*, 8(15), 1786–1796. https://doi.org/10.5377/ribcc.v8i15.13986.

Zurano-Cervelló, P., Pozo, C., Mateo-Sanz, J.M., Jiménez, L., & Guillén-Gosálbez, G. (2018). Eco-efficiency assessment of EU manufacturing sectors combining input-output tables and data envelopment analysis following production and consumption-based accounting approaches. *Journal of Cleaner Production*, 174, 1161–1189.

Zúniga-González, C.A., Durán Zarabozo, O., Dios Palomares, R., Sol Sánchez, A., Guzman Moreno, M.A., Quiros, O., & Montoya Gaviria, G.D.J. (2014). Estado del arte de la bioeconomía y el cambio climático (No. 1133-2016-92457, pp. 20–329). https://doi.org/10.22004/ag.econ.168356.

Part II

Data Envelopment Analysis (DEA) in Banking and Finance

https://doi.org/10.1142/9781800615786_0008

Chapter 8

Assessing the Efficiency Impact of COVID-19 on Australian Private Health Insurers Using DEA

Lan Nguyen[*,‡] **and Andrew C. Worthington**[†,§]

[*]*Centre for the Business and Economics of Health, University of Queensland, Australia*

[†]*Department of Accounting, Finance and Economics, Griffith University, Australia*

[‡]*lan.nguyen@uq.edu.au*

[§]*a.worthington@griffith.edu.au*

Abstract

The COVID-19 pandemic has had a pronounced effect on the Australian private health insurance (PHI) industry. With government-imposed restrictions limiting access to hospital and nonhospital treatments, fund benefits paid have declined or grown only slowly since 2020. At the same time, PHI insurers have pursued technological innovations in service delivery, including work-from-home arrangements under mandated lock-downs, resulting in declining fund administration and claims handling expenses, while the share of Australians with PHI increased for the first time since 2015 given the increased community focus on health, with a corresponding increase in premium revenue. Investment revenue has been and still is extremely volatile. This chapter examines the impacts of these on the technical and scale efficiency of Australia's 31 largest PHI insurers during the period 2016–2022 using data envelopment analysis. We find that mean pure technical efficiency initially worsened and then quickly improved during and after 2020, while scale efficiency also declined but took longer to recover. However, the PHI industry continues to run at lower levels of technical and scale efficiency than before the pandemic, with fewer best-practice insurers, corresponding to the incredibly significant disruption of COVID-19 to this industry.

Keywords: Data envelopment analysis, private health insurance, COVID-19, pure technical efficiency, technical efficiency, scale efficiency, value-added, inputs and outputs.

1. Introduction and Background

Australia has a mixed public and private healthcare system, consisting of Medicare — a mandatory universal healthcare scheme introduced in 1984 — and voluntary private health insurance (PHI). While Medicare accounts for most healthcare expenditure in Australia, PHI plays an essential role in the sustainability of the healthcare system, offering a flexible choice of healthcare for consumers, injecting funds into the health system, increasing healthcare capacity, and enhancing the responsibility of individuals for good health. It does this by providing cover for healthcare not covered by Medicare, including hospital cover to help with the cost of treatment as a private patient in the hospital and extras (or general or ancillary) cover to help pay for out-of-hospital health services that Medicare does not cover (including dental, optical, obstetrics, physiotherapy. and ambulance levies where applicable). As one consequence, the PHI industry in Australia is not only highly regulated but also strongly supported through government policy aimed at improving the affordability of PHI for policyholders and thereby the viability and sustainability of the PHI industry, especially given the demands of strongly growing healthcare pricing and demand (and therefore PHI benefit claims). Nguyen & Worthington (2021) discuss these policies and their impact on individual insurers and the industry.

Prior to the outbreak of COVID-19 in Australia in January 2020, the PHI industry was already facing tough questions about its ongoing viability. The industry is highly concentrated, with the five largest funds accounting for three-quarters of all industry assets, but with a considerable number of small- and medium-sized funds making up the balance. The Australian Prudential Regulation Authority (APRA), as the industry regulator, has long expressed its concerns about the stability of the industry given the growth of benefit claims, and there have been calls for a sector wide inquiry by the Productivity Commission. A 2017 Senate inquiry into the affordability of PHI for policymakers made several recommendations, and some rule changes in 2018 were implemented to make PHI simpler and more attractive, especially to younger cohorts. However, these reforms, alongside existing PHI carrot-and-stick incentives (rebates) and penalties (lifetime cover), have not had the desired impact on membership. As shown in Table 1, the population share of Australians with hospital cover continuously declined from September 2017 to September 2019, representing only 44.0%, as did general (or nonhospital or ancillary or extras) cover, for 53.4% of the population.

Table 1. Hospital and general cover, number and % share of population, September 2017–September 2022.

	Hospital cover		**General cover**	
Qtr.	***N***	**%**	***N***	**%**
Sep-17	11,318,588	45.8	13,531,883	54.9
Dec-17	11,306,294	45.7	13,534,821	54.8
Mar-18	11,316,775	45.5	13,581,613	54.7
Jun-18	11,259,263	45.1	13,543,193	54.6
Sep-18	11,254,007	44.9	13,564,249	54.1
Dec-18	11,241,637	44.7	13,567,720	54.0
Mar-19	11,256,108	44.6	13,612,007	53.9
Jun-19	11,227,569	44.3	13,593,547	53.7
Sep-19	11,242,726	44.2	13,628,326	53.6
Dec-19	11,233,123	44.0	13,629,205	53.4
Mar-20	11,225,962	43.8	13,637,241	53.2
Jun-20	11,197,395	43.7	13,599,782	53.0
Sep-20	11,301,501	44.1	13,742,486	53.6
Dec-20	11,336,302	44.2	13,798,905	53.8
Mar-21	11,396,181	44.4	13,886,876	54.1
Jun-21	11,442,584	44.6	13,952,256	54.3
Sep-21	11,506,349	44.8	14,023,893	54.6
Dec-21	11,564,806	44.9	14,102,979	54.7
Mar-22	11,621,039	44.9	14,186,527	54.8
Jun-22	11,678,283	45.0	14,260,742	54.9
Sep-22	11,763,117	45.2	14,357,579	55.1

Notes: This table lists the number and percentage share of the population of Australians with hospital and general (extras) cover for PHI at the end of each quarter from September 2017 to September 2022 (the latest quarter available).

Source: APRA, Quarterly Private Health Insurance Membership and Benefits Trends, September 2022.

This changed abruptly and largely unexpectedly with the outbreak of COVID-19 in Australia in January 2020. At once, government-imposed health restrictions in response to the pandemic began to limit PHI policyholders' access to elective surgery and extra treatments (including most dental, optical, and other health services). Paradoxically, despite these significant restrictions on the use of cover, the number of Australians taking up PHI cover grew strongly. A greater focus on health prompted by the pandemic was an obvious contributor, as was the restraint placed by PHI insurers on premium increases. After an initial fall in membership, as shown in Table 1, the number of Australians with hospital and general cover, respectively, grew by 103,179 and 169,700 persons in the first calendar year of the pandemic and by another 228,204 and 304,074 persons in the second, placing the shares of the population with PHI hospital (45.2%) and general (55.1%) cover at highs not seen since 2017.

The impact on the PHI industry is clear. As shown in Table 2, insured days in private hospitals, private day facilities, and public hospitals have all declined since 2019/20 (taken up only in part by the increase in insured days for hospital substitutes), with total insured days and the ratio of total insured days to average insured persons also falling. Clearly, COVID-19 restrictions resulted in fewer claims, and this reduced the benefits paid out by insurers for both hospital and extras treatments during part of 2019/2020, all of 2020/2021, and much of 2021/2022. For example, it was not until October 5, 2021, that NSW private hospitals were given permission to recommence some non-urgent day surgery, with overnight elective surgery suspended again on January 10, 2022, to ensure sufficient staffing and bed capacity to meet demand during the Omicron wave, followed by a further staged resumption of elective surgery on February 1, 2022. Likewise, dental care, being the largest part of extras cover by expenditure, did not fully recommence until October 22, 2021, in Victoria, some 22 months after the outbreak of COVID-19.

Table 3 summarizes the impact of these circumstances on the financial performance and position of the industry. As shown, while premium revenue continued to grow by 1.4% in 2019/2020 and 3.2% in 2020/2021, the growth in fund benefits paid slowed to almost zero, growing by 3.8% in 2019/2020, but only 0.1% in 2020/2021 and just 0.7% in 2021/2022. For its part, health insurance business (HIB) expenses, including claims handling, grew in the period 2019/2020–2021/2022 by between 4.4% and 8.1%, but declined by

Table 2. Privately insured hospital and average days per privately insured person, inclusive of nursing home patients, 2017/18–2021/2022.

	2017/2018		2018/2019		2019/2020		2020/2021		2021/2022	
Measure	*N*	Δ%	*N*	Δ%	*N*	Δ%	*N*	Δ%	*N*	Δ%
Insured days in private hospitals	8,093,276	1.7	8,167,202	0.9	8,052,533	−1.4	8,034,878	−0.2	7,769,011	−3.3
Insured days in private day hospital facilities	629,301	0.0	619,442	−1.6	583,198	−5.9	626,737	7.5	640,469	2.2
Insured days in public hospitals	3,030,807	0.3	2,982,434	−1.6	2,931,983	−1.7	2,624,180	−10.5	2,537,126	−3.3
Insured days for hospital-substitute	234,179	3.7	286,669	22.4	313,780	9.5	319,154	1.7	416,223	30.4
Total days	11,987,563	1.3	12,055,747	0.6	11,881,494	−1.4	11,604,949	−2.3	11,362,829	−2.1
Average hospital insured persons	11,307,665	−0.2	11,248,792	−0.5	11,228,573	−0.2	11,338,493	1.0	11,563,157	2.0
Ratio of total days to average insured persons	1.06	1.5	1.07	1.1	1.06	−1.3	1.02	−3.3	0.98	−4.0

Notes: This table details the number of PHI insured days in private, private day, and public hospitals and hospital substitutes as a percentage of total PHI insured days from 2017/2018 to 2021/2022. Insured days (in hospital) are: (a) intended to manage a disease, injury, or condition; (b) provided to a person: (i) by a person who is authorized by a hospital to provide treatment, (ii) under the management or control of such a person; (c) either: (i) provided at a hospital, (ii) provided or arranged with the direct involvement of a hospital. Insured days in hospital substitute is general treatment that: (a) substitutes for an episode of hospital treatment, (b) is any of, or any combination of, nursing, medical, surgical, pediatric surgical, diagnostic, therapeutic, prosthetic, pharmacological, pathology, or other services intended to manage a disease, injury, or condition. Average hospital insured persons is the number of persons for hospital cover (less than total insured days for multiple-day stays), and the ratio of total days to averaged insured persons is a measure of the relative length of hospital stay, with a ratio of 1.0 showing an average stay of one day. The percentage change is the change from the previous year (data for 2016/2017 are not shown).

Source: APRA, Quarterly Private Health Insurance Membership and Benefits Trends, various years.

Table 3. PHI industry financial performance and position, 2017/18–2021/2022.

	2017/2018		2018/2019		2019/2020		2020/2021		2021/2022	
	$	Δ%	$	Δ%	$	Δ%	$	Δ%	$	Δ%
Premium revenue	23,899,411	—	24,561,953	2.8	24,893,599	1.4	25,679,086	3.2	26,586,551	3.5
Investment revenue	412,318	—	464,700	12.7	149,788	−67.8	627,118	318.7	(293,822)	−146.9
Other revenue	202,203	—	203,348	0.6	276,195	35.8	69,604	−74.8	(3,390)	−104.9
Total revenue	**24,513,932**	—	**25,230,001**	**2.9**	**25,319,582**	**0.4**	**26,375,808**	**4.2**	**26,289,339**	**−0.3**
Fund benefits	20,279,843	—	20,884,820	3.0	21,676,089	3.8	21,688,962	0.1	21,837,587	0.7
State ambulance levies	227,390	—	233,634	2.7	238,397	2.0	244,484	2.6	256,760	5.0
Total fund benefits	**20,507,233**	—	**21,118,453**	**3.0**	**21,914,487**	**3.8**	**21,933,446**	**0.1**	**22,094,347**	**0.7**
HIB expenses	1,733,472	—	1,828,324	5.5	1,887,468	3.2	2,017,520	6.9	2,197,476	8.9
HIB claims handling	397,064	—	413,671	4.2	412,833	−0.2	422,464	2.3	430,613	1.9
Other expenses	66,643	—	51,324	−23.0	95,778	86.6	85,574	−10.7	103,073	20.4
Total expenses	**2,197,179**	—	**2,293,318**	**4.4**	**2,396,079**	**4.5**	**2,525,558**	**5.4**	**2,731,162**	**8.1**
Surplus/(deficit) before tax	1,809,521	—	1,818,229	0.5	1,009,016	−44.5	1,916,804	90.0	1,463,830	−23.6
Taxation expense	444,716	—	437,424	−1.6	286,509	−34.5	440,607	53.8	440,768	0.0
Surplus/(deficit) after tax	**1,364,805**	—	**1,380,806**	**1.2**	**722,507**	**−47.7**	**1,476,197**	**104.3**	**1,023,062**	**−30.7**
Gross margin	14.2%	—	14.00%	−1.4	12.0%	−14.5	14.6%	21.9	16.9%	15.8
Management expenses	8.9%	—	9.10%	2.1	9.2%	1.5	9.5%	2.8	9.9%	4.0
Net margin	5.3%	—	4.90%	−7.2	2.7%	−44.3	5.1%	86.5	7.0%	37.9
Total assets	14,335,152	—	15,176,910	5.9	16,591,401	9.3	17,724,263	6.8	19,750,997	11.4
Total liabilities	6,122,852	—	6,487,363	6.0	7,951,221	22.6	8,223,467	3.4	9,869,104	20.0
Health benefits fund capital	8,212,300	—	8,689,547	5.8	8,640,180	−0.6	9,500,795	10.0	9,881,893	4.0

Notes: This table details the PHI industry aggregate financial performance and position from 2017/2018 to 2021/2022. Percentage change is the change from the previous year (data for 2016/2017 are not available, so not shown for 2017/2018). Total revenue – total fund benefits – total expenses = Surplus/(deficit) after tax. Gross margin is the difference between total premium revenue and the total cost of benefits, including state ambulance levies, expressed as a percentage of total premium revenue. Health insurance business (HIB) is the business of undertaking liability by way of insurance or an employee health benefits scheme that relates to hospital treatment and general treatment. Net margin is gross margin less management expenses, expressed as a percentage of premium revenue. Values in $000s.
Source: APRA, Operations of Private Health Insurers Annual Report, various years.

0.2% in 2019/2020, and other expenses at first increased by 86.6% in 2019/2020 and then fell by 10.7% in 2020/2021.

These changes in premium revenue, fund benefit payments, and HIB expenses reflect the varying effects of the disruption to benefit claims from government restrictions on healthcare services and insurer initiatives in improving the value of PHI (including short-stay hospital care and services provided via telehealth, such as psychology, physiotherapy, and speech pathology), and treatment outside of hospital (such as chemotherapy at home). They also mirror responses by insurers to the pandemic aimed at policyholders (including hardship measures such as premium waivers and the refund of unclaimed benefits) and changes to insurer business operations and the handling of claims. Lastly, investment revenue varied directly with the initially negative and then positive effects of COVID-19 on volatile global financial markets, growing by 12.7% in 2018/2019 and falling by 67.8% in 2019/2020, before surging in 2020/2021 by 318.7% and settling back at 146.7% in 2021/2022.

Finally, the summary financial performance measures in Table 3 reflect these changes in revenues, fund benefits, and fund expenses. As shown, while PHI insurer gross margins fell by 1.4% and 14.5% on average in 2018/2019 and 2019/2020, they improved by 21.9% and 15.8% in 2020/2021 and 2021/2022, respectively. Net margins improved even more, rising by 86.5% and 37.9% in 2020/2021 and 2021/2022 after declining by 7.2% and 44.3% in 2018/2019 and 2019/2029, respectively. Of course, univariate and bivariate financial statement analysis of dollar values, percentages, and ratios only go so far in explaining changes in the way in which PHI insurers changed their operations during and after the COVID-19 pandemic. This is especially the case of PHI as a highly regulated industry where there are significant constraints on profit-making through controls on premiums, policies restrict coverage to only certain healthcare services and not Medicare-funded services, and the industry consists of few open-for-profit entities and relatively many restricted not-for-profit entities.

To address these conditions, we employ data envelopment analysis (DEA), a nonparametric frontier efficiency measurement technique, in the rest of the chapter. Section 2 outlines the method and the data employed. Section 3 presents the results. Section 4 supplies a summary and the implications of the analysis for policy and practice.

2. Method and Data

In this chapter, we employ DEA, a technique widely used in the analysis of efficiency in the insurance industry, to analyze the efficiency of the Australian PHI industry over the period 2016–2022. Nguyen & Worthington (2021) present the DEA model and discuss the selection of the inputs and outputs used in the model for this industry in detail. We employ identical inputs and outputs as justified at length there and specify premium and investment revenue as outputs, and equity capital, liabilities, and management expenses as inputs in our value-added approach to insurer behavior.

In terms of the application of DEA in insurance studies, Worthington & Hurley (2002), Luhnen (2009), Cummins *et al.* (2010), Barros *et al.* (2010), Biener & Eling (2012), Cummins & Xie (2013), Huang & Eling (2013), Yaisawarng *et al.* (2014), Alhassan & Biekpe (2015), Biener *et al.* (2016), and Eling & Schaper (2017) consider life and nonlife (general) insurance, and Carrington *et al.* (2011) and Nguyen & Worthington (2021) private health insurance, while Boubaker *et al.* (2022a) apply DEA to other financial institutions during COVID-19. Boubaker *et al.* (2022b) and Le *et al.* (2021) present some useful recent extensions to DEA in financial products and services. Of relevance, as there are some negative values for investment revenue given the financial market volatility surrounding COVID-19, we transform negative values into positive values using the base point method introduced by Tone *et al.* (2020).

We collect data from the Operations of Private Health Insurers Annual Reports from 2018/2019 to 2021/2022, as produced by APRA under Section 167 of the Private Health Insurance (Prudential Supervision) Act 2015 (2018, 2019, 2020, 2021, 2022a). The reports summarize the financial performance and financial position of registered PHI insurers (and funds) for the year ending June 30. We deflate all monetary values (2016/2017 = 100) using the medical care price index from the Australian Bureau of Statistics.

Of the 34 registered private health insurers registered on June 30, 2022, we include only the 31 insurers operating continuously over the sample period, representing 99.7% of the market and thus reflective of the Australian private health insurance industry as a whole. Table 4 lists the insurers and whether they are for-profit (alternatively not-for-profit) or restricted access (alternatively unrestricted

Table 4. Sampled PHIs.

Insurer's name	Code	For-profit	Restricted access	Ten largest	Ten smallest
ACA Health Benefits Fund Ltd.	ACA	—	√	—	√
Australian Unity Health Ltd.	AUHL	√	—	√	—
BUPA HI Pty Ltd.	BUPA	√	—	√	—
CBHS Health Fund Ltd.	CBHS	—	√	√	—
CBHS Corporate Health Pty Ltd.	CBHSCH	—	—	—	√
Cessnock District Health Benefits Fund Ltd.	CDH	—	—	—	√
CUA Health Ltd.	CUA	—	—	—	√
Defence Health Ltd.	Defence	—	√	√	—
Doctors' Health Fund Pty Ltd.	DHF	√	√	—	√
GMHBA Ltd.	GMHBA	—	—	√	—
HBF Health Ltd.	HBF	—	—	√	—
Hospitals Contribution Fund of Australia Ltd.	HCF	—	—	√	—
Health Care Insurance Ltd.	HCI	—	—	—	√
Health Insurance Fund of Australia Ltd.	HIF	—	—	—	—
Health Partners Ltd.	HP	—	—	—	—
Latrobe Health Services Ltd.	Latrobe	—	—	—	—
Mildura District Hospital Fund Ltd.	Mildura	—	—	—	—
Medibank Private Ltd.	MPL	√	—	√	—
Navy Health Ltd	Navy	—	√	—	—
National Health Benefits Australia Pty Ltd.	NHBA	√	—	—	√
NIB Health Funds Ltd.	NIB	√	—	√	—
Peoplecare Health Ltd.	Peoplecare	—	—	—	—
Phoenix Health Fund Ltd.	Phoenix	—	—	—	√
Police Health Ltd.	Police	—	√	—	√
Queensland Country Health Fund Ltd.	QCH	—	—	—	—
Queensland Teachers' Union Health Fund Ltd.	QTUH	—	√	—	—
Reserve Bank Health Society Ltd.	RBHS	—	√	—	√
St Luke's Medical and Hospital Benefits Assoc.	St Luke's	—	—	—	—
Teachers Federation Health Ltd.	TFH	—	√	√	—
Transport Health Pty Ltd.	Transport	√	—	—	√
Westfund Ltd.	Westfund	—	—	—	—

Table 5. Descriptive statistics, 2017/18–2021/22.

Statistic	Variable	Type	2017/2018	2018/2019	2019/2020	2020/2021	2021/2022
Mean	Premium revenue	Output	751.43000	758.01000	756.160000	770.42000	788.89000
	Investment revenue	Output	12.89452	14.36129	4.599226	19.00852	−8.76081
	Equity capital	Input	257.00390	267.69870	263.418000	284.97860	293.55830
	Liabilities	Input	191.48580	199.05710	242.231500	246.48190	293.42770
	Management expenses	Input	66.47290	68.48774	69.523540	73.99037	79.01195
Std. dev.	Premium revenue	Output	1,597.51300	1,588.59100	1,567.378000	1,586.62300	1,589.60100
	Investment revenue	Output	22.50721	24.94098	9.426011	32.71372	17.67469
	Equity capital	Input	435.41100	452.05590	431.247000	453.15630	459.48920
	Liabilities	Input	395.86000	416.99000	500.730000	492.88000	610.07000
	Management expenses	Input	131.05750	132.03030	131.037000	137.09100	146.54530
Min.	Premium revenue	Output	1.65000	3.38000	4.727300	4.46380	4.74770
	Investment revenue	Output	0.10000	0.12000	−0.917700	−0.17870	−96.90200
	Equity capital	Input	4.66000	3.59000	5.742600	5.58610	9.12480
	Liabilities	Input	0.84000	1.88000	2.861200	3.15800	3.58500
	Management expenses	Input	1.47000	1.77000	1.554100	1.04520	0.68110
Max.	Premium revenue	Output	6,498.50000	6,364.28000	6,193.723000	6,241.57300	6,204.34600
	Investment revenue	Output	92.88000	97.48000	37.822000	139.59920	2.41130
	Equity capital	Input	1,669.66000	1,779.65000	1,670.267000	1,762.51900	1,777.11100
	Liabilities	Input	1,639.03000	1,648.11000	2,051.506000	1,905.17900	2,429.91300
	Management expenses	Input	514.89000	512.83000	490.764600	538.78360	601.86040

Notes: This table provides means, standard deviations, minimums, and maximums for the two outputs (premium revenue and investment revenue) and three inputs (equity capital, liabilities, and management expenses) specified in the DEA analysis for each year of the same from 2017/2018 to 2021/2022. All values are real dollar values (2017/2018 = 100) in $000s.

Source: APRA, Operations of Private Health Insurers Annual Report, various years.

access) insurers (2022b). Unlike not-for-profit insurers, for-profit insurers pay income tax but have greater flexibility in the use of fund assets under existing regulation and potentially improved access to capital. Likewise, under existing regulations, all PHI funds cannot refuse membership on the grounds of health status, age, or claims history, nor if the individual is outside lifetime cover conditions, and must charge the same premium for the same policy to all policyholders, even if they are a restricted fund available only to those employed in a particular industry or profession.

Table 5 details descriptive statistics (means, standard deviation, minimum, and maximum) for the three inputs and two outputs for each year of the sample period. Even in the absence of COVID-19, we can clearly see using the values of the standard deviation that there is a wide dispersion in the sizes of these insurers. With the mean values, we can see that premium revenue has grown in real terms throughout the period, while investment revenue has declined, increased, declined again, and then became negative. For their part, liabilities in real terms increased dramatically from 2019/2020 onward, as did management expenses. Equity capital (as variously measured) has remained stable throughout in real terms.

3. Results & Discussion

Table 6 provides the overall technical efficiency (OTE), pure technical efficiency (PTE), and scale efficiency (SE) scores for each fund for each year from 2017/208 to 2021/2022. OTE is made up of PTE and SE, such that OTE $=$ PTE $\times$ SE. A maximum score of 1.000 represents an efficient fund, with anything less representing an inefficient fund. We also include whether a fund is experiencing increasing (irs), constant (crs), or decreasing (drs) returns to scale (RTS). For example, a fund may have an efficiency of less than one, but this may be because it is either too small or too large, and irs or drs returns to scale shows whether it needs to increase or reduce its scale to become fully efficient, respectively.

Consider, as an example, ACA in the first row in 2019/2020, which has an OTE of 0.789, a PTE of 0.831, an SE of 0.949, and a RTS of irs. Overall, this fund is 78.9% efficient (or 11.1% inefficient) in terms of its OTE input–output relation to the best-practice frontier and could

Table 6. Technical and scale efficiencies and returns to scale, 2017/18–2021/2022.

Insurer	2017/2018				2018/2019				2019/2020				2020/2021				2021/2022			
	OTE	PTE	SE	RTS	OTE	PTE	SE	RTS	OTE	PTE	SE	RTS	OTE	PTE	SE	RTS	OTE	PTE	SE	RTS
ACA	0.961	0.968	0.992	drs	0.996	1.000	0.996	irs	0.789	0.831	0.949	irs	0.802	0.921	0.871	irs	0.905	0.905	1.000	crs
AUHL	1.000	1.000	1.000	crs	0.939	0.943	0.996	irs	0.804	0.828	0.970	drs	0.741	0.742	0.999	irs	0.771	0.794	0.971	drs
BUPA	1.000	1.000	1.000	crs	1.000	1.000	1.000	crs	1.000	1.000	1.000	crs	1.000	1.000	1.000	crs	1.000	1.000	1.000	crs
CBHS	1.000	1.000	1.000	crs	1.000	1.000	1.000	crs	0.889	0.982	0.905	drs	0.956	1.000	0.956	drs	0.715	0.805	0.889	drs
CBHSCH	0.501	1.000	0.501	irs	0.419	1.000	0.419	irs	1.000	1.000	1.000	crs	0.335	1.000	0.335	irs	1.000	1.000	1.000	crs
CDH	0.620	1.000	0.620	irs	0.591	1.000	0.591	irs	1.000	1.000	1.000	crs	0.769	1.000	0.769	irs	1.000	1.000	1.000	crs
CUA	0.876	0.881	0.993	drs	0.931	0.931	1.000	crs	0.866	0.867	0.999	drs	0.892	0.909	0.981	irs	0.888	0.889	0.999	crs
Defense	1.000	1.000	1.000	crs	1.000	1.000	1.000	crs	1.000	1.000	1.000	crs	0.858	0.966	0.888	drs	0.717	0.807	0.889	drs
DHF	0.847	0.975	0.869	irs	0.925	1.000	0.925	irs	0.903	1.000	0.903	drs	0.752	0.856	0.878	irs	1.000	1.000	1.000	crs
GMHBA	0.819	0.832	0.984	drs	0.832	0.968	0.860	drs	0.807	1.000	0.807	drs	0.697	0.817	0.853	drs	0.666	0.670	0.995	drs
HBF	0.772	1.000	0.772	drs	0.745	1.000	0.745	drs	0.692	1.000	0.692	drs	0.580	0.851	0.682	drs	0.497	0.980	0.507	drs
HCF	0.831	1.000	0.831	drs	0.818	1.000	0.818	drs	0.729	1.000	0.729	drs	0.822	1.000	0.822	drs	0.694	1.000	0.694	drs
HCI	0.799	0.939	0.851	irs	0.804	1.000	0.804	irs	0.822	0.956	0.860	irs	0.695	0.944	0.736	irs	0.972	0.973	0.999	irs
HIF	1.000	1.000	1.000	crs	0.816	0.824	0.991	irs	0.762	0.763	0.999	drs	1.000	1.000	1.000	crs	0.665	0.665	1.000	crs
HP	1.000	1.000	1.000	crs	1.000	1.000	1.000	crs	0.809	0.822	0.984	drs	0.878	0.883	0.995	irs	0.794	0.795	0.999	crs
Latrobe	0.751	0.774	0.970	irs	0.639	0.660	0.968	irs	0.657	0.670	0.981	drs	0.584	0.588	0.994	irs	0.487	0.487	1.000	crs
Mildura	0.807	0.943	0.856	irs	1.000	1.000	1.000	crs	0.962	1.000	0.962	drs	0.851	0.960	0.886	irs	0.899	0.900	0.999	irs
MPL	0.971	1.000	0.971	drs	0.955	1.000	0.955	drs	0.999	1.000	0.999	drs	1.000	1.000	1.000	crs	1.000	1.000	1.000	crs
Navy	1.000	1.000	1.000	crs	0.946	0.984	0.961	irs	0.745	0.753	0.990	irs	1.000	1.000	1.000	crs	0.814	0.814	1.000	crs
NHBA	1.000	1.000	1.000	crs	0.999	1.000	0.999	irs	1.000	1.000	1.000	crs	0.816	1.000	0.816	irs	1.000	1.000	1.000	crs
NIB	1.000	1.000	1.000	crs	1.000	1.000	1.000	crs	0.785	0.924	0.850	drs	1.000	1.000	1.000	crs	0.947	1.000	0.947	drs
Peoplecare	0.808	0.808	1.000	crs	0.853	0.854	0.999	irs	0.849	0.852	0.996	irs	0.777	0.789	0.985	irs	0.697	0.697	1.000	crs
Phoenix	0.866	0.967	0.896	irs	0.893	1.000	0.893	irs	0.836	0.836	1.000	crs	0.972	1.000	0.972	irs	0.863	0.863	0.999	irs
Police	1.000	1.000	1.000	crs	1.000	1.000	1.000	crs	1.000	1.000	1.000	crs	1.000	1.000	1.000	crs	1.000	1.000	1.000	crs
QCH	0.897	0.902	0.995	irs	0.800	0.802	0.998	irs	0.795	1.000	0.795	drs	0.788	0.795	0.991	irs	0.690	1.000	0.690	drs
QTUH	0.829	0.831	0.998	drs	0.821	0.822	0.999	irs	0.927	0.928	0.999	irs	1.000	1.000	1.000	crs	0.858	0.858	0.999	drs

RBHS	1.000	1.000	1.000	crs	1.000	1.000	1.000	crs	1.000	1.000	1.000	crs	0.740	1.000	0.740	irs	1.000	1.000	1.000	crs
St Luke's	0.919	0.924	0.994	irs	0.914	0.915	0.999	irs	0.777	0.779	0.998	irs	0.955	0.963	0.992	irs	0.638	0.638	1.000	crs
TFH	0.943	0.969	0.972	drs	0.884	0.918	0.963	drs	0.891	0.917	0.972	drs	0.992	0.993	0.998	drs	0.853	0.949	0.899	drs
Transport	0.871	1.000	0.871	irs	0.781	1.000	0.781	irs	0.986	1.000	0.986	drs	0.585	0.904	0.647	irs	0.893	0.895	0.998	irs
Westfund	0.599	0.600	0.998	drs	1.000	1.000	1.000	crs	0.660	0.724	0.912	drs	0.790	0.790	1.000	crs	0.566	0.701	0.807	drs
Mean	0.880	0.946	0.933	—	0.881	0.956	0.925	—	0.863	0.917	0.943	—	0.827	0.925	0.896	—	0.822	0.874	0.945	—
Std.dev.	0.132	0.092	0.119	—	0.140	0.082	0.136	—	0.111	0.102	0.086	—	0.162	0.102	0.150	—	0.159	0.139	0.117	—
Min.	0.501	0.600	0.501	—	0.419	0.660	0.419	—	0.657	0.670	0.692	—	0.335	0.588	0.335	—	0.487	0.487	0.507	—
Max.	1.000	1.000	1.000	—	1.000	1.000	1.000	—	1.000	1.000	1.000	—	1.000	1.000	1.000	—	1.000	1.000	1.000	—
# eff.	11	17	12	—	9	20	10	—	7	15	8	—	7	14	8	—	8	11	14	—
% eff.	35.48	54.84	38.71	—	29.03	64.52	32.26	—	22.58	48.39	25.81	—	22.58	45.16	25.81	—	25.81	35.48	45.16	—
#irs	—	—	—	10	—	—	—	16	—	—	—	6	—	—	—	17	—	—	—	4
#crs	—	—	—	12	—	—	—	10	—	—	—	8	—	—	—	8	—	—	—	16
#drs	—	—	—	9	—	—	—	5	—	—	—	17	—	—	—	6	—	—	—	11

Notes: This table provides the measures of efficiency for each PHI fund for each year from 2017/18 to 2021/22, obtained using DEA. OTE: overall technical efficiency, PTE: pure technical efficiency, SE: scale efficiency, RTS: returns to scale, crs: constant returns to scale, drs: decreasing returns to scale, irs: increasing returns to scale. # eff. and % eff. are the funds with an efficiency score of 1.000 (fully efficient) by number and as a percentage of all funds (32 in each year of the sample), respectively.

theoretically improve its efficiency by increasing (decreasing) its outputs (inputs) relative to its inputs (outputs) by this percentage. In turn, this includes a PTE of 83.1% efficiency (16.9% inefficiency) and a SE of 94.9% efficiency (5.1% inefficiency). An improvement in PTE would be possible with the existing size but a different combination of inputs and outputs, while an improvement in SE would only be possible with more or less of both inputs and outputs. The irs value for RTS says this would be by increasing the size (or scale) of the fund, with correspondingly more inputs and outputs.

For the most part, it is easier and quicker to improve PTE in PHI funds as at least some inputs can be cut through better managing expenses (as was the case during COVID-19). Decreasing liabilities and equity capital is more difficult, although a decrease in liabilities is possible through repaying debt and refunding already paid premiums (as well as during COVID-19). Equity capital is exceedingly difficult to decrease very quickly for all insurance funds, apart from listed for-profit funds that can repurchase stock or issue dividends. Output alternatives in terms of increasing premium revenue and/or investment revenue are also possible. Increasing premium revenue requires either higher premiums or more customers, and this can be difficult to attain in the short term in a mature and highly competitive market (especially during COVID-19 when the payout of benefits was constrained). Investment revenue will, of course, largely depend on external financial market conditions.

For their part, any improvements in SE are ordinarily limited to the longer term when funds can increase the size of both inputs and outputs, usually through a combination of substantial asset, liability, or capital restructuring. The simplest and most effective way to improve scale is through merger with another fund, although this may also require some post-merger adjustment of inputs and/or outputs to secure any hoped-for synergies in revenue, costs, or the cost of capital. The alternative to long-term growth is slower and can involve cost-reducing technologies, such as investment in the automation of core processes, self-servicing tools for customers, regulatory technology, outsourcing, talent acquisition and retention, product promotion and simplification through marketing investment, and improvements in investment portfolio planning and management. It might also be possible to achieve rapid scale-enhancing growth through removing restrictions placed upon the fund by itself, including not-for-profit status (and thereby opening itself to more equity capital and fewer restrictions on its use, but at the cost of higher taxes) or as a restricted fund (by allowing it to attract new policyholders from other professions or industries). However, either of these would be more difficult, costly, and potentially provocative for existing policyholders.

Finally, the years 2017/2018 and 2018/2019 correspond fully to the period before the COVID-19 pandemic, 2019/2020 to the first six months of the pandemic — the World Health Organization (WHO) declared COVID-19 a worldwide pandemic on March 11, 2020 — and 2020/2021 and 2021/2022 to the next 24 months of the pandemic. In practical terms, PHI insurers began to implement relief measures (including delays in increasing premiums) in March 2020, and this also corresponded with periodic suspensions of lower-urgency elective surgery. The Commonwealth and state and territory governments eased these in 2020/2021; however, as outbreaks became more widespread, suspensions again took hold, causing further disruptions to the delivery of healthcare services. While these restrictions on elective surgery and services such as dental and physiotherapy were removed in late 2020/2021, as late as July 2022, state health departments were still requiring hospitals to delay and reschedule some categories of elective surgery.

Focusing on the means, standard deviations, minimums, maximums, and the number and percentage of efficient PHI funds at the bottom of Table 6, we can see that the mean OTE, which had been 0.880 in 2017/2018 and 0.881 in 2018/2019, at first improved to 0.863 in 2019/2020, being the first year of the pandemic, but then declined to 08.27 in 2020/2021 and 0.822 in 2021/2022. This reflects similar changes in the PTE and SE components. The standard deviations of OTE, PTE, and SE also increased, suggesting a much greater range of inefficiencies among the 32 insurers after the pandemic than before. Lastly, while some PHI funds were fully efficient across all years and all measures (as always the case with DEA as a relative measure of efficiency), the minimum efficiencies declined majorly, with a minimum OTE of just 0.335 in 2020/2021 (CBHSCH), a PTE of only 0.487 in 2021/2022 (LaTrobe), and an SE of merely 0.507 in 2021/2022 (HBF). So, in the years of the pandemic, at least some PHI funds were only 33.5%, 48.7%, and 50.7% as overall technically, pure technically, and scale efficient as the best-practice funds. This points to a significant disruption resulting from COVID-19.

Table 6 also provides details on the number and percentage of efficient funds. As shown, in 2018/2019, the last full year before the outbreak of the pandemic, nine funds (29.03%) were fully OTE, twenty funds (6.4.52%) were fully PTE, and ten funds (32.26%) were fully SE. However, by the first full year of the pandemic (2020/2021), these had fallen to seven (22.58%), fourteen (45.16%), and eight (25.81%) funds being fully OTE, PTE, and SE, respectively. As shown, there was a modest recovery of these in 2021/2022, but only in relation to SE, where an additional six funds became fully SE.

Finally, Table 6 also contains information on the nature of the scale efficiency improvements possible. It is clear from the earlier discussion that the unexpected arrival of COVID-19 and the public health policy responses of the state and territory governments caught most PHI funds unaware and unprepared. Investment revenues declined with global investment volatility from the pandemic and remained unpredictable and premium revenue remained stable and even grew, but liabilities and expenses surged as funds grappled with how to best service their policyholders in an environment where it was simply not possible for them to use their policies anywhere near their best effect. This entailed both worsening technical efficiency (funds were using the wrong combinations of inputs and outputs) and scale efficiency (funds were simply too large in terms of both inputs and outputs for what was required).

We can see this best in the number of irs, crs, and drs funds. In 2018/2019, the last full year before the pandemic, 16, or precisely half of all funds, had irs, indicating that increasing their size (scale) would yield an average efficiency improvement of 7.5% across the entire PHI industry. This accords with other research indicating that most PHI funds are simply too small in light of the economies of scale possible and practiced in the industry, and longstanding calls for structural reform. However, in 2019/2020, 17 or more than half of all funds were now experiencing drs and only six irs, such that reducing their size would have been the best possible strategy. By 2020/2021, there was a return to the situation where about half of all funds could again derive efficiency improvements by increasing, if possible, their scale of operations.

While it is not of great benefit to tediously analyze the fund-by-fund efficiencies, it may be interesting to gauge the impact of COVID-19 on the five-largest PHI funds by assets (BUPA, HBF, HCF, MPL, and NIB). As discussed, the PHI industry in Australia is highly concentrated, with the five largest funds accounting for some 73% of industry assets, and even the fifth-largest fund is twice as large as the sixth-largest fund. The largest PHI fund is almost three hundred times larger than the smallest! In the case of BUPA, MPL, and NIB, these funds mostly remained fully efficient (OTE, PTE, and SE) throughout the pandemic, with a temporary fall in efficiency for MPL and NIB only during the pandemic's first six months. This points to the resilience of these institutions in meeting the operational and financial challenges of COVID-19. However, in the case of HBF and HCF, while both were fully PTE in 2017/2018, 2018/2019, and 2019/2020, this deteriorated in the case of HBF. It is also of note that both HBF and HCF were not fully SE before, during, and after the COVID-19 pandemic, and that measured SE has deteriorated further throughout. This is of some interest, as in February 2018, HBF and HCF announced plans to merge, later called off in June of that same year after not coming to an agreement in the interest of their non-profit policy holding members. This suggests that the shock of COVID-19 served to worsen several pre- existing industry conditions.

4. Conclusion

This chapter examined the changes in technical and scale efficiency for the Australian PHI industry resulting from the COVID-19 pandemic using DEA. As discussed, the outbreak of the COVID-19 pandemic in January 2020 resulted in significant disruption to the PHI industry. Premium revenue increased in real terms as Australians became more conscious of possible impacts on their health and took up PHI cover, countering a decade-long decline in the share of the population with PHI. Investment revenue became volatile with the disruption to global financial markets. But most significantly, state and territory restrictions on elective surgery and the use of services typically covered by PHI extras, such as dental and physiotherapy, saw a significant decline in benefits paid, and this, along with the pandemic itself, resulted in PHI funds delaying premium increases and even refunding premiums to policyholders. At the same time, PHI funds set about reorganizing their operations to streamline claims online, facilitate online services for their members, and enable working from home for their employees.

The measures used show that all PHI funds suffered efficiency losses from COVID-19. While some were technical efficiency losses that could be addressed by reducing and rearranging inputs, others resulted from an inefficient scale that suddenly resulted from the inputs and outputs used by the fund being greatly in excess of their requirements. In this situation, it would normally be a recommended strategy for a fund to scale back to an optimal size, but given the temporary nature of the pandemic, funds were required to maintain an inefficient size. While both technical and scale efficiency have now both improved with the relaxation of healthcare controls and the resumption of elective surgery, the levels are still lower than they were before the pandemic. Of particular concern is the extremely low levels of efficiency of some smaller and medium-sized funds.

It is clear that COVID-19 has caused considerable disruption to any industry, for which questions were already raised about its long-term sustainability given the falling numbers of policyholders and rapidly increasing healthcare costs. As a highly regulated industry, there are already significant constraints placed on PHI funds in terms of setting premium increases as well as what healthcare services PHI funds can cover. This certainly limited the ability of the industry to respond to the changes in the healthcare sector associated with COVID-19. For this reason, it would be timely to undertake an industry wide review of the PHI industry, particularly that relating to premium rebates, coverage, and the potential for future structural reform.

References

Alhassan, A.L. & Biekpe, N. (2015). Efficiency, productivity and returns to scale economies in the non-life insurance market in South Africa. *Geneva Papers on Risk and Insurance — Issues and Practice*, 40, 493–515.

Australian Prudential Regulation Authority (APRA). (2016, 2017, 2018, 2019, 2020, 2021, 2022a). Operations of private health insurers annual report. https://www.apra.gov.au/operations-of-private-health-insurers-annual-report; https://webarchive.nla.gov.au/awa/19981206130713; http://www.apra.gov.au/.

Australian Prudential Regulation Authority (APRA) (2022b, September). Statistics: Private health insurance membership and benefits. https://www.apra.gov.au/quarterly-private-health-insurance-statistics.

Barros, C.P., Nektarios, M., & Assaf, A. (2010). Efficiency in the Greek insurance industry. *European Journal of Operational Research*, 205, 431–436.

Biener, C. & Eling, M. (2012). Organization and efficiency in the international insurance industry: A cross-frontier analysis. *European Journal of Operational Research*, 221, 454–468.

Biener, C., Eling, M., & Wirfs, J.H. (2016). The determinants of efficiency and productivity in the Swiss insurance industry. *European Journal of Operational Research*, 248, 703–714.

Boubaker, S., Le, T.D.Q., & Ngo, T. (2022a). Managing bank performance under COVID-19: A novel inverse DEA efficiency approach. *International Transactions in Operational Research*, 30(2023), 2436–2452. https://doi.org/10.1111/itor.13132.

Boubaker, S., Do, D.T., Hammami, H., & Ly, K.C. (2022b). The role of bank affiliation in bank efficiency: A fuzzy multiobjective data envelopment analysis approach. *Annals of Operations Research*, 311, 611–639. https://doi.org/10.1007/s10479-020-03817-z.

Carrington, R., Coelli, T., & Prasada Rao, D. (2011). Australian private health insurance productivity growth: Is there scope to limit premium increases? *Economic Record*, 87, 125–139.

Cummins, J.D. & Xie, X. (2013). Efficiency, productivity, and scale economies in the U.S. property-liability insurance industry. *Journal of Productivity Analysis*, 39, 141–164.

Cummins, J.D., Weiss, M.A., Xie, X., & Zi, H. (2010). Economies of scope in financial services: A DEA efficiency analysis of the US insurance industry. *Journal of Banking and Finance*, 34, 1525–1539.

Eling, M. & Schaper, P. (2017). Under pressure: How the business environment affects productivity and efficiency of European life insurance companies. *European Journal of Operational Research*, 258, 1082–1094.

Huang, W. & Eling, M. (2013). An efficiency comparison of the non-life insurance industry in the BRIC countries. *European Journal of Operational Research*, 226, 577–591.

Le, T.D.Q., Ho, T.H., Nguyen, D.T., & Ngo, T. (2021). Fintech credit and bank efficiency: International evidence. *International Journal of Financial Studies*, 9(3), 44. https://www.mdpi.com/2227-7072/9/3/44.

Luhnen, M. (2009). Determinants of efficiency and productivity in German property-liability insurance: Evidence for 1995–2006. Geneva Papers on Risk and Insurance Issues and Practice, 34, 483–505.

Nguyen, L. & Worthington, A.C. (2021). Industry regulation, fund characteristics, and the efficiency of Australian private health insurers. *Accounting and Finance*, 61(1), 781–801.

Tone, K., Chang, T.-S., & Wu, C.-H. (2020). Handling negative data in slacks-based measure data envelopment analysis models. *European Journal of Operational Research*, 282(3), 926–935.

Worthington, A.C. & Hurley, E.V. (2002). Cost efficiency in Australian general insurers: A non-parametric approach. *British Accounting Review*, 34, 89–108.

Yaisawarng, S., Asavadachanukorn, P., & Yaisawarng, S. (2014). Efficiency and productivity in the Thai non-life insurance industry. *Journal of Productivity Analysis*, 41, 291–306.

https://doi.org/10.1142/9781800615786_0009

Chapter 9

Devising the Best Prospective Merger Plan for a Banking Sector Through a Hybrid DEA-Based Methodology: An Inverse DEA Perspective

Amar Oukil

College of Economics & Political Science,
Sultan Qaboos University, Al Khoud, Oman
aoukil@squ.edu.om

Abstract

Mergers & acquisitions (M&A) are strategic decisions that have long been associated with the banking sector, entailing the consolidation of assets among a group of banks through various types of financial transactions. Though a lot of research has been dedicated to the application of data envelopment analysis (DEA) to multiple aspects of M&A within this particular sector, little or even no significant attention has been paid to investigating the optimal matchings among banks, i.e., what should be the best partners of prospective bank mergers that are more likely to maximize the overall performance of the whole banking sector? To answer this question, we propose a hybrid DEA methodology that operates over two levels. The first level entails solving an inverse DEA (IDEA) model to evaluate the optimal gains that could potentially be generated out of pairwise consolidations among banks. As a result, all productive post-merger banks, i.e., those mergers that have real potential for gains' generation, are duly discerned. In the second level, a DEA procedure integrating a standard DEA model with a greedy heuristic is devised to select the best pairs of merging banks based on the post-merger banks' expected outcomes. Here, the best prospective merger plan is derived for the whole banking sector out of the entire sample of banks. Using data from the Office of the Superintendent of Financial Institutions (OSFI) database, the pertinence of the proposed methodology is shown by evaluating the potential merger gains of 28 Canadian banks prior to building the associated best prospective merger plan.

Keywords: Data envelopment analysis (DEA), inverse DEA, heuristics, mergers and acquisitions, gains, partnerships, alliances, strategic decision making, merger plan, collective performance, bank efficiency, financial sector.

1. Introduction

The intercessional role of the banking sector is undeniably vital for all markets, which certainly stimulates a broad spectrum of studies on the evaluation of various banking aspects, including efficiency (Doumpos *et al.*, 2022). A review of the literature reveals that the largest share of efficiency studies are based on data envelopment analysis (DEA). The popularity of DEA stems partly from the fact that it can assess decision-making units (DMUs) that are defined with multiple inputs and multiple outputs (Sow *et al.*, 2016). Moreover, DEA does not require prior assumptions on the functional relationships between inputs and outputs (Oukil & Zekri, 2021).

The application of DEA spans broad business areas, such as production scheduling (Oukil & El-Bouri, 2021), open field agriculture (Branco *et al.*, 2021; Kamiyama *et al.*, 2021; Zalaghi *et al.*, 2021), material handling selection (Hassan & Oukil, 2021; Oukil, 2020a; Oukil, 2022b), greenhouse production (Al-Mezeini *et al.*, 2020; Nourani & Bencheikh, 2020, 2021), performance and productivity of airlines (Ngo & Tsui, 2022), water quality assessment (Oukil *et al.*, 2022c), hospitality management (Fang *et al.*, 2020), ranking football players (Oukil & Govindaluri, 2017), evaluation of faculty academic performance (Oral *et al.*, 2014, 2015), and energy optimization (Oukil *et al.*, 2022a).

In the banking sector, the most prevailing research stream is concerned with the application of DEA to estimate efficiency at the bank level (e.g., Marjanović *et al.*, 2018; Henriques *et al.*, 2018; Husejinovic, 2019; Kamarudin *et al.*, 2019; Zhu *et al.*, 2020; Antil *et al.*, 2020; Cvetkoska & Čiković, 2020; Akhtar *et al.*, 2021; Učkar &

Petrović, 2021; Maity & Sahu, 2021; Kundu & Banerjee, 2022; Yeh *et al.*, 2022). The DEA Malmquist index has been used to measure the banks' productivity change over the time horizon (e.g., Tziogkidis *et al.*, 2018; Nguyen, 2021; Zaman *et al.*, 2022). More advanced DEA models have been developed to handle technological heterogeneity among banks (e.g., Liu *et al.*, 2020a; Yu *et al.*, 2021) or the relationships among major indicators within the banks (e.g., Wanke *et al.*, 2019; Dia *et al.*, 2020; de Mello Valerio *et al.*, 2021). In another direction, DEA models have also been employed to gauge the influence of specific contextual factors on efficiency, such as controlling shareholders (Boubaker *et al.*, 2021), capital (Zhou *et al.*, 2019), currency hedging (Boubaker *et al.*, 2022b), liquidity (Fernandes *et al.*, 2018), affiliation (Boubaker *et al.*, 2022c), CEO compensation (Matousek & Tzeremes, 2016), corporate governance (Thaker *et al.*, 2022), operational behavior (Pournader *et al.*, 2017), market structure (Karlsson *et al.*, 2021), regulations (Gaganis *et al.*, 2020), funding structure (Ofori-Sasu *et al.*, 2019), technology development (Le *et al.*, 2021), ownership structure (Liu *et al.*, 2020b), competition and stability (Phan *et al.*, 2019), and risk credit (Simper *et al.*, 2017). The two-stage DEA technique has been adopted to further investigate the influence of more factors over time (e.g., Bou-Hamad *et al.*, 2017; Sukmana *et al.*, 2020; Nguyen *et al.*, 2021; Antunes *et al.*, 2022; Cobas *et al.*, 2022; Zaman & Bhandari, 2022). See the work of Henriques *et al.* (2020) for an extensive discussion of the application of the two-stage DEA approach to banking. A broader review of the literature relating to bank efficiency has been carried out by Alber *et al.* (2019) and de Abreu *et al.* (2019). The reader is also referred to Cvetkoska & Savic (2021) for a bibliometric review of the literature relating to the application of DEA in banking.

Operating within a global environment of speedy technological progress, perpetual reformatting of customer preferences, and incessant product and service innovations, banks do face intensive market competition, where merging with peers becomes a premium option for growth (Montgomery *et al.*, 2014; Kowalik *et al.*, 2015; Chiorazzo *et al.*, 2018). Per se, market power growth, technology enhancement, and efficiency improvement are seen as the key motivators of mergers and acquisitions (M&A) in the banking sector (Fraisse *et al.*, 2018). The literature dealing with the efficiency of M&A in the banking sector keeps growing due to the substantial expansion of consolidations

(Avinadav *et al.*, 2017). The effect of consolidations on the bank's efficiency has been investigated in many DEA-based studies. See, for example, Vo (2018), Martín *et al.* (2018), Hsu *et al.* (2018), Vo & Nguyen (2018), Edward & Manoj (2019), Nguyen & Vân Anh (2019), Al-Khasawneh *et al.* (2020), Borisov (2020), Coccorese & Ferri (2020), Duong *et al.* (2020), Akbar *et al.* (2021), Blanco-Oliver (2021), Galariotis *et al.* (2021), Barkhordari *et al.* (2022), and Čiković *et al.* (2022).

Practically, the consolidation process, which underpins M&A, constitutes the ultimate mirror for potential growth, competition dynamics, and prospective efficiency gains (Larkin & Lyandres, 2019; Rao & Mishra, 2020; Huhtilainen *et al.*, 2022). Here, DEA models have also been developed to estimate potential merger gains (e.g., Bogetoft & Wang, 2005; Lozano & Villa, 2010; Halkos & Tzeremes, 2013; Xiao *et al.*, 2017; Shi *et al.*, 2017; Li *et al.*, 2018; Nguyen & Pham, 2020; Xie *et al.*, 2020; Chiu *et al.*, 2021; Ray & Sethia, 2022; Khoveyni & Eslami, 2022; Al Tamimi *et al.*, 2022). However, these models are exclusively focused on the post-merger stage, where the decision-maker (DM) is attentive only to possible cost and profit efficiency gains given the on-hand levels of inputs and/or outputs. In other words, these models cannot inform the DM earlier, i.e., at the pre-merger stage, about the levels of inputs and/or outputs that would enable the intended merger to achieve a desired preset efficiency target. Such decision problems can be addressed through inverse DEA (IDEA) (Wei *et al.*, 2000; Yan *et al.*, 2002).

While the purpose of traditional DEA is basically to estimate the efficiency scores of DMUs whose input and output data are known, IDEA aims to determine the optimal levels of inputs and/or outputs needed for reaching an efficiency target that is set *a priori*. Recent applications of IDEA at the level of individual banks include performance evaluation after a financial crisis (Ghiyasi & Zhu, 2020), resilience appraisal in the aftermath of a pandemic (Boubaker *et al.*, 2022a), ranking efficient banks (Soleimani-Chamkhorami *et al.*, 2020), assessing the expected yields of consolidations (Venkatesh & Narasimhan, 2021), resource planning (An *et al.*, 2019), fund management (Kazemi & Galagedera, 2022), and sensitivity analysis (Ghobadi, 2019).

In M&A, except for a few studies in agriculture production (Oukil & Zekri, 2015; Oukil, 2022a, 2023; Oukil *et al.*, 2022b), all known applications are limited to the banking sector. The pioneering study on the application of IDEA in an M&A setting is due to Gattoufi *et al.* (2014), with the introduction of the first related IDEA model. The latter was extended by Amin *et al.* (2017b) to generalized firm restructuring, where both consolidations and splits can be handled. With the same restructuring objectives, Gerami *et al.* (2021) proposed an approach that enables integrating the DM's preferences besides determining the worst-case scenario for restructuring. The DM's preferences were also incorporated into the combined IDEA and goal programming approach of Amin *et al.* (2019). Guijarro *et al.* (2020) also adopted a combined scheme with a genetic algorithm to identify the merging units. Amin & Al-Muharrami (2018) developed a new IDEA model for M&A in the presence of negative data. A similar case was modeled by Soltanifar *et al.* (2023) on top of the potential existence of ratio data. An IDEA model for merging units defined with interval data was presented by Ghobadi (2021). In the approach of Amin *et al.* (2017a), DMs can anticipate whether a prospective merger is a major or a minor consolidation. Kamyab *et al.* (2019) addressed the same issue in a two-stage production system through a network IDEA model. Amin & Ibn Boamah (2021) formulated other network IDEA models specifically for M&A within the US banking sector. Amin & Ibn Boamah (2020) also proposed an IDEA cost-efficiency model for restructuring scenarios, where cost minimization may be required. Instead of using a series of models for studying multiple scenarios of M&A, Soltanifar *et al.* (2022) devised a single IDEA model based on a common set of weights.

Interestingly, regardless of the IDEA model adopted, all the existing studies assume implicitly that the merging DMUs are known *a priori*, and the only concern is to estimate the merger's potential gains for a preset efficiency target. Yet, the ideal scenario for a DMU that is planning to embark on a merger is certainly to be able to choose the best partner among the candidate DMUs. As such, the current study aims to answer the following question: What should be the best partners in prospective bank mergers that would qualify a selected sample of banks to achieve maximum performance collectively rather than individually? In other words, assuming that each bank is a candidate to merge with another bank within the selected sample, the objective is to find the optimal matchings among these banks.

To answer the above question, the proposed methodology operates over two levels. The IDEA model is initially solved to evaluate the optimal gains that are likely to be generated as a consequence of pairwise consolidations among banks. As a result, all possibly productive post-merger banks, i.e., those mergers that have real potential for gains' generation, are duly discerned. In the second stage, a DEA procedure integrating a standard DEA model with a greedy heuristic is devised to select the best pairs of merging banks based on the post-merger banks' expected outcomes. Here, the best prospective merger plan is derived for the entire sample of banks.

To the author's best knowledge, there are no previous studies that attempted to build the best merger plan for a sample of banks by identifying the best partners in potential post-mergers. Therefore, the contribution of this chapter to the banking literature is twofold: (1) It is the first study that investigates comprehensively all prospective mergers within a selected sample of banks; (2) a new hybrid procedure is developed for identifying the best merging partners based on the post-merger's expected outcomes.

Using data from the Office of the Superintendent of Financial Institutions (OSFI) database, the pertinence of the proposed methodology is shown through the evaluation of the potential merger gains of 28 Canadian banks prior to building the associated best prospective merger plan.

The application of the standard DEA approach to evaluate these banks' performance produced an expense savings index of only 3.88% for an average technical efficiency of 0.7797, with 35.71% of the banks declared strongly efficient, i.e., not requiring input savings.

The optimal input levels for post-merger banks are derived through the IDEA approach for all possible pairwise consolidations among the banks. The results revealed that the potential savings per post-merger bank can reach proportions as high as 84.82% and 71.27% for the input deposits and man-hours, respectively. Nonetheless, one of the most notable results is undeniably the occurrence of productive post-merger banks that comprise solely strongly efficient banks. In other words, despite a status that denies any input savings under the individual banks' assessment scheme for strongly efficient banks, it is found that savings become possible after merging.

The integration of the DEA model with a greedy heuristic enabled the selection of the best non-overlapping pairs of banks from the sample of productive post-merger banks and, hence, the merger plan. The latter does not necessarily involve all the banks since only 11 productive post-merger banks are selected. The expense savings index of the resulting merger plan is 11.95%, i.e., an improvement by a factor of more than three in comparison with its pre-merger value of 3.88%.

The remainder of the chapter proceeds as follows: Section 2 presents the methodological background, which comprises DEA concepts and an IDEA model for the merger process. Section 3 introduces briefly a case study prior to a discussion of the results. In Section 4, the approach to selecting the best partners is described and implemented. We end the chapter with concluding remarks.

2. Methodological Background

In line with the literature relating to the application of DEA in the banking industry, this section starts with the basic concepts for the evaluation of a bank's efficiency, including the slack values, which are practically important to measure the pre-merger savings. The following section presents the modeling tools for the merger of two banks, with an emphasis on the optimal savings that are expected from potential mergers. The comparison of the pre-merger and post-merger savings at the application stage will highlight one side of the benefits of a merger's decision process.

2.1. *Bank efficiency evaluation*

Let K be the number of banks to be evaluated, with each bank S_k represented with I inputs x_{ik} and J outputs y_{ik}, where $i = 1, \ldots, I$ and $j = 1, \ldots, J$. The following BCC (Banker *et al.*, 1984) input-oriented model is used to estimate the efficiency score e_o^* of $S_o = (x_o y_o)$ and set targets for inefficient banks:

$$\text{(BCC)} \qquad e_o^* = \min \pi$$

$$\text{s.t.:} \quad \sum_{k=1}^{K} \alpha_k x_{ik} \leq \pi x_{io}, \quad i = 1, \ldots, I,$$

$$\sum_{k=1}^{K} \alpha_k y_{jk} \geq y_{jo}, \quad j = 1, \ldots, J,$$
$$\sum_{k=1}^{K} \alpha_k = 1,$$
$$\alpha_k \geq 0, \quad k = 1, \ldots, K.$$

The optimal efficiency score e_o^* of S_o represents the minimum inputs reduction that is required to achieve full efficiency for given outputs levels (Oukil & Govindaluri, 2020). Thus, S_o is efficient if $e_o^* = 1$; otherwise, it is inefficient, i.e., it is not using its inputs optimally.

The intensity value α_k measures the contribution of each bank S_k toward the efficiency evaluation of S_o (Soltani *et al.*, 2021). An optimum $\alpha_k^* > 0$ indicates that S_k is a potential benchmark for S_o (Oukil *et al.*, 2016).

To achieve full efficiency, an inefficient bank S_o can decrease its ith input by an amount S_{io}^* while keeping its outputs unchanged:

$$s_{io}^* = e_o^* x_{io} - \sum_{k=1}^{K} \alpha_k^* x_{ik} \quad i = 1, \ldots, I,$$

where S_{io}^* is the slack value of the ith input. S_o is strongly efficient if $S_{io}^* = 0$ for all inputs $i = 1, \ldots, I$. Otherwise, S_o is weakly efficient if $e_0^* = 1$ and $S_{io}^* > 0$ for at least one input $i = 1, \ldots, I$ (Oukil & Al-Zidi, 2018). In other words, S_{io}^* represents the savings of input i that an inefficient or a weakly efficient bank S_o is required to achieve if it is willing to upgrade to efficient.

2.2. *Inverse DEA model for merger*

We assume, for modeling simplicity, that two pre-merger banks, S_A and S_B, are merging into a larger bank M_m. Meanwhile, this assumption does not exclude the fact that the proposed methodology applies to the merger of more than two banks.

Let $\bar{\pi}$ be the efficiency target set *a priori* for the post-merger bank M_m. Under an input-oriented DEA setting, bank M_m keeps the entire outputs of the merging banks S_A and S_B while consuming the minimum levels of their current inputs with, as an objective, achieving a preset efficiency target $\bar{\pi}$.

Let ∂_{iA} and ∂_{iB} denote the minimum levels of ith input associated with banks S_A and S_B, respectively. Hence, the level of ith input for bank M_m is $\partial_{im} = \partial_{iA} + \partial_{iB}$, which is expected to be less than or equal to $x_{im} = x_{iA} + x_{iB}$, while the level of its jth output remains $y_{jm} = y_{jA} + y_{jB}$, for $i = 1, \ldots, I$ and $j = 1, \ldots, J$. The problem, as stated, can be formulated using the following IDEA model (Amin & Ibn Boamah, 2020):

$$
\begin{aligned}
&\text{(InvDEA)}\ \min \sum_{i=1}^{I} (\partial_{iA} + \partial_{iB}) \\
&\text{s.t.:} \quad \sum_{k \in P} \alpha_k x_{ik} + \alpha_m x_{im} \\
&\qquad\qquad - (\partial_{iA} + \partial_{iB}) \times \bar{\pi} \leq 0, \qquad i = 1, \ldots, I, \\
&\qquad \sum_{k \in P} \alpha_k y_{jk} + \alpha_m y_{jm} \geq y_{jm}, \qquad j = 1, \ldots, J, \\
&\qquad \sum_{k \in P} \alpha_k + \alpha_m = 1, \\
&\qquad 0 \leq \partial_{iA} \leq x_{iA}, \quad 0 \leq \partial_{iB} \leq x_{iB}, \\
&\qquad\qquad \alpha_k \geq 0, \quad k \in P, \quad \alpha_m \geq 0, \qquad i = 1, \ldots, I.
\end{aligned}
$$

Model (InvDEA) is an IDEA model because, here, the efficiency score $\bar{\pi}$ is known and the objective is to find the minimum required levels of corresponding inputs, i.e., the values of the $2I$ variables, ∂_{iA} and ∂_{iB}, $i = 1, \ldots, I$.

With P denoting the set of peers available for the evaluation of the post-merger bank $M_m P$ may contain one of the merging banks, S_A and S_B, or none of them, depending on the form of the merger. If the merger is a *survival*, only the acquiring bank will continue to operate under its former name (Amin & Oukil, 2019a). Otherwise, if the merger is a *consolidation*, the merging banks will be fusing into a new entity with a new name, denoted by M_m. The objective of our study being specifically to boost banks' efficiency through consolidation, none of the two banks will be considered in P. The reader is referred to Oukil (2022a) for a detailed discussion of possible merger forms.

Regardless of the merger's form, the intensity variables α_k are now restricted to $k \in P$, which includes α_m, the intensity variable associated with M_m. If $\alpha_m^* = 1$ and $\bar{\pi} = 1$ is the only solution to model (InvDEA), the merger will be a *major consolidation*, where the post-merger bank M_m uses the totality of the inputs, i.e., $\partial_{im} = x_{im}$, and no input gains are expected from the merger.

3. Application and Practical Scope

In this chapter, we use the 2017 consolidated financial statements of 28 domestic Canadian banks, which were presented by Amin & Ibn Boamah (2020). Following the latter study, we consider two inputs: interest expenses (x_1) and non-interest expenses (x_2) (e.g., Niepmann & Schmidt-Eisenlohr, 2017). Interest expenses represent the expenses that the bank pays on deposits and other borrowed funds (Ibn Boamah, 2018). Non-interest expenses include all the operating and overhead costs of the bank, such as employee salaries and benefits, professional services, administration, equipment, and other expenses. Meanwhile, we use two outputs: interest income (y_1) and non-interest income (y_2). Interest income comprises interest on personal and commercial loans, mortgages, and government securities. Non-interest income consists of charges on transactions and deposits, such as monthly cheque account service charges, cheque and deposit slip fees, annual credit card fees, and other bank operation income. To measure the total savings that could result from a potential merger, Amin & Ibn Boamah (2020) suggest converting the interest and non-interest expenses to deposits and man-hours, respectively, assuming unit input prices of \$0.02 and \$33. The full dataset is presented in Table A1.

Without loss of generality, we assume that the sample of 28 banks represents the whole Canadian banking sector.

3.1. *Banks performance evaluation*

Using IBM-ILOG CPLEX version 12.6, the BCC models are solved to compute the optimal efficiency scores e_k^* for all banks S_k, $k = 1, \ldots, 28$. The results are summarized in Table 1.

The average efficiency score is 0.7797, with 10 banks out of 28 (35.71%) strongly efficient, i.e., not requiring input savings. The remaining banks are inefficient ($e_k^* < 1$) and need to benchmark one of the strongly efficient banks to achieve better performance (see, e.g.,

Table 1. DEA efficiency results for 28 banks.

Bank	e_k^*	Bank	e_k^*	Bank	e_k^*	Bank	e_k^*
S_{01}	0.507	S_{08}	0.572	S_{15}	1.000	S_{22}	0.972
S_{02}	0.938	S_{09}	0.490	S_{16}	0.509	S_{23}	0.648
S_{03}	1.000	S_{10}	1.000	S_{17}	0.550	S_{24}	1.000
S_{04}	0.457	S_{11}	1.000	S_{18}	0.913	S_{25}	1.000
S_{05}	0.478	S_{12}	1.000	S_{19}	0.821	S_{26}	0.512
S_{06}	0.946	S_{13}	0.628	S_{20}	1.000	S_{27}	0.687
S_{07}	1.000	S_{14}	0.724	S_{21}	1.000	S_{28}	0.479

Notes: This table reports the efficiency score e_k^* of each bank S_k. Bank S_k is declared efficient if $e_k^* = 1$, otherwise it is inefficient, i.e., it is not using its inputs (expenses) efficiently, and hence, it has potential for savings.

Table 2. Expense savings per input ($).

Input	Present expenses	Target expenses	Expense savings	Contribution to savings (%)
Deposits	43,321,267	41,398,438	1,922,829	54.62
Man-hours	47,433,033	45,835,595	1,597,438	45.38
Total	90,754,300	87,234,033	3,520,267	100.00

Notes: This table reports the distribution of the inputs' totals (expenses), where **Present expenses** refer to the current input consumptions in the whole sample of 28 banks, while **Expense savings** represent the input savings that can potentially be achieved at the banks' individual levels, and **Target expenses** is the difference between the two sets of expenses. **Contribution to savings** computes the proportion of each input in the expense savings.

Oukil *et al.*, 2021). To reach the efficient status, an inefficient bank S_o needs to shrink its inputs by the slack values S_{io}^* $(i = 1, \ldots, 2)$ while keeping its outputs unaffected. Table A2 shows the slack values for the inefficient banks. Thus, the input savings pattern for the entire sample of banks is shown in Table 2.

The results reveal that the total expense savings are $3,520,267, with the highest share of 54.62% for deposits, followed by 45.38% for man-hours. As the current expenses are $90,754,300, the expense savings index is 3.88%. In other words, under the assumption that all inefficient banks in the sample succeed in improving their operating conditions from deposit and man-hour perspectives, the overall banking sector's savings cannot exceed 3.88%.

3.2. *Mergers of banks*

Using the sample of banks, the InvDEA model is solved for each potential pairwise bank merger $M_m = (S_A, S_B)$, $A \neq B$, with the efficiency target as $\bar{\pi} = 1$. With $K = 28$ banks, 378 pairwise mergers, M_m, are possible. The optima of InvDEA reveal that 131 mergers out of 378 are major consolidations, with the post-merger banks M_m retaining the entire inputs of the merging banks S_A and S_B, i.e., $\partial^*_{iA} = x_{iA}, \partial^*_{iB} = x_{iB}$, for all $i = 1, \ldots, I$, and $\alpha^*_m = 1$. In other words, no input gains are expected from nearly 34.65% of the mergers, which is not surprising with the high levels of efficiency noted in the sample. The remaining 247 post-merger banks are potentially productive.

Definition 9.1. A merger $M_m = (S_A, S_B)$, $A \neq B$, is productive if and only if $\alpha^*_m = 0$ and $\partial^*_{iA} < x_{iA}$ or $\partial^*_{iB} < x_{iB}$ for at least one input $i (i = 1, \ldots, I)$.

Given the large number of productive mergers, Table 3 shows only the first 10 pairs for illustrative purposes.

Consider, for instance, the consolidation of banks S_{01} and S_{10}, whose efficiency scores are $e^*_1 = 0.507$ and $e^*_{10} = 1$, respectively. With $\overline{\pi} = 1$ as the efficiency target for the merger, the minimum levels of deposits and man-hours required to create the newly merged bank M_7 are $\partial^*_{11} = \$2,307,628, \partial^*_{110} = \$0, \partial^*_{21} = 1351\,\text{h}$, and $\partial^*_{210} = 0\,\text{h}$. With $\partial^*_{110} = 0$ and $\partial^*_{210} = 0$, it appears that the contribution of bank S_{10} to the merger M_7 is not needed, despite being an efficient bank, and all the expenses are provided exclusively by bank S_{01}.

The most striking result of the IDEA application is certainly the fact that, among the productive post-merger banks, there are pairs that involve exclusively strongly efficient banks, namely, $M_{136} = (S_{10}, S_{12})$, $M_{145} = (S_{10}, S_{25})$, $M_{149} = (S_{11}, S_{12})$, $M_{154} = (S_{11}, S_{25})$, $M_{160} = (S_{12}, S_{15})$, and $M_{197} = (S_{15}, S_{25})$. Interestingly, $S_{10}, S_{11}, S_{12}, S_{15}$, and S_{25} have all been identified as strongly efficient,

Table 3. Bank mergers with minimum input usages for 10 productive mergers.

Banks			Deposits ($)		Man-hours (h)	
M_m	S_A	S_B	∂^*_{1A}	∂^*_{1B}	∂^*_{2A}	∂^*_{2B}
M_1	S_{01}	S_{02}	9,086,901	266,458,044	1,837	217,519
M_2	S_{01}	S_{04}	2,584,417	0	1,484	0
M_3	S_{01}	S_{05}	3,012,259	0	1,689	0
M_4	S_{01}	S_{06}	9,086,901	208,026,669	1,837	156,124
M_5	S_{01}	S_{08}	9,086,901	3,611,425	1,837	5,955
M_6	S_{01}	S_{09}	3,664,907	0	1,837	166
M_7	S_{01}	S_{10}	2,307,628	0	1,351	0
M_8	S_{01}	S_{12}	2,200,288	0	1,299	0
M_9	S_{01}	S_{13}	2,284,787	0	1,340	0
M_{10}	S_{01}	S_{14}	2,466,440	0	1,427	0

Notes: This table reports the minimum input usages ∂^*_{iA} and ∂^*_{iB} ($i = 1$ for deposits and $i = 2$ for man-hours) for potential mergers M_m, of banks S_A and S_A. The results are presented for a sample of 10 potential mergers taken from the set of 247 mergers that are potentially productive, i.e., are likely to generate input gains after merging.

i.e., no savings are required from each bank while operating individually. Yet, after merging, it appears that there is still room for savings.

Definition 9.2. Given a merger $M_m = (S_A, S_B)$, $A \neq B$, the potential gains of S_A and S_B for input i are $\gamma_{iA} = x_{iA} - \partial^*_{iA}$ and $\gamma_{iB} = x_{iB} - \partial^*_{iB}$, respectively.

Noting that the pre-merger inputs of S_{01} and S_{10} are $x_{11} =$ \$9,086,901, $x_{110} =$ \$596,651, $x_{21} = 1837\,\text{h}$ and $x_{210} = 47\,\text{h}$ the potential gains for deposits comprise $\gamma_{11} = x_{11} - \partial^*_{11} = 9{,}086{,}901 - 2{,}307{,}628 = \$6{,}779{,}273$ and $\gamma_{110} = x_{110} - \partial^*_{110} = 596{,}651 - 0 = \$596{,}651$, while the potential gains for man-hours consist of $\gamma_{21} = x_{21} - \partial^*_{21} = 1837 - 1351 = 486\,\text{h}$ and $\gamma_{210} = x_{210} - \partial^*_{210} = 47 - 0 = 47\,\text{h}$.

Definition 9.3. The cumulative potential gain of merger $M_m = (S_A, S_B)$, $A \neq B$, for input i is $\mathfrak{g}_{im} = \gamma_{iA} + \gamma_{iB}$, and the corresponding proportion of energy gains is $g_{im}/x_{im} \times 100\%$.

Table 4. Cumulative gains for 10 bank mergers with related proportions.

	Deposits ($)		**Man-hours (h)**	
M_m	g_{1m}	%	g_{2m}	%
M_1	275,544,945	8.27	219,356	4.77
M_2	2,584,417	76.06	1,484	33.09
M_3	3,012,259	72.06	1,689	55.06
M_4	217,113,570	9.49	157,961	1.85
M_5	12,698,326	55.18	7,792	9.22
M_6	3,664,907	78.62	2,002	32.57
M_7	2,307,628	76.17	1,351	28.31
M_8	2,200,288	75.79	1,299	32.28
M_9	2,284,787	75.24	1,340	34.10
M_{10}	2,466,440	75.89	1,427	27.46
Avg		51.79		17.92
Max		84.82		71.27

Notes: This table reports the **Cumulative gains** g_{im} ($i = 1$ for deposits and $i = 2$ for man-hours) for potential mergers M_m of banks S_A and S_A. These gains are the differences between the original input usages of M_m and the corresponding minimum input usages ∂^*_{iA} and ∂^*_{iB}. The proportion (%) is the ratio of the Cumulative gains over the original input usages. The results are presented for a sample of 10 potential mergers taken from the set of 247 mergers that are potentially productive, i.e., are likely to generate input gains after merging. The averages **Avg** and the maxima **Max** are determined based on the full set of 247 mergers.

Hence, the cumulative gains resulting from the merger M_7 are $g_{11} = \gamma_{11} + \gamma_{110} = \$7,375,924$ for deposits, accounting for not less than 76.17% savings of the joint deposits allocated separately to the banks S_{01} and S_{10}. Similarly, the proportion of cumulative gains for man-hours is 28.31%. Table 4 presents these proportions with the cumulative gains g_{im} for the selected set of 10 post-merger banks M_m, $m = 1, \ldots, 10$.

Though Table 4 presents the results for only 10 post-merger banks, the rows Avg and Max refer to the average and maximum proportions of gains recorded over the full sample, i.e., 247 productive post-merger banks. Even if the average gain proportions are 51.79% for deposits and only 17.92% for man-hours, the corresponding maxima attain values as high as 84.82% (M_{169}) and 71.27% (M_{81}), respectively.

Under a standard DEA slack analysis, $e_1^* = 0.507$ means that bank S_{01} must shrink its current expenses by 49.3% before it can aspire for full efficiency. After merging with S_{10}, the derived merger M_7 could attain an efficiency target of $\bar{\pi} = 1$ by retaining only fractions of the combined inputs of the merging banks, as small as 23.83% of deposits. Figure 1 presents the frequency distributions of the gains' proportions for the post-merger banks.

Figure 1 reveals that there is a remarkable domination of deposits' gains, which preserve the cap of 50% and above for 160 post-merger banks out of 247 (64.77%), with 40% of these gains exceeding 70%. In other words, most of the merged entities are using less than 30% of the joined deposits inherited from the merging banks; yet, full efficiency is attained. The picture is almost reversed for man-hours, where 137 post-merger banks (55.46%) are only able to reach 10%

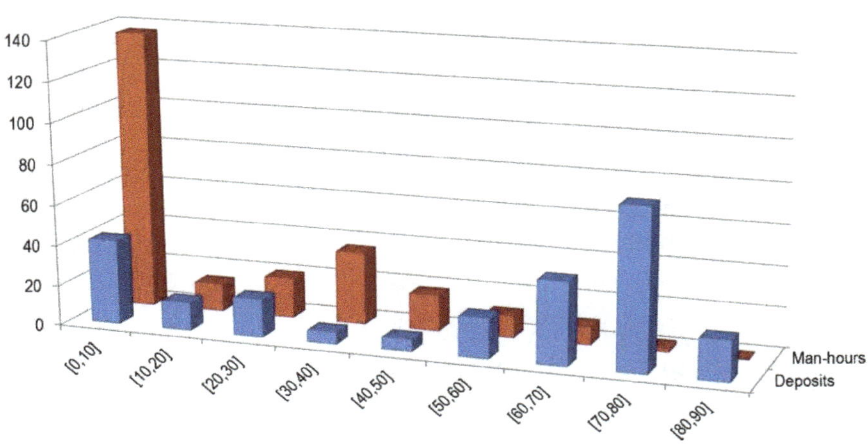

Figure 1. Frequency distributions of the input gains' proportions for post-merger banks.

Notes: This figure shows the number of post-merger banks that achieve proportions of input gains (**Deposits** and **Man-hours**) within the class [a, b] of width 10%, a minimum, and a maximum 90%. Here, the whole set of 247 post-merger banks that are potentially productive is considered.

gains; yet, there are still gains of more than 30% for at least 76 mergers (30.77%).

In the meantime, it is noteworthy that most of the 247 post-merger banks M_m are pairs with possibly a common partner. For instance, all mergers presented in Table 3 are pairs involving S_{01} alongside 10 different banks. A merger cannot occur simultaneously with more than one partner, and hence, it is the responsibility of the DM to select the best match for each bank from its list of potential candidates, i.e., the best set of pairs (S_A, S_B). Practically, such a step entails ranking the alternative post-merger banks $M_m = (S_A, S_B)$ and selecting the best set of non-overlapping mergers, viz. pairs involving distinct banks S_A and S_B.

4. Selecting the Best Partners for a Merger

An important step toward building the best merger plan is the selection of the best set of post-merger banks in the sense that (1) the collective performance of the merging process is maximized and (2) there is no overlapping among the pairs of post-merger banks that are included in the merger plan. Since each productive post-merger bank is defined with two cumulative gains, i.e., deposits and man-hours, measuring the collective performance of the merging process can be treated as a multi-criteria problem. Hence, a DEA model is first proposed to rank the post-merger banks based on the associated relative efficiency. Next, a greedy heuristic is run to select the best non-overlapping pairs of post-merger banks, as explained in the following two sections.

4.1. *Ranking the post-merger banks*

The cumulative expense gains g_{im} of inputs i $(i = 1, \ldots, 2)$ are plausible post-merger outcomes for each pair of banks, $M_m = (S_A, S_B)(m = 1, \ldots, 247)$. It is evident that the higher the value of g_{im}, the more attractive is M_m. As such, g_{im} abides by the dictum "more is better," and hence, each gain g_{im} can be treated as an output of the post-merger production technology. Subsequently, each post-merger bank M_m can be regarded as a DMU that produces the cumulative expense gains g_{im} $(i = 1, \ldots, 2)$ as outputs, and DEA could be a suitable approach to ranking the post-merger banks M_m. The following "best partners" model (BPM) is used:

$$\text{(BPM)} \qquad E_o^* = \max \psi$$

$$\text{s.t.:} \quad \sum_{m=1}^{247} \beta_m g_{im} \geq \psi g_{io}, \quad i = 1, \ldots, 2,$$

$$\sum_{m=1}^{247} \beta_m = 1,$$

$$\beta_m \geq 0, \qquad m = 1, \ldots, 247.$$

BPM is a variable returns to scale (VRS) output-oriented DEA model for the evaluation of post-merger bank M_o, defined with two outputs and no inputs. β_m is the intensity variable associated with the post-merger bank M_m. Hence, M_o is efficient if $E_o^* = 1$, otherwise it is inefficient $(E_o^* > 1)$.

Even with $M = 247$ post-merger banks and two outputs $(I = 2)$ only, model (BPM) does not enable full ranking. To discriminate among the three efficient post-merger banks that BPM produces, we use the associated benchmarking powers P_o, as shown in Table 5, where P_o is the number of times the efficient post-merger bank M_o appears in the reference sets of inefficient post-merger banks.

Table 5. Ranking of the efficient post-merger banks.

M_o	S_A	S_B	P_o	**Rank**
M_{25}	S_{02}	S_{08}	49	3
M_{36}	S_{02}	S_{19}	77	2
M_{211}	S_{17}	S_{19}	197	1

Notes: This table reports the ranks, **Rank**, of the three efficient post-merger banks M_o of banks S_A and S_A. The ranks are determined based on the highest benchmarking power P_o of each post-merger bank M_o.

Based on the benchmarking powers, there is no doubt that M_{211} is the leading post-merger bank, followed by M_{36} and M_{25}. The remaining 244 inefficient post-merger banks are ranked based on the corresponding scores, E_k^*.

4.2. *Building a merger plan*

Let $\mathfrak{R}$ denote the list of ranked post-merger banks, where $\mathfrak{R} = \{M^{(r)} = (S_A^{(r)}, S_B^{(r)}) \mid r = 1, \ldots, 247\}$, and let $\mathfrak{I}$ the list of banks that have already been paired.

The following greedy heuristic is implemented to select the list $\aleph$ of post-merger banks that are best suited for the merger plan:

Iteration 0: $\aleph = \varphi$, $\mathfrak{I} = \varphi$
$r = 1$
Iteration r: while $(r \leq 274)$ do {
If $(S_A^{(r)} \notin \mathfrak{I}$ and $S_B^{(r)} \notin \mathfrak{I})$ {
$\aleph \leftarrow \aleph \cup M^{(r)} = (S_A^{(r)}, S_B^{(r)})\}$
$\mathfrak{I} \leftarrow \mathfrak{I} \cup \{S_A^{(r)}, S_B^{(r)}\}$
$\mathfrak{R} \leftarrow \mathfrak{R} \backslash \{M^{(r)}\}$
}
$r \leftarrow r + 1$
}

Thus, M_{211} the lead post-merger bank, is selected in iteration $r = 1$, i.e., $\aleph = \{M_{211}\} \Rightarrow \mathfrak{I} = \{S_{17}, S_{19}\}$ Although M_{36} ranks in the second position, it is discarded from the list $\aleph$ in iteration $r = 2$ because it is overlapping on S_{19} with M_{211}. In iteration $r = 3$, M_{25} enters $\aleph$, i.e., $\aleph = \{M_{211}, M_{25}\} \Rightarrow \mathfrak{I} = \{S_{17}, S_{19}, S_{02}, S_{08}\}$.

The above algorithm is run again for $r = 4, \ldots, 247$ based on the rank succession of the remaining 244 pairs of banks. The final list of post-merger banks $\aleph$, which involves all the banks except $S_{03}, S_{07}, S_{12}, S_{20}, S_{21}$, and S_{24} is $\aleph = \{M_{211}, M_{25}, M_{89}, M_{131}, M_{14}, M_{79}, M_{56}, M_{198}, M_{191}, M_{137}, M_{243}\}$. Thus, the best merger plan is presented in Table 6.

Hence, the total savings on deposits and man-hours together can reach \$10,848,663.27 with the proposed merger plan. In proportional terms, 78.83% of these savings are due on deposits and the rest (21.17%) on man-hours.

Considering the initial total of \$90,754,300 input expenses, the post-merger saving index will be 11.95%. In comparison with its

Table 6. Merger plan with pairs of banks and corresponding gains.

M_m	S_A	S_B	Deposits ($)	Man-hours (h)
M_{211}	S_{17}	S_{19}	44,045,201	5,511
M_{25}	S_{02}	S_{08}	43,746,323	11,647
M_{89}	S_{06}	S_{11}	27,745,970	13,529
M_{131}	S_{09}	S_{23}	34,012,665	12,508
M_{14}	S_{01}	S_{18}	42,253,468	3,309
M_{79}	S_{05}	S_{22}	42,163,303	3,859
M_{56}	S_{04}	S_{16}	41,680,149	3,624
M_{198}	S_{15}	S_{26}	19,996,874	11,942
M_{191}	S_{14}	S_{28}	19,858,101	11,847
M_{137}	S_{10}	S_{13}	32,130,767	10,856
M_{243}	S_{25}	S_{27}	37,778,506	6,532
Gains			385,411,327	95,165

Notes: This table reports the best mergers among the 28 banks, with the total gains that are likely to be generated for each input (**Deposits** and **Man-hours**) from each potential post-merger $\boldsymbol{M}_{\mathbf{m}}$. The total gains for the whole merger plan are shown in the last row.

pre-merger value of 3.88%, there is an improvement by a factor of more than three.

5. Concluding Remarks

In this chapter, we introduced a hybrid DEA-based methodology to build the best prospective merger plan for the banking sector on the grounds of the potential gains of individual post-merger banks. As such, the objective of the proposed methodology is to identify the partners of the post-merger banks that would ultimately qualify the whole banking sector, which is hypothetically represented by the selected sample of banks, to achieve maximum collective performance. Specifically, if we assume that each bank is willing to merge with another bank, within the selected sample, all pairwise consolidations among banks are evaluated through an IDEA model. Based on the optimal gains resulting from the merger, only possibly productive post-merger banks, i.e., those mergers that have the potential to generate gains, are further considered. In the next stage, a hybrid procedure that incorporates a DEA model with a greedy heuristic enables the best pairs of productive post-merger banks to be selected in such a way that the collective performance of the ultimate selection is maximized.

To assess the practical scope of the proposed methodology, we considered a case study involving 28 domestic Canadian banks, each defined with two inputs, namely man-hours and deposits, and two outputs, i.e., interest income and non-interest income.

The performance analysis carried out on the selected sample of banks produced an average technical efficiency of 0.7797, with 35.71% of the banks strongly efficient, i.e., not requiring input savings. As a result of the high relative efficiency level, the expense savings index was only 3.88%. In other words, based on the standard DEA approach, if each inefficient bank manages to upgrade its respective status to strongly efficient, this will generate an overall savings of 3.88% on the expenses of the whole sample of banks.

With a sample of 28 banks, the IDEA approach was applied to investigate 378 pairwise consolidations, among which 247 potential post-merger banks have been identified as productive, i.e., likely to generate expense savings.

One of the most salient results at this stage is undoubtedly the fact that, among the productive post-merger banks, there are pairs that involve exclusively strongly efficient banks, which are not supposed to require any savings based on the individual banks' production possibility set. Yet, it appears that, after merging, there is still room for savings. Moreover, a thorough analysis of the outcomes of these mergers revealed that the proportions of potential savings per post-merger bank can reach 84.82% and 71.27% for deposits and man-hours, respectively. These proportions, put in the banking context, can already be considered strong evidence of the benefits of mergers as potential contributors to enhancing gains.

An important step toward devising the best merger plan is the selection of the best non-overlapping pairs of banks from the sample of productive post-merger banks. With each productive post-merger bank being defined with two cumulative gains, i.e., deposits and man-hours, the problem is again multi-criteria, which requires primarily a DEA model as a ranking device. Though the ranking pattern that the DEA model produces allows for clear discrimination among the post-merger banks, developing a greedy heuristic was required for selecting the best non-overlapping pairs of banks. The selection process unveiled a final list of 11 productive post-merger banks, which include all the banks except $S_{03}, S_{07}, S_{12}, S_{20}, S_{21}$, and S_{24}. Accordingly, 22 out of 28 banks have been paired with the best partner among the existing candidates. The six banks that have been excluded from the list can always merge with each other, but all consolidations of two or more banks among these banks are likely to be unproductive.

The new expense saving index for the anticipated merger plan has been evaluated at 11.95%, which shows an improvement by a factor of more than three, in comparison with its pre-merger value of 3.88%. These figures are strategically important for supporting new policies that would promote mergers among banks as an established option for collective benefits.

Nonetheless, the present methodology may suffer limitations from both methodological and practical perspectives. In terms of methodology, the adoption of a standard DEA model for ranking the productive post-merger banks may not be sufficient to attain full discrimination if there are more inputs and outputs. To overcome such a difficulty, one could resort to developing more advanced ranking techniques, such as DEA cross-efficiency (see, e.g., Oukil & Amin, 2015; Amin & Oukil, 2019b), where a DM's preferences could be integrated as a key factor in the decision-making process (Oukil, 2018, 2019, 2020b, 2021). On the other hand, it is apparent that a comprehensive investigation of all possible pairwise consolidations might be computationally costly if the number of banks under scrutiny is much higher. Hence, another future research avenue may consider developing models that would duly integrate the merger's evaluation as well as the best partners' selection stages under the same framework.

In practice, the concept of a "merger plan" might not be feasible or even realistic in some business contexts. Therefore, a future study may focus on investigating various contextual settings of horizontal cooperation as well as corporate restructuring strategies where developing a large-scale merger plan may be conceptually appropriate and practically relevant.

Appendix

Table A1. Data for 28 Canadian banks (year 2017).

	Inputs		Outputs	
Bank	Deposits	Man-hours	Interest income	Non-interest income
S_{01}	9,086,901	1,836.92961	366,548	49,622
S_{02}	291,290,201	228,513.93	15,833,418	10,714,905
S_{03}	444,602,101	225,711.05	23,927,247	12,119,904
S_{04}	1,709,301	380.4032135	63,971	6,399
S_{05}	1,695,801	1,921.428462	135,206	41,972
S_{06}	230,792,651	159,096.58	13,592,577	7,303,171
S_{07}	5,588,401	2,926.17215	930,678	222,957
S_{08}	19,244,701	6,746.707728	1,027,285	65,951
S_{09}	8,058,401	1,132.843917	243,871	23,238
S_{10}	596,651	47.12088753	17,886	1,565
S_{11}	20,074,351	2,017.587605	717,506	33,611
S_{12}	2,501	81.42846213	14	5,129
S_{13}	141,501	196.10329	14,083	3,822
S_{14}	1,142,301	130.1813313	44,328	781
S_{15}	1,777,701	42.83626626	77,231	1,777
S_{16}	3,147,751	483.3259373	124,493	918
S_{17}	30,872,751	11,049.23	1,255,546	354,938
S_{18}	14,367,351	2,025.942617	652,659	70,503
S_{19}	117,432,701	72,068.02	5,580,044	3,377,188
S_{20}	2,758,801	1,011.711553	392,617	506,482
S_{21}	488,211,851	407,965.32	26,904,175	20,475,659
S_{22}	417,551	953.04	67,788	627
S_{23}	20,976,751	3,474.664881	765,464	81,060
S_{24}	449,259,351	306,581.72	29,832,307	13,505,543
S_{25}	6,401	245.5906656	1,283	31,553
S_{26}	1,389,001	418.9342005	71,766	307
S_{27}	43,551	119.0719204	1,461	33
S_{28}	1,376,051	186.7689365	34,765	7,074

Table A2. Slack analysis of the inefficient banks.

Bank	Deposits s^*_{1k}	Man-hours s^*_{2k}	Bank	Deposits s^*_{1k}	Man-hours s^*_{2k}	Bank	Deposits s^*_{1k}	Man-hours s^*_{2k}
S_{01}	4,478,343	905	S_{09}	4,112,162	578	S_{19}	20,995,323	12,885
S_{02}	18,146,506	14,236	S_{13}	52,645	73	S_{22}	11,717	27
S_{04}	928,390	207	S_{14}	315,510	36	S_{23}	7,375,803	1,222
S_{05}	885,050	1,003	S_{16}	1,545,656	237	S_{26}	677,559	204
S_{06}	12,510,808	8,624	S_{17}	13,902,988	4,976	S_{27}	13,647	37
S_{08}	8,228,284	2,885	S_{18}	1,244,644	176	S_{28}	716,401	97

References

Akbar, I., Sebastian Herman, S., & Rusydiana, A.S. (2021). Efficiency of pre-merged state-owned Sharia Banks in Indonesia using 2-stage DEA. *Ekonomi Islam Indonesia, 3*(2). https://doi.org/10.58968/eii.v3i2.39.

Akhtar, S., Alam, M., & Ansari, M.S. (2021). Measuring the performance of the Indian banking industry: Data envelopment window analysis approach. *Benchmarking: An International Journal*, 29(9), 2842–2857.

Alber, N., Elmofty, M., Kishk, I., & Sami, R. (2019). Banking efficiency: Concepts, drivers, measures, literature and conceptual model. https://ssrn.com/abstract=3310982; http://dx.doi.org/10.2139/ssrn.3310982Drivers.

Al-Khasawneh, J.A., Essaddam, N., & Hussain, T. (2020). Total productivity and cost efficiency dynamics of US merging banks: A non-parametric bootstrapped analysis of the fifth merger wave. *The Quarterly Review of Economics and Finance*, 78, 199–211.

Al-Mezeini, N.K., Oukil, A., & Al-Ismaili, A.M. (2020). Investigating the efficiency of greenhouse production in Oman: A two-stage approach based on Data Envelopment Analysis and double bootstrapping. *Journal of Cleaner Production*, 247, 119160. https://doi.org/10.1016/j.jclepro.2019.119160.

Al Tamimi, H.A.H., Duqi, A., Kanas, A., & Zervopoulos, P.D. (2022). Directional distance function DEA estimators for evaluating efficiency gains from possible mergers and acquisitions. *Journal of the Operational Research Society*, 73(6), 1240–1257.

Amin, G.R. & Al-Muharrami, S. (2018). A new inverse data envelopment analysis model for mergers with negative data. *IMA Journal of Management Mathematics*, 29(2), 137–149.

Amin, G.R. & Ibn Boamah, M. (2020). A new inverse DEA cost efficiency model for estimating potential merger gains: A case of Canadian banks. *Annals of Operations Research*, 295, 21–36.

Amin, G.R. & Ibn Boamah, M. (2021). A two-stage inverse data envelopment analysis approach for estimating potential merger gains in the US banking sector. *Managerial and Decision Economics*, 42(6), 1454–1465.

Amin, G.R. & Oukil, A. (2019a). Flexible target setting in mergers using inverse data envelopment analysis. *International Journal of Operational Research*, 35(3), 301–317.

Amin, G.R. & Oukil, A. (2019b). Gangless cross-evaluation in DEA: An application to stock selection. *RAIRO-Operations Research*, 53(2), 645–655.

Amin, G.R., Al-Muharrami, S., & Toloo, M. (2019). A combined goal programming and inverse DEA method for target setting in mergers. *Expert Systems with Applications*, *115*, 412–417.

Amin, G.R., Emrouznejad, A., & Gattoufi, S. (2017a). Minor and major consolidations in inverse DEA: Definition and determination. *Computers & Industrial Engineering*, 103, 193–200.

Amin, G.R., Emrouznejad, A., & Gattoufi, S. (2017b). Modelling generalized firms' restructuring using inverse DEA. *Journal of Productivity Analysis*, 48(1), 51–61.

An, Q., Liu, X., Li, Y., & Xiong, B. (2019). Resource planning of Chinese commercial banking systems using two-stage inverse data envelopment analysis with undesirable outputs. *PLoS ONE*, 14(6), e0218214.

Antil, S., Kumar, M., & Swain, N. (2020). Evaluating the efficiency of regional rural banks across the Indian states during different phases of structural development. *Applied Economics*, 52(41), 4457–4473.

Antunes, J., Hadi-Vencheh, A., Jamshidi, A., Tan, Y., & Wanke, P. (2022). Bank efficiency estimation in China: DEA-RENNA approach. *Annals of Operations Research*, 315(2), 1373–1398.

Avinadav, T., Chernonog, T., & Perlman, Y. (2017). Mergers and acquisitions between risk-averse parties. *European Journal of Operational Research*, 259(3), 926–934.

Banker, R.D., Charnes, A., & Cooper, W.W. (1984). Some models for estimating technical and scale inefficiencies in data envelopment analysis. *Management Science*, 9, 1078–1092.

Barkhordari, F., Dehghan Dehnavi, M.A., Sargolzaei, M., & Rajaei Bagh Siaei, M. (2022). The effect of potential mergers on liquidity creation and efficiency in Iranian banks. *Quarterly Journal of Fiscal and Economic Policies*, 10(38), 141–160.

Blanco-Oliver, A. (2021). Banking reforms and bank efficiency: Evidence for the collapse of Spanish savings banks. *International Review of Economics & Finance*, 74, 334–347.

Bogetoft, P. & Wang, D. (2005). Estimating the potential gains from mergers. *Journal of Productivity Analysis,* 23, 145–171.

Borisov, L.D. (2020). The impact of mergers and acquisitions on the efficiency of banks in Bulgaria. *Economic Thought Journal*, 2020(3), 66–85.

Boubaker, S., Le, T.D.Q., & Ngo, T. (2022a). Managing bank performance under COVID-19: A novel inverse DEA efficiency approach. *International Transactions in Operational Research*, 30(2023), 2436–2452. https://doi.org/10.1111/itor.13132.

Boubaker, S., Manita, R., & Mefteh-Wali, S. (2022b). Foreign currency hedging and firm productive efficiency. *Annals of Operations Research*, 313, 833–854.

Boubaker, S., Manita, R., & Rouatbi, W. (2021). Large shareholders, control contestability and firm productive efficiency. *Annals of Operations Research*, 296(1), 591–614.

Boubaker, S., Do, D.T., Hammami, H., & Ly, K.C. (2022c). The role of bank affiliation in bank efficiency: A fuzzy multi-objective data envelopment analysis approach. *Annals of Operations Research*, 311, 611–639. https://doi.org/10.1007/s10479-020-03817-z.

Bou-Hamad, I., Anouze, A.L., & Larocque, D. (2017). An integrated approach of data envelopment analysis and boosted generalized linear mixed models for efficiency assessment. *Annals of Operations Research*, 253(1), 77–95.

Branco, J.E.H., Bartholomeu, D.B., Junior, P.N.A., & Caixeta Filho, J.V. (2021). Mutual analyses of agriculture land use and transportation networks: The future location of soybean and corn production in Brazil. *Agricultural Systems*, 194, 103264.

Chiorazzo, V., D'Apice, V., DeYoung, R., & Morelli, P. (2018). Is the traditional banking model a survivor? *Journal of Banking & Finance*, 97, 238–256.

Chiu, Y.H., Lin, T.Y., Chang, T.H., Lin, Y.N., & Chiu, S.Y. (2021). Prevaluating efficiency gains from potential mergers and acquisitions in the financial industry with the Resample Past–Present–Future data envelopment analysis approach. *Managerial and Decision Economics*, 42(2), 369–384.

Čiković, K.F., Lozić, J., & Guzovski, M. (2022). Do mergers and acquisitions improve bank efficiency? Evidence from North Macedonia. *Journal of International Studies*, 15(2), 77–93.

Cobas, A., Maziotis, A., & Villegas, A. Measurement of Efficiency and its Drivers in the Chilean Banking Industry. Available at *SSRN* 4266049. https://ssrn.com/abstract=4266049.

Coccorese, P. & Ferri, G. (2020). Are mergers among cooperative banks worth a dime? Evidence on efficiency effects of M&As in Italy. *Economic Modelling*, 84, 147–164.

Cvetkoska, V. & Čiković, K.F. (2020). Assessing the relative efficiency of commercial banks in the Republic of North Macedonia: DEA window analysis. *Croatian Operational Research Review*, 11, 217–227.

Cvetkoska, V. & Savic, G. (2021). DEA in banking: Analysis and visualization of bibliometric data. *Data Envelopment Analysis Journal*, 5(2), 455–485.

de Abreu, E.S., Kimura, H., & Sobreiro, V.A. (2019). What is going on with studies on banking efficiency? *Research in International Business and Finance*, 47, 195–219.

de Mello Valerio, V.E., de Oliveira Pamplona, E., Fonseca, M.N., Junior, P.R., Rocha, L.C.S., & Peruchi, R.S. (2021). Efficiency and productivity of the Brazilian banks: A new approach based on two-stage network DEA. *Brazilian Review of Finance*, 19(4), 130–159.

Dia, M., Golmohammadi, A., & Takouda, P.M. (2020). Relative efficiency of Canadian banks: A three-stage network bootstrap DEA. *Journal of Risk and Financial Management*, 13(4), 68.

Doumpos, M., Zopounidis, C., Gounopoulos, D., Platanakis, E., & Zhang, W. (2022). Operational research and artificial intelligence methods in banking. *European Journal of Operational Research*. https://doi.org/10.1016/j.ejor.2022.04.027.

Duong, T.T.N., Phan, H.T., Hoang, T.N., & Vo, T.T.T. (2020). The effect of financial restructuring on the overall financial performance of the commercial banks in Vietnam. *The Journal of Asian Finance, Economics and Business*, 7(9), 75–84.

Edward, A. & Manoj, J. (2019). Analysis of pre and post-merger financial performance of SBI associate banks. *International Journal of Research and Analytical Reviews*, 6(2), 630–635.

Emrouznejad, A., Yang, G.L., Khoveyni, M., & Michali, M. (2022). Data envelopment analysis: Recent developments and challenges. In: S. Salhi & J. Boylan (Eds.) *The Palgrave Handbook of Operations Research* (pp. 307–350). Cham, Switzerland: Palgrave Macmillan.

Fang, D., Hsueh, M.C., & Chiu, C.R. (2020). An investigation of international hotel productivity considering quasifixed inputs and negative outputs. *Managerial & Decision Economics*, 41(3), 380–388.

Fernandes, F.D.S., Stasinakis, C., & Bardarova, V. (2018). Two-stage DEA-truncated regression: Application in banking efficiency and financial development. *Expert Systems with Applications*, 96, 284–301.

Fraisse, H., Hombert, J., & Lé, M. (2018). The competitive effect of a bank megamerger on credit supply. *Journal of Banking & Finance*, 93, 151–161.

Gaganis, C., Galariotis, E., Pasiouras, F., & Staikouras, C. (2020). Bank profit efficiency and financial consumer protection policies. *Journal of Business Research*, 118, 98–116.

Galariotis, E., Kosmidou, K., Kousenidis, D., Lazaridou, E., & Papapanagiotou, T. (2021). Measuring the effects of M&As on Eurozone bank efficiency: An innovative approach on concentration and credibility impacts. *Annals of Operations Research*, 306(1), 343–368.

Gattoufi, S., Amin, G.R., & Emrouznejad, A. (2014). A new inverse DEA method for merging banks. *IMA Journal of Management Mathematics*, 25, 73–87.

Gerami, J., Mozaffari, M.R., Wanke, P.F., & Correa, H.L. (2021). A generalized inverse DEA model for firm restructuring based on value efficiency. *IMA Journal of Management Mathematics*. dpab043, https://doi.org/10.1093/imaman/dpab043.

Ghiyasi, M. & Zhu, N. (2020). An inverse semi-oriented radial data envelopment analysis measure for dealing with negative data. *IMA Journal of Management Mathematics*, 31(4), 505–516.

Ghobadi, S. (2019). A generalized DEA model for inputs (outputs) estimation under inter-temporal dependence. *RAIRO-Operations Research*, 53(5), 1791–1805.

Ghobadi, S. (2021). Merging decision-making units with interval data. *RAIRO-Operations Research*, 55, S1605–S1631.

Guijarro, F., Martínez-Gómez, M., & Visbal-Cadavid, D. (2020). A model for sector restructuring through genetic algorithm and inverse DEA. *Expert Systems with Applications*, 154, 113422.

Halkos, G.E. & Tzeremes, N.G. (2013). Estimating the degree of operating efficiency gains from a potential bank merger and acquisition: A DEA bootstrapped approach. *Journal of Banking & Finance*, 37(5), 1658–1668.

Hassan, M. & Oukil, A. (2021). Design of efficient systems of commercial material handling equipment for supply chain and logistics facilities using DEA. *International Journal of Logistics Systems & Management*, 39(2), 241–272.

Henriques, I.C., Sobreiro, V.A., Kimura, H., & Mariano, E.B. (2018). Efficiency in the Brazilian banking system using data envelopment analysis. *Future Business Journal*, 4(2), 157–178.

Henriques, I.C., Sobreiro, V.A., Kimura, H., & Mariano, E.B. (2020). Two-stage DEA in banks: Terminological controversies and future directions. *Expert Systems with Applications*, 161, 113632.

Hsu, S.Y., Lu, C.C., Chou, K.W., & Chiu, Y.H. (2018). Dynamic performance evaluation of banking industry in Taiwan after mergers. *International Journal of Organizational Innovation*, 11(1), 1–25.

Huhtilainen, M., Saastamoinen, J., & Suhonen, N. (2022). Determinants of mergers and acquisitions among Finnish cooperative and savings banks. *Journal of Banking Regulation*, 23(3), 339–349.

Husejinovic, A. (2019). Efficiency of commercial banks operating in Federation of Bosnia and Herzegovina using DEA method. *Sustainable Engineering and Innovation*, 1(2), 106–111. https://ssrn.com/abstract=3519131.

Ibn Boamah, M.I. (2018). Real interest parity: Evidence from trade partnerships. *Review of Financial Economics*, 36(3), 199–205.

Kamarudin, F., Sufian, F., Nassir, A.M., Anwar, N.A.M., & Hussain, H.I. (2019). Bank efficiency in Malaysia a DEA approach. *Journal of Central Banking Theory and Practice*, 8(1), 133–162.

Kamiyama, H., Kefi, M., & Kashiwagi, K. (2021). Irrigation water use efficiency in olive trees in Kairouan, Tunisia. *Asian Journal of Agriculture and Rural Development*, 11(3), 255–261.

Kamyab, P., Mozaffari, M.R., & Gerami, J. (2019). Determining minor and major consolidations in network inverse data envelopment analysis. *Journal of Mathematical Extension*, 14, 195–223.

Karlsson, L., Häggqvist, H., & Hedberg, P. (2021). Market structure and efficiency in Swedish commercial banking, 1912–1938. *Scandinavian Economic History Review*, 69(2), 103–123.

Kazemi, A. & Galagedera, D.U. (2022). An inverse DEA model for intermediate and output target setting in serially linked general two-stage processes. *IMA Journal of Management Mathematics*, dpab041. https://doi.org/10.1093/imaman/dpab041.

Khoveyni, M. & Eslami, R. (2022). Merging two-stage series network structures: A DEA-based approach. *OR Spectrum*, 44(1), 273–302.

Kowalik, M., Davig, T., Morris, C.S., & Regehr, K. (2015). Bank consolidation and merger activity following the crisis. *Economic Review*, 1612387(100), 31–49.

Kundu, S. & Banerjee, A. (2022). Operational and policy efficiency: A comparison between public and private Indian banks. *International Journal of Productivity and Performance Management*, 71(4), 1537–1558.

Larkin, Y. & Lyandres, E. (2019). Inefficient mergers. *Journal of Banking & Finance*, 108, 105648.

Le, T.D.Q., Ho, T.H., Nguyen, D.T., & Ngo, T. (2021). Fintech credit and bank efficiency: International evidence. *International Journal of Financial Studies*, 9(3), 44.

Li, F., Liang, L., Li, Y., & Emrouznejad, A. (2018). An alternative approach to decompose the potential gains from mergers. *Journal of the Operational Research Society*, 69(11), 1793–1802.

Liu, X., Yang, F., & Wu, J. (2020a). DEA considering technological heterogeneity and intermediate output target setting: The performance analysis of Chinese commercial banks. *Annals of Operations Research*, 291(1), 605–626.

Liu, X., Sun, J., Yang, F., & Wu, J. (2020b). How ownership structure affects bank deposits and loan efficiencies: An empirical analysis of Chinese commercial banks. *Annals of Operations Research*, 290(1), 983–1008.

Lozano, S. & Villa, G. (2010). DEA-based pre-merger planning tool. *Journal of the Operational Research Society*, 61, 1485–1497.

Maity, S. & Sahu, T.N. (2021). How far the Indian banking sectors are efficient? An empirical investigation. *Asian Journal of Economics and Banking*, 6(3), 413–431.

Marjanović, I., Stanković, J.J., & Popović, Ž. (2018). Efficiency estimation of commercial banks based on financial performance: Input oriented DEA CRS/VRS models. *Economic Themes*, 56(2), 239–252.

Martín, E., Bachiller, A., & Bachiller, P. (2018). The restructuring of the Spanish banking system: Analysis of the efficiency of financial entities. *Management Decision*, 56(2), 474–487.

Matousek, R. & Tzeremes, N.G. (2016). CEO compensation and bank efficiency: An application of conditional nonparametric frontiers. *European Journal of Operational Research*, 251(1), 264–273.

Montgomery, H., Harimaya, K., & Takahashi, Y. (2014). Too big to succeed? Banking sector consolidation and efficiency. *Journal of International Financial Markets, Institutions and Money*, 32, 86–106.

Ngo, T. & Tsui, K.W.H. (2022). Estimating the confidence intervals for DEA efficiency scores of Asia-Pacific airlines. *Operational Research*, *22*(4), 3411–3434.

Nguyen, M.H. (2021). Productivity growth of Vietnamese commercial Banks: An application of non-parametric analysis. *The Journal of Asian Finance, Economics and Business*, 8(9), 177–187.

Nguyen, P.A. & Pham, L.D. (2020). Non-parametric analysis of bank merger gains: The case of Vietnam. *Cogent Business & Management*, 7(1), 1823582.

Nguyen, N.T. & Vân Anh, T.T. (2019). Has merger and acquisition been considered as a method of dealing with weak banks? Evidence from the third bank restructuring process in Vietnam. *Banks and Bank Systems*, 14(1), 55–64.

Nguyen, P.A., Tran, B.L., & Simioni, M. (2021). Optimal capital adequacy ratio: An investigation of Vietnamese commercial banks using two-stage DEA. *Cogent Business & Management*, 8(1), 1870796.

Niepmann, F. & Schmidt-Eisenlohr, T. (2017). International trade, risk and the role of banks. *Journal of International Economics*, 107, 111–126.

Nourani, A. & Bencheikh, A. (2020). Energy requirement optimization of greenhouse vegetable production using Data Envelopment Analysis (DEA) Method in Algeria. *Acta Technologica Agriculturae*, 23(2), 60–66.

Nourani, A. & Bencheikh, A. (2021). Applying the DEA approach to optimize the energy required and GHG emission for greenhouse vegetable production in Algeria. *Agricultural Engineering International: CIGR Journal*, 23(3), 156–164.

Ofori-Sasu, D., Abor, J.Y., & Mensah, L. (2019). Funding structure and technical efficiency: A data envelopment analysis (DEA) approach for banks in Ghana. *International Journal of Managerial Finance*, 15(4), 425–443.

Oral, M., Amin, G.R., & Oukil, A. (2015). Cross-efficiency in DEA: A maximum resonated appreciative model. *Measurement*, 63, 159–167.

Oral, M., Oukil, A., Malouin, J.-L., & Kettani, O. (2014). The appreciative democratic voice of DEA: A case of faculty academic performance evaluation. *Socio-Economic Planning Sciences*, 48, 20–28.

Oukil, A. (2018). Ranking via composite weighting schemes under a DEA cross-evaluation framework. *Computers & Industrial Engineering*, 117, 217–224.

Oukil, A. (2019). Embedding OWA under preference ranking for DEA cross-efficiency aggregation: Issues and procedures. *International Journal of Intelligent Systems*, 34(5), 947–965.

Oukil, A. (2020a). Designing efficient material handling systems: A two-stage approach based on DEA cross-efficiency. In *2020 5th International Conference on Logistics Operations Management (GOL)* (pp. 1–5). https://doi.org/10.1109/GOL49479.2020.9314754.

Oukil, A. (2020b). Exploiting value system multiplicity and preference voting for robust ranking. *Omega, International Journal of Management Science*, 94, 102048. https://doi.org/10.1016/j.omega.2019.03.006.

Oukil, A. (2021). A two-level Induced OWA procedure for ranking DMUs under a DEA cross-efficiency framework. In *Intelligent and Transformative Production in Pandemic Times: Proceedings of the 26th International Conference on Production Research (ICPR)*, 19–21 July 2021, Taichung, Taiwan.

Oukil, A. (2022a). Investigating prospective gains from mergers in the agricultural sector through Inverse DEA. *IMA Journal of Management Mathematics*, In press. https://doi.org/10.1093/imaman/dpac004.

Oukil, A. (2022b). Selecting material handling equipment through a market weight scheme based DEA cross-efficiency approach. *International Journal of Management Science and Engineering Management*. https://doi.org/10.1080/17509653.2022.2116122.

Oukil, A. (2023). Merging farms to enhance agricultural sustainability: An energy efficiency perspective. *International Conference on Innovation and Technological Advances for Sustainability* (ITAS 2023) 1–3 March 2023, Doha, Qatar.

Oukil, A. & Al-Zidi, A. (2018). Benchmarking the hotel industry in Oman through a three-stage DEA-based procedure. *Journal of Arts and Social Sciences*, 9(2), 5–23.

Oukil, A. & Amin, G.R. (2015). Maximum appreciative cross-efficiency in DEA: A new ranking method. *Computers & Industrial Engineering*, 81, 14–21.

Oukil, A. & El-Bouri, A. (2021). Ranking dispatching rules in multi-objective dynamic flow shop scheduling: A multi-faceted perspective. *International Journal of Production Research*, 59(2), 388–411.

Oukil, A. & Govindaluri, S.M. (2017). A systematic approach for ranking football players within an integrated DEA-OWA framework. *Managerial & Decision Economics*, 38(8), 1125–1136.

Oukil, A. & Govindaluri, S.M. (2020). A hybrid multi-attribute decision-making procedure for ranking project proposals: A historical data perspective. *Managerial and Decision Economics*, 41(3), 461–472.

Oukil, A. & Zekri, S. (2015). Inverse optimization: An efficient tool for resource reallocation in the farming sector. In *Operational Research Practice in Africa* (ORPA), 20–22 April 2015, Algiers, Algeria.

Oukil, A. & Zekri, S. (2021). Investigating farming efficiency through a two stage analytical approach: Application to the agricultural sector in Northern Oman. Preprint, *arXiv preprint* arXiv:2104.10943.

Oukil, A., Channouf, N., & Al-Zidi, A. (2016). Performance evaluation of the hotel industry in an emerging tourism destination: Case of Oman. *Journal of Hospitality & Tourism Management*, 29, 60–68.

Oukil, A., El-Bouri, A., & Emrouznejad, A. (2022a). Energy-aware job scheduling in a multi-objective production environment — An integrated DEA-OWA model. *Computers & Industrial Engineering*, 168, 108065.

Oukil, A., Nourani, A. Soltani, A.A., & Benchikh, A. (2022b). Using inverse data envelopment analysis to investigate potential impact of mergers on energy use optimization — Application in agricultural production. *Journal of Cleaner Production*, 381(1), 135199.

Oukil, A., Soltani, A.A., Boutaghane, H., Abdalla, O., Bermad, A., Hasbaia, M., & Boulassel, M.R. (2021). A surrogate water quality index to assess groundwater using a unified DEA-OWA framework. *Environmental Science and Pollution Research*, 28, 56658–56685.

Oukil, A., Soltani, A.A., Zeroual, S., Boutaghane, H., Abdalla, O., Bermad, A., & Boulassel, M.R. (2022c). A DEA cross-efficiency inclusive methodology for assessing water quality: A Composite Water Quality Index. *Journal of Hydrology*, 612, 128123.

Phan, H.T., Anwar, S., Alexander, W.R.J., & Phan, H.T.M. (2019). Competition, efficiency and stability: An empirical study of East Asian commercial banks. *The North American Journal of Economics and Finance*, 50, 100990.

Pournader, M., Kach, A., Hajiagha, S.H.R., & Emrouznejad, A. (2017). Investigating the impact of behavioral factors on supply network efficiency: Insights from banking's corporate bond networks. *Annals of Operations Research*, 254(1), 277–302.

Rao, U. & Mishra, T. (2020). Posterior analysis of mergers and acquisitions in the international financial market: A re-appraisal. *Research in International Business and Finance*, 51, 101062.

Ray, S.C. & Sethia, S. (2022). Nonparametric measurement of potential gains from mergers: An additive decomposition and application to Indian bank mergers. *Journal of Productivity Analysis*, 57(2), 115–130.

Shi, X., Li, Y., Emrouznejad, A., Xie, J., & Liang, L. (2017). Estimation of potential gains from bank mergers: A novel two-stage cost efficiency DEA model. *Journal of the Operational Research Society*, 68(9), 1045–1055.

Simper, R., Hall, M.J., Liu, W., Zelenyuk, V., & Zhou, Z. (2017). How relevant is the choice of risk management control variable to non-parametric bank profit efficiency analysis? The case of South Korean banks. *Annals of Operations Research*, 250(1), 105–127.

Soleimani-Chamkhorami, K., Hosseinzadeh Lotfi, F., Jahanshahloo, G.R., & Rostamy-Malkhalifeh, M. (2020). Preserving cost and revenue efficiency through inverse data envelopment analysis models. *INFOR: Information Systems and Operational Research*, 58(4), 561–578.

Soltani, A.A., Oukil, A., Boutaghane, H., Bermad, A., & Boulassel, M.R. (2021). A new methodology for assessing water quality, based on data envelopment analysis: Application to Algerian dams. *Ecological Indicators*, 121, 106952.

Soltanifar, M., Ghiyasi, M., & Sharafi, H. (2023). Inverse DEA-R models for merger analysis with negative data. *IMA Journal of Management Mathematics*, 34(3), 491–510.

Soltanifar, M., Ghiyasi, M., Emrouznejad, A., & Sharafi, H. (2022). Novel model for merger analysis and target setting: A CSW-Inverse DEA approach. Available at *SSRN* 4115552.

Sow, O. Oukil, A. Ndiaye, B.M. Marcos, A. (2016). Efficiency analysis of public transportation subunits using data envelopment analysis (DEA) and bootstrap: Dakar Dem Dikk case study. *Journal of Mathematics Research*, 8, 114–127.

Sukmana, R., Ajija, S.R., Salama, S.C.U., & Hudaifah, A. (2020). Financial performance of rural banks in Indonesia: A two-stage DEA approach. *Heliyon*, 6(7), e04390.

Thaker, K., Charles, V., Pant, A., & Gherman, T. (2022). A DEA and random forest regression approach to studying bank efficiency and corporate governance. *Journal of the Operational Research Society*, 73(6), 1258–1277.

Tziogkidis, P., Matthews, K., & Philippas, D. (2018). The effects of sector reforms on the productivity of Greek banks: A step-by-step analysis of the pre-Euro era. *Annals of Operations Research*, 266(1–2), 531–549.

Učkar, D. & Petrović, D. (2021). Efficiency of banks in Croatia. *Zbornik radova Ekonomskog fakulteta u Rijeci: časopis za ekonomsku teoriju i praksu*, 39(2), 349–379.

Vo, X.V. (2018). M&As in the process of banking consolidation — Preliminary evidence from Vietnam. *Asian Journal of Law and Economics*, 9(2), 20170032.

Vo, X.V. & Nguyen, H.H. (2018). Bank restructuring and bank efficiency — The case of Vietnam. *Cogent Economics & Finance*, 6(1), 1520423.

Wanke, P., Azad, M.A.K., Emrouznejad, A., & Antunes, J. (2019). A dynamic network DEA model for accounting and financial indicators: A case of efficiency in MENA banking. *International Review of Economics & Finance*, 61, 52–68.

Wei, Q., Zhang, J., & Zhang, X. (2000). An inverse DEA model for inputs/outputs estimate. *European Journal of Operational Research*, 121(1), 151–163.

Xiao, S., Li, Y., Emrouznejad, A., Xie, J., & Liang, L. (2017). Estimation of potential gains from bank mergers: A novel two-stage cost efficiency DEA model. *Journal of the Operational Research Society*, 68(9), 1045–1055.

Xie, J., Zhu, X., & Liang, L. (2020). A multiplicative method for estimating the potential gains from two-stage production system mergers. *Annals of Operations Research*, 288(1), 475–493.

Yan, H., Wei, Q., & Hao, G. (2002). DEA models for resource reallocation and production input/output estimation. *European Journal of Operational Research*, 136(1), 19–31.

Yeh, L.T., Chang, D.S., & Li, H.M. (2022). Developing a network data envelopment analysis model to measure the efficiency of banking with the governance, innovation, and operations. *Managerial and Decision Economics*, 43(7), 2863–2874.

Yu, M.M., Lin, C.I., Chen, K.C., & Chen, L.H. (2021). Measuring Taiwanese bank performance: A two-system dynamic network data envelopment analysis approach. *Omega*, 98, 102145.

Zalaghi, A.H., Lotfalian Dehkordi, A., Abedi, A., & Taki, M. (2021). Applying data envelopment analysis (DEA) to improve energy efficiency of apple fruit, focusing on cumulative energy demand. *Energy Equipment and Systems*, 9(1), 37–52.

Zaman, M.S. & Bhandari, A.K. (2022). Stressed assets, off-balance sheet business activities and performance of Indian banking sector: A DEA double bootstrap approach. *Studies in Economics and Finance*, 39(4), 572–592.

Zaman, M.S., Valiyattoor, V., & Bhandari, A.K. (2022). Dynamics of total factor productivity growth: An empirical analysis of Indian commercial banks. *The Journal of Economic Asymmetries*, 26, e00268.

Zhou, X., Xu, Z., Chai, J., Yao, L., Wang, S., & Lev, B. (2019). Efficiency evaluation for banking systems under uncertainty: A multi-period three-stage DEA model. *Omega*, 85, 68–82.

Zhu, N., Hougaard, J.L., Yu, Z., & Wang, B. (2020). Ranking Chinese commercial banks based on their expected impact on structural efficiency. *Omega*, 94, 102049.

https://doi.org/10.1142/9781800615786_0010

Chapter 10

Data Envelopment Analysis in Microfinance Research: A Systematic Review

Md Imran Hossain[*,†,‡] **and Md Aslam Mia**[*,§]

[*]*School of Management, Universiti Sains Malaysia, Penang, Malaysia*

[†]*Department of Finance, Jagannath University, Dhaka, Bangladesh*

[‡]*imranhossain@student.usm.my*

[§]*aslammia@usm.my*

Abstract

Numerous studies in the microfinance context have been observed to employ the data envelopment analysis (DEA) technique for the estimation of efficiency and/or productivity at the organizational level. This chapter aims to synthesize the existing DEA literature in the microfinance context to provide an overview of the methodological development over the past two decades and identify future research directions. To this end, literature data spanning the period 2006–2022 have been gathered from the Scopus database, and after filtering them based on inclusion and exclusion criteria, a total of 70 research articles relevant to this topic were considered for the study. With respect to methodological development, although most of the studies relied on the classical DEA, a recent trend revealed that studies are also employing various forms of network DEA (NDEA) techniques that integrate complex and multiple stages of the production process to help capture the diverse objectives of microfinance institutions. Furthermore, most studies utilize either regional or global data in their analysis of efficiency/productivity, with only a few exceptions. Lastly, bibliometric processing software (e.g., VOSViewer) is utilized for the identification of influential authors, institutions, networking, and countries. The study identifies important prospects for future research.

Keywords: Data envelopment analysis, microfinance institutions, productivity, efficiency, outreach, financial sustainability, microfinance, social outreach, VOSViewer, systematic review, MIX market, stochastic frontier analysis.

1. Introduction

Microfinance institutions (MFIs) are considered one of the major economic tools for poverty alleviation. Although MFIs primarily aim to promote financial inclusion of the poor and marginalized in a society, they have, over time, recognized the need for financial sustainability to achieve their social goal — reaching out to the maximum number of poor people (Efendić & Hadžić, 2017). This is further buttressed by the rapid expansion and proliferation of MFIs globally, which has resulted in increased competition in the industry. Consequently, efficient operation has become the rallying point of practitioners, policymakers, and researchers to ensure the sustainability of MFIs (Bassem, 2014).

Numerous studies have emphasized the significance of MFIs' productivity or efficiency as a primary driver of their success (Mia & Soltane, 2016; Rana *et al.*, 2021; Wijesiri & Meoli, 2015).[1] According to Isik & Hassan (2003), the evaluation of productivity and its components is crucial since it can greatly enhance how well a firm utilizes its limited resources. Moreover, the microfinance industry is faced with several changes involving the industry environment, policy reforms, technological advancements, financial technology (fintech), ownership structures, etc., all of which might influence the productivity of MFIs in terms of efficiency or technological change (Khan & Gulati, 2022). Therefore, in this modern era, the attainment of efficiency and productivity may be a challenge for MFIs, and this underscores the need for further research in this area.

[1] "Productivity" and "efficiency" are interchangeably used in this study. However, the estimation procedures and concept subtly varies between them.

To investigate the efficiency of MFIs, many researchers relied on the data envelopment analysis (DEA) technique, making it one of the most popular non-parametric approaches in the microfinance context. For example, Khan & Gulati (2022), Piot-Lepetit & Tchakoute Tchuigoua (2022), Li *et al.* (2019), Al-Awlaqi and Aamer (2019), Muneer Babu & Kulshreshtha (2014), Mia & Soltane (2016), etc., investigated the efficiency of MFIs in different geographic regions, and many more are still under investigation. As an instance, Soldátková & Černý (2022), Bharti & Malik (2022), Kolloju & Meoli (2022), and Hussain *et al.* (2022) investigated the financial and social efficiencies of MFIs alongside the determinants of efficiency. Lastra *et al.* (2022) employed the DEA technique to determine the efficiency of MFIs and compared their efficiency scores with those of commercial banks in 18 countries. Similarly, Mia *et al.* (2022) divided efficiency into three dimensions, namely operational, financial, and social, and by applying the network DEA (NDEA) — a more advanced type of the conventional DEA — to a global dataset, they identified several institutional and macroeconomic factors influencing the efficiency of MFIs.

Also, in analyzing the efficiency of MFIs, some studies, including those conducted by Hermes *et al.* (2018), Riaz (2015), and Servin *et al.* (2012), utilized the parametric stochastic frontier analysis (SFA) approach. While a growing number of studies have employed both the SFA and DEA approaches in analyzing the efficiency and productivity of MFIs, the DEA approach is more commonly used. The major challenge to applying the SFA approach is its requirement for the specification of production functions, which is quite complex in the microfinance context (Fall *et al.*, 2018). Furthermore, stochastic boundaries and other inherent weaknesses of SFA make the approach even less popular among microfinance researchers. The DEA approach, on the other hand, is a linear programming function that does not require a specific production function but rather calculates the relative efficiency based on the frontier (Kabir Hassan *et al.*, 2012). The DEA approach can also cater to multiple inputs and outputs, which is more relevant in capturing the dual objectives of MFIs. Therefore, this chapter focuses on research works that have used the DEA approach to estimate efficiency/productivity at the organizational level.

Although several studies have applied the DEA approach in the microfinance context, there seems to be an inadequate or limited systematic review that summarizes their overall findings. One of the exceptions was the meta-analysis by Fall *et al.* (2018), which investigated MFIs' efficiency based on 262 observations extracted from 38 relevant research papers that consider both DEA and SFA approaches. The authors highlighted that the measurement of efficiency varies as a result of differences in the approaches, methodologies, sample sizes, and dimensions of efficiency (e.g., social or financial), among others. However, as the study was conducted based on limited resources, the authors have recommended further research in this context. Moreover, the research applied a meta-analysis approach based on the mean technical efficiency (MTE) score of the existing literature, as opposed to a bibliometric review and a network analysis, in order to help uncover the research trend. Also, Fall *et al.* (2018) mostly excluded works of literature on MFIs' productivity that were estimated based on the Malmquist productivity index (MPI), an extension of the original DEA, which allowed researchers to estimate productivity rather than efficiency. While efficiency can be generally estimated for a single point in time using unbalanced panel data, a minimum of two time periods and balanced panel data are required to estimate productivity. Therefore, this chapter considers both the efficiency and productivity dimensions in the analysis of MFI literature applying the DEA framework in the past two decades.

This chapter contributes to the current literature in several ways: First, it consolidates the existing studies from two broad areas, i.e., productivity and efficiency, under one parameter, i.e., the DEA approach. Second, this chapter guides future researchers on the selection of appropriate parameters, such as orientation (input/output), returns to scale, and DEA type, to estimate the efficiency and productivity of the microfinance industry globally or for a single country. Third, future researchers can identify available opportunities that can help expand investigations on the efficiency of MFIs through a network of co-authors, co-occurrences, and influential papers, all of which were extensively discussed in the chapter.

The organization of the chapter is as follows: Section 2 reviews the concept of DEA and its key features. Section 3 discusses the methodology and features of the data used in this chapter. Section 4 sets forth the basic results and features of the collected data. Section 5 advances the different forms of network analysis conducted using the VOSViewer application, and Section 6 presents conclusions and guidance for future research.

2. DEA and Its Features

DEA is a popular tool for efficiency analysis in various fields, such as banking and finance (Boubaker *et al.*, 2022a, 2022b; Le, 2018; Tripe, 2010) and manufacturing (Becchetti & Sierra, 2003; Boubaker *et al.*, 2023; Vu, 2016). However, in the context of microfinance literature, researchers have used several measures to estimate the efficiency and productivity of the MFIs, with DEA being the most popular approach due to its advantage over other methods (e.g., SFA), particularly for service-oriented organizations, where inputs (funding) are generated from multiple sources, including market and non-market (Segun & Anjugam, 2013). DEA is a non-parametric mathematical linear programming technique for measuring the performance of decision-making units (DMU) based on the association between different inputs and outputs (Cooper *et al.*, 2011). The concept of DEA evolved from the study of Farrell (1957), which suggests the division of efficiency into two dimensions, namely technical and allocative efficiencies. In general, technical efficiency means achieving maximum outputs with fewest inputs, whereas allocative efficiency refers to how different sets of resource inputs are utilized to generate different sets of intended outputs (Akazilli *et al.*, 2008).

Based on the concept proposed by Farrell (1957), several researchers have attempted to develop a mathematical linear programming technique that can measure the efficiency or productivity of firms. Charnes *et al.* (1978) were the first to coin the term "DEA" for their linear programming method, which was built on two assumptions (input orientation and constant return to scale, or CRS) and later expanded by researchers (Banker *et al.*, 1984; Färe *et al.*, 1983) to include a variable return to scale (VRS) assumption.

In applying the DEA approach for measuring efficiency or productivity, among others, three properties should be initially determined. The first is the input and output combination. As data concerning allocative efficiency (prices of the inputs) are often not available in MFIs, most researchers considered multiple inputs and outputs for the estimation of efficiency/technical efficiency/productivity. Second, the return to scale (RTS), which explains the nature of the fluctuations of the inputs and outputs utilized in the production frontier, must be decided to estimate the efficiency of a firm. Third, the researcher must ascertain the orientation of the DMU, whether it is input- or output-oriented. However, it is argued that orientation may not influence the efficiency of the DMUs significantly (Coelli, 1996), and for efficient DMUs, both input and output orientations will lead to the same frontier. The only aspect of the two models (based on orientation) that may differ is the efficiency measures related to inefficient DMUs. A detailed explanation of these aspects is discussed in Section 4 of this chapter.

3. Methodology

While there are a plethora of ways to conduct a systematic review, we have adopted Tranfield *et al.*'s (2003) approach for this review. The processes involved in this approach are depicted in Figure 1.

3.1. *Research objectives*

The objective of this research is to synthesize existing literature works that consider the DEA framework in assessing the efficiency and/or productivity of MFIs. This review is instrumental

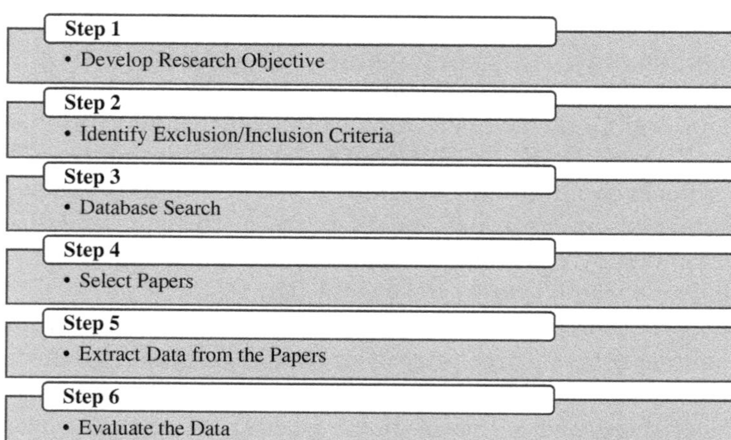

Figure 1. Systematic review process.

in providing an understanding of the methodological and empirical development of the DEA approach in microfinance research over the past two decades.

3.2. *Inclusion/Exclusion criteria*

The selection and identification of relevant papers from existing literature works is the most crucial part of conducting a systematic review. Owing to the availability of extant research papers from different databases, it is advisable to select a specific database first to narrow down the search. Due to its credibility and wider coverage, the Scopus database, as opposed to others such as the Web of Science (WOS), Google Scholar, and EBSCO, has been considered the source of relevant papers for this review. The Scopus database is generally popular and renowned among academics because it possesses high-quality (as the articles indexed with Scopus search engines are peer-reviewed) indexed journals (Khan & Khan, 2021) and also features the most number of journals compared to other databases, such as WOS or EBSCO (Falagas *et al.*, 2008). Further to the selection of the preferred database, we have established the following inclusion/exclusion criteria after consulting several past studies:

- The paper must be written in English.
- It must be focused on the microfinance industry.
- It must consider the DEA technique in its methodology/analysis.
- The DEA application should be utilized for the estimation of either efficiency and/or productivity.
- The paper must focus on the institutional level of efficiency and/or productivity analysis.
- It should not be a conference proceeding or a working paper.

3.3. *Data search*

The search and collection of data began in early 2022 and were concluded in October 2022. To obtain as many studies as possible for the review, a Scopus search was conducted without period limits. Consequently, the earliest literature relating to efficiency/productivity was discovered to be published in 2006. Prior to 2006, our search results failed to identify any paper that had exclusively

Table 1. Paper search in Scopus.

Strings	Number of papers found in Scopus
"Data envelopment analysis" OR "DEA" OR "efficiency" OR "productivity" AND "microfinance" OR "microfinance institutions" OR "microcredit"	506
Data AND envelopment AND analysis OR DEA AND microfinance OR microcredit	90
Efficiency AND DEA AND microfinance OR microcredit	61
Productivity AND DEA AND microfinance OR microcredit	14
Total	**671**

focused on the application of the DEA approach in the microfinance industry. In conducting the search, keywords such as "Data Envelopment Analysis or DEA and Microfinance or Microcredit" were used. However, since the DEA approach was employed in the methodology of this paper to determine the efficiency or productivity of MFIs, matching papers were searched using keywords such as "Efficiency of Microfinance or Microcredit" and "Productivity of Microfinance or Microcredit" and, more specifically, using search strings highlighted in Table 1.

3.4. *Selection of papers*

Using the keywords and string equations highlighted above, a total of 671 papers were discovered during the online search, and after screening their titles based on the exclusion/inclusion criteria, 544 papers were dropped, while 127 were retained for further scrutiny. As the title could be misleading, an in-depth study of the abstract was performed, resulting in the selection of 90 relevant papers. However, after examining the content of these papers, only 70 papers were considered suitable for the review based on their fulfillment of the specified criteria. The process of paper selection is depicted in Figure 2.

Figure 3 highlights the number of published papers related to DEA in microfinance. Interestingly, a growing trend was observed in the number of publications, particularly after 2012. The highest

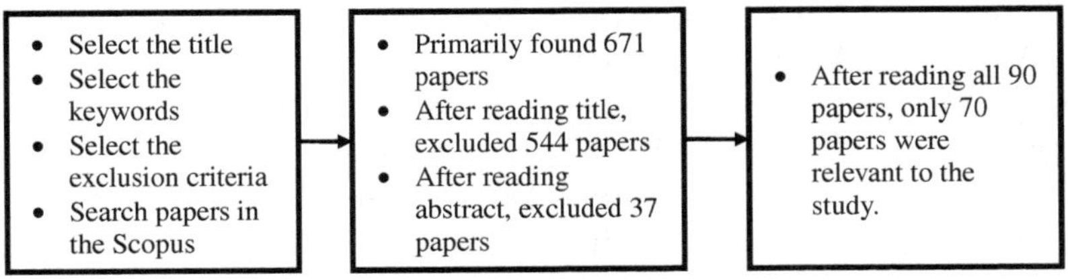

Figure 2. Paper selection process.

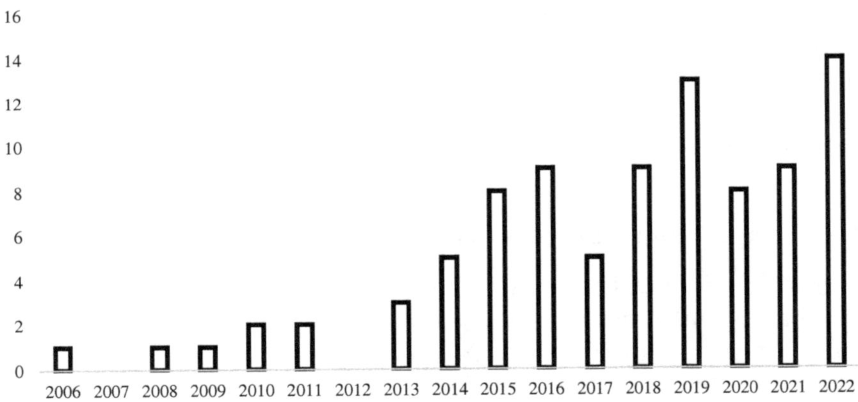

Figure 3. Year-wise number of papers published on DEA and microfinance in the Scopus database.

number of published papers was observed in 2022, and it is expected to increase in the near future.

4. Results and Findings

Based on the data obtained from Scopus (see Figure 3), we have divided our analysis into two main segments: First, we discuss the input/output selection process, RTS and orientation, and types of DEA used by the researchers, and second, we demonstrate the varying network analysis of the selected papers based on the VOSViewer application.

Based on the diversified measures of financial and social outputs, various authors have observed different scores of efficiencies and productivity. Overall, the majority of the researchers have found that

MFIs from the South Asian region and the Eastern Europe and Central Asia (EECA) region performed better in terms of financial efficiency compared to social efficiency (Bibi *et al.*, 2018; Hussain *et al.*, 2020; Khan & Gulati, 2021; Khan & Shireen, 2020; Zainal *et al.*, 2020a, 2020b). Whereas, Collins (2019), based on 238 Latin American MFIs, suggested that banking MFIs were more efficient compared to other MFI types (e.g., NGOs, cooperatives, etc.). Similarly, Ayayi & Wijesiri (2018) also discovered that non-profit MFIs were efficient before the financial crisis but became otherwise afterward. On the other hand, the study conducted by Ben Abdelkader & Mansouri (2019) on the Middle East and North Africa (MENA) region argued that the efficiency score varied across the countries due to the age and regulation of MFIs. Consequently, the obtained efficiency and/or productivity results are inconclusive.

4.1. *Input and output selection*

A crucial aspect of the DEA technique is the identification of potential inputs/outputs that can truly capture the types of services rendered by the institutions. A poor reflection of these values will result in either an under- or overestimation of efficiency/productivity. Therefore, careful attention should be paid to the selection of the input and output variables for DEA. Also, the identification of these variables mostly depends on how we classify the sector under investigation. Moreover, diverse models from different philosophical perspectives exist concerning the definition of financial institutions and their activities and the precise interpretation of their efficiency (Berger & Mester, 1997). Generally, the specification of inputs and outputs to analyze efficiency and/or productivity in the banking industry is a challenging task (Ahn & Le, 2014; do Castelo Gouveia *et al.*, 2018); however, reliance on an appropriate approach could assist researchers in the selection of suitable inputs and outputs for DEA.

A summary of the input and output variables utilized in the literature to measure the efficiency and productivity of the MFIs is presented in Table A1. Most researchers selected these variables using the production concept. Accordingly, authors were observed to frequently utilize total assets, the number of employees or staff, and operating expenses as input variables (see Table A1). However, in defining the output variables, researchers have classified the output into two categories, namely financial and social outputs, in line with MFIs' dual objectives (Gutiérrez-Nieto *et al.*, 2009; Serrano-Cinca & Gutiérrez-Nieto, 2014; Serrano-Cinca *et al.*, 2011). The gross loan portfolio, financial revenue, operating income, interest income, ROA, and ROE are some of the commonly used financial outputs, while the average loan balance, average loan balance over GNI per capita, number of active borrowers, female borrowers, and poor borrowers, among others, are popular social outputs (see Table A1).

Another method to ensure that the inputs and outputs are correctly specified is for the researcher to perform an "isotonicity" test, which examines the correlation between the inputs and outputs (Hong & Jeong, 2019). The assumption is that the relationship between the inputs and outputs is not erratic and that increasing the value of any input while keeping other factors constant should not decrease any output but rather lead to an increase in the value of at least one output (DEA Zone, 2022). Therefore, the existing correlation coefficients between the inputs and the outputs must remain positive to fulfill the "isotonicity" assumption prior to the application of the DEA approach.

4.2. *Types of DEA used in the analysis*

Although most earlier literature works on microfinance utilized classical models of DEA developed by Charnes *et al.* (1978) and Banker *et al.* (1984), some of the recent literature relied on more comprehensive and advanced models of DEA, such as the network slack-based measure (NSBM), developed by Tone & Tsutsui (2009). The NSBM permits more complex and comprehensive stages of the production process to accurately estimate the efficiency of DMUs (Mia *et al.*, 2022; Wijesiri & Tremblay, 2020).

Apart from that, it has been observed from the existing literature that the two most prominent models used by researchers are the production and intermediation approaches. According to the production model, MFIs are regarded as producers that utilize inputs, such as labor (e.g., number of employees) and capital (e.g., total assets or equity), to produce outputs in the form of returns (e.g., gross loan portfolio or the number of active borrowers) (Jaiyeoba *et al.*, 2018).

It was also observed that most of the microfinance literature considers the production approach in evaluating efficiency or productivity (Hussain *et al.*, 2022; Khan & Gulati, 2021; Kolloju & Meoli, 2022; Wijesiri *et al.*, 2015; Zainal *et al.*, 2022).

Under the intermediation model, MFIs are regarded as financial intermediaries that channel funds from surplus to deficit units (Bassem, 2014). Several studies relied on the intermediary approach in estimating the efficiency or productivity of MFIs (Bassem, 2008; Gutierrez-Nieto *et al.*, 2007; Haq *et al.*, 2010). Meanwhile, few studies have tried to combine and compare both approaches in selecting inputs and outputs or utilized a mixed approach in selecting inputs and outputs (Mia *et al.*, 2022; Piot-Lepetit & Nzongang, 2014; Segun & Anjugam, 2013). It is worth mentioning that under the intermediation approach, MFIs are assumed to act as intermediaries, utilizing the savings (deposit) of clients to cater to the loan demands of borrowers. However, there are several MFIs, especially in the sub-Saharan African and Latin American regions, that are not allowed to take deposits from savers due to regulatory restrictions but accept short-term credit facilities from other institutions (Segun & Anjugam, 2013). This limitation may justify the high preference among scholars for the production approach compared to the intermediation approach (Mia, 2017).

4.3. *Return to scale and orientation of DEA*

Orientation dictates the direction of DEA to estimate efficiency or productivity. The initial DEA model developed by Charnes *et al.* (1978) assumed a CRS. However, the main limitation of this approach is that in the real world, not all firms or DMUs run at an optimal scale due to issues of imperfect competition, financing constraints, and others (Coelli, 1996). A VRS was consequently integrated into the DEA extension by Banker *et al.* (1984) to enable a more realistic estimation of efficiency. However, the existing literature on microfinance revealed that equal numbers of papers (25) were observed to have adopted the CRS and VRS orientations, while a total of 12 papers employed both.

In applying the DEA, researchers are required to decide on the use of either an input- or output-oriented model. According to Segun & Anjugam (2013), "the input-oriented model identifies technical

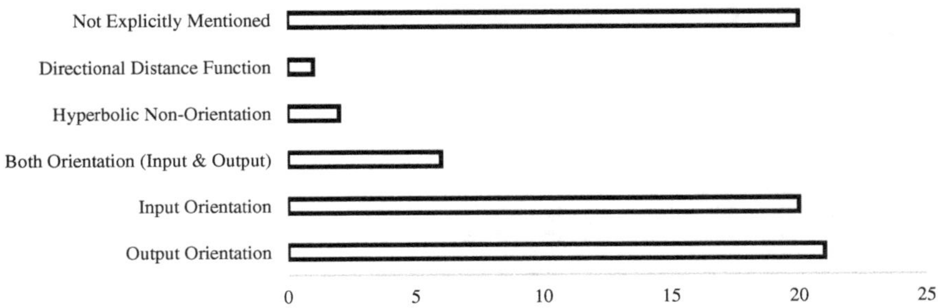

Figure 4. Types of orientation of the DEA and number of papers.

efficiency as a proportional reduction in input usage, whereas the output-oriented Model identifies technical efficiency as a proportional increase in output production" (p. 404). Although the efficiency/productivity results might change marginally between these two orientations, the production frontiers remain the same in both models (Coelli & Perelman, 1999; Coelli, 1996). For this reason, the efficiency/productivity scores only change for inefficient DMUs. That being established, output orientation remains the preferred choice among researchers in the microfinance industry (see Figure 4), since MFIs aim to optimally expand their outreach and generate financial revenue to cover their costs. Although a couple of papers have utilized the directional distance function (Sai Mohini & Vilvanathan, 2021) and hyperbolic non-orientation approaches (Widiarto *et al.*, 2017) to estimate the efficiency/productivity of MFIs, their usage remains rather limited.

4.4. *Trend of DEA in the microfinance literature*

Researchers have relied on the DEA approach over the past few decades to estimate the efficiency or productivity of MFIs. However, the use of DEA methods was observed to be more frequent for estimating the efficiency of MFIs (89% of papers published in the past two decades) than for estimating their productivity (11% of papers published in the same period).

The trend and evolution of DEA in the microfinance literature paint an intriguing picture. To illustrate, it was observed that available research works on the efficiency or productivity of MFIs did not employ diversified dimensions of DEA until after the year 2013.

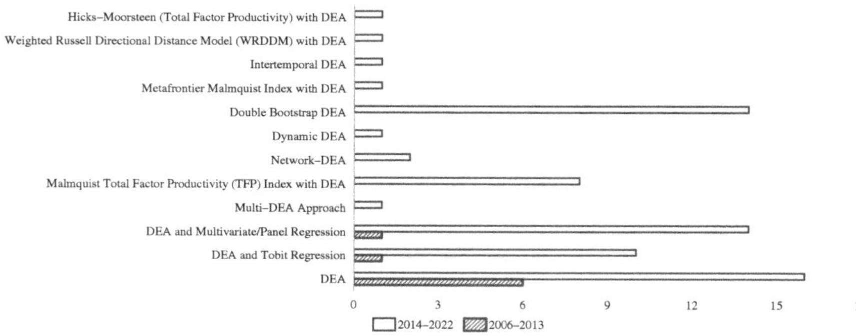

Figure 5. Different types of DEA used in the microfinance literature.
Note: Based on the available literature in Scopus between 2006 and 2022.

Therefore, in this chapter, the entire microfinance literature can be classified into two phases: one spanning the period 2006–2013 and the other covering the period 2014–2022 (see Figure 5).

From Figure 5, it can be observed that in the first phase (2006–2013), the use of DEA was limited to two-stage analyses with tobit regression and multivariate or panel regression. However, in phase two (2014–2022), different forms and variations of DEA were applied, such as network DEA (Mia *et al.*, 2022; Nourani *et al.*, 2021), bootstrap DEA (Al-Awlaqi & Aamer, 2019; Bibi *et al.*, 2018; Khan & Shireen, 2020; Sinha & Pandey, 2019), dynamic DEA (Mia *et al.*, 2019), DEA based on the Malmquist total factor productivity (TFP) index (Jaiyeoba *et al.*, 2018; Khan & Gulati, 2022; Mia & Soltane, 2016; Muneer Babu, 2016; Wijesiri & Meoli, 2015), DEA based on the pseudo-panel Malmquist index (Piot-Lepetit & Tchakoute Tchuigoua, 2022), DEA based on the meta-frontier Malmquist index (Azad *et al.*, 2016), intertemporal DEA (Li *et al.*, 2019), DEA based on the weighted Russell directional distance model (WRDDM) (Sai Mohini & Vilvanathan, 2021), and DEA based on the Hicks–Moorsteen total factor productivity (TFP) index (Al-Awlaqi & Aamer, 2019). Among the researchers, the most popular techniques to identify the determinants of efficiency or productivity of MFIs were truncated bootstrap DEA and two-stage DEA with tobit regression or a multivariate panel regression approach. However, in recent literature, the use of advanced and expanded models of DEA, such as the WRDDM and Hicks–Moorsteen TFP index, is gaining momentum.

5. Network Analysis

5.1. *Co-authorship network analysis*

It is recognized that international collaboration contributes greatly to the comprehension of the geographic and demographic disparities observed in scientific studies and allows authors to create a network of organizations and nations in the field of study (Cisneros *et al.*, 2018). To comprehend the contribution of researchers to the literature's structure, the co-authorship network analysis was conducted (Khanra *et al.*, 2021), which is one of the most popular tools utilized in the development of bibliometric network analyses (Cavalcante *et al.*, 2021; Oladinrin *et al.*, 2023; Shah *et al.*, 2019).

Although 138 authors collaborated worldwide to form the co-authorship network, only a few of the authors were observed to have conducted research in the same area more than twice. We adopted a threshold of two as the minimum number of documents published by the authors and discovered that only 26 authors attained this threshold. Among these authors, 15 were considered to be productive based on their number of publications and citations (see Table 2). However, Kamarudin was observed to be the most productive author, publishing the highest number of papers (8) on DEA and microfinance research in the past two decades. With respect to the number of citations, which gauges the quality of the publications and influence of the researcher, Gutierrez-Nieto, Mar Molinero, and Serrano-Cinca were the most cited authors in the usage of DEA for the estimation of efficiency and productivity in the microfinance industry.

The network visualization among authors is shown in Figure 6, and eight clusters were discovered in the VOSViewer based on the threshold of documents published by the authors and the number of citations. From Figure 6, it is observed that cluster 1 comprises nine documents (upper-left network), and cluster 8 represents only one publication (in the lower-right frame of the figure) by the author (Hermes). Based on the network of co-authorships, it may be said that international collaboration among the researchers is minimal in the efficiency and productivity literature.

Furthermore, Figure 7 has been included to provide more insight into the top 15 countries where most of the studies on DEA and microfinance efficiency and productivity were conducted. The finding

Table 2. Most productive authors based on publications and citations.

Author	Publications	Rank (Document)	Citations	Rank (Citations)
Kamarudin F.	8	1	75	7
Wijesiri M.	6	2	237	4
Meoli M.	5	3	216	5
Hussain H.I	5	4	64	9
Zainal N.	5	5	19	12
Mia M.A	4	6	55	10
Khan A.	4	7	19	13
Sufian F.	4	8	13	14
Gutierrez-nieto B.	3	9	378	1
Mar Molinero C.	3	10	378	2
Serrano-Cinca C.	3	11	378	3
Piot-Lepetit I.	3	12	70	8
Widiarto I.	2	13	98	6
Van Damme P.	2	14	24	11
Hook I.S	2	15	10	15

Note: Citation counts are based on Scopus.

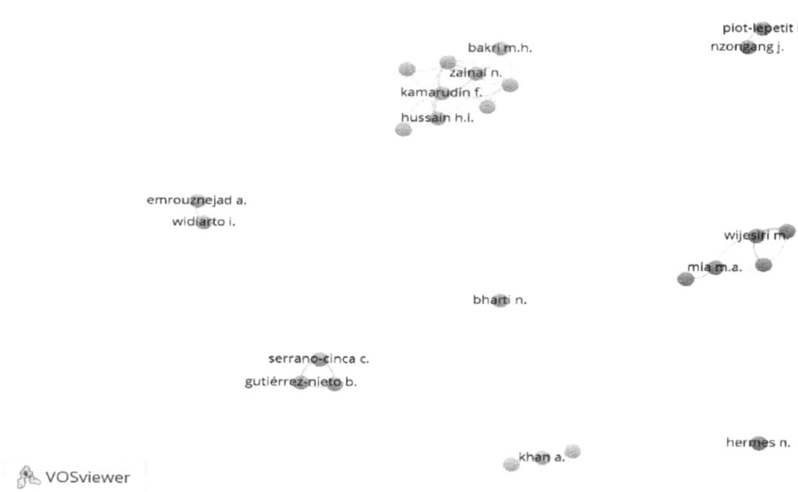

Figure 6. Network visualization among the co-authors based on citations. *Note*: Based on the available literature in Scopus between 2006 and 2022.

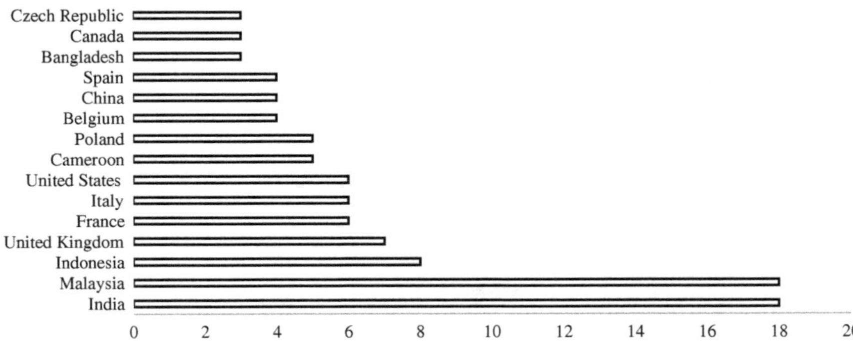

Figure 7. Number of papers published by country (top 15).

Notes: Based on the available literature in Scopus between 2006 and 2022. The countries are determined based on the authors affiliation and not the sample data that they have used in the analysis.

indicated that Malaysia and India had the highest (18) numbers of published papers (18), followed by Indonesia with a total of eight papers. Meanwhile, the number of papers from the African region remained limited, as none of the countries in the region were listed in the top 15 countries.

5.2. *Most influential paper: Citation network analysis*

It is popularly argued in academia that the most cited paper in any field is considered the most influential (Merigó *et al.*, 2016). In a bid to identify the most influential paper in the field of microfinance efficiency or productivity, we have used the VOSViewer application to highlight the overlay density networks of authors (Figure 8). We have also set the minimum number of papers by the authors to two in order to enable the incorporation of a maximum number of authors (based on citation) into the network. Consequently, a total of 26 items attained the minimum threshold and were divided into three clusters based on the citation network (see Figure 8). The paper titled "Microfinance Institutions and Efficiency" by Gutierrez-Nieto *et al.* (2007) was also observed to be the most influential in the area, followed by the research conducted by Serrano-Cinca *et al.* (2011) and then the study by Wijesiri *et al.* (2015). It was also observed that the authors Wejisiri and Meoli had the most number of links with other authors with respect to collaborative research. These authors had links with three different clusters, as highlighted in Figure 8.

Moreover, we have presented Figure 9 to spotlight the top 15 affiliated organizations with the most number of papers on microfinance

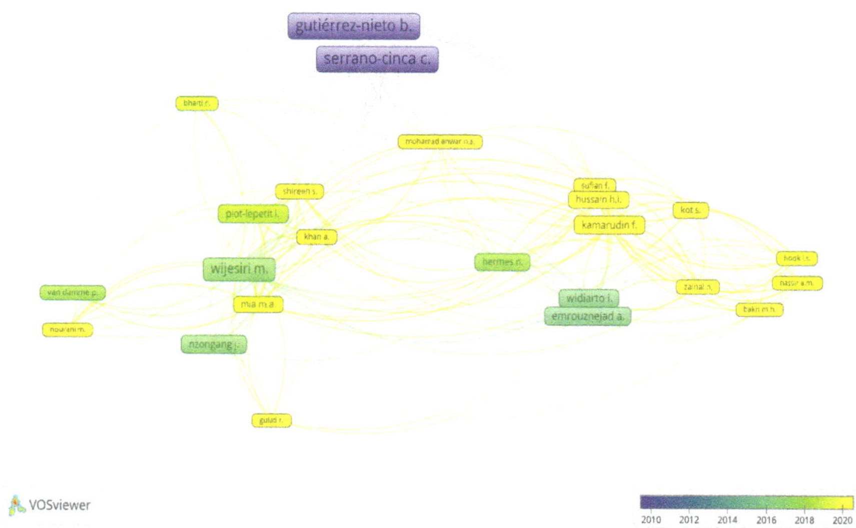

Figure 8. Overlay density network of authors based on citations.

Note: Based on the available literature from Scopus between 2006 and 2022.

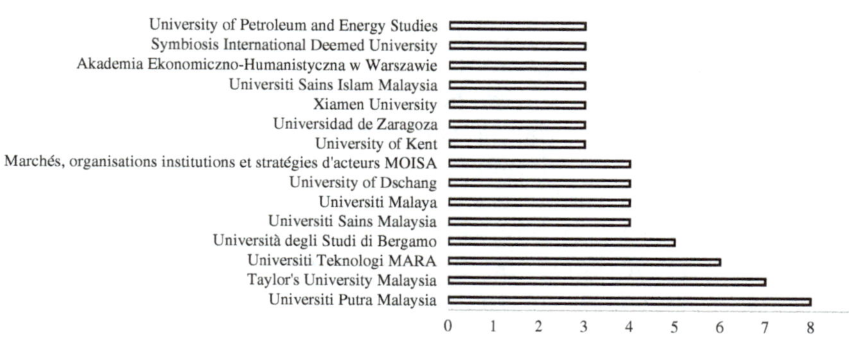

Figure 9. Number of papers published from the top 15 affiliated organizations of the authors.

Note: Based on the available literature in Scopus between 2006 and 2022.

and DEA research. Universities from Malaysia were observed to have produced the highest number of papers (see Figure 9), indicating that the use of DEA in analyzing the efficiency and productivity of microfinance is one of the most interesting areas of research for researchers based in Malaysia. Among the universities, Universiti Putra Malaysia (UPM), where the most productive author (Kamarudin) is based, ranked first. Another top-publishing academic organization outside of Malaysia is the Università Degli Studi di Bergamo, which is located in Italy.

5.3. *Co-occurrence network analysis*

We also made attempts to investigate the correlation between the keywords highlighted by the researchers in this research dimension. This analysis is essential because keyword co-occurrence offers a dimension of literature that exists on a certain topic (Khanra *et al.*, 2020). Co-occurrence network analysis provides directions for future research by identifying already trending keywords in academia, indicating the network between keywords in the existing literature, and emphasizing the development of research across the keywords. Again, the VOSViewer application has been used to analyze the network of co-occurrences of the keywords, and in so doing, we have set the minimum number of co-occurrences of a keyword to three and discovered that 30 of 229 keywords attained the required thresholds.

According to Figure 10, we observed that keywords such as "microfinance," "data envelopment analysis," "financial performance," "social performance," and "productivity" fit into cluster 1 and possess the maximum number of co-occurrences with other keywords. On the other hand, keywords such as "microfinance institutions," "financial efficiency," "social efficiency," and "bootstrap truncated regression" were categorized under cluster 2. This indicates that authors who have adopted the DEA approach in ascertaining the social and financial efficiency of MFIs are likely to utilize the bootstrap truncated regression model alongside the DEA. This finding is interesting for future researchers in the sense that in evaluating efficiency (e.g., social, financial, or technical), second-stage regression techniques such as bootstrap truncated regression might provide a better outcome in identifying determinants of efficiency.

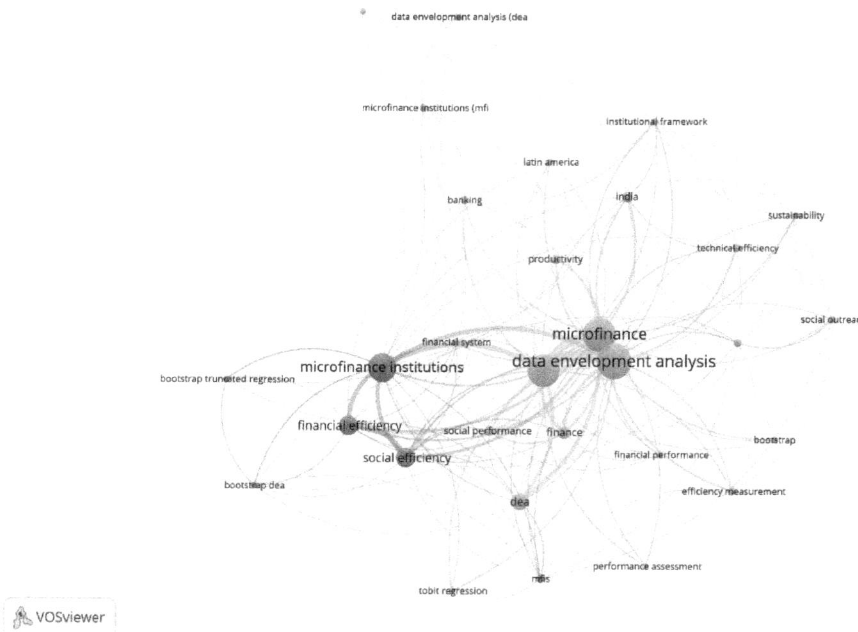

Figure 10. Network visualization of co-occurrence of the keywords.
Note: Based on the available literature in Scopus between 2006 and 2022.

6. Conclusion and Guidance for Future Research

This research aims to synthesize the existing literature works that consider the DEA framework in the assessment of MFIs' efficiency and productivity. This review is significant for an understanding of the empirical development over the past two decades. To conduct the systematic review, this study explored the Scopus database and found 70 relevant papers that were published between 2006 and 2022 and exclusively focused on estimating the efficiency and/or productivity of MFIs using the DEA approach. The review was conducted in two dimensions. First, the papers were synthesized based on input/output selection, RTS and orientation selection, and the type of DEA. Second, a comprehensive network analysis was adopted in the study based on co-authorship networks, co-occurrence networks, and the most influential paper or citation networks.

From the conducted methodological analysis, this chapter concludes that the efficiency and productivity score of MFIs vary across the study, owing to the inconsistency in selecting inputs and outputs, RTS and orientation, and the selection of different forms of DEA approaches. As the selection of methodology depends on the focus of the researcher and the perspective of the research, diversified methodological implications tend to result in inconclusive findings. Overall, the trend of DEA in the microfinance literature showed the adoption of diverse models, including more robust and sound methodological models, such as network DEA, WRDDM, Hicks–Moorsteen TFP index, and meta-frontier Malmquist TFP index. However, most of the recent studies relied on the two-stage double bootstrap DEA to investigate efficiency as well as determinants of the efficiency of MFIs. Furthermore, it can be concluded from the network analysis that Kamarudin F. was the most productive author, publishing the highest number of papers, while Gutierrez-Nieto B. was the most influential author in this research area based on the number of citations. Moreover, Asian countries, such as Malaysia, India, China, and Bangladesh, were discovered to be the most productive based on the co-authorship networks. The co-occurrence network analysis visualizes that the DEA framework is increasingly favored in the determination of the social and financial efficiencies of MFIs, along with second-stage regression techniques, such as bootstrap truncated regression.

Most research have focused on the production approach to analyze the efficiency and productivity of MFIs, and only a limited number of studies were observed to have simultaneously considered both production and intermediation approaches. Thus, to understand the global trend of efficiency and productivity in the microfinance industry, we recommend the concurrent implementation of both approaches and a comparison of results. Furthermore, advanced models, such as network DEA, WRDDM, Hicks–Moorsteen TFP index, and meta-frontier Malmquist TFP index, can be used in future studies for robust findings. Also, the most recent Euclidean common set of weights (ECSW) in DEA can be applied to the microfinance studies, in line with the banking literature (Hammami *et al.*, 2022). Existing research has shown that researchers are selective in their choice of input and output variables, which may generate inconsistent results in the literature. Thus, it is recommended that future studies conduct an "isotonicity test," which inspects the correlation between the inputs and the outputs, prior to the selection of variables. Furthermore, the RTS and orientation of the DEA approach might marginally affect the efficiency score. Limited research has simultaneously considered both CRS and VRS alongside input and output orientations. To compare efficiency/productivity results, we suggest the utilization of scale and orientation to observe the changes.

From the network analysis, we observed that most research works in this area were conducted in the Asian region, while the least number of studies emerged from the African region. Therefore, we recommend that researchers from the African region explore the methodological framework of the DEA model to shed more light on the efficiency and productivity of MFIs. Moreover, the co-authorship network is dispersed across the countries, indicating the potentiality of future collaboration among researchers, especially from South Asia and the African region. The co-occurrence network indicates that only a limited number of studies had compared the efficiency and productivity between commercial banks and MFIs, despite their similarities in the provision of services, such as loans. Hence, future research could be focused on the comparative study of efficiency and productivity.

Since we solely relied on the Scopus search engine, we may have omitted high-quality papers published in the Web of Science (WoS), Australian Business Deans Council (ABDC), Chartered Association of Business Schools (ABS), and other indexed and abstracted journals. Therefore, a comprehensive systematic review comparing major indexed and abstract search engines can be conducted in the future. Lastly, our systematic review was limited to the first stage of productivity/efficiency of MFIs. However, future reviews could investigate the determinants of productivity/efficiency of MFIs using a second-stage non-parametric approach. In this regard, investigating the role of corporate governance (e.g., board independence, board gender diversity), technology (fintech), and macroprudential policies in the efficiency/productivity of MFIs would be valuable additions to the existing literature.

Acknowledgments

The authors greatly appreciate constructive feedback and suggestions from the anonymous reviewer and the editor(s) of the book on the earlier version of the text. We believe that the quality of the chapter has improved after addressing the recommended revisions.

Appendix

Table A1. Summary of the input/output variables used in the existing literature.

Input variables	Study	Output variables (Financial)	Study	Output variables (Social)	Study
Administrative expense	(Bharti & Malik, 2022), (Farida *et al.*, 2018)	Gross loan portfolio	(Bharti & Malik, 2022), (Khan & Gulati, 2022), (Lastra *et al.*, 2022), (Mia *et al.*, 2022), (Ahamad *et al.*, 2023), (Sai Mohini & Vilvanathan, 2021), (Khan *et al.*, 2021), (Deb & Sinha, 2022), (Khan & Gulati, 2021), (Tan & Anh, 2021), (Nourani *et al.*, 2021), (Li *et al.*, 2019), (Al-Awlaqi & Aamer, 2019), (Balammal *et al.*, 2019), (Ayayi & Wijesiri, 2018), (Bibi *et al.*, 2018), (Jaiyeoba *et al.*, 2018), (Efendic & Hadziahmetovic, 2017), (Wijesiri *et al.*, 2017), (Goyal & Agarwal, 2017), (Bharti & Chitnis, 2016), (Muneer Babu, 2016), (Van Damme *et al.*, 2016), (Kaur, 2016), (Lebovics *et al.*, 2016), (Kablan & Boussemart, 2015), (Alinsunurin, 2014), (Muneer Babu & Kulshreshtha, 2014),	Number of active borrowers/ per staff	(Bharti & Malik, 2022), (Khan & Gulati, 2022), (Hussain *et al.*, 2022), (Mia *et al.*, 2022), (Ahamad *et al.*, 2023), (Piot-Lepetit & Tchakoute Tchuigoua, 2022), (Hussain *et al.*, 2021b), (Sai Mohini & Vilvanathan, 2021), (Khan *et al.*, 2021), (Deb & Sinha, 2022), (Hussain *et al.*, 2021), (Khan & Gulati, 2021), (Tan & Anh, 2021), (Nourani *et al.*, 2021), (Hussain *et al.*, 2020), (Mohamad Anwar *et al.*, 2021), (Purwanto *et al.*, 2020), (Zainal *et al.*, 2020b), (Piot-Lepetit & Nzongang, 2019), (Zainal *et al.*, 2019), (Ayayi & Wijesiri, 2018), (Bibi *et al.*, 2018), (Khanam *et al.*, 2018), (Jaiyeoba *et al.*, 2018), (Efendic & Hadziahmetovic, 2017), (Widiarto *et al.*, 2017), (Wijesiri *et al.*, 2017),

(*Continued*)

Table A1. (*Continued*)

Input variables	**Study**	**Output variables (Financial)**	**Study**	**Output variables (Social)**	**Study**
			(Asawaruangpipop & Suwunnamek, 2014), (Segun & Anjugam, 2013), (Nzongang & Nishimikijimana, 2013), (Serrano-Cinca *et al.*, 2011), (Haq *et al.*, 2010), (Gutiérrez-Nieto *et al.*, 2009), (Gutierrez-Nieto *et al.*, 2007)		(Bharti & Chitnis, 2016), (Muneer Babu, 2016), (Mia & Soltane, 2016), (Wu, Escalante, & Li, 2016), (Wijesiri & Meoli, 2015), (Widiarto & Emrouznejad, 2015), (Alinsunurin, 2014), (Muneer Babu & Kulshreshtha, 2014), (Nzongang & Nishimikijimana, 2013), (Haq *et al.*, 2010), (Sedzro *et al.*, 2010)
Personnel expense	(Bharti & Malik, 2022), (Hussain *et al.*, 2021b), (Piot-Lepetit & Nzongang, 2019), (Sinha & Pandey, 2019), (Zainal *et al.*, 2019), (Farida *et al.*, 2018), (Piot-Lepetit & Nzongang, 2014), (Segun & Anjugam, 2013)	Net operating income/Net income	(Bharti & Malik, 2022), (Mia *et al.*, 2022), (Nourani *et al.*, 2021), (Khan & Shireen, 2020), (Collins, 2019), (Cabanda & Domingo, 2014)	Female borrowers	(Kolloju & Meoli, 2022), (Ahamad *et al.*, 2023), (Khan *et al.*, 2021), (Deb & Sinha, 2022), (Li *et al.*, 2019), (Al-Awlaqi & Aamer, 2019), (Piot-Lepetit & Nzongang, 2019), (Sinha & Pandey, 2019), (Balammal *et al.*, 2019), (Van Damme *et al.*, 2016), (Kaur, 2016), (Wijesiri *et al.*, 2015), (Sarma, 2015), (Serrano-Cinca *et al.*, 2011), (Gutiérrez-Nieto *et al.*, 2009)

Total assets	(Zainal *et al.*, 2022), (Khan & Gulati, 2022), (Kolloju & Meoli, 2022), (Hussain *et al.*, 2022), (Lastra *et al.*, 2022), (Mia *et al.*, 2022), (Ahamad *et al.*, 2023), (Piot-Lepetit & Tchakoute Tchuigoua, 2022), (Hussain *et al.*, 2021b), (Sai Mohini & Vilvanathan, 2021), (Khan *et al.*, 2021), (Hussain *et al.*, 2021a), (Khan & Gulati, 2021), (Nourani *et al.*, 2021), (Zainal *et al.*, 2020a), (Khan & Shireen, 2020), (Hussain *et al.*, 2020), (Mohamad Anwar *et al.*, 2021), (Purwanto *et al.*, 2020), (Zainal *et al.*, 2020b), (Li *et al.*, 2019), (Al-Awlaqi & Aamer, 2019), (Piot-Lepetit & Nzongang, 2019), (Sinha & Pandey, 2019), (Collins, 2019), (Ben Abdelkader & Mansouri, 2019), (Balammal *et al.*, 2019), (Zainal *et al.*, 2019), (Ayayi & Wijesiri, 2018), (Jaiyeoba *et al.*, 2018), (Efendic & Hadziahmetovic, 2017), (Widiarto *et al.*, 2017), (Bharti & Chitnis, 2016), (Muneer Babu, 2016),	Financial revenue/ Total income	(Zainal *et al.*, 2022), (Khan & Gulati, 2022), (Kolloju & Meoli, 2022), (Hussain *et al.*, 2022), (Lastra *et al.*, 2022), (Ahamad *et al.*, 2023), (Piot-Lepetit & Tchakoute Tchuigoua, 2022), (Hussain *et al.*, 2021b), (Hussain *et al.*, 2021a), (Khan & Gulati, 2021), (Zainal *et al.*, 2020a), (Khan & Shireen, 2020), (Hussain *et al.*, 2020), (Mohamad Anwar *et al.*, 2021), (Purwanto *et al.*, 2020), (Zainal *et al.*, 2020b), (Al-Awlaqi & Aamer, 2019), (Sinha & Pandey, 2019), (Ben Abdelkader & Mansouri, 2019), (Balammal *et al.*, 2019), (Farida *et al.*, 2018), (Bibi *et al.*, 2018), (Efendic & Hadziahmetovic, 2017), (Wijesiri *et al.*, 2017), (Mia & Soltane, 2016), (Kaur, 2016), (Lebovics *et al.*, 2016), (Wijesiri & Meoli, 2015), (Widiarto & Emrouznejad, 2015), (Wijesiri *et al.*, 2015), (Alinsunurin, 2014), (Piot-Lepetit & Nzongang, 2014), (Nzongang & Nishimikijimana, 2013), (Serrano-Cinca *et al.*, 2011),	Number of poor borrowers	(Piot-Lepetit & Tchakoute Tchuigoua, 2022), (Zainal *et al.*, 2020a), (Balammal *et al.*, 2019), (Kaur, 2016), (Serrano-Cinca *et al.*, 2011)

(*Continued*)

Table A1. (*Continued*)

Input variables	**Study**	**Output variables (Financial)**	**Study**	**Output variables (Social)**	**Study**
	(Kaur, 2016), (Gebremichael & Gessesse, 2016), (Wijesiri & Meoli, 2015), (Widiarto & Emrouznejad, 2015), (Wijesiri *et al.*, 2015), (Sarma, 2015), (Muneer Babu & Kulshreshtha, 2014), (Piot-Lepetit & Nzongang, 2014), (Asawaruangpipop & Suwunnamek, 2014), (Nzongang & Nishimikijimana, 2013), (Serrano-Cinca *et al.*, 2011), (Gutiérrez-Nieto *et al.*, 2009), (Bassem, 2008)		(Gutiérrez-Nieto *et al.*, 2009)		
Operating expense	(Zainal *et al.*, 2022), (Khan & Gulati, 2022), (Hussain *et al.*, 2022), (Lastra *et al.*, 2022), (Mia *et al.*, 2022), (Ahamad *et al.*, 2023), (Piot-Lepetit & Tchakoute Tchuigoua, 2022), (Hussain *et al.*, 2021b), (Sai Mohini & Vilvanathan, 2021), (Khan *et al.*, 2021), (Hussain *et al.*, 2021a), (Khan & Gulati, 2021), (Tan & Anh, 2021), (Zainal *et al.*, 2020a), (Khan & Shireen, 2020), (Hussain *et al.*, 2020), (Mohamad Anwar *et al.*, 2021),			Average loan balance/ per borrower	(Hussain *et al.*, 2022), (Hussain *et al.*, 2021b), (Zainal *et al.*, 2020a), (Zainal *et al.*, 2019), (Wijesiri *et al.*, 2017), (Widiarto & Emrouznejad, 2015), (Muneer Babu & Kulshreshtha, 2014), (Nzongang & Nishimikijimana, 2013)

	(Zainal *et al.*, 2020b), (Al-Awlaqi & Aamer, 2019), (Piot-Lepetit & Nzongang, 2019), (Ben Abdelkader & Mansouri, 2019), (Zainal *et al.*, 2019), (Ayayi & Wijesiri, 2018), (Farida *et al.*, 2018), (Bibi *et al.*, 2018), (Jaiyeoba *et al.*, 2018), (Widiarto *et al.*, 2017), (Wijesiri *et al.*, 2017), (Bharti & Chitnis, 2016), (Muneer Babu, 2016), (Van Damme *et al.*, 2016), (Kaur, 2016), (Gebremichael & Gessesse, 2016), (Wijesiri & Meoli, 2015), (Widiarto & Emrouznejad, 2015), (Kablan & Boussemart, 2015), (Amersdorffer *et al.*, 2015), (Alinsunurin, 2014), (Muneer Babu & Kulshreshtha, 2014), (Cabanda & Domingo, 2014), (Piot-Lepetit & Nzongang, 2014), (Asawaruangpipop & Suwunnamek, 2014), (Nzongang & Nishimikijimana, 2013), (Serrano-Cinca *et al.*, 2011), (Haq *et al.*, 2010), (Gutiérrez-Nieto *et al.*, 2009), (Gutierrez-Nieto *et al.*, 2007)				

(Continued)

Table A1. (*Continued*)

Input variables	**Study**	**Output variables (Financial)**	**Study**	**Output variables (Social)**	**Study**
Number of personnel/ staff/ employees	(Khan & Gulati, 2022), (Hussain *et al.*, 2022), (Mia *et al.*, 2022), (Ahamad *et al.*, 2023), (Sai Mohini & Vilvanathan, 2021), (Khan *et al.*, 2021), (Deb & Sinha, 2022), (Hussain *et al.*, 2021a), (Khan & Gulati, 2021), (Tan & Anh, 2021), (Nourani *et al.*, 2021), (Zainal *et al.*, 2020a), (Hussain *et al.*, 2020), (Mohamad Anwar *et al.*, 2021), (Purwanto *et al.*, 2020), (Zainal *et al.*, 2020b), (Li *et al.*, 2019), (Mia *et al.*, 2019), (Al-Awlaqi & Aamer, 2019), (Collins, 2019), (Ben Abdelkader & Mansouri, 2019), (Balammal *et al.*, 2019), (Ayayi & Wijesiri, 2018), (Bibi *et al.*, 2018), (Khanam *et al.*, 2018), (Efendic & Hadziahmetovic, 2017), (Widiarto *et al.*, 2017), (Wijesiri *et al.*, 2017), (Goyal & Agarwal, 2017),	Interest income	(Lastra *et al.*, 2022), (Mia *et al.*, 2022), (Nourani *et al.*, 2021), (Farida *et al.*, 2018), (Widiarto *et al.*, 2017), (Sarma, 2015), (Asawaruangpipop & Suwunnamek, 2014), (Gutierrez-Nieto *et al.*, 2007)	Health	(Bharti & Malik, 2022)

	(Muneer Babu, 2016), (Van Damme *et al.*, 2016), (Mia & Soltane, 2016), (Lebovics *et al.*, 2016), (Gebremichael & Gessesse, 2016), (Wijesiri & Meoli, 2015), (Widiarto & Emrouznejad, 2015), (Kablan & Boussemart, 2015), (Sarma, 2015), (Alinsunurin, 2014), (Muneer Babu & Kulshreshtha, 2014), (Cabanda & Domingo, 2014), (Segun & Anjugam, 2013), (Nzongang & Nishimikijimana, 2013), (Serrano-Cinca *et al.*, 2011), (Haq *et al.*, 2010), (Sedzro *et al.*, 2010), (Gutiérrez-Nieto *et al.*, 2009), (Bassem, 2008)				
				Poverty index/ Alleviation	(Bharti & Malik, 2022), (Kablan & Boussemart, 2015)
Cost per borrower	(Kolloju & Meoli, 2022), (Sinha & Pandey, 2019), (Balammal *et al.*, 2019), (Goyal & Agarwal, 2017), (Azad *et al.*, 2016), (Wijesiri *et al.*, 2015), (Sarma, 2015), (Segun & Anjugam, 2013), (Haq *et al.*, 2010)	ROA	(Lastra *et al.*, 2022), (Collins, 2019), (Kablan & Boussemart, 2015), (Bassem, 2008)	Education	(Bharti & Malik, 2022)

(*Continued*)

Table A1. (*Continued*)

Input variables	**Study**	**Output variables (Financial)**	**Study**	**Output variables (Social)**	**Study**
Number of loan officer	(Kolloju & Meoli, 2022), (Sinha & Pandey, 2019), (Collins, 2019), (Kaur, 2016), (Wijesiri *et al.*, 2015), (Segun & Anjugam, 2013), (Gutierrez-Nieto *et al.*, 2007)	ROE	(Lastra *et al.*, 2022), (Collins, 2019), (Wu *et al.*, 2016)	Target market	(Bharti & Malik, 2022)
Equity	(Lastra *et al.*, 2022), (Mia *et al.*, 2022), (Piot-Lepetit & Tchakoute Tchuigoua, 2022), (Nourani *et al.*, 2021), (Piot-Lepetit & Nzongang, 2019), (Collins, 2019), (Lebovics *et al.*, 2016), (Kablan & Boussemart, 2015)	Operating income	(Lastra *et al.*, 2022), (Farida *et al.*, 2018), (Piot-Lepetit & Nzongang, 2014)	Benefits to the poor	(Ben Abdelkader & Mansouri, 2019), (Gutiérrez-Nieto *et al.*, 2009)
Deposit	(Mia *et al.*, 2022), (Nourani *et al.*, 2021), (Farida *et al.*, 2018), (Lebovics *et al.*, 2016)	Total investment	(Lastra *et al.*, 2022)		
Total borrowings	(Mia *et al.*, 2022), (Deb & Sinha, 2022), (Nourani *et al.*, 2021), (Purwanto *et al.*, 2020)			Average loan balance/GNI per capita	(Ahamad *et al.*, 2023), (Khan *et al.*, 2021), (Deb & Sinha, 2022), (Hussain *et al.*, 2021a), (Hussain *et al.*, 2020), (Mohamad Anwar *et al.*, 2021), (Zainal *et al.*, 2020b), (Widiarto *et al.*, 2017), (Mia & Soltane, 2016)

Fixed assets	(Lastra *et al.*, 2022), (Wu *et al.*, 2016), (Cabanda & Domingo, 2014), (Sedzro *et al.*, 2010)	Portfolio risk par 30 days	(Sai Mohini & Vilvanathan, 2021)		
Interest expense	(Lastra *et al.*, 2022), (Khan & Shireen, 2020), (Farida *et al.*, 2018)	Loan outstanding	(Mia *et al.*, 2019), (Piot-Lepetit & Nzongang, 2019), (Farida *et al.*, 2018), (Khanam *et al.*, 2018), (Gebremichael & Gessesse, 2016), (Sarma, 2015), (Amersdorffer *et al.*, 2015), (Piot-Lepetit & Nzongang, 2014), (Segun & Anjugam, 2013), (Gutierrez-Nieto *et al.*, 2007)		
Total expense	(Mia *et al.*, 2019)	Deposit	(Mia *et al.*, 2019), (Piot-Lepetit & Nzongang, 2019), (Cabanda & Domingo, 2014), (Piot-Lepetit & Nzongang, 2014), (Asawaruangpipop & Suwunnamek, 2014), (Haq *et al.*, 2010)		
Financial expense	(Bibi *et al.*, 2018), (Lebovics *et al.*, 2016), (Piot-Lepetit & Nzongang, 2014)	Depositors per staff member	(Azad *et al.*, 2016), (Sedzro *et al.*, 2010)		
Cost per loan	(Azad *et al.*, 2016)	Borrowers per staff member	(Azad *et al.*, 2016)		

(Continued)

Table A1. (*Continued*)

Input variables	**Study**	**Output variables (Financial)**	**Study**	**Output variables (Social)**	**Study**
Operating expense over average GLP	(Mia & Soltane, 2016)	Borrowers per loan officer	(Azad *et al.*, 2016)		
Debt to equity	(Wu *et al.*, 2016)	Operational self sufficiency	(Sarma, 2015)		
Gross loan portfolio	(Wu *et al.*, 2016)	Equity capital	(Amersdorffer *et al.*, 2015)		
Loan loss provision	(Wu *et al.*, 2016)				
Portfolio risk 30 days	(Widiarto & Emrouznejad, 2015)				
Number of branches	(Cabanda & Domingo, 2014)				

Notes: Gross loan portfolio (GLP) can be used as both financial and social outreach indicators. However, since the interest income is directly dependent on the amount of GLP, we have considered it in the financial output of MFIs.

References

Ahamad, S., Al-Jaifi, H.A.A., & Ehigiamusoe, K.U. (2023). Impact of intellectual capital on microfinance institutions' efficiency: The moderating role of external governance. *Journal of the Knowledge Economy*, 14, 691–717.

Ahn, H. & Le, M.H. (2014). An insight into the specification of the input-output set for DEA-based bank efficiency measurement. *Management Review Quarterly*, 64(1), 3–37.

Al-Awlaqi, M.A. & Aamer, A.M. (2019). Financial and social efficiency in the Yemen microfinance institutions. *International Journal of Islamic and Middle Eastern Finance and Management*, 12(3), 322–345. DOI: 10.1108/IMEFM-10-2017-0276.

Alinsunurin, M.K. (2014). Efficiency of microfinance institutions in the Philippines. *Enterprise Development & Microfinance*, 25(4), 311–324.

Amersdorffer, F., Buchenrieder, G., Bokusheva, R., & Wolz, A. (2015). Efficiency in microfinance: Financial and social performance of agricultural credit cooperatives in Bulgaria. *Journal of the Operational Research Society*, 66(1), 57–65.

Asawaruangpipop, P. & Suwunnamek, O. (2014). Analysis on savings and credit cooperatives efficiency in Thailand: A data envelopment analysis (DEA) approach. *Research Journal of Business Management*, 8(3), 242–253.

Ayayi, A.G. & Wijesiri, M. (2018). Better with age? The relationship between longevity and efficiency dynamics of nonprofit microfinance institutions. *Quality & Quantity*, 52(5), 2331–2343.

Azad, M.A., Munisamy, S., Masum, A.K., & Wanke, P. (2016). Do African microfinance institutions need efficiency for financial stability and social outreach? *South African Journal of Science*, 112(9–10), 1–8.

Balammal, A., Madhumathi, R., & Ganesh, M. (2019). Performance evaluation frameworks in the context of Indian microfinance institutions. *Foundations of Management*, 11(1), 209–228.

Banker, R.D., Charnes, A., & Cooper, W.W. (1984). Some models for estimating technical and scale inefficiencies in data envelopment analysis. *Management Science*, 30(9), 1078–1092. DOI: 10.1287/mnsc.30.9.1078.

Bassem, B.S. (2008). Efficiency of microfinance institutions in the Mediterranean: An application of DEA. *Transition Studies Review*, 15(2), 343–354.

Bassem, B.S. (2014). Total factor productivity change of MENA microfinance institutions: A Malmquist productivity index approach. *Economic Modelling*, 39, 182–189.

Becchetti, L. & Sierra, J. (2003). Bankruptcy risk and productive efficiency in manufacturing firms. *Journal of Banking & Finance*, 27(11), 2099–2120.

Ben Abdelkader, I. & Mansouri, F. (2019). Performance of microfinance institutions in the MENA region: A comparative analysis. *International Journal of Social Economics*, 46(1), 47–65. DOI: 10.1108/IJSE-06-2017-0242.

Berger, A.N. & Mester, L.J. (1997). Inside the black box: What explains differences in the efficiencies of financial institutions? *Journal of Banking & Finance*, 21(7), 895–947.

Bharti, N., & Chitnis, A. (2016). Size and efficiency of MFIs: A data envelopment analysis of Indian MFIs. *Enterprise Development & Microfinance*, 27(4), 255–272.

Bharti, N. & Malik, S. (2022). Financial inclusion and the performance of microfinance institutions: Does social performance affect the efficiency of microfinance institutions? *Social Responsibility Journal*, 18(4), 858–874. DOI: 10.1108/SRJ-03-2020-0100.

Bibi, U., Balli, H.O., Matthews, C.D., & Tripe, D.W. (2018). Impact of gender and governance on microfinance efficiency. *Journal of International Financial Markets, Institutions and Money*, 53, 307–319.

Boubaker, S., Le, T.D.Q., & Ngo, T. (2022a). Managing bank performance under COVID-19: A novel inverse DEA efficiency approach. *International Transactions in Operational Research*, 30(2023), 2436–2452. https://doi.org/10.1111/itor.13132.

Boubaker, S., Do, D.T., Hammami, H., & Ly, K.C. (2022b). The role of bank affiliation in bank efficiency: A fuzzy multi-objective data envelopment analysis approach. *Annals of Operations Research*, 311, 611–639. https://doi.org/10.1007/s10479-020-03817-z.

Boubaker, S., Le, T.D.Q., Ngo, T., & Manita, R. (2023). Predicting the performance of MSMEs: A hybrid DEA-machine learning approach. *Annals of Operations Research*. DOI: 10.1007/s10479-023-05230-8.

Cabanda, E. & Domingo, E.C. (2014). A production approach to performance of banks with microfinance operations. *International Journal of Information Systems in the Service Sector (IJISSS)*, 6(2), 18–35.

Cavalcante, W.Q.D.F., Coelho, A., & Bairrada, C.M. (2021). Sustainability and tourism marketing: A bibliometric analysis of publications between 1997 and 2020 using vosviewer software. *Sustainability*, 13(9), 4987.

Charnes, A., Cooper, W.W., & Rhodes, E. (1978). Measuring the efficiency of decision making units. *European Journal of Operational Research*, 2(6), 429–444.

Cisneros, L., Ibanescu, M., Keen, C., Lobato-Calleros, O., & Niebla-Zatarain, J. (2018). Bibliometric study of family business succession between 1939 and 2017: Mapping and analyzing authors' networks. *Scientometrics*, 117(2), 919–951.

Coelli, T.J. (1996). A Guide to DEAP Version 2.1: A data envelopment analysis (computer) program. Center for Efficiency and Productivity Analysis. University of New England. Working paper, 96(8). Available at https://economics.uq.edu.au/cepa/software.

Coelli, T.J. & Perelman, S. (1999). A comparison of parametric and non-parametric distance functions: With application to European railways. *European Journal of Operational Research*, 117(2), 326–339. https://doi.org/10.1016/S0377-2217(98)00271-9.

Collins, T. (2019). Pursuing efficiency: A data envelopment analysis of MFIs in Latin America. *Applied Economics Letters*, 26(6), 480–484.

Cooper, W.W., Seiford, L.M., & Zhu, J. (2011). Data Envelopment Analysis: History, Models, and Interpretations. In: W.W. Cooper, L.M. Seiford & J. Zhu (Eds.), *Handbook on Data Envelopment Analysis*. International Series in Operations Research & Management Science (Vol. 164). Boston, MA: Springer. https://doi.org/10.1007/978-1-4419-6151-8_1.

Deb, J. & Sinha, R.P. (2022). Impact of competition on efficiency of microfinance institutions: Cross country comparison of India and Bangladesh. *International Journal of Rural Management*, 18(2), 250–270.

do Castelo Gouveia, M., Duarte Neves, E., Cândido Dias, L., & Henggeler Antunes, C. (2018). Performance evaluation of Portuguese mutual fund portfolios using the value-based DEA method. *Journal of the Operational Research Society*, 69(10), 1628–1639.

Efendic, V. & Hadziahmetovic, N. (2017). The social and financial efficiency of microfinance institutions: The case of Bosnia and Herzegovina. *South East European Journal of Economics and Business*, 12(2), 85–101.

Efendić, V. & Hadžić, F. (2017). Business model and operational efficiency of microfinance in Bosnia and Herzegovina: Lessons for Islamic microfinance. In *Critical Issues and Challenges in Islamic Economics and Finance Development* (pp. 191–211). Cham: Palgrave Macmillan. https://doi.org/10.1007/978-3-319-45029-2_9.

Falagas, M.E., Pitsouni, E.I., Malietzis, G.A. & Pappas, G. (2008). Comparison of PubMed, Scopus, web of science, and Google scholar: Strengths and weaknesses. *The FASEB Journal*, 22(2), 338–342.

Fall, F., Akim, A.-M., & Wassongma, H. (2018). DEA and SFA research on the efficiency of microfinance institutions: A meta-analysis. *World Development*, 107, 176–188.

Färe, R., Grosskopf, S., & Logan, J. (1983). The relative efficiency of Illinois electric utilities. *Resources and Energy*, 5(4), 349–367.

Farida, F., Osman, I.R., Kurniawan Lim, A., & Wahyuni, N. (2018). The efficiency of formal microfinance in Indonesia: Using data envelopment analysis application. *Iranian Economic Review*, 22(3), 787–810.

Farrell, M.J. (1957). The measurement of productive efficiency. *Journal of the Royal Statistical Society: Series A (General)*, 120(3), 253–281.

Gebremichael, B.Z. & Gessesse, H.T. (2016). Technical efficiency of Microfinance Institutions (MFIs). *International Journal of Development Issues*, 15(3), 224–239. DOI: 10.1108/IJDI-04-2016-0026.

Goyal, S. & Agarwal, S. (2017). Technical efficiency of microfinance institutions in India: Data envelopment analysis. *Journal of Rural Development*, 36(1), 83–96.

Gutierrez-Nieto, B., Serrano-Cinca, C., & Molinero, C.M. (2007). Microfinance institutions and efficiency. *Omega*, 35(2), 131–142.

Gutiérrez-Nieto, B., Serrano-Cinca, C., & Molinero, C.M. (2009). Social efficiency in microfinance institutions. *Journal of the Operational Research Society*, 60(1), 104–119.

Hammami, H., Ngo, T., Tripe, D., & Vo, D.-T. (2022). Ranking with a Euclidean common set of weights in data envelopment analysis: With application to the Eurozone banking sector. *Annals of Operations Research*, 311(2), 675–694. DOI: 10.1007/s10479-020-03759-6.

Haq, M., Skully, M., & Pathan, S. (2010). Efficiency of microfinance institutions: A data envelopment analysis. *Asia-Pacific Financial Markets*, 17(1), 63–97.

Hermes, N., Lensink, R., & Meesters, A. (2018). Financial development and the efficiency of microfinance institutions. In *Research Handbook on Small Business Social Responsibility: Global Perspectives*. Cheltenham, UK: Edward Elgar Publishing.

Hong, J.-D. & Jeong, K.-Y. (2019). Combining data envelopment analysis and multi-objective model for the efficient facility location–allocation decision. *Journal of Industrial Engineering International*, 15(2), 315–331. DOI: 10.1007/s40092-018-0294-2.

Hussain, H.I., Kot, S., Kamarudin, F., & Wong, C.M. (2020). The nexus of competition freedom and the efficiency of microfinance institutions. *Journal of Competitiveness*, 12(2), 67.

Hussain, H.I., Kot, S., Kamarudin, F., & Yee, L.H. (2021a). Impact of rule of law and government size to the microfinance efficiency. *Economic Research-Ekonomska Istraživanja*, 34(1), 1870–1895.

Hussain, H.I., Szczepańska-Woszczyna, K., Kamarudin, F., Anwar, N.A.M., & Saudi, M.H.M. (2021b). Unboxing the black box on the dimensions of social globalisation and the efficiency of microfinance institutions in Asia. *Oeconomia Copernicana*, 12(3), 557–592.

Hussain, H.I., Kamarudin, F., Anwar, N.A.M., Sufian, F., Ali, A., & Saudi, M.H. (2022). Social globalisation and efficiency of microfinance institutions nexus: Empirical evidence on financial and social efficiency. *Engineering Economics*, 33(1), 27–46.

Isik, I. & Hassan, M.K. (2003). Financial deregulation and total factor productivity change: An empirical study of Turkish commercial banks. *Journal of Banking & Finance*, 27(8), 1455–1485.

Jaiyeoba, H.B., Adewale, A.A., & Ibrahim, K. (2018). Measuring efficiencies of Bangladeshi and Indonesian microfinance institutions. *International Journal of Bank Marketing*, 36(2), 305–321. DOI: 10.1108/IJBM-01-2017-0009.

Kabir Hassan, M., Sanchez, B., & Ngene, G. (2012). Scales and technical efficiencies in Middle East and North African (MENA) micro financial institutions. *International Journal of Islamic and Middle Eastern Finance and Management*, 5(2), 157–170. DOI: 10.1108/17538391211233434.

Kablan, S.A. & Boussemart, J.-P. (2015). Bootstrapping efficiency scores of microfinance institutions in West African economic and monetary union. *Economics Bulletin*, 35(4), 2220–2232.

Kaur, P. (2016). Efficiency of microfinance institutions in India: Are they reaching the poorest of the poor? *Vision*, 20(1), 54–65.

Khan, A. & Gulati, R. (2021). Efficiency of microfinance institutions of South Asia: A bootstrap DEA approach. *International Journal of Computational Economics and Econometrics*, 11(1), 84–104.

Khan, A. & Gulati, R. (2022). Productivity growth, catching-up and technology innovation in microfinance institutions in India: Evidence using a bootstrap Malmquist Index approach. *Benchmarking: An International Journal*, 29(3), 878–904. DOI: 10.1108/BIJ-10-2020-0535.

Khan, A. & Shireen, S. (2020). Drivers of financial and operational efficiency of MFIs: Empirical evidences from Eastern Europe and Central Asia. *Benchmarking: An International Journal*, 27(9), 2679–2697. DOI: 10.1108/BIJ-11-2019-0515.

Khan, N.F. & Khan, M.N. (2021). A bibliometric analysis of peer-reviewed literature on smartphone addiction and future research agenda. *Asia-Pacific Journal of Business Administration*, 14(2), 199–222. https://doi.org/10.1108/APJBA-09-2021-0430.

Khan, A., Ahmad, A., & Shireen, S. (2021). Ownership and performance of microfinance institutions: Empirical evidences from India. *Cogent Economics & Finance*, 9(1), 1930653.

Khanam, D., Parvin, S.S., Mohiuddin, M., Hoque, A., & Su, Z. (2018). Financial sustainability of non-governmental microfinance institutions (MFIs): A cost-efficiency analysis of BRAC, ASA, and

PROSHIKA from Bangladesh. *Review of Economics & Finance*, 12(2), 43–56.

Khanra, S., Dhir, A., & Mäntymäki, M. (2020). Big data analytics and enterprises: A bibliometric synthesis of the literature. *Enterprise Information Systems*, 14(6), 737–768.

Khanra, S., Dhir, A., Parida, V., & Kohtamäki, M. (2021). Servitization research: A review and bibliometric analysis of past achievements and future promises. *Journal of Business Research*, 131, 151–166.

Kolloju, A.K. & Meoli, M. (2022). Efficiencies of faith and secular microfinance institutions in regions of Asia, Africa, and Latin America: A two-stage dual efficiency bootstrap DEA approach. *Economies*, 10(3), 66.

Lastra, M.G.U., Salgado, R.J.S., & Castro, H.J.F. (2022). A data envelopment analysis of the global microfinance industry. *Contaduría y Administración*, 67(1), 10.

Le, T.D.Q. (2018). Bank risk, capitalisation and technical efficiency in the Vietnamese banking system. *Australasian Accounting, Business and Finance Journal*, 12(3), 41–61.

Lebovics, M., Hermes, N., & Hudon, M. (2016). Are financial and social efficiency mutually exclusive? A case study of Vietnamese microfinance institutions. *Annals of Public and Cooperative Economics*, 87(1), 55–77.

Li, L.Y., Hermes, N., & Meesters, A. (2019). Convergence of the performance of microfinance institutions: A decomposition analysis. *Economic Modelling*, 81, 308–324. https://doi.org/10.1016/j.econmod.2019.05.014.

Merigó, J.M., Cancino, C.A., Coronado, F., & Urbano, D. (2016). Academic research in innovation: A country analysis. *Scientometrics*, 108(2), 559–593.

Mia, M.A. & Soltane, B.I.B. (2016). Productivity and its determinants in microfinance institutions (MFIs): Evidence from South Asian countries. *Economic Analysis and Policy*, 51, 32–45.

Mia, M.A., Dalla Pellegrina, L., Van Damme, P., & Wijesiri, M. (2019). Financial inclusion, deepening and efficiency in microfinance programs: Evidence from Bangladesh. *The European Journal of Development Research*, 31(4), 809–835.

Mia, M.A., Rangel, G.J., Nourani, M., & Kumar, R. (2022). Institutional factors and efficiency performance in the global microfinance industry. *Benchmarking: An International Journal*(ahead-of-print). DOI: 10.1108/BIJ-06-2021-0326.

Mohamad Anwar, N.A., Iqbal Hussain, H., Kamarudin, F., Sufian, F., Zainal, N., & Wong, C.M. (2021). Impact of regulatory efficiency and market openness to social and financial efficiency: Empirical evidence from microfinance institutions. *Society and Business Review*, 16(3), 374–397. DOI: 10.1108/SBR-04-2020-0056.

Muneer Babu, M. (2016). Macroeconomic determinants of technological change in Indian microfinance institutions. *Global Business Review*, 17(5), 1179–1195.

Muneer Babu, M. & Kulshreshtha, P. (2014). Productivity change and technical efficiency in Indian microfinance institutions. *Studies in Microeconomics*, 2(2), 165–200.

Nourani, M., Malim, N.A.K., & Mia, M.A. (2021). Revisiting efficiency of microfinance institutions (MFIs): An application of network data envelopment analysis. *Economic Research-Ekonomska Istraživanja*, 34(1), 1146–1169.

Nzongang, J. & Nishimikijimana, E. (2013). Social and financial performance of microfinance institutions: A multi-stage data envelopment analysis application. In: R. Manos, J.P. Gueyié & J. Yaron (Eds.) *Promoting Microfinance*. London: Palgrave Macmillan. https://doi.org/10.1057/9781137034915_8.

Oladinrin, O.T., Arif, M., Rana, M.Q., & Gyoh, L. (2023). Interrelations between construction ethics and innovation: A bibliometric analysis using VOSviewer. *Construction Innovation*, 23(3), 505–523. doi:10.1108/CI-07-2021-0130.

Piot-Lepetit, I. & Nzongang, J. (2014). Financial sustainability and poverty outreach within a network of village banks in Cameroon: A multi-DEA approach. *European Journal of Operational Research*, 234(1), 319–330.

Piot-Lepetit, I. & Nzongang, J. (2019). Performance assessment and definition of improvement paths for microfinance institutions: An application to a network of village banks in Cameroon. *International Transactions in Operational Research*, 26(4), 1188–1210.

Piot-Lepetit, I. & Tchakoute Tchuigoua, H. (2022). Ownership and performance of microfinance institutions in Latin America: A pseudo-panel malmquist index approach. *Journal of the Operational Research Society*, 73(5), 1167–1180. DOI: 10.1080/01605682.2021.1895683.

Purwanto, Primiana I., Masyita D., Febrian, E. (2020). Alleviating poverty through Islamic microfinance: Outreach and efficiency (Study on BMT and Islamic Cooperatives in Indonesia). *Ekonomika regiona (Economy of Region)*, 16(4), 1362–1376. https://doi.org/10.17059/ekon.reg.2020-4-24.

Rana, M.S., Banna, H., Mia, M.A., Ismail, I.B., & Ismail, M.N.B. (2021). How productive are the microfinance institutions in Bangladesh? An application of malmquist productivity index. *Studies in Microeconomics*. https://doi.org/10.1177/23210222211024445.

Riaz, M. (2015). In competency aspects of microfinance industry: Via SFA approach. *Journal of Economics and Behavioral Studies*, 7(1 (J)), 1–12.

Sai Mohini, M. & Vilvanathan, L. (2021). Efficiency assessment of microfinance institutions: Using DEA with weighted Russell directional distance model. *Benchmarking: An International Journal*, 28(3), 769–791. DOI: 10.1108/BIJ-06-2020-0288.

Sarma, G.K. (2015). Efficiency of microfinance institutions and branch level determinants: Empirical evidence from Assam, India. *International Journal of Economics and Business Research*, 9(3), 337–354.

Sedzro, K., Keita, M., & Assogbavi, T. (2010). The impact of the organizational structure on the performance of microfinance institutions using data envelopment analysis. In *Applications in Multicriteria Decision Making, Data Envelopment Analysis, and Finance* (Vol. 14, pp. 293–308). Bingley, England: Emerald Group Publishing Limited.

Segun, K.R.S. & Anjugam, M. (2013). Measuring the efficiency of sub-Saharan Africa's microfinance institutions and its drivers. *Annals of Public and Cooperative Economics*, 84(4), 399–422.

Serrano-Cinca, C. & Gutiérrez-Nieto, B. (2014). Microfinance, the long tail and mission drift. *International Business Review*, 23(1), 181–194.

Serrano-Cinca, C., Gutiérrez-Nieto, B., & Molinero, C.M. (2011). Social and financial efficiency of microfinance institutions. In: *The Handbook of Microfinance* (pp. 397–418). Singapore: World Scientific.

Servin, R., Lensink, R., & Van den Berg, M. (2012). Ownership and technical efficiency of microfinance institutions: Empirical evidence from Latin America. *Journal of Banking & Finance*, 36(7), 2136–2144.

Shah, S.H.H., Lei, S., Ali, M., Doronin, D., & Hussain, S.T. (2019). Prosumption: Bibliometric analysis using HistCite and VOSviewer. *Kybernetes*, 49(3), 1020–1045. https://doi.org/10.1108/K-12-2018-0696.

Sinha, R.P. & Pandey, P. (2019). Efficiency of microfinance institutions in India: A two-stage DEA approach. *International Journal of Rural Management*, 15(1), 49–77.

Soldátková, N. & Černý, M. (2022). Microfinance in Sub-Saharan Africa: Social efficiency, financial efficiency and institutional factors. *Central European Journal of Operations Research*, 30(2), 449–477.

Tan, N.N. & Anh, L.H. (2021). Performance of Microfinance Institutions in Vietnam. In: N. Ngoc Thach, V. Kreinovich, &

N.D. Trung (Eds.) *Data Science for Financial Econometrics*. Studies in Computational Intelligence (Vol. 898). Cham: Springer. https://doi.org/10.1007/978-3- 030-48853-6_12.

Tone, K. & Tsutsui, M. (2009). Network DEA: A slacks-based measure approach. *European Journal of Operational Research*, 197(1), 243–252. https://doi.org/10.1016/j.ejor.2008.05.027.

Tranfield, D., Denyer, D., & Smart, P. (2003). Towards a methodology for developing evidence-informed management knowledge by means of systematic review. *British Journal of Management*, 14(3), 207–222.

Tripe, D. (2010). Using DEA to investigate bank safety and soundness — Which approach works best? *Journal of Financial Economic Policy*, 2(3), 237–250.

Van Damme, P., Wijesiri, M., & Meoli, M. (2016). Governance and efficiency of microfinance institutions: Empirical evidence from Sri Lanka. *South Asia Economic Journal*, 17(2), 236–247.

Vu, H.D. (2016). Technical efficiency of FDI firms in the Vietnamese manufacturing sector. *Review of Economic Perspectives*, 16(3), 205–230.

Widiarto, I. & Emrouznejad, A. (2015). Social and financial efficiency of Islamic microfinance institutions: A data envelopment analysis application. *Socio-Economic Planning Sciences*, 50, 1–17.

Widiarto, I., Emrouznejad, A., & Anastasakis, L. (2017). Observing choice of loan methods in not-for-profit microfinance using data envelopment analysis. *Expert Systems with Applications*, 82, 278–290.

Wijesiri, M. & Meoli, M. (2015). Productivity change of microfinance institutions in Kenya: A bootstrap Malmquist approach. *Journal of Retailing and Consumer Services*, 25, 115–121.

Wijesiri, M. & Tremblay, D.-G. (2020). Corporate social responsibility and financial performance in microfinance institutions. In: N. Nemoto & P.J. Morgan (Eds.), *Environmental, Social, and Governance Investment* (pp. 175–186). Tokyo, Japan: Asian Development Bank Institute.

Wijesiri, M., Viganò, L., & Meoli, M. (2015). Efficiency of microfinance institutions in Sri Lanka: A two-stage double bootstrap DEA approach. *Economic Modelling*, 47, 74–83.

Wijesiri, M., Yaron, J., & Meoli, M. (2017). Assessing the financial and outreach efficiency of microfinance institutions: Do age and size matter? *Journal of Multinational Financial Management*, 40, 63–76.

Wu, Y., Escalante, C.L., & Li, X. (2016). Technical efficiency and business maturity: Evidence from Chinese and Indian microfinance institutions. *Enterprise Development & Microfinance*, 27(2), 97–114.

Zainal, N., Nassir, A.M., Kamarudin, F., & Hook, L.S., (2020b). Does bank regulation and supervision impedes the efficiency of microfinance institutions to eradicate poverty? Evidence from ASEAN-5 countries. *Studies in Economics and Finance*, 38(2), 272–302.

Zainal, N., Bakri, M.H., Sanusi, Z.M., Razak, N.A., Roslan, S., & Harjito, D.A. (2022). Financial efficiency of microfinance institutions in the event of pre and post financial crisis. *Quality — Access to Success*, 23(187), 180–191.

Zainal, N., Kamarudin, F., Hook, L.S., Bakri, M.H., Sufian, F., & Nassir, A.M. (2020a). Regulation, supervision and social and financial efficiency of microfinance institutions in ASEAN-5 countries. *Malaysian Journal of Economic Studies*, 57(2), 253–273.

Zainal, N., Nassir, A.M., Kamarudin, F., Hook, L.S., Sufian, F., & Hussain, H.I. (2019). The social role of microfinance institutions in poverty eradication: Evidence from ASEAN-5 countries. *International Journal of Innovation, Creativity and Change*, 5(2), 1551–1576.

https://doi.org/10.1142/9781800615786_0011

Chapter 11

Technical Efficiency Determinants of the Indonesian Employer Pension Funds: A Double-Bootstrap DEA Analysis

Paskalis Seran[*,‡], **Apriani Dorkas Rambu Atahau**[†,§], **Supramono Supramono**[†,¶], and **Usil Sis Sucahyo**[†,‖]

[*]*Universitas Katolik Widya Mandira, Indonesia*
[†]*Universitas Kristen Satya Wacana, Indonesia*
[‡]*seranpascalsvd@gmail.com*
[§]*apriani@staff.uksw.edu*
[¶]*supramono.supramono@uksw.edu*
[‖]*usil.sucahyo@uksw.edu*

Abstract

The world's aging population requires a robust and effective pension system, yet research on its efficiency is limited. This research examines the technical efficiency determinants of pension funds using financial data on 40 Indonesian Employer Pension Funds (EPFs) during 2011–2017 (280 firm-year observations). We use the double-bootstrap data envelopment analysis (DEA) method, as described by Simar & Wilson (2007), to analyze the data. The findings indicate that size, ownership, and pension plan type affect pension funds' efficiency. Size positively affects operating and investment efficiency, presumably due to economies of scale. In addition, pension funds sponsored by state-owned enterprises exhibit higher operating and investment technical efficiency due to their larger size and better financial knowledge/skills. On the other hand, pension plan type negatively affects operating management but not investment efficiency. For macroeconomic factors, only the interest rate significantly and negatively affects pension funds' investment efficiency, while GDP growth does not significantly affect pension funds' efficiency.

Keywords: Pension funds' performance, pension funds' efficiency, employer pension funds, Indonesian pension funds, pension funds determinants, technical efficiency, two-stage DEA, double-bootstrap DEA, Simar–Wilson, truncated regression.

1. Introduction

The United Nations (UN) forecasts that in 2050, around 16% of the global population will be over 65 years old (United Nations Population Division, 2019). Even though Indonesia is now experiencing a demographic boom, this will only endure until 2035. Statistics Indonesia projects that in 2045, the number of elderly Indonesians will reach 63.3 million (19.8%) (Badan Pusat Statistik Indonesia, 2018).

This demographic trend necessitates a robust pension system to support the elderly. In this context, one of the most important concerns in today's global pension business is efficiency, or the capacity of pension funds to offer participants optimal pension benefits (Chybalski, 2015). Internal inefficiency can result in high operating expenses and low investment returns, leading to lower pension benefits accrued when employees reach retirement age (Bikker & Dreu, 2009; Bateman & Mitchell, 2004). Internal inefficiencies might even cause pension funds to dissolve (Casey & Yiallouros, 2013). Thus, efficiency is crucial for ensuring that pensioners have sufficient pension income.

Nevertheless, the issue of pension fund efficiency remains relatively understudied (Ambachtsheer *et al.*, 2008). Previous studies focus on developed countries, such as the Netherlands (Alserda *et al.*, 2018; Bikker, 2017) and Australia (Bui *et al.*, 2016; Galagedera & Watson, 2015). Similar studies using the Indonesian context have only appeared recently. For instance, Hasanudin *et al.* (2017) investigated the impacts of market timing abilities, stock selection skills, and firm size on the portfolio performance of 64 employer pension funds (EPFs) during 2014–2016. Nanda & Atahau (2020) and Putri *et al.* (2020) analyze the impact of pension fund size on their efficiency and investment performance. Further, Sunaryo *et al.* (2020) examine the effects of financial ratios, such as return on assets, investment expense ratio, and operating expense ratio, on the fund adequacy ratio of 20 defined-benefit program pension funds during 2008–2017.

These studies only focus on the effects of a few pension fund characteristics on efficiency. Meanwhile, studies in other countries document that other specific characteristics, such as pension plan type (Bikker & Dreu, 2009) and ownership (Kowalewski, 2012), as well as macroeconomic factors, such as interest rates (Wanjiku, 2014) and GDP growth (Akwimbi, 2020), significantly affect pension fund efficiency. Hence, our study evaluates not only the influence of size on pension fund efficiency but also the effects of other specific characteristics and macroeconomic variables in the Indonesian setting. Specifically, this research seeks to investigate the determinants of Indonesian EPFs' technical efficiency during 2011–2017, including the impact of the specific characteristics of pension funds (size, plan type, and sponsor type or ownership) and macroeconomic factors (interest rate and GDP growth).

Methodologically, we employ data envelopment analysis (DEA), a widely used non-parametric technique (Emrouznejad & Yang, 2018) in the financial sector to measure and evaluate the efficiency and performance of banks (Defung *et al.*, 2016; Boubaker *et al.*, 2022a, 2022b), insurance firms (Kaffash *et al.*, 2020; Abdin *et al.*, 2022; Zhao *et al.*, 2021), mutual funds (Basso & Funari, 2017; Vidal-Garcia *et al.*, 2018; Bilbao-Terol *et al.*, 2021), or pension funds (Bui *et al.*, 2016; Demirtaş & Kececi, 2020; Seran *et al.*, 2022). DEA measures the relative efficiency of a set of similar production units (decision-making units, or DMUs) that utilize the same inputs and produce the same outputs during certain periods. DEA identifies a DMU's inefficiency by comparing it with other efficient DMUs (Coelli *et al.*, 2005). DEA can handle several inputs and outputs and operate with a limited sample size. Hence, the DEA procedure is preferable in the Indonesian context, where pension data are limited.

Along with the widespread use of DEA as a method of measuring the efficiency and performance of organizations in various industrial sectors, new DEA models are also emerging. More advanced and sophisticated DEA models are being developed and are becoming prominent in efficiency and performance studies. These include network DEA (Färe & Grosskopf, 2000; Cook *et al.*, 2010; Seran *et al.*, 2023), bootstrap DEA (Simar & Wilson, 2007; Zaman & Bhandari, 2022), slack-based DEA (Tone & Tsutsui, 2010; Xue & Ma, 2022), inverse DEA (Wei *et al.*, 2000; Boubaker *et al.*, 2022a), and common weight DEA (Hammami *et al.*, 2020; Bánhidi & Dobos, 2023). Even the recent developments in artificial intelligence (AI) and machine learning (ML) in "Industry 4.0" have also contributed to the development of the DEA methodology (Revuelta *et al.*, 2021; Thaker *et al.*, 2022; Zhu, 2022).

In this study, we investigate the determinants of Indonesian EPFs' technical efficiency using the two-stage DEA method. In the literature, the term "two-stage DEA" is used interchangeably for two different approaches (Henriques *et al.*, 2020). First, the term "two-stage DEA" refers to two-stage network DEA models, in which the outputs that are generated using available inputs in the first stage become inputs to the second stage to produce the final outputs (Kao & Hwang, 2008). Second, the term "two-stage DEA" refers to "two-stage analysis," which includes estimating efficiency scores using DEA in the first stage, and the estimated efficiency scores are then regressed on contextual variables in the second stage using regression techniques, such as ordinary least squares (OLS) or maximum likelihood (ML)-based probit, logit, and truncated regressions (Banker & Natarajan, 2008; Simar & Wilson, 2007). Henriques *et al.* (2020) distinguish the two approaches by naming the first approach (two-stage network DEA) as "internal two-stage DEA," while the second approach is called "external two-stage DEA."

We use the "external two-stage" approach in the current study. Accordingly, we initially estimate the efficiency scores of DMUs (EPFs) using the conventional DEA model (input-oriented variable return to scale, or VRS). Further, we regress the DEA efficiency scores against exogenous variables (Simar & Wilson, 2007; Banker & Natarajan, 2008; Ngo & Tsui, 2021), including EPFs' specific characteristics (size, pension plan type, and sponsor/ownership) and macroeconomic factors (Bank Indonesia's interest rate and GDP growth). The traditional DEA model (Charnes *et al.*, 1978) is arguably inadequate to explain DMUs' complex production processes that are affected by not only endogenous factors (managerial efficiency) but also exogenous factors, such as DMUs' production environments (Fried *et al.*, 2002). Thus, the two-stage procedure is necessary to overcome the problem.

We specifically follow Simar & Wilson (2007), SW in short, who adopt the two-stage DEA with the double bootstrapping DEA approach. SW argue that the conventional two-stage DEA, where the efficiency scores of estimation results in the first phase are directly regressed (using OLS or tobit regression) against exogenous variables, is not statistically valid due to an ambiguous data generating process (DGP) and serial correlation problems between efficiency scores and estimation results. The double bootstrapping procedure likely mitigates the problems and provides more statistically valid inference results. Several studies on pension fund efficiency have utilized the two-stage method (Sathye, 2011; Njuguna & Arnolds, 2012; Galagedera, 2017). However, double bootstrapping remains underutilized in this research stream.

Our contributions are twofold. First, this study extends the knowledge of the determinants of Indonesian pension funds' technical efficiency by incorporating several firm-specific variables (size, pension plan type, and ownership) and macroeconomic variables that likely affect pension funds' efficiency but are relatively understudied. Prior studies have only focused on the effect of size on portfolio returns or the impact of financial ratios on capital adequacy (Sunaryo *et al.*, 2017; Nanda & Atahau, 2020; Hasanudin *et al.*, 2017). Second, this study advises pension fund managers to take strategic actions to improve their pension funds' efficiency by highlighting the effects of size, pension plan type, ownership, and macroeconomic conditions on their efficiency.

Our results demonstrate that Indonesian EPFs are not optimal in their operating and investment efficiency. Size (i.e., total asset) positively affects efficiency, where larger EPFs exhibit greater operating and investment efficiency, which is thus in line with the economies of scale concept. In a similar vein, sponsor type or ownership affects efficiency, where EPFs sponsored by state-owned enterprises (SOEs) are more efficient in operating and investment activities than those sponsored by non-SOEs. SOE-sponsored pension funds are arguably larger and possess better skills and knowledge. However, plan type only affects operating efficiency, where those offering defined-benefit plans exhibit lower operating efficiency than those offering defined-contribution plans, likely due to higher operating costs. Besides, changes in BI rates during 2011–2017 negatively affect investment efficiency, whereas GDP growth does not significantly affect operating and investment efficiency. In sum, size, plan type, sponsor type/ownership, and interest rate (i.e., Bank Indonesia's rate) crucially determine Indonesian EPFs' technical efficiency, either positively or negatively.

This chapter is organized as follows: After the introduction, Section 2 presents the theoretical framework and hypothesis development. Section 3 presents the methodology. Section 4 discusses and analyzes the results, and Section 5 concludes.

2. Theoretical Framework and Hypothesis Development

2.1. *Pension funds' technical efficiency*

Technical efficiency refers to a production unit's managerial ability to select and combine a number of inputs to minimize costs and produce optimal outputs (Coelli *et al.*, 2005). Regarding pension funds, *efficiency* can be defined as the ability to manage resources (contributions and investments) to produce optimal outputs to ensure optimal pension payments (Chybalski, 2015).

The ability to provide optimal pension benefits is affected by various factors, including pension fund-specific characteristics and macroeconomic conditions. Several prior studies demonstrate that pension fund-specific characteristics, such as size, plan type, and sponsor (ownership), affect pension funds' efficiency (Sathye, 2011; Bateman & Mitchell, 2004; Bikker & Dreu, 2009). Macroeconomic factors, such as interest rates and GDP growth, are also likely to affect pension funds' efficiency (Bohl *et al.*, 2011; Wanjiku, 2014; Akwimbi, 2020).

2.2. *The influence of specific characteristics on efficiency*

Size and Efficiency: Pension funds' size likely affects their efficiency. Larger pension funds are arguably more efficient due to their economies of scale (Bikker & Dreu, 2009). Bikker (2017) documents declining unused scale economies in operating and investment activities when pension funds' assets increase. Larger pension funds can spread their fixed operating costs (such as IT infrastructure-related ones) due to their larger asset values (Cummings, 2015). In terms of investments, larger asset values offer more flexibility, increase bargaining power, and facilitate economies of scale that lead to lower costs and higher investment income (Dyck & Pomorsky, 2011). Galagedera & Watson (2015) find a positive correlation between asset values and technical efficiency.

Nonetheless, diseconomies of scale can emerge if, as pension funds grow, their operations become more involved and complex, resulting in rising costs that are disproportionate to their size (Chen *et al.*, 2004). Alserda *et al.* (2018) observe that Dutch pension funds exhibit larger economies of scale in administrative activities but diseconomies of scale in investment activities. In a similar vein, Bauer *et al.* (2010) reveal that although larger US pension funds enjoy lower operating costs, they generate lower investment returns than smaller pension funds due to domestic security markets' limited liquidity. Thus, we propose the following hypotheses to predict the effect of size on EPFs' technical efficiency during 2011–2017:

H1a: Larger EPFs exhibit higher operating technical efficiency.
H1b: Larger EPFs exhibit higher investment technical efficiency.

Pension Plan Type and Efficiency: EPFs can offer defined-benefit or defined-contribution plans. Defined-benefit programs promise predetermined benefits with uncertain contributions depending on actuaries' estimations (Kadarisman & Wahyuni, 2010). Conversely, defined-contribution programs predetermine contributions with the amount of pension benefits depending on investment returns (Bodie *et al.*, 1988). In defined-benefit (defined-contribution) programs, employers (employees) bear the investment risk. These differences likely affect the operating and investment efficiencies of both pension programs. Defined-benefit programs incur greater operating costs because they have to hire actuaries or guarantee fund adequacy (Bateman & Mitchell, 2004). Bikker & Dreu (2009) demonstrate that Dutch defined-benefit programs exhibit higher operating costs than their defined-contribution counterparts.

Nevertheless, defined-benefit plans arguably exhibit better investment performance because owners/sponsors who bear the financial repercussions of all investment activities will be encouraged to closely monitor management (Jang & Wu, 2021), whereas defined-contribution programs' sponsors are less motivated to do so because participants bear the investment risks (agency problem) (Aslan, 2022). Besides, defined-benefit pension funds' top management teams collectively possess better financial knowledge than defined-contribution programs' individual participants. Several studies observe that defined-contribution pension funds have lower investment performance due to their managers' or participants' lower financial knowledge or agency problems (Andonov & Mao, 2019;

Clark *et al.*, 2015). We then formulate the following hypotheses predicting the effect of program type on EPFs' technical efficiency:

H2a: Defined-benefit EPFs exhibit lower operating technical efficiency than defined-contribution EPFs.

H2b: Defined-benefit EPFs exhibit higher investment technical efficiency than defined-contribution EPFs.

Sponsor/Ownership and Efficiency: Ownership is closely related to governance, which is crucial in explaining firm performance (Giannetti & Laeven, 2009). Ownership is associated with key personnel (shareholders and the board of trustees) who control firms' strategic decisions (Clark & Urwin, 2008). Pension funds controlled by foreign institutions arguably exhibit different governance from domestic ones. Similarly, pension funds controlled by SOEs differ from those owned by private firms. The Indonesian EPF industry identifies two sponsorship/ownership types: those sponsored/owned by SOEs or private firms. SOE-sponsored EPFs are typically managed by larger firms or state-owned banks, while privately sponsored EPFs are typically managed by private firms and smaller foundations or schools. These differences likely affect EPFs' efficiency levels.

Empirical studies in the banking and pension fund industries on the relationship between ownership and efficiency indicate different results. For example, Defung *et al.* (2016) find that state-owned banks are more efficient than private and other banks. Conversely, Shaban & James (2018) investigate the relationship between ownership and the efficiency of Indonesian banks during 2005–2012 and indicate that private banks are more efficient than state-owned banks. In the pension fund industry, Pasqualeto *et al.* (2014) show that SOE-sponsored Brazilian pension funds incur higher administrative costs than privately sponsored ones. Given the characteristics of SOE-sponsored and privately sponsored Indonesian EPFs, we formulate the following hypotheses:

H3a: SOE-sponsored EPFs exhibit higher operating technical efficiency than privately sponsored EPFs.

H3b: SOE-sponsored EPFs exhibit higher investment technical efficiency than privately sponsored EPFs.

2.3. *The impact of macroeconomic factors on pension funds' efficiency*

Interest Rate and Efficiency: Changes in interest rates may affect pension funds' performance depending on their asset and liability structures, investment portfolio structures, and pension plan types (Antolin *et al.*, 2011). A prolonged period of lower interest rates might result in negative duration gaps (i.e., debt/liability duration is greater than asset duration) and higher debts/liabilities. However, lower interest rates will increase bond portfolio values and deposit interest rates and reduce credit risks in the investment portfolios, allowing defined-benefit and defined-contribution pension programs to generate higher returns (Antolin *et al.*, 2011). On the other hand, higher interest rates may reduce assets' market values, especially bond portfolios (Fleuriet & Lubochinsky, 2005). Stock portfolio returns also decline when interest rates increase (Jensen *et al.*, 2019).

Changes in interest rates affect risky and asset allocation decisions that will eventually affect investment returns (Boubaker *et al.*, 2018). Andonov *et al.* (2017) document that US public pension funds exhibit negative performance because they switch their portfolios to highly risky assets when interest rates are lower. Wanjiku (2014) finds a negative correlation between interest rates and Kenyan pension funds' investment returns. Based on these arguments, we propose the following hypotheses that predict the impact of interest rates set by Bank Indonesia (BI) on Indonesian EPFs' technical efficiency during 2011–2017:

H4a: BI interest rates positively affect EPFs' operating technical efficiency.

H4b: BI interest rates positively affect EPFs' investment technical efficiency.

GDP Growth and Efficiency: GDP growth may negatively or positively affect pension funds' efficiency. Conducive macroeconomic conditions, as indicated by higher GDP growth, increase pension funds' investment portfolios, which will positively affect pension funds' efficiency (Mazreku *et al.*, 2020). Conversely, lower GDP growth due to economic crises or natural economic cycles will reduce pension funds' efficiency performance, and lower investment activities will inhibit pension funds' investment portfolio growth.

Besides, especially for EPFs, their sponsors will experience poorer performance, which negatively affects their ability to provide pension contributions on time. Bohl *et al.* (2011) indicate that poor macroeconomic conditions explain pension funds' lower performance in Hungary and Poland. Similarly, Najafi *et al.* (2021) observe that fluctuations in overall macroeconomic variables likely increase Iranian social security institutions' total costs. Accordingly, we propose the following hypotheses that predict the effect of GDP on Indonesian EPFs' technical efficiency during 2011–2017:

H5a: GDP growth positively affects EPFs' operating technical efficiency.

H5b: GDP growth positively affects EPFs' investment technical efficiency.

3. Methodology

3.1. *Data, sample, and variables*

Our secondary data are EPFs' financial statements during 2011–2017 that are manually generated from Asosiasi Dana Pensiun Indonesia (the Association of Indonesian Pension Funds, or ADPI) and Badan Kerja Sama Dana Pensiun Kristen Indonesia (the Cooperation Agency for Christian Pension Funds in Indonesia, or BKS Dapen-KI). After screening for data completeness, we generate 40 EPFs (280 firm-year observations) as the final sample, consisting of 31 defined-benefit pensions, nine defined-contribution pensions, and 21 SOE-sponsored and 19 privately sponsored pension funds.

Our first research phase analyzes input and output variables selected by assuming that pension funds' production activities consist of operating and investment ones (Galagedera, 2017). We investigated the input and output variables used in previous studies and selected them for our study based on the data available to us. Specifically, our input and output variables are almost identical to those selected by Galagedera (2017), who differentiates the process of pension fund management into operational management and investment management. Table 1 reports the input and output variables used

Table 1. Summary of input and output variables in the previous literature.

Author(s)	Units	Inputs	Outputs
Galagedera (2017)	Australian superannuation funds	Operational Management: (1) Net assets beginning of period (2) Total contributions (3) Total operating expenses	(1) Total benefit payment (2) Net assets end of period
		Portfolio Management: (1) Total net asset end period (2) Total risks (3) Total investment expenses	(1) Returns
Alserda *et al.* (2018)	Dutch occupational pension funds	(1) Administrative costs (2) Investment costs	(1) The number of members (2) Total assets
Sathye (2011)	Australian retail superannuation funds	(1) Operating Expenses (2) Contribution received	(1) Value of the fund (2) Benefits paid
Bui *et al.* (2016)	183 Australian superannuation funds	(1) Investment expenses (2) Operating expenses (3) Management, administration, and director fees (4) Total expenses (5) Volatility/SD investment return	(1) Average net assets (2) Member account (3) Return before tax/multiple period return
Barrientos & Boussofiane (2005)	Pension fund management companies in Chille, 1982–1999)	(1) Marketing expenses (2) Staff's salaries (3) Administrative costs	(1) Total income (2) Number of participant

Table 2. Descriptive statistics of input and output variables.

Variable	Obs.	Mean	SD	Min	Max
A: Operating Management					
Net assets, beginning of period (x_1)	280	2,140,000	3,620,000	4,010	17,800,000
Operating costs (x_2)	280	11,700	20,400	49.7	121,000
Net assets, end of period (y)	280	2,290,000	3,870,000	4,340	19,400,000
B: Investment Management					
Average investments (x_1)	280	2,140,000	3,640,000	4,150	18,300,000
Investment costs (x_2)	280	8,740	20,100	11	159,000
Investment income (y)	280	216,000	398,000	80.4	2,670,000

Notes:
Input Variables (x):
Net assets, beginning of period = The value of net asset at the beginning of each period.
Operating costs = Total operational costs, including staff's salaries and other administrative expenses.
Average investment = Total investment beginning of period + Total investment end of period divided by two.
Investment costs = Investment manager's compensation and other investment expenses.
Output Variables (y):
Net asset, end of period = Total net asset at the end of each period.
Investment income = Total income resulted from investment activities.

in previous literature. Table 2 lists the descriptive statistics of the input and output variables used in this study.

Operational management includes activities such as collecting contributions, paying pension benefits, accounting, and other administrative activities that belong to operating activities. In these activities, pension funds utilize their beginning-of-period assets and contributions received to pay pension benefits and other administrative activities. Operating activities' final outputs are end-of-period net asset values. Therefore, we select the following input and output variables for model A (operating management): net asset values at the beginning of the period (x_1), total operating costs (x_2), and end-of-period net asset values (y). The ratio of net asset value at the end of the period to net asset value at the beginning of the period and total management expenses reflect operating efficiency.

The input and output variables for model B (investment activities) are the average investment values (x_1), investment costs (x_2), and investment income (y). We include the average investment values as an input variable (x_1) because investment activities consist of

Table 3. Second-stage variables.

Dependent variables		
Technical efficiency	TE(A) TE(B)	Technical efficiency for operational management (model A), and investment management (model B)
Independent variables		
Size		Natural log of total assets
Type		A dummy variable, 1 for D.B. and 0 for D.C.
Ownership		A dummy variable, 1 for SOE-sponsored EPF and 0 for privately sponsored EPF
BIrate		Average annual Bank Indonesia (BI) interest rates (%)
GDP		Annual GDP growth (%)
Control variable		
INFrate		Annual inflation rate (%)

Note: DB: Defined benefit; DC: Defined contribution; SOE: State-owned enterprises; EPF: Employer pension fund.

portfolio management that involves a willingness to take risks and diversify to generate optimal returns. However, the ability to take risks and diversify is also affected by the amount of available funds (Gorter & Bikker, 2011). Hence, the average investment values and costs (consisting of investment managers' remunerations and other investment costs) are selected as the input variables, while the output variables are investment income. The ratio of investment income to average investment value and investment cost measures the efficiency of investing in a pension fund.

Next, the second phase determines several contextual/environmental variables that are related to employer funds' efficiency, including EPF-specific characteristics (size, pension plan type, and ownership) and macroeconomic conditions (average BI interest rate and the percentage of GDP growth). Table 3 lists the second stage variables.

3.2. *The approach model*

This study utilizes double bootstrap DEA analysis following SW's procedure (2007). According to SW, estimating the efficiency determinants with conventional two-stage DEA analysis lacks a clear theoretical foundation for the data generation process and is problematic owing to invalid inferences caused by finite-sample bias and serial correlation problems.

Hence, SW apply the bootstrapping technique to generate robust efficiency scores and mitigate serial correlations in efficiency scores. Following SW, this study estimates pension funds' efficiency using the following equation:

$$\theta_i = z_i\beta + \varepsilon_i, \tag{1}$$

where θ_i is the efficiency of EPF i, as stated by SW as the function of the vector of the external/environmental variables z_i experienced by each pension fund, consisting of pension fund-specific characteristics (i.e., size, pension program type, and ownership) and macroeconomic variables (i.e., BI interest rates and GDP growth); and β is the estimation coefficient vector. The estimation of efficiency, θ_i, and its determinants are obtained in two interrelated stages.

The efficiency score θ_i of each decision-making unit (DMU), i.e., an EPF, is estimated using the input-based VRS DEA. Technically, the following explains the calculation of the efficiency score for each DMU. Each DMU from the DMU set N has k inputs and m outputs, where for each ith DMU, each input and output are represented by input vectors of $x_i = x_{i1}, \ldots, x_{ik}$, and output vectors of $y_i = y_{ir}, \ldots, y_{im}$. The θ_i efficiency score of each DMU is generated by solving the following linear programming problem:

$$\min_{\theta\lambda} \theta$$

$$\begin{aligned}
\text{s.t.:}\quad & \sum_{j=1}^{n} \lambda_j y_{jr} \geq y_{ir},\ r = 1, \ldots, m;\\
& \sum_{j=1}^{n} \lambda_j x_{jl} \leq \theta x_{il},\ l = 1, \ldots, k;\\
& \sum_{j=1}^{n} \lambda_j = 1,\ \lambda_j \geq 0,\ j = 1, \ldots, n,
\end{aligned} \tag{2}$$

where θ (technical efficiency) is a scalar value, λ is a constant vector $N \times 1$. The optimal value of θ is the efficiency score of a certain DMU (the ith DMU) that satisfies $0 < \theta \leq 1$.

Referring to SW (2007), we run the following procedures to analyze EPFs' technical efficiency scores and their determinants according to the second algorithm:

(1) Based on Equation (2), we calculate efficiency scores for all DMUs, $i = 1, \ldots, N$.
(2) We then use M DMUs ($M < N$), i.e., all DMUs that satisfy $0 < \hat{\theta}_i < 1$ in a two-sided truncated regression of $\hat{\theta}_i$ against z_i to generate the estimation coefficient $\hat{\beta}$ and an estimation for variance parameter $\hat{\sigma}$ through the maximum likelihood method.
(3) Next, we repeat steps 3a–3d for $B1$ times ($B1 = 2000$) to generate a set of bootstrap estimates $\hat{\theta}_i^b$ for each DMU $i = 1, \ldots, N$, where $b = 1, \ldots, B1$.
   (3a) For each DMU, $i = 1, \ldots, N$, we randomly select a dummy error term $\tilde{\varepsilon}_i$ from the two-sided truncated normal distribution $N(0, \hat{\sigma})$, with the left-hand side at $-z_i\hat{\beta}$ and right-hand side at $1 - z_i\hat{\beta}$.
   (3b) For each DMU, $i = 1, \ldots, N$, we also calculate a dummy efficiency score $\tilde{\theta}_i$ as $\tilde{\theta}_i = z_i\hat{\beta} + \tilde{\varepsilon}_i$.
   (3c) We then create dummy DMUs, $i = 1, \ldots, N$, with the number of inputs $\tilde{x}_i = x_i$ and outputs $\tilde{y}_i = \left(\frac{\hat{\theta}_i}{\tilde{\theta}_i}\right) y_i$.
   (3d) The dummy DMUs N created in 3c are used as reference data in Equation (2) to generate $\hat{\theta}_i^b$, for each original DMU $i = 1, \ldots, N$.
(4) We then calculate a bias-corrected efficiency score $\hat{\theta}_i^{bc}$ as $\hat{\theta}_i - \left(\frac{1}{B1}\right) \sum_{b=1}^{B1} \hat{\theta}_i^b - \hat{\theta}_i$ for each DMU $i = 1, \ldots, N$.
(5) Next, we run the two-sided truncated regression (the left-hand side at 0 and the right-hand side at 1) of $\hat{\theta}_i^{bc}$ against z_i to generate the estimation coefficient $\hat{\hat{\beta}}$ and an estimation for variance parameter $\hat{\hat{\sigma}}$ through maximum likelihood.
(6) Further, we repeat steps 6a–6c for $B2$ times ($B2 = 2000$) to generate the bootstrap estimation sets ($\hat{\hat{\beta}}^b$, $\hat{\hat{\sigma}}^b$), where $b = 1, \ldots, B2$.
   (6a) For each DMU $i = 1, \ldots, N$, we randomly select a dummy error term $\tilde{\tilde{\varepsilon}}_i$ from the two-sided truncated normal distribution of $(0, \hat{\hat{\sigma}})$, with the left-hand side at $-z_i\hat{\hat{\beta}}$ and right-hand side at $1 - z_i\hat{\hat{\beta}}$.

(6b) We then calculate a dummy efficiency score $\tilde{\tilde{\theta}}_i$ as $\tilde{\tilde{\theta}}_i = z_i\hat{\hat{\beta}} + \tilde{\tilde{\varepsilon}}_i$ for each DMU $i = 1, \ldots, N$.

(6c) We run the two-sided truncated regression (the left-hand side at 0 and right-hand side at 1) of $\tilde{\tilde{\theta}}_i$ against z_i to generate a bootstrap estimation $\hat{\hat{\beta}}^b$ and $\hat{\hat{\sigma}}^b$ through the maximum likelihood method.

(7) We then calculate the confidence intervals and standard errors for $\hat{\hat{\beta}}$ and $\hat{\hat{\sigma}}$ from the bootstrap distribution $\hat{\hat{\beta}}^b$ and $\hat{\hat{\sigma}}^b$.

We run the entire procedure using the STATA I/C 16 software based on Badunenko & Tauchmann's (2019) contribution.

4. Results and Discussion

4.1. *Indonesian EPFs' technical efficiency, 2011–2017*

Table 4 presents the average of EPFs' efficiency scores during 2011–2017 under two approaches (operating and investment management models).

This table indicates that Indonesian EPFs exhibit relatively low efficiency levels for 2011–2017. The average operating efficiency is 48.2%, while investment efficiency is only 49%. The operating and investment efficiency levels become lower after correcting for bias

Table 4. Average EPFs' efficiency scores, 2011–2017.

	A: Operating management			**B: Investment management**		
Year	**Original scores**	**Bias-corrected**	**Biases**	**Original scores**	**Bias-corrected**	**Biases**
2011	0.459	0.306	0.154	0.442	0.382	0.060
2012	0.468	0.336	0.132	0.452	0.383	0.069
2013	0.491	0.360	0.131	0.541	0.404	0.137
2014	0.472	0.355	0.117	0.512	0.423	0.090
2015	0.497	0.378	0.120	0.410	0.232	0.178
2016	0.488	0.367	0.121	0.534	0.427	0.107
2017	0.496	0.358	0.137	0.536	0.445	0.091
Mean	0.482	0.351	0.130	0.490	0.385	0.105

(only 35.2% and 38.5%, respectively). Indonesian EPFs' investment efficiency levels are slightly higher (49% and 38.5%) than their operating efficiency (48.2% and 35.2%).

Nevertheless, Figure 1 suggests that investment performance fluctuates more than operating efficiency. A possible explanation for these results is that EPFs' management can control operating performance better than their investment performance, which is highly affected by external factors, such as market conditions and government regulations.

Figure 1 indicates that 2015 saw the lowest investment performance (41% and 23.2%). High stock price fluctuations due to rising US interest rates and a slowing Chinese economy may have contributed to Indonesian EPFs' low investment income in 2015. Besides, in 2015, the Indonesian government enacted several new regulations concerning risk management (Financial Services Authority Regulation No. 1/2015) and Indonesian pension funds' investment portfolio allocation (Financial Services Authority Regulation No. 3/2015). These new regulations likely affected pension funds' investment decisions and, eventually, their investment returns.

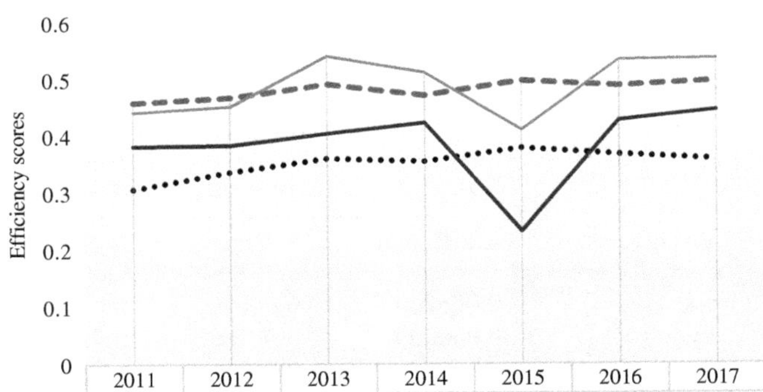

	2011	2012	2013	2014	2015	2016	2017
A: Operational Efficiency (Original scores)	0.459	0.468	0.491	0.472	0.497	0.488	0.496
A: Operational Efficiency (Bias-corrected)	0.306	0.336	0.36	0.355	0.378	0.367	0.358
B: Investment Efficiency (Original scores)	0.442	0.452	0.541	0.512	0.41	0.534	0.536
B: Investment Efficiency (Bias-corrected)	0.382	0.383	0.404	0.423	0.232	0.427	0.445

Figure 1. Trend of Indonesian EPFs' efficiency during 2011–2017.

4.2. The determinants of EPFs' technical efficiency

Table 5 presents the results of regressing pension fund-specific characteristics and macroeconomic conditions against EPFs' technical efficiency for 2011–2017. As can be seen in Table 5, size strongly and positively affects operating and investment technical efficiency, empirically supporting our hypotheses that greater asset values are beneficial for EPFs. EPFs with greater assets are more efficient. Large-scale pension funds can exploit economies of scale in their operating and investment activities.

In other words, Indonesian EPFs exhibit economies of scale in their operating and investment management. Our results are inconsistent with those of Hasanudin *et al.* (2017) and Putri *et al.* (2020), who do not find the effect of size on Indonesian EPFs' performance. However, our findings are consistent with prior studies on large pension funds' economies of scale in their operating and investment activities (Bikker & Dreu, 2009; Bikker, 2017). Large pension funds can distribute their operating costs to their fixed assets (Cummings, 2015) and enjoy greater bargaining power and flexibility (Dyck & Pomorsky, 2011), facilitating economies of scale in operating and investment activities.

Further, our regression results suggest a strong and negative effect (at the 1% level) of pension plan type (defined benefit vs. defined contribution) on operating technical efficiency but not on investment

Table 5. The determinants of Indonesian EPFs' technical efficiency.

	Model A (Operational)		**Model B (Investment)**	
Variables	**Coefficient**	**SE**	**Coefficient**	**SE**
Size	0.065200***	0.00526	0.040800***	0.00977
Type	−0.066600***	0.02180	0.026500	0.04140
Ownership	0.045100**	0.02210	0.229000***	0.04210
BIrate	0.004700	0.00961	−0.051300***	0.01800
GDP	−0.004760	0.00730	−0.006570	0.01380
INFrate	0.000799	0.00419	0.022900***	0.00805
sigma	0.131000***	0.00631	0.216000***	0.01420
Constant	−1.366000***	0.16000	−0.590000**	0.29600
Observations	273	273	275	275

Note: SE: standard errors; ***, **, denote significance level at 1% and 5% levels.

technical efficiency. The results imply that defined-benefit pension funds exhibit lower operating technical efficiency than their defined-contribution counterparts, as predicted by our hypothesis. Nevertheless, defined-benefit pension funds do not exhibit higher investment technical efficiency than defined-contribution pension funds. In other words, different pension program types are only relevant in explaining operating efficiency but not investment efficiency.

The results support prior studies, such as those by Bateman & Mitchell (2004) and Bikker & Dreu (2009), who observe that defined-benefit pension funds exhibit lower operating efficiency than defined-contribution pension funds due to higher actuarial and other administrative costs. However, the results are inconsistent with those reported by Jang & Wu (2021), who argue that defined-benefit pension funds invest more efficiently because of stricter monitoring and their management's better collective financial skills than those of defined-contribution pension funds. A likely explanation for our results is that Indonesian EPFs, both defined-benefit and defined-contribution ones, have a board of trustees whose members are selected by their founders or participants. Furthermore, management makes investment decisions based on the founders and supervisors' guidance, thus mitigating agency problems. Meanwhile, both pension funds apparently have management with similar knowledge and skill levels, resulting in similar investment efficiency for both pension program types.

Table 5 also demonstrates that the ownership variable (SOE vs. non-SOE) strongly and positively affects operating and investment technical efficiency. Thus, SOE-sponsored EPFs are more efficient than privately sponsored EPFs. The results are inconsistent with those of Pasqualeto *et al.* (2014), who find that Brazilian privately sponsored pension funds are more efficient than SOE-sponsored ones. It is worth noting that Indonesian SOE-sponsored EPFs are typically sponsored by large banks with sufficient skilled personnel and better access to technology. Conversely, privately sponsored EPFs are typically organized by small firms or foundations that manage their investments conventionally by their in-house part-time managers. In this respect, SOE-sponsored EPFs enjoy two advantages: their relatively larger size and the availability of experts. These factors likely cause SOE-sponsored EPFs to exhibit better performance than privately sponsored ones.

We recognize macroeconomic conditions with average BI interest rate changes and GDP growth. Our regression results in Table 5 reveal that BI interest rates negatively affect EPFs' investment technical efficiency but not operating efficiency. Changes in BI interest rates are more associated with EPFs' investment activities than operating ones. Antolin *et al.* (2011) suggest that changes in interest rates likely affect pension funds' investment efficiency in various scenarios, depending on pension program types and asset structures. In this respect, changes in BI interest rates during 2011–2017 negatively affected Indonesian EPFs. Our results are similar to those presented by Wanjiku (2014), who observes a negative correlation between changes in interest rates and Kenyan pension funds' investment returns. Changes in central bank interest rates are closely associated with general macroeconomic conditions. Unstable economic conditions tend to negatively affect firm performance (Najafi *et al.*, 2021).

GDP growth does not affect operating and investment technical efficiency. Nevertheless, the regression coefficients of GDP growth are negative for both operating and investment technical efficiency. Our results are in line with those of Bohl *et al.* (2011) and Najafi *et al.* (2021), who find that less conducive macroeconomic conditions negatively affect pension funds' efficiency. In the Indonesian context, 2011–2017 marked a period of economic recovery following the 2008 financial crisis. The statistical data indicate that the average GDP growth in 2011–2017 was lower than in the prior period (Badan Pusat Statistik Indonesia, 2019). The condition likely contributes to pension funds' lower operating and investment technical efficiency.

We included the annual inflation rate as a control variable (INFrate). It exhibits no significant effect on operational management but a significant positive effect on investment efficiency.

5. Conclusions

This study empirically analyzed the determinants of Indonesian EPFs' technical efficiency. We used the semi-parametric two-stage DEA analysis with the double bootstrapping method to test the effects of pension fund-specific characteristics (i.e., size, pension plan type, and sponsorship/ownership) and macroeconomic factors (i.e., average BI interest rates and GDP growth) on EPFs' operating and investment technical efficiency.

Data from 40 EPFs operational during 2011–2017 reveal that size (i.e., total asset value) strongly and positively affects operating and investment technical efficiency. Consistent with the economies of scale concept, larger pension funds are more efficient. Larger EPFs can exploit their economies of scale for their operating and investment activities. Based on plan type, defined-benefit plans exhibit lower operating technical efficiency than defined-contribution plans. However, these two pension plans do not exhibit different investment technical efficiency levels. In other words, pension plan type, as an efficiency determinant, is only relevant for operating activities but not for investment ones, likely because the regulatory framework for Indonesian EPF management has minimized agency problems that largely explain investment efficiency differences between defined-benefit and defined-contribution plans (Jang & Wu, 2021). Consequently, investment performance differences between Indonesian defined-benefit and defined-contribution EPFs are negligible. Further, SOE-sponsored EPFs exhibit higher operating and investment technical efficiency than privately sponsored EPFs, likely because SOE-sponsored EPFs are larger and employ more skilled managerial personnel. Besides, this study also highlights that changes in BI rates during 2011–2017 negatively affect investment technical efficiency, while GDP growth does not significantly affect EPFs' investment and operating technical efficiency.

In sum, this study concludes that size, pension plan type, sponsor/ownership type, and changes in BI interest rates crucially determine Indonesian EPFs' technical efficiency, either negatively or positively.

Our empirical evidence offers several academic and managerial contributions. First, this study extends the literature on the determinants of EPFs' technical efficiency by including several variables that are understudied in the Indonesian pension fund literature, such as pension program type, sponsor/ownership type, BI interest rates, and GDP growth. Second, our reliance on the double bootstrapping DEA method empirically demonstrates that pension fund-specific characteristics, such as size, pension plan type, and sponsor/ownership type, and macroeconomic factors, such as BI interest rates, influence EPFs' technical efficiency positively or negatively.

Besides, this study has several practical implications. First, this study empirically documents that size is a major determinant of operating efficiency. Hence, EPFs need to enlarge their size, such as through mergers or consolidation. Second, this study shows that defined-benefit EPFs exhibit lower operating performance than their defined-contribution counterparts due to higher operating costs and similar investment performance. In other words, defined-contribution EPFs enjoy lower operating costs (as suggested by various scholars) and exhibit similar investment performance to their defined-benefit counterparts. The findings practically imply that EPFs' management or potential participants should consider defined-contribution EPFs as they are more beneficial for their pension plans. Third, privately sponsored EPFs need to benchmark their performance against SOE-sponsored EPFs because SOE-sponsored EPFs exhibit better operating and investment performance.

This study is subject to several caveats. First, this study does not investigate Indonesian EPFs' optimal scale points. Second, this study does not precisely inform but merely suggests that skills/knowledge and regulation crucially determine EPFs' technical efficiency. Further analysis is needed to investigate the effect of these factors on EPFs' efficiency. Third, this study cannot inform whether the negative impacts of BI interest rates and GDP growth are solely attributable to declining BI interest rates and lower GDP growth during the observation period or whether they are also related to EPFs' asset and portfolio structures, as suggested by prior studies. Future studies that investigate our caveats in detail are necessary to expand our knowledge of this topic.

References

Abdin, Z., Prabantarikso, R., Fahmy, E., & Farhan, A. (2022). Analysis of the efficiency of insurance companies in Indonesia. *Decision Science Letters*, 11(2), 105–112.

Akwimbi, W. (2020). Effect of corporate governance, investment strategy and macroeconomic factors on financial performance of pension schemes in Kenya. *SSRN Electronic Journal*. http://dx.doi.org/10.2139/ssrn.3586384.

Alserda, G.A.G., Bikker, J.A., & van der Lecq, F.S.G. (2018). X-efficiency and economies of scale in pension fund administration and investment.

Applied Economics, 50(48), 5164–5188. https://doi.org/10.1080/00036846.2018.1486011.

Ambachtsheer, K., Cappele, R., & Lum, H. (2008). The pension governance deficit: Still with us. *Rotman International Journal of Pension Management*, 1(1), 14–22.

Andonov, A. & Mao, M.Q. (2019). Financial sophistication and conflicts of interest: Evidence from 401(k) investment menus (November 22, 2019). http://dx.doi.org/10.2139/ssrn.3485462.

Andonov, A., Bauer, R.M.M.J., & Cremers, M.K.J. (2017). Pension fund asset allocation and liability discount rates. *Review of Financial Studies*, 30(8), 2555–2595.

Antolin, P., Schich, S., & Yermo, J. (2011). The economic impact of protracted low interest rates on pension funds and insurance companies. *OECD Journal: Financial Market Trends*, 2011(1), 237–256.

Aslan, R.G. (2022). Agency problems in the private pension system of Turkey: Pension sector employee perspectives. *Qualitative Research in Financial Markets* (ahead-of-print). https://doi.org/10.1108/QRFM-07-2021-0120.

Badan Pusat Statistik Indonesia (2018). *Proyeksi Penduduk Indonesia 2015–2045 Hasil SUPAS 2015 (Edisi Revisi)*. Jakarta: BPS-Statistics Indonesia.

Badunenko, O. & Tauchmann, H. (2019). Simar and Wilson two-stage efficiency analysis for Stata. *The Stata Journal*, 19(4), 950–988. DOI: 10.1177/1536867X19893640.

Bánhidi, Z. & Dobos, I. (2023). A data envelopment analysis model for ranking digital development in the countries of the European Union without explicit inputs and common weights analysis. *Decision Analytics Journal*, 6, 100167. https://doi.org/10.1016/j.dajour.2023.100167.

Banker, R.D. & Natarajan, R. (2008). Evaluating contextual variables affecting productivity using data envelopment analysis. *Operations Research*, 56(1), 48–58. DOI: 10.1287/opre.1070.0460.

Barrientos, A. & Boussofiane, A. (2005). How efficient are pension fund managers in Chile? *Revista de Economia Contemporanea*, 9(2), 289–311.

Basso, A. & Funari, S. (2017). The role of fund size in the performance of mutual funds assessed with DEA models. *The European Journal of Finance*, 23(6), 457–473.

Bateman, H. & Mitchell, O.S. (2004). New evidence on pension plan design and administrative expenses: The Australian experience. *Journal of Pension Economics and Finance*, 3(1), 63–76. DOI: 10.1017/S1474747204001465.

Bauer, R.M.M.J., Cremers, M.K.J., & Frehen, R.G.P. (2010). Pension fund performance and costs: Small is beautiful. (April 30, 2010). https://dx.doi.org/10.2139/ssrn.965388.

Bikker, J.A. (2017). Is there an optimal pension fund size? A scale-economy analysis of administrative and investment costs. *The Journal of Risk and Insurance*, 84(2), 739–769. DOI: 10.1111/jori.12103.

Bikker, J.A. & Dreu, J.D. (2009). Operating costs of pension funds: The impact of scale, governance, and plan design. *Journal of Pension Economics and Finance*, 8(1), 63–89. DOI: 10.1017/S1474747207002995.

Bilbao-Terol, A., Arenas-Parra, M., & Bilbao-Terol, C. (2021). Measuring the overall efficiency of SRI and conventional mutual funds by a diversification-consistent DEA model. *International Transactions in Operating Research.* 0(2021), 1–33. DOI: 10.1111/itor.12974.

Bodie, Z., Marcus, A.J., & Merton, R.C. (1988). Defined benefit versus defined contribution pension plans: What are the real trade-offs? In: Z. Bodie, J.B. Shoven & D.A. Wise (Eds.), *Pensions in the US Economy* (pp. 139–162). http://www.nber.org/chapters/c6047.

Bohl, M.T., Lischewski, J., & Voronkova, S. (2011). Pension funds' performance in strongly regulated industries in central Europe: Evidence from Poland and Hungary. *Emerging Markets Finance and Trade*, 47(3), 80–94. DOI: 10.2753/REE1540-496X4704S305.

Boubaker, S., Le, T.D.Q., & Ngo, T. (2022a). Managing bank performance under COVID-19: A novel inverse DEA efficiency approach. *International Transactions in Operating Research*, 30(2023), 2436–2452. https://doi.org/10.1111/itor.13132.

Boubaker, S., Do, D.T., Hammami, H., & Ly, K.C. (2022b). The role of bank affiliation in bank efficiency: A fuzzy multi-objective data envelopment analysis approach. *Annals of Operations Research*, 311, 611–639. DOI: 10.1007/s10479-020-03817-z.

Boubaker, S., Gounopoulos, D., Nguyen, D.K., & Paltalidis, N. (2018). Reprint of: Assessing the effects of unconventional monetary policy and low interest rates on pension fund risk incentives. *Journal of Banking and Finance*, 92, 340–357. https://doi.org/10.1016/j.jbankfin.2018.03.003.

Bui, Y.H., Sarath, D., & Ahmed, A.D. (2016). Efficiency of Australian superannuation funds: A comparative assessment. *Journal of Economic Studies*, 43(6), 1022–1038. http://dx.doi.org/10.1108/JES-05-2015-0088.

Casey, B.H. & Yiallouros, P. (2013). The slow growth and sudden demise of supplementary pension provision in Cyprus. *Cyprus Economic Policy Review*, 7(2), 25–51.

Charnes, A., Cooper, W.W., & Rhodes, E. (1978). Measuring the efficiency of decision-making units. *European Journal of Operating Research*, 2, 429–444.

Chen, J., Hong, H.G., Huang, M., & Kubik, J.D. (2004). Does fund size Erode mutual fund performance? The role of liquidity and organization. *American Economic Review*, 94(5), 1276–1302.

Chybalski, F. (2015). The multidimensional efficiency of pension system: Definition and measurement in cross-country studies. *Social Indicators Research.* DOI: 10.1007/s11205-015-1017-3.

Clark, G.L. & Urwin, R. (2008). Best-practice pension fund governance. *Journal of Asset Management*, 9(1), 2–21. DOI: 10.1057/jam.2008.1.

Clark, G.L., Luzardi, A., & Mitchell, O.S. (2015). Financial knowledge and 401(k) investment performance: A case study. *Journal of Pension Economics and Finance*, 16(03), 324–347. https://doi.org/10.1017/S 1474747215000384.

Coelli, T., Rao, P.D.D., O'Donnell, C., & Battese, G.E. (2005). *An Introduction to Efficiency and Productivity Analysis* (2nd edn.). New York: Springer Science + Business Media, Inc.

Cook, W.D., Zhu, J., Bi, G., & Yang, F. (2010). Network DEA: Additive efficiency decomposition. *European Journal of Operational Research*, 207(2), 1122–1129.

Cummings, J.R. (2015). Effect of fund size on the performance of Australian superannuation funds. *Accounting and Finance.* DOI: 10.1111/acfi.12110.

Defung, F., Salim, R., & Bloch, H. (2016). Has regulatory reform had any impact on bank efficiency in Indonesia? A two-stage analysis. *Applied Economics*, 1–15. DOI: 10.1080/00036846.2016.1170934.

Demirtaş, Y.E. & Keçeci, N.F. (2020). The efficiency of private pension companies using dynamic data envelopment analysis. *Quantitative Finance and Economics*, 4(2), 204–219.

Dyck, A. & Pomorsky, L. (2011 June 1). Is bigger better? Size and performance in pension plan management [Working Paper]. DOI: 10.2139/ssrn.1690724.

Emrouznejad, A., & Yang, G.-L. (2018). A survey and analysis of the first 40 years of scholarly literature in DEA: 1978–2016. *Socio-Economic Planning Sciences*, 61, 4–8.

Färe, R. & Grosskopf, S. (2000). Network DEA. *Socio-Economic Planning Sciences*, 34, 35–49.

Fleuriet, V. & Lubochinsky, C. (2005 June). Interest rate risk management by life insurance companies and pension funds. *Financial Stability Review*, 2005(6), 95–111.

Galagedera, D.U.A. (2017). Modeling superannuation fund management functions as a two-stage process for overall and stage-level performance appraisal. *Applied Economics*. DOI: 10.1080/00036846.20171400649.

Galagedera, D.U.A. & Watson, J. (2015). Benchmarking superannuation funds based on relative performance. *Applied Economics*, 47(28), 2959–2973. DOI: 10.1080/00036846.2015.1011315.

Giannetti, M. & Laeven, L. (2009). Pension reform, ownership structure, and corporate governance: Evidence from a natural experiment. *The Review of Financial Studies*, 22(10). DOI: 10.1093/rfs/hhn091.

Gorter, J. & Bikker, J. (2011). Investment risk taking by institutional investors. *Applied Economics*, 45(33), 4629–4640. DOI: 10.1080/00036846.2013.795282.

Hammami, H., Ngo, T., Tripe, D., & Vo, D.-T. (2020). Ranking with a Euclidean common set of weights in data envelopment analysis: With application to the Eurozone banking sector. *Annals of Operations Research*, 311, 675–694 (2022). DOI: 10.1007/s10479-020-03759-6.

Hasanudin, Wahyudi, S. & Pangestuti, I.R.D. (2017). Managing the pension fund to improve portfolio performance: An empirical study on EPFs in Indonesia. *International Journal of Civil Engineering and Technology*, 8(8), 714–723.

Henriques, I.C., Sobreiro, V.A., Kimura, H., & Mariano, E.B. (2020). Two-stage DEA in banks: Terminological controversies and future directions. *Expert Systems with Applications*, 161, 113632. https://doi.org/10.1016/j.eswa.2020.113632.

Jang, D. & Wu, Y. (2021). Size and investment performance: Defined benefit vs. Defined contribution pension plans. (21 August 2021). http://dx.doi.org/10.2139/ssrn.3697711.

Jensen, T.K., Johnson, R.R., & McNamara, M.J. (2019). Funding conditions and insurance stock returns: Do insurance stocks really benefit from rising interest rate regimes? *Risk Management and Insurance Review*, 22(4), 367–391. DOI: 10.1111/rmir.12133.

Kadarisman & Wahyuni, S. (2010). *Manajemen Dana Pensiun Indonesia* [*The Management of Indonesian Pension Funds*]. Jakarta: PT. Mediantara Semesta.

Kao, C. & Hwang, S.-N. (2008). Efficiency decomposition in two-stage data envelopment analysis: An application to non-life insurance companies in Taiwan. *European Journal of Operational Research*, 185(1), 418–429. DOI: 10.1016/j.ejor.2006.11.041.

Kowalewski, O. (2012). Corporate governance and pension fund performance. *Contemporary Economics*, 6(1), 14–44.

Mazreku, I., Morina, F., & Curraj, E. (2020). Evaluation of the financial performance of pension funds. Empirical evidence: Kosovo, Albania and North Macedonia. *European Journal of Sustainable Development*, 9(1), 161. DOI: 10.14207/ejsd.2020.v9n1p161.

Najafi, H., Damankeshideh, M., Mohammadi, T., Shojaee, M., & Daghighiasli, A. (2021). Fluctuations in macroeconomic variables on the funding of pension funds in Iran (Combined data pattern method with different frequency (MIDAS)). *International Journal of Nonlinear Analysis and Applications*. DOI: 10.22075/ijnaa.2021.23252.2589.

Nanda, Y.F. & Atahau, A.D.R. (2020). An empirical study on pension funds' portfolio and investment performance: The effect of pension funds' size. *Journal Manajemen dan Kewirausahaan*, 22(2), 115–121.

Ngo, T. & Tsui, K.W.H. (2021). Estimating the confidence intervals for DEA efficiency scores of Asia-Pacific airlines. *Operational Research*, 1–24. https://doi.org/10.1007/s12351-021-00667-w.

Njuguna, A.G. & Arnolds, C. (2012). Determinants of pension fund efficiency in Kenya: An exploratory study. *African and Asian Studies*, 11(2012), 182–218.

Pasqualeto, J.C.M., Mangoni, L.A., Da Silva, M.H., & Teixeira, B.D.M. (2014). A influência dos patrocinadores nas despesas administrativas dos fundos de pensão brasileiros. In paper presented at the *Congresso internacional de administração*, Ponta Grossa. http://www.admpg.com.br/2014/down.php?id=949&q=1.

Putri, L., Sakti, I.M., & Atahau, A.D.R. (2020). Does ownership moderate the effects of size on pension funds' efficiency and investment performance? *Journal Keuangan dan Perbankan*, 24(3), 253–266. http://jurnal.unmer.ac.id/index.php/jkdp.

Revuelta, I., Santos-Arteaga, F.J., Montagud-Marrahi, E., Ventura-Aguiar, P., Di Caprio, D., Cofan, F., Cucchiari, D., Torregrosa, V., Piñeiro, G.J., Esforzado, N. (2021). A hybrid data envelopment analysis — Artificial neural network prediction model for COVID-19 severity in transplant recipients. *Artificial Intelligence Review*, 54(6), 4653–4684.

Sathye, M. (2011). The impact of the financial crisis on the efficiency of superannuation funds: Evidence from Australia. *Journal of Law and Financial Management*, 10(2), 16–27.

Seran, P., Sucahyo, U.S., Atahau, A.D.R., & Supramono, S. (2022). Investigating the efficiency of Indonesian employee pension funds. *Romanian Economic Journal*, 25(83), 74–87.

Seran, P., Sucahyo, U.S., Atahau, A.D.R., & Supramono, S. (2023). The efficiency of Indonesian pension funds: A two-stage additive network DEA approach. *International Journal of Financial Studies*, 11(1), 28.

Shaban, M. & James, G.A. (2018). The effects of ownership change on bank performance and risk exposure: Evidence from Indonesia. *Journal of Banking & Finance*, 88, 483–497. https://doi.org/10.1016/j.jbankfin.2017.02.002.

Simar, L. & Wilson, P.W. (2007). Estimation and inference in two-stage, semi-parametric models of production processes. *Journal of Econometrics*, 136(1), 31–64. https://doi.org/10.1016/j.jeconom.2005.07.009.

Sunaryo, S., Santoni, A., Endri, E., & Harahap, M.N. (2020). Determinants of capital adequacy ratio for pension funds: A case study in Indonesia. *International Journal of Financial Research*, 11(4), 203–213. https://doi.org/10.5430/ijfr.v11n4p203.

Thaker, K., Charles, V., Pant, A., & Gherman, T. (2022). A DEA and random forest regression approach to studying bank efficiency and corporate governance. *Journal of the Operational Research Society*, 73(6), 1258–1277.

Tone, K. & Tsutsui, M. (2010). Dynamic DEA: A slacks-based measure approach. *Omega*, 38(3–4), 145–156.

United Nation Population Division. (2019). World population prospect 2019. https://population.un.org/wpp/.

Vidal-García, J., Vidal, M., Boubaker, S., & Hassan, M. (2018). The efficiency of mutual funds. *Annals of Operations Research*, 267(1), 555–584.

Wanjiku, E. (2014). The effect of macroeconomic variables on portfolio returns of the pension industry in Kenya (Master Thesis). University of Nairobi. http://erepository.uonbi.ac.ke/handle/11295/75724.

Wei, Q., Zhang, J., & Zhang, X. (2000). An inverse DEA model for inputs/outputs estimate. *European Journal of Operational Research*, 121(1), 151–163.

Xue, Z. & Ma, Z. (2022). Performance evaluation of China's basic pension insurance based on a three-stage superefficient SBM-DEA model. *Discrete Dynamics in Nature and Society*, 2022. 1–10. DOI: 10.1155/2022/2429927.

Zaman, M.S. & Bhandari, A.K. (2022). Stressed assets, off-balance sheet business activities and performance of Indian banking sector: A DEA double bootstrap approach. *Studies in Economics and Finance*, 39(4), 572–592.

https://doi.org/10.1142/9781800615786_0012

Chapter 12

Efficiency Analysis of the Banking Sector in Pakistan: An Application of the Malmquist Index Approach

Farah Naz*, **Alishba Naseer**†, **and Tooba Lutfullah**‡

Kinnaird College for Women University, Pakistan

*farah.naz@kinnaird.edu.pk

†alishba.6511@gmail.com

‡tooba.lutfullah@gmail.com

Abstract

The purpose of this chapter is to analyze the efficiency of the banking sector in Pakistan in order to provide sufficient evidence for the improvement of the banks. This chapter determines the efficiency levels of 25 sample Pakistani banks over the period of 2010–2021 in order to investigate the factors which affect the effectiveness of the banks. Data envelopment analysis (DEA) and the Malmquist index approach have been used to determine the efficiency of the banking sector, and an overall regression analysis provides the results of the model. This determines the factors affecting the efficiency levels of the bank and how significant an impact they have on the banking sector. Empirical results have shown that internal and external determinants have a significant impact on the efficiency levels of the banks. However, some variables did not have a significant impact on the total efficiency of the banks due to specific factors. This research contributes to the most important scope of efficiency in the banking sector in Pakistan following the financial crisis. Moreover, the research will help in analyzing the factors that affect the efficiency levels of the banks, particularly in Pakistan.

Keywords: Banking sector, efficiency, Malmquist index approach, data envelopment analysis, Pakistani banks, economic growth, financial institutions, internal determinants, external determinants, banking performance.

1. Introduction

Financial Institutions are a source of economic development, with their main aim to reduce costs in a financial system of a country. This can be done with the help of financial contracts, markets, and intermediaries that constantly focus on the process of reducing the costs of information asymmetry and transactions (Ibrahim & Vo, 2021). Thus, the financial sector provides efficient allocation of resources, which can be helpful in economic advancement, and provides new opportunities. Developing economies, such as Pakistan, are required to advance their financial sector in order to bring about changes, as financial development is a key source in boosting economic development (Zhu *et al.*, 2021). According to the State Bank of Pakistan (SBP), the banking sector has shown healthy growth and signs of development in the private sector until June 30, 2017 (Arifeen, 2018). With the rise of globalization and competition around the world, banks are becoming the most important component of financial institutions in order to implement sound financial management practices (Bayiley, 2022). Thereby, the banking sector is an essential component of the financial system of a country: a perfect economy is directly related to the well-being of the banking sector of a country.

The performances of financial institutions can be analyzed through the efficiency and stability of the banking sector, which conceptualize efficiency by calculating the difference between the inputs and outputs that can be observed alongside ideal values (Haralayya & Aithal, 2021). Efficiency scores serve as indicators to evaluate the overall performance of the banking industry, determining the efficiency of banks. These scores play a crucial role in measuring the effects of government regulations and policies on the economic performance of the country (Nhan *et al.*, 2021). The financial sector is of utmost importance, primarily due to the essential role played by the banking sector as an intermediary between borrowers and creditors. This intermediary function significantly contributes to the smooth facilitation of economic activities within the financial sector.

Consequently, the banking sector holds prime significance for owners, depositors, policymakers, and potential investors, as banks actively implement the government's monetary policy (Sadalia *et al.*, 2018).

A bank's efficiency is considered an indicator of the efficiency of the financial system of the respective country. An inefficient and unproductive banking system may cause severe consequences for the overall economic performance of a country and may ultimately lead to financial crisis with increased default tendencies. The valuation of risk and risk management also become more challenging due to the increasing rate of uncertainties and their associated risks. Whereas, the process of managing risks and securing the stability of banks is necessary for the economic stability of a country, as financial instability of banks is directly proportional to the financial condition of a country (Sangeetha, 2020). The banking stability is measured by the default risk of banks because it is defined as the chances that a bank can default to fulfill financial commitments. Usually, a bank defaults only when it falls short of resources which are necessary or required to fulfill its financial obligations. But whenever a bank defaults, not only do its customers and shareholders suffer financially, but the bank itself goes through destabilization and also bears a huge loss of competition, which ultimately results in a banking and financial crisis. Moreover, macroeconomic stability can only be achieved through an efficient and well-organized working system of financial institutions. Therefore, the objective of determining the efficiency and profitability of the banks is a topic worthy of much consideration.

Measuring the efficiency and profitability of the banking sector is necessary in order to determine the strength of the banks as a main support to the country affected by the financial crisis of 2008. This financial crisis cautioned policymakers to not be rigid but rather confident enough to face risk — efficiency, profitability, and risks are closely related and have a huge impact (Soedarmono *et al.*, 2013). Measuring the efficiency of banks has been carried out in many countries. However, this study uses the value-added approach for the selection of variables to evaluate the efficiency of the banks. Moreover, this research is performed to fill the research gap of measuring profitability in Pakistan.

The aim of this research is to determine the efficiency and profitability of the banks of Pakistan and examine their progress in the given time period. Therefore, this chapter aims to contribute to the literature on measuring the efficiency and profitability of banks, which can provide a practical vision for the banking system and clarify avenues for more efficient economic growth. This chapter contributes to the two most important scopes of the banking sector, namely the efficiency and profitability of the banks, in Pakistan since the financial crisis to the present day by evaluating the profitability, scale efficiency, technical efficiency, and overall efficiency in the context of Pakistan's banking sector. For the task of evaluating efficiency, data envelopment analysis (DEA) is used, which is a comparatively new data-oriented approach used for evaluating efficiency and performance by converting multiple inputs into outputs. This approach has been widely used in recent years in a huge variety of contexts and countries in studies of the banking industry, along with ratio analysis in order to check the uniformity of results (Dar *et al.*, 2021b). In Section 2 of the chapter, a thorough literature review is presented, referencing previous studies that are pertinent to the main objective of the current study. This section provides valuable insights and contextual background for the research being conducted. Section 3 provides details for the relevant methodology, while Section 4 provides a detailed analysis of the results of this research. Last but not least, Section 5 provides the conclusion to this chapter.

2. Literature Review

The banking sector is considered to be the backbone of the financial structure of any country; therefore, it is necessary to examine the overall progress and developments made in this sector. For this, it is crucial to know the process of evaluating bank performance. The presence of a robust and resilient financial sector is indispensable for promoting economic growth in every country. The banking sector of Pakistan comprises five types of banks: commercial banks, Islamic banks, foreign banks, specialized banks, and microfinance banks. The central bank of Pakistan, SBP, is responsible for regulations and monitoring (Zhu *et al.*, 2021).

For the development of the condition of commercial banks, considerable effort has been made on research and improvement throughout the world for the better efficiency of the banking sector. The overall performance of the banking industry improved due to technological improvements, deregulation, and some financial reforms and modernization. The management of banks usually has a major focus on the procedure of getting a favorable and satisfactory return on investments, considering the intensity of the risk exposure the bank may face. In the new era of technological advancements, both conventional banking and online banking play a vital role in growing the banking sector of a country (Sadeghi & Farokhian, 2011). The process of checking the profitability of banks is also necessary because the major party responsible for providing funds for development or any other purpose has always been a bank.

The banking industry has faced some very radical changes in the past decade. The main reasons for such considerable changes are, most importantly, changes in customer needs and demands. Second, the process of globalization signified a huge step in declining the barriers to entering the banking sector. Finally, the tightening of laws and technological advancements played a vital role (Otaviya & Rani, 2020). Due to all this, the banking sector of Pakistan faced tough competition and pressure, both internally and externally. Efficiency measurement can be used in order to improve the overall efficiency of procedures. The liberalization of prices, changes at the social, political, and economic levels, improvements in the circulation of better-quality services and goods, globalization, and deregulation affect the development of the banking industry (Abdul-Majid *et al.*, 2011). Therefore, some major reforms were taken by the banks of developing countries, including Pakistan, for the purpose of making their banking institutions more efficient, sound, and capable of boosting economic growth (Estrada *et al.*, 2010).

The relationship between the efficiency of commercial banks and risk was analyzed by determining the values of non-performing loans and financial capital. It was found that financial capital had a sizeable impact on the values and efficiency of banks (Alandejani, 2022). For the evaluation of the profitability factor, it is stated that private banks make more profit as compared to banks owned by the governments of each respective country (Iannotta *et al.*, 2007). This study conducted an investigation into the risk and quality dimensions of banks, revealing that both private and public banks outperformed. Moreover, in the context of competition within the banking industry, it was observed that higher market power emerged as a critical factor affecting a bank's risk-taking capacity (Bayiley, 2022; Soedarmono *et al.*, 2013). Therefore, in order to achieve higher levels of efficiency and effectively navigate financial crises, banks must prioritize the enhancement of service quality provided to their customers.

The banks which are well capitalized usually face a comparatively lower bankruptcy cost than others, and this factor benefits the profitability of the bank. In the year 2002, a study was conducted in European countries, seeking to evaluate the determinants which are responsible for affecting the profitability and interest of banks, where it was stated that the rate of unemployment and the inflation rate are somehow relevant to the profitability of the banks (Abreu & Mendes, 2002). There was initially a decline in the efficiency of the Singapore banking industry, followed by a dramatic and rapid positive change in later years (Sufian, 2009). In Taiwan, the study was done for the period 1994–2004 on 44 banks to evaluate their technical and allocative efficiency using the DEA method, with the intermediate approach chosen for the selection of input and output variables. This study found that a decline in the efficiency of the banks was experienced during the financial crisis; hence, the banks in Taiwan showed a poorer performance during that phase. The main cause of such a case of decline and depreciation were bad debts and a worse rate of return (Ho & Zhu, 2004).

> $\boldsymbol{H}_{\boldsymbol{\alpha}}$ = The profitability of banks is affected by external determinants.

Few empirical studies have focused on the efficiency of banks in the South Asian region and have examined the overall efficiency of the banking sector in Pakistan using simplistic and basic models, majorly focusing on the period before the financial crisis (Chowdhury & Haron, 2021). Burki & Niazi (2010) investigated the significant effect of financial reforms on local and foreign banks in Pakistan for the period 1991–2000. Soedarmono *et al.* (2013) analyzed the influence of competition between various banks and the financial stability in Pakistan on the efficiency of the banks. A study conducted by Kolapo *et al.* (2012) in Nigeria evaluated about 13 commercial banks through the application of DEA, where only 3–4 banks were considered efficient. Furthermore, a study was conducted by Ahmad *et al.* (2009) in

Pakistan for the period 1990–2005. According to the study, about 20 Pakistani domestic banks were chosen, which were divided into three phases. The authors concluded that the banks of Pakistan showed a huge improvement following all the financial reforms.

DEA is one of the most common and popular techniques or methods applied in various studies in order to measure the efficiency of a company or bank (Boubaker *et al.*, 2022a; Zhu, 2020). Regarding the countries which are new entrants and recognized as new members of the European Union, only a few papers were found that investigated using the DEA approach. The diversity and variety of the various models differ because of the choices and variety of the combinations of the input and output variables DEA offers (Bahadji & Cheikh, 2021). Ariff & Can (2009) investigated the cost and profit efficiency of 28 Chinese banks using DEA and concluded that more efficient banks have a higher profitability rate.

Mercan *et al.* (2003) investigated the banking performance of Turkey for the period 1989–1999 in order to measure the performance in terms of the size of the bank and ownership. It was determined that government-owned banks were inefficient, displaying worse performance when compared to foreign or even private banks. In 2014, further research was done on 26 commercial banks in Bangladesh to evaluate their profitability for the years 2008–2011. Through that paper, it became clear that there was a direct relation between the size of the bank, capital adequacy, and profitability (Abdullah *et al.*, 2014). Basri *et al.* (2018) analyzed the importance of Malaysian banks performing efficiently in order to remain competitive, which can be determined using the DEA approach of highlighting government initiatives.

One of the points of interest for this chapter is to measure the efficiency of the banking sector in Pakistan with the use of a very popular non-frontier approach named data envelopment analysis (DEA). It is one of the most widely used techniques for the measurement of the banking sector through the relative measure of banks within a sector (Dar *et al.*, 2021a). A large amount of research has used this technique and agreed that DEA is the most suitable for evaluation purposes in the banking industry (Boubaker *et al.*, 2022b; Emrouznejad, 2017; Le, 2018; Rusydiana & Rahmawati, 2021; Vidal-García *et al.*, 2018).

3. Methodology

For the purpose of this research, the sample data consists of the years 2010–2021, which are taken directly from the annual financial reports of 25 Pakistani banks currently operational in the market. This was the time period in which the Islamic banks were emerging in Pakistan and providing competition to the country's commercial banks, as well as determining the dynamic forces of efficiency and profitability after the financial crisis.

The objective of this study was to examine all of the commercial banks in Pakistan; however, unfortunately, because of either the unavailability of data or a few limitations in them, the sample size of this study was reduced. DEA and the Malmquist index approach were simultaneously employed, encompassing all the input and output variables of diverse decision-making units (DMUs) in the calculation and analysis procedure. This approach facilitated a comprehensive comparison of individual units within the sample to assess their relative performance. This study is based on two objectives: evaluating the efficiency of the banks in Pakistan and determining the factors that affect their profitability.

3.1. *Efficiency analysis*

Evaluating the efficiency of banks requires determining the relationship between input and output variables. The results can be used as the performance standard and can also assist or help managers in knowing and understanding the performance of internal resources and determining ways for maximizing efficiency (Ngo & Tsui, 2022). The estimation of productive efficiency in this study is based on the utilization of relative efficiency measurements. For the purpose of evaluating the efficiency of the banks in Pakistan, we need to authenticate the selection of variables for the DEA model used and applied in the Pakistani banking sector through the value-added approach. For processing all the sampled data, the DEAP computer program was used (see Table 1).

3.1.1. *Malmquist productivity indexes*

There are two accounting practices which are popularly used for the approximation of the growth of total factor productivity: the parametric approach (SFA) and the non-parametric approach

Table 1. Variables measurement for DEA analysis.

	Measured by	**Sources**
Input variables		
Fixed assets	Total no. of long-term, tangible assets	Titko *et al.* (2014)
Interest expense	Total non-operating expenses	
Labor	Total no. of full-time employees	Favero & Papi (1995)
Non-performing loans	Total loans/advances that are in default or are close to being in default	Ataullah *et al.* (2004)
Total assets	Total amount of assets owned by the bank	Dietsch & Lozano-Vivas (2000)
Output variables		
Deposits	Total amount of deposits	Titko *et al.* (2014)
Non-interest income	Total income derived from the deposit, transaction fees, NSF fees, annual fees, etc.	Karray & Eddine Chichti (2013)
Investment	Total amount of investments of the banks	Favero & Papi (1995)
Liquid assets	Total amount of the assets that are readily convertible into cash	Akmal & Saleem (2008)

(the Malmquist index). In this study, the Malmquist index was used. It is likely possible to approximate the Malmquist index by applying the DEA methodology. It computes productivity growth, technical efficiency, and changes in terms of geometric means. It then calculates the relative efficiency of the sample, whereas there is no limitation for the frontier.

The equation for the Malmquist index is as follows:

$$M(x_l^{t+1}, y_l^{t+1}, x_l^t, y_l^t) = \sqrt{\left(\frac{D^t(x_l^{t+1}, y_l^{t+1})D^{t+1}(x_l^{t+1}, y_l^{t+1})}{D^t(x_l^t, y_l^t)D^{t+1}(x_l^t, y_l^t)}\right)}.$$

In this equation, x_l^t and y_l^t indicate the input and output vectors, respectively, and l denotes the unit used in time period t and $t+1$. This equation is used for measuring the productivity of l unit at (x_l^{t+1}, y_l^{t+1}) relative to (x_l^t, y_l^t).

3.2. *Efficiency performance*

This chapter is mainly based on research which is quantitative in nature and therefore uses secondary panel data. For the second objective of this chapter — testing the efficiency performance of the banking sector in Pakistan — variables which are considered include the total factor productivity of banks (TFPB) as the dependent variable, reflecting the ability of the management to utilize the finances of the banks and the ways in which and resources with which the profits are generated. Whereas, the internal determinants include the natural log of total assets, LNTA, which is used as a substitute for the size; loan loss provisions/total loans (LLPTL), included as a substitute of credit risk; non-interest income/total assets (NIITA); non-interest expense/total assets (NIETA) for providing the operating costs of the banks; and equity/total assets (EQTA). These are internal factors which are used to examine the relationship between a bank's profitability and its capitalization. The external factors include the natural log of gross domestic products (LNGDP), inflation (INF), and stock market capitalization (StockMktC) in order to measure the relationship between three factors: bank efficiency, economic conditions, and market conditions (see Table 2).

Table 2. Variables measurement for regression analysis.

	Measurement
Dependent variable	
Total Factor Productivity (TFPB)	Efficiency variable generated through the efficiency analysis of input and output variables mentioned in DEA
Independent variables	
Internal determinants	
LNTA	Natural log of total assets
LLPTL	Loan loss provisions divided by total loans
NIITA	Non-interest income divided by total assets
NIETA	Non-interest expense divided by total assets
EQTA	Equity divided by total assets
External determinants	
LNGDP	Natural log of gross domestic product
INF	Annual inflation rate
StockMktC	Stock market capitalization % of GDP

3.3. *Model specification*

To determine the relationship between the internal and external determinants considered in this study and the profitability of the banks, a linear regression model is implemented in the following way:

$$y_{jt} = \delta_j + \alpha' X_{ijt} + \beta' X_{et} + \varepsilon_{jt}.$$

In this equation, y_{jt} denotes the return on assets (TFPB) generated after efficiency analysis of the variables mentioned above and used as a dependent variable for our research for the observation of an individual bank j in a specific year; j denotes an individual bank or financial institution; t denotes year; X_i refers to the internal determinants of the financial institution; X_e refers to the external determinants of the financial institution; ε_{jt} represents a random variable which is normally distributed. When Equation (1) is extended, it gets the following form:

$$\begin{aligned}\text{TFPB}_{jt} = {} & \delta_0 + \alpha_1\text{LLP/TL}_{jt} + \alpha_2\text{NII/TA}_{jt} + \alpha_3\text{NIE/TA}_{jt} \\ & + \alpha_4\text{LNTA}_{jt} + \alpha_5\text{EQ/TA}_{jt} + \beta_1\text{LNGDP}_t \\ & + \beta_2\text{INF}_t + \beta_3\text{StockMktC}_t + \varepsilon_{jt}.\end{aligned}$$

4. Results & Discussion

4.1. *Descriptive statistics*

The descriptive statistics results displayed in Table 3 show the values of the mean, standard deviation, minimum, and maximum of the sample data. The mean values show the averages of the data, which are used to describe the sample with a single value, and that value denotes the center of the sampled data. If the mean values appear to be unusual, then it means that there is a presence of outliers in the data, as the mean is affected by them.

The column containing the values of standard deviations shows the dispersion of the data and the spread of the data from the mean. If the value of the standard deviation is higher, then it means that there is a greater spread in the sampled data. And if the value is close to zero, it means that the values are closer to the mean.

Table 3. Descriptive statistics.

Variable	Obs	Mean	Std. dev.	Min	Max
TFPB	300	3.346	8.132	0.000	63.381
LNTA	300	8.433	0.652	6.724	9.610
LLPTL	300	0.145	0.640	−0.212	10.032
NIITA	300	0.011	0.010	−0.002	0.141
NIETA	300	0.030	0.017	0.003	0.120
EQTA	300	0.097	0.116	−0.495	0.785
GDP	300	26.329	0.215	25.900	26.599
Inflation	300	7.858	4.640	0.400	19.600
StockMktCap	300	22.003	4.784	15.713	29.830

From Table 3, it can be seen that total factor productivity has the highest standard deviation among all the variables, whereas other variables have a standard deviation closer to 0, showing a lower deviation in the data, except stock market capitalization. This has a deviation of 4.784, the highest after total factor productivity.

The GDP saw a continued increase from 2010. The reason for the difference in stock market capitalization is that in 2008, an uncertain political environment was observed after the general elections. Inflation started mounting, and a steep decline was observed on the Karachi Stock Exchange. Due to all this, in 2009, the market bounced back, and this continued until 2011. Afterward, in 2014, the five major banks of Pakistan changed their positions and became stable, when the KSE 100 Index jumped.

4.2. *Correlation matrix*

A correlational matrix provides information regarding the degree of correlation among selected explanatory variables, which are used in the regression analysis in this study. The correlation matrix does not show a strong relationship, which suggests that the problem of multi-collinearity does not exist, as it is an issue which occurs when the relationship between the variables is above 0.8 (Kennedy, 2008). Another test consisting of the variance inflation factor (VIF) was performed to detect the problem of multi-collinearity, and it was seen that the value of VIF was below 10 for the variables.

From the correlation results in Table 4, it can be seen that total TFPB has a negative correlation with LLPTL, NIITA, NIETA,

Table 4. Correlation matrix results

Variables	(1)	(2)	(3)	(4)	(5)	(6)	(7)	(8)	(9)
(1) TFPB	1.000								
(2) LNTA	0.101	1.000							
(3) LLPTL	−0.039	−0.303	1.000						
(4) NIITA	−0.033	0.096	−0.033	1.000					
(5) NIETA	−0.056	−0.762	0.402	−0.078	1.000				
(6) EQTA	−0.023	−0.348	0.139	0.252	0.314	1.000			
(7) GDP	0.135	0.292	−0.154	−0.167	−0.141	−0.199	1.000		
(8) Inflation	0.060	−0.097	0.022	−0.030	0.053	0.027	−0.436	1.000	
(9) StockMktCap	−0.239	0.009	0.003	0.035	−0.001	0.059	0.268	−0.646	1.000

EQTA, and StockMktCap and is closer to zero, showing a weaker correlation. Nevertheless, the variables LNTA, GDP, and INF exhibit a positive correlation, although the correlation coefficient is below 0.8. This suggests that there is no significant concern regarding multicollinearity among the data. Therefore, from the results, we can see that there is no high correlation issue among the independent variables as well.

4.3. *Efficiency scores analysis*

Evaluating the efficiency of banks means determining the relationship between their input and output variables. The results can be used as performance standards and can also assist or help managers in knowing and understanding the performance of internal resources and determining ways to maximize efficiency. This study applies the relative efficiency to measure productive efficiency for the estimations. This approach to productive efficiency includes technical, overall, and allocative efficiency (Farrell, 1957). Technical efficiency is further divided into scale efficiency, congestion efficiency, and pure technical efficiency (Bjurek *et al.*, 1990).

The banks exhibiting total factor productivity change (TFPCH) to be 1 or above, meaning 100% productivity level or more, show that the particular bank has increased its productivity in the selected time period (Alandejani, 2022). Tables 4 and 5 consist of the values of technical efficiency change (EFFCH), technology change (TECHCH), pure technical efficiency change (PECH), and scale efficiency change (SECH), along with TFPCH for annual means in Table 4 and firm means in Table 5.

Table 5. Malmquist index summary of annual means.

Year	EFFCH	TECHCH	PECH	SECH	TFPCH
2011	1.253	0.577	1.101	1.138	0.723
2012	0.980	0.000	1.000	0.980	0.730
2013	1.030	0.000	1.038	0.992	0.740
2014	0.951	0.799	0.937	1.014	0.759
2015	0.723	0.000	0.893	0.809	0.640
2016	1.385	0.000	1.181	1.173	0.000
2017	1.047	0.000	1.018	1.029	0.000
2018	1.041	0.208	0.992	1.049	0.216
2019	0.821	0.901	0.862	0.953	0.740
2020	1.174	6.415	1.168	1.005	7.529
2021	0.891	2.923	0.903	0.987	2.605

Table 5 provides the annual means for the 25 sampled banks in this research, where the results indicate an increase in TFPCH during the years 2011–2014. However, a decreasing pattern can be seen until 2019. But later, the efficiencies of the banks have been increasing due to impact of COVID-19 that shift the banking operations to online services which ultimately increased the pressure on the banks to improve their banking services. The other efficiency scales show fluctuations in the results between each year. However, the values of the efficiencies are closest to 1.

Table 6 provides the mean efficiency scores of the banks for the period 2010–2021. It can be seen from the table that all banks have TFPCH of close to, equal to, or greater than 1, which shows higher efficiencies of the banks over the years. However, Bank Alfalah Ltd., Faysal Bank Ltd., and Habib Bank Ltd. show an average TFPCH of 0.7, which is the lowest among all the banks' values. This is because of their lowest TECHCH, which decreased the TFPCH average over the period.

From Table 6, it is very clear that the banks are performing efficiently over time in order to attract more customers. Efficiency analysis alone does not determine the value of a bank in the market, however. There are other factors too which provide detailed growth analyses of the banks. In this chapter, one of the analyses conducted pertains to profitability analysis, focusing on the examination of the relationship between efficiency and bank profitability.

Table 6. Malmquist index summary of bank means.

Banks	EFFCH	TECHCH	PECH	SECH	TFPCH
Al Baraka Bank Ltd.	1.107	0.936	1.086	1.019	1.036
Allied Bank Ltd.	0.970	1.048	0.948	1.023	1.016
Askari Bank Ltd.	1.100	0.876	1.023	1.075	0.963
Bank Alfalah Ltd.	0.896	0.882	0.892	1.004	0.790
Bank Islami Pak Ltd.	1.074	1.032	1.020	1.053	1.108
Dubai Islamic Bank Pak Ltd.	1.003	1.000	1.000	1.003	1.000
Faysal Bank Ltd.	0.974	0.000	1.001	0.973	0.752
First Women Bank Ltd.	1.175	1.224	1.147	1.024	1.439
Habib Bank Ltd.	0.932	0.000	0.953	0.978	0.737
Standard Chartered Bank Ltd.	1.000	0.986	1.000	1.000	0.986
Habib Metropolitan Bank Ltd.	1.021	1.052	0.999	1.022	1.074
JS Bank Ltd.	1.067	1.174	1.036	1.029	1.252
MCB Bank Ltd.	1.000	1.063	1.000	1.000	1.063
Meezan Bank Ltd.	1.000	2.004	1.000	1.000	2.004
National Bank of Pakistan	1.000	1.246	1.000	1.000	1.246
S.M.E Bank Ltd.	0.996	0.985	0.998	0.998	0.981
Silk Bank Ltd.	1.000	1.457	1.000	1.000	1.457
Soneri Bank Ltd.	1.000	1.030	1.000	1.000	1.030
Summit Bank Ltd.	1.000	1.240	1.000	1.000	1.240
The Bank of Khyber	0.999	1.000	1.000	0.999	1.000
The Bank of Punjab	1.000	0.913	1.000	1.000	0.913
United Bank Limited	0.996	0.997	1.000	0.996	0.993
Punjab Provincial Cooperative	1.000	1.261	1.000	1.000	1.261
Bank Al Habib	1.000	1.362	1.000	1.000	1.362
Samba Bank Ltd.	1.000	1.897	1.000	1.000	1.897

TFPCH is utilized as the dependent variable to assess the impact of efficiency on bank profitability.

4.4. *Regression results*

The results of DEA contain a collection of scores of efficiencies. These efficiency scores represent the current costs of the banks, which should be capable enough to produce the current output of the firm. The TFPB values are taken as dependent variables in order to analyze the efficiency of the banks and their performance in the stock market.

In this study, the generalized least squares (GLS) method is applied to a random effects (RE) model. The RE model is restricted by maximum likelihood estimation and does not require the dependent variable to be linearly transformed. Additionally, the RE model

Table 7. Regression results for the model of the study.

TFPB	Coef.	Std. err.	t-value	p-value	[95% conf]	Interval	Sig.
LNTA	0.896	1.141	0.79	0.432	−1.341	3.133	
LLPTL	−0.443	0.074	−5.98	0.000	−0.589	−0.297	***
NIITA	−11.649	47.677	−0.24	0.807	−105.094	81.796	
NIETA	0.659	0.332	1.98	0.049	0.003	1.314	**
EQTA	4.322	4.405	0.98	0.326	−4.312	12.957	
GDP	6.905	2.540	2.72	0.007	1.927	11.882	***
Inflation	−0.134	0.137	−0.98	0.329	−0.402	0.135	
StockMktCap	−0.580	0.124	−4.67	0.000	−0.823	−0.337	***
Constant	−172.596	65.337	−2.64	0.008	−300.654	−44.539	***

Mean dependent var	3.346	SD dependent var	8.132
Overall R-squared	0.107	Number of obs	300.000
Chi-square	35.041	Prob > chi^2	0.000
R-squared within	0.114	R-squared between	0.011

Note: $^{***}p < 0.01$, $^{**}p < 0.05$, $^{*}p < 0.1$.

assumes normality as a fundamental characteristic (Gardiner *et al.*, 2009). However, it has been processed due to the existence of the heteroscedasticity of the data, and GLS is best suited to cater to such types of data. Table 7 shows the variable coefficients of this study and their associated standard errors, which would help in analyzing the accuracy of the sample. The column of t-values helps in testing the hypothesis that the individual coefficient is dissimilar from the value of zero. The results have been analyzed at the 10%, 5%, and 1% levels of significance, where LLPTL, NIITA, GDP, and StockMktCap show significance at these levels.

Among the internal determinants, LLPTL and NIETA have a significant impact on the efficiency of the banks. On the other hand, GDP and StockMktCap have a significant impact on the efficiency of the banks. This shows that among all the variables considered, the significant variables have an effect on the efficiency of the banks for the period 2010–2021. The results depict that the economic and market conditions of Pakistan, except the inflation rate, have an insignificant impact on the efficiency of the banks, which is consistent with the results of Gul *et al.* (2011), showing significant results. The macroeconomic variables considered in this study are more likely to be affected by the efficiency of the banking sector due to exerting a stronger influence on financial stability rather than having an impact on the profitability, productivity, and efficiency of banks

(Klein & Weill, 2022). A study conducted in Pakistan evaluated that internal determinants had a strong impact on profitability irrespective of the external determinants (Bukhari & Qudous, 2012). Nevertheless, in this study, it was observed that out of the various internal determinants examined, only two (i.e., LLPTL and NIETA) demonstrated a statistically significant influence on both the efficiency and profitability of the firms. The annual rate of inflation in Pakistan has been seeing an increasing trend. However, the inflation decreased in 2011 but showed an increasing pattern after 2017 that increased the chances of risk and loan defaults. This impacted the rise in interest rates and reduction in money supply. Therefore, the inflation rate affected the profitability and efficiency of the banks in this way. Stock market capitalization has a significant impact on the profitability of the banks, according to the regression model. This is due to the stock market of Pakistan emerging into the world but being greatly affected during the pandemic crisis of COVID-19. The results depict that this has a significant impact on the efficiency and productivity of the banks. Moreover, the banks have been affected by the political instability in the country, creating high economic instability. Moreover, the PSX was unable to attract foreign investors because of political instability and the huge tension between civilians and the military.

One of the reasons for an improvement in the banking sector in Pakistan after the financial crisis may be the prudent and wise regulatory policies of SBP. Its policies regarding the number of branches to be increased helped in increasing technological growth, performances, operations, and the quality of services. It was found that all the sampled banks are enthusiastic enough to increase their investments and are expanding their activities. Another reason may be the energetic and spirited response of both foreign and local shareholders. Because of this, quality of services was achieved, which somehow promoted increased and healthy competition among the banks in Pakistan (Zhu *et al.*, 2021). The concern and anxiousness of the customers regarding the quality of services and their rights, requirements, and needs play a very important role in the progress of the banking sector (Klein & Weill, 2022). They expect and demand more because of the increased awareness related to the banking system. This forces and encourages the management to put in more efforts for the betterment of their banks (Thaker *et al.*, 2022).

5. Conclusion

The aim of the research is to analyze the relationship between multiple determinants in the context of performance along with efficiency and profitability for the sample taken from the banking sector in Pakistan over the 2010–2021 period. From the point of view of a bank, customer satisfaction is a factor of key importance. In order to achieve that, it is very important to determine the efficiency of banks and the factors which are responsible for affecting their profitability. Therefore, this study comprises the recent reports regarding the performances of the sampled banks in Pakistan and concludes that these banks are giving their best in both of these terms through their constant and untiring struggle.

On the basis of the results, it can be concluded that the efficiency of the banks in Pakistan considered for this study showed an overall increase throughout the period. The influencing factors included financial reforms, prudent and wise regulatory policies by the State Bank of Pakistan, expansion in size, increased awareness of the rights and needs of the customers, enhancements in structure, an energetic response from stakeholders, and a wide spread of the bank networks. The internal determinants of LLPTL and NIETA have a significant impact on profitability. Moreover, the external determinants GDP and StockMktCap show a significant impact on the profitability of banks. It is not necessary for the banks with better efficiency to be profitable, as profitability is mainly linked with higher returns obtained on the money the bank lends, the amount of interest they earn on the loans, the amount of interest they pay on the deposits which customers make, and the interest payments they get on debt securities.

Bank efficiency is concerned with the continuous efforts of the managers of the banks and of the personnel utilizing resources effectively, as well as the determined efforts they put in in order to control the costs of all aspects, including non-interest expenses, fixed costs, and salary expenses. Undoubtedly, some of the factors, such as the rate of illiteracy, poverty, unemployment, political instability, and terrorism, have affected Pakistan's economy significantly; however, as the overall efficiency of the sampled banks in Pakistan is increasing and the factors which affect their profitability are giving progressive results, a ray of hope can be predicted.

This research may be helpful for banks to analyze how efficiency increases can lead to higher growth of the firm using internal and external factors as determinants to address issues related to the profitability of the banks. The government can also take steps through SBP to ensure economic growth and the future of Pakistan. The higher the efficiency of a bank, the more satisfied the clients of that bank are. Thus, bank efficiency is key to profitability because the investors who want to take loans from the banks obtain them easily, while banks can only provide loans when they attain enough profits.

This study performs the analysis for the period 2010–2021, whereas future research can be done on efficiency analysis, comparing the pre-financial crisis and post-financial crisis situations in order to understand the difference the banking sector in Pakistan experienced under the financial crisis. Additionally, when examining profitability, it is essential to broaden the scope beyond quantitative data. Future research endeavors could enrich the analysis by incorporating qualitative measures, such as evaluating the banks' service quality, technological advancements, and provision of additional customer facilities, to provide a more comprehensive understanding of the factors influencing profitability. Other variables, such as exchange rates and taxation, can also be taken into consideration. Further research should consider comparing the determinants of profitability of the banks according to their sizes, small and large. These extensions would add some meaningful content to the current literature.

References

Abdullah, M.N., Parvez, K., & Ayreen, S. (2014). Bank specific, industry specific and macroeconomic determinants of commercial bank profitability: A case of Bangladesh. *World*, 4(3), 82–96.

Abdul-Majid, M., Saal, D.S., & Battisti, G. (2011). Efficiency and total factor productivity change of Malaysian commercial banks. *The Service Industries Journal*, 31(13), 2117–2143.

Abreu, M. & Mendes, V. (2002). Commercial bank interest margins and profitability: Evidence from EU countries. Porto Working paper series.

Ahmad, U., Farooq, S., & Jalil, H.H. (2009). Efficiency dynamics and financial reforms: Case study of Pakistani banks. *International Research Journal of Finance and Economics*, 25, 172–182.

Akmal, M. & Saleem, M. (2008). Technical efficiency of the banking sector in Pakistan. *SBP Research Bulletin*, 4(1), 61–80.

Alandejani, M. (2022). An overview of efficiency and profitability in Islamic banking: A comparative study between Islamic banking and conventional banking. *Social Management Research Journal*, 19(1), 209–234.

Arifeen, M. (2018). Banking sector performance in 2016 and 2017. Retrieved from *The Financial Daily International*. https://thefinancialdaily.com/banking-sector-performance-2016-2017/.

Ariff, M. & Can, L. (2009). IMF bank-restructuring efficiency outcomes: Evidence from East Asia. *Journal of Financial Services Research*, 35(2), 167–187.

Ataullah, A., Cockerill, T., & Le, H. (2004). Financial liberalization and bank efficiency: A comparative analysis of India and Pakistan. *Applied Economics*, 36(17), 1915–1924.

Bahadji, K. & Cheikh, S. (2021). Measurement of the productivity of Algerian Banks: Using DEA-based Malmquist productivity index approach, *Journal of Economic Integration*, 9(2), 607–620.

Basri, M.F., Muhamat, A.A., & Jaafar, M.N. (2018). The efficiency of Islamic banks in Malaysia: Based on DEA and Malmquist productivity index. *Journal of Emerging Economies Islamic Research*, 6(3), 1–14.

Bayiley, Y.T. (2022). Assessing bank performance using Malmquist productivity index approach and one-step system GMM dynamic panel data model. *Open Journal of Business Management*, 10(2), 798–821.

Bjurek, H., Hjalmarsson, L., & Forsund, F.R. (1990). Deterministic parametric and nonparametric estimation of efficiency in service production: A comparison. *Journal of Econometrics*, 46(1–2), 213–227.

Boubaker, S., Le, T.D.Q., & Ngo, T. (2022a). Managing bank performance under COVID-19: A novel inverse DEA efficiency approach. *International Transactions in Operational Research*, 30(2023), 2436–2452. https://doi.org/10.1111/itor.13132.

Boubaker, S., Do, D.T., Hammami, H., & Ly, K.C. (2022b). The role of bank affiliation in bank efficiency: A fuzzy multi-objective data envelopment analysis approach. *Annals of Operations Research*, 311, 611–639. https://doi.org/10.1007/s10479-020-03817-z.

Bukhari, S.A.J. & Qudous, R.A. (2012). Internal and external determinants of profitability of banks: Evidence from Pakistan. *Interdisciplinary journal of contemporary research in business*, *3*(9), 1037–1058.

Burki, A.A. & Niazi, G. (2010). Impact of financial reforms on efficiency of state-owned, private and foreign banks in Pakistan. *Applied Economics*, *42*(24), 3147–3160.

Chowdhury, M.A.M. & Haron, R. (2021). The efficiency of Islamic Banks in the Southeast Asia (SEA) region. *Future Business Journal*, 7(1), 1–16.

Dar, A.H., Mathur, S.K., & Mishra, S. (2021a). The efficiency of Indian Banks: A DEA, Malmquist and SFA analysis with bad output. *Journal of Quantitative Economics*, 19(4), 653–701.

Dar, Q.F., Ahn, Y.-H., & Dar, G.F. (2021b). The impact of international trade on central bank efficiency: An application of DEA and Tobit regression analysis. *Statistics, Optimization Information Computing*, 9(1), 223–240.

Dietsch, M. & Lozano-Vivas, A. (2000). How the environment determines banking efficiency: A comparison between French and Spanish industries. *Journal of Banking & Finance*, 24(6), 985–1004.

Emrouznejad, A.Y. (2017). A survey and analysis of the first 40 years of scholarly literature in DEA: 1978–2016. *Socio-Economic Planning Sciences*, 61, 4–8.

Estrada, G.B., Park, D., & Ramayandi, A. (2010). Financial development and economic growth in developing Asia. Asian Development Bank Economics Working Paper (233).

Farrell, M.J. (1957). The measurement of productive efficiency. *Journal of the Royal Statistical Society: Series A*, 120(3), 253–281.

Favero, C.A. & Papi, L. (1995). Technical efficiency and scale efficiency in the Italian banking sector: A non-parametric approach. *Applied Economics*, 27(4), 385–395.

Gardiner, J.C., Luo, Z., & Roman, L.A. (2009). Fixed effects, random effects and GEE: What are the differences? *Statistics in Medicine*, 28(2), 221–239.

Gul, S., Irshad, F., & Zaman, K. (2011). Factors affecting bank profitability in Pakistan. *Romanian Economic Journal*, 14(39), 61–87.

Haralayya, D. & Aithal, P. (2021). Factors determining the efficiency in Indian banking sector: An Tobit regression analysis. *International Journal of Scientific Development and Research*, 6(6), 1–6.

Ho, C.T. & Zhu, D.S. (2004). Performance measurement of Taiwan's commercial banks. *International Journal of Productivity Performance Management*, 53, 425–434.

Iannotta, G., Nocera, G., & Sironi, A. (2007). Ownership structure, risk and performance in the European banking industry. *Journal of Banking Finance*, 31(7), 2127–2149.

Ibrahim, M. & Vo, X.V. (2021). Exploring the relationships among innovation, financial sector development and environmental pollution in selected industrialized countries. *Journal of Environmental Management*, 284, 112057.

Karray, S.C. & eddine Chichti, J. (2013). Bank size and efficiency in developing countries: Intermediation approach versus value added approach and impact of non-traditional activities. *Asian Economic Financial Review*, 3(5), 593–613.

Kennedy, P. (2008). *A Guide to Econometrics.* Chichester: John Wiley & Sons.

Klein, P.-O. & Weill, L. (2022). Bank profitability and economic growth. *The Quarterly Review of Economics & Finance*, Elsevier, 84(C), 183–199.

Kolapo, T.F., Ayeni, R.K., & Oke, M.O. (2012). Credit risk and commercial banks performance in Nigeria: A panel model approach. *Australian Journal of Business Management Research*, 2(2), 31.

Le, T.D.Q. (2018). Bank risk, capitalisation and technical efficiency in the Vietnamese banking system. *Australasian Accounting, Business & Finance Journal*, 12(3), 41–61.

Mercan, M., Reisman, A., Yolalan, R., & Emel, A.B. (2003). The effect of scale and mode of ownership on the financial performance of the Turkish banking sector: Results of a DEA-based analysis. *Socio-Economic Planning Sciences*, 37(3), 185–202.

Ngo, T. & Tsui, K.W.H. (2022). Estimating the confidence intervals for DEA efficiency scores of Asia-Pacific airlines. *Operational Research*, 22(4), 3411–3434.

Nhan, D.T.T., Pho, K.-H., Anh, D.T.V., & McAleer, M. (2021). Evaluating the efficiency of Vietnam Banks using data envelopment analysis. *Annals of Financial Economics*, *16*(02), 2150010.

Otaviya, S.A. & Rani, L.N. (2020). Productivity and its determinants in Islamic banks: Evidence from Indonesia. *Journal of Islamic Monetary Economics Finance*, 6(1), 189–212.

Rusydiana, A.S. & Rahmawati, R. (2021). DEA on Islamic Banking: A bibliometric study and critical perspective. *Library Philosophy & Practice*, 1–27.

Sadalia, I., Kautsar, M.H., Irawati, N., & Muda, I. (2018). Analysis of the efficiency performance of Sharia and conventional banks using stochastic frontier analysis. *Banks & Bank Systems*, 13(2), 27–38.

Sadeghi, T. & Farokhian, S. (2011). The role of behavioral adoption theories in online banking services. *Middle-East Journal of Scientific Research*, 7(3), 374–380.

Sangeetha, R. (2020). How efficient are public sector banks in India? A non-parametric approach. *Banks Bank Systems*, 15(4), 108.

Soedarmono, W., Machrouh, F., & Tarazi, A. (2013). Bank competition, crisis and risk taking: Evidence from emerging markets in Asia. *Journal of International Financial Markets, Institutions & Money*, 23, 196–221.

Sufian, F. (2009). Determinants of bank efficiency during unstable macroeconomic environment: Empirical evidence from Malaysia. *Research in International Business & Finance*, 23(1), 54–77.

Thaker, K., Charles, V., Pant, A., & Gherman, T. (2022). A DEA and random forest regression approach to studying bank efficiency and corporate governance. *Journal of the Operational Research Society*, 73(6), 1258–1277.

Titko, J., Stankevičienė, J., & Lāce, N. (2014). Measuring bank efficiency: DEA application. *Technological Economic Development of Economy*, 20(4), 739–757.

Vidal-García, J., Vidal, M., Boubaker, S., & Hassan, M. (2018). The efficiency of mutual funds. *Annals of Operations Research*, 267, 555–584.

Zhu, J. (2020). DEA under big data: Data enabled analytics and network data envelopment analysis. *Annals of Operations Research*, 309, 1–23.

Zhu, N., Shah, W.U.H., Kamal, M.A., & Yasmeen, R. (2021). Efficiency and productivity analysis of Pakistan's banking industry: A DEA approach. *International Journal of Finance Economics*, 26(4), 6362–6374.

https://doi.org/10.1142/9781800615786_0013

Chapter 13

Antecedents of Efficiency: Empirical Evaluation of the Life Insurance Sector in India

Aparna Bhatia* **and Megha Mahendru†**

University School of Financial Studies,
Guru Nanak Dev University, Amritsar, India
**aparnamohindru@yahoo.co.in*
†mahendru.megha30@gmail.com

Abstract

The purpose of this chapter is to identify the attributes affecting the cost, revenue, and profit efficiency of life insurance companies in India from 2013–2014 to 2018–2019. A two-phase analysis is applied in the study. In the first phase, the cost, revenue, and profit efficiency scores of all the life insurance companies are calculated using the technique of data envelopment analysis. In the second phase, a panel tobit regression model is run to estimate the antecedents of efficiency. The results of the study emphasize that capital adequacy, asset quality, reinsurance and actuarial issues, management soundness, and liquidity have a positive relationship with cost, revenue, and profit efficiency. However, "earning and profitability" has a negative impact on all the efficiency scores, depicting that Indian life insurance companies are not getting much return from their investments, which is their major source of revenue. Low revenues do not seem to be sufficient to cover the cost of insurance and, consequently, generate low profits. In order to improve efficiency, insurers should focus on balancing the input–output mix, taking into consideration their prices. Also, modern virtual platforms should be adopted, which can lead to cost savings and higher profitability.

Keywords: Cost efficiency, revenue efficiency, profit efficiency, CARAMEL framework, capital adequacy, asset quality, reinsurance and actuarial issues, management soundness, earnings and profitability, liquidity, data envelopment analysis, life insurance companies, India.

1. Introduction

The life insurance business serves dual objectives for the masses. It is a contract of assurance wherein, on the one hand, if the policyholder passes away during the policy tenure, the insurance company will offer a lump sum amount to his/her nominee. On the other hand, in the case of the survival of the policyholder, the insurance company has to pay a sum assured on maturity along with a bonus. Thus, in both cases, the life insurance company is required to pay back the policy amount to the policyholder. Consequently, working with efficiency is of prime importance for a life insurance company, as against general insurers, for whom indemnification arises only in the case of the occurrence of an event. Hence, life insurance companies must gauge their efficiency and identify the factors that affect it. There are three main concepts of economic efficiencies for analyzing financial performance, namely cost, revenue, and profit efficiency. Cost efficiency reveals the scope for a company to cut down its costs if it operates efficiently. Cost efficiency measurement alone is not adequate, as revenue generation is also important. Revenue efficiency discloses the extent to which a company can maximize its revenue from the available resources. However, on the whole, if a company wants to perform completely efficiently in terms of profits, it needs to focus on both cost efficiency and revenue efficiency concurrently.

Prolific literature suggests that researchers have worked on several factors affecting the efficiency of insurance companies. Variables such as premium, reinsurance rates, size, solvency ratio, profitability, earnings, age, liquidity, market share, management expenses and commission ratio, conservation ratio, persistency ratio, and retention ratio (Diacon *et al.*, 2002; Ansah-Adu *et al.*, 2012; Kwadjo *et al.*, 2014; Ilyas & Rajasekaran, 2019; Sen, 2020) have been widely used to study the impact on the efficiency of insurance companies. But a notable observation suggests that these variables have been picked up randomly by the researchers, either as per the availability of data in their respective countries or according to the best of their knowledge. Convenience dominates the logic and rationale in the selection of variables. It is important to formalize the selection of variables in order to remove randomness.

The CARAMEL model is suggested by the World Bank and the International Monetary Fund to evaluate the financial soundness, stability, and performance of the insurance sector. It was developed considering Insurance Core Principles. It is capable of providing a myriad amount of information by using limited ratios (International Association of Insurance Supervisors, 2000). A CARAMEL analysis allows for quantitative monitoring and analysis of the financial performance of insurance companies. Under the CARAMEL model, C stands for "Capital Adequacy," A for "Asset Quality," R for "Reinsurance and Actuarial Issues," M for "Management Soundness," E for "Earnings and Profitability," and L for "Liquidity" of the company. The current study selects the variables affecting the efficiency of life insurance companies on the basis of the CARAMEL framework and evaluates the impact of these variables on the cost, revenue, and profit efficiency of life insurance companies operating in India. The current study, based on the CARAMEL framework, would in fact be the first of its kind to thread the drivers of efficiency into a structured framework. Hence, the study would contribute significantly to the extant literature on efficiency in several ways. First, there would be no discretion or randomness in choosing company-specific variables affecting efficiency scores. Second, the study would be comprehensive enough to identify variables affecting all three concepts of efficiency in the life insurance sector, namely cost, revenue, and profit efficiency. Third, the study is based on all 24 life insurance companies operating in India, thus providing a holistic analysis of performance of the life insurance business in India. The study could contribute to the insurance sector as well. India has the 10th largest life insurance market in terms of insurance penetration. It recorded a total premium of INR 5.73 trillion in the financial year 2020–2021 (Invest India, 2022). Overall, insurance penetration in India increased from 2.7% in 2001 to 3.7% in 2020. Hence, in order to maintain the escalated trend, the current study presents some concrete factors affecting the efficiency of the insurance business in India and thus serves as a beacon for managers in the insurance business. The study involves a two-phase analysis: First, it calculates the cost, revenue, and profit efficiency scores of all the life insurance companies in India using the data envelopment analysis (DEA) technique, and then, in the second phase, it brings out the drivers affecting these efficiency scores with the help of tobit panel regression.

The chapter is divided into six sections. The introductory Section 1 brings out the importance of the life insurance business in India. Section 2 reviews the existing literature on factors affecting the efficiency of life insurance companies. Section 3 discusses the database and research methodology. Section 4 outlines hypotheses development. Section 5 focuses on results and discussion. Finally, Section 6 delineates the implications, brings out the conclusions, and opens the road for future scope of the study.

2. Review of Literature

The literature on the measurement of "efficiency performance" in the insurance sector is inexhaustible. A review of existing literature suggests that voluminous research papers are based on evaluating the efficiency of insurance companies (Kaffash *et al.*, 2019). Researchers around the globe have been wrestling to identify the drivers of efficiency in the insurance business. Meador *et al.* (1997) suggested that diversification across multiple insurance and investment product lines resulted in greater X-efficiency among the US life insurance industry during 1990–1995. Donni & Fecher (1997) concluded that reinsurance rates and market shares had a favorable impact on technical efficiency, while the life share of the total insurance business and the mean size of companies had a negative impact across 15 OECD countries over the period 1983–1991. Cummins (1999) examined the relationship between mergers and acquisitions, cost, revenue efficiency, and scale economies on 750 US life insurers per year between 1988 and 1995. The results of the study found that mergers and acquisitions in the life insurance industry had a beneficial effect on efficiency. Diacon (2001) examined the impact of explanatory environmental variables on the value-based technical efficiency of 431 general insurers in six European countries for the year 1999. The increase in the concentration of an insurer's assets and a higher proportion of premiums in life and pension insurance businesses were associated with higher efficiency. Diacon *et al.* (2002) exhibited that technical and scale efficiencies were strongly associated with insurer size. Solvency ratio and profitability lead to higher technical efficiency, whereas there was a modest relationship between liquidity and efficiency in 450 companies from 15 European countries for the period 1996–1999. Karim & Jhantasana (2005) showed that age had no significant relationship with inefficiency. Size was negatively related to inefficiency. The study also showed that efficient firms had a higher return on equity and on assets in Thailand's life insurance firms during the period 1997–2002. Luhnen (2009) found a positive relationship between size and efficiency in the German property-liability insurance industry from 1995 to 2006. The relationship between premium growth and efficiency was negative. Kader *et al.* (2010) tested the influence of corporate characteristics on the cost efficiency of 26 Takaful insurance companies operating in 10 Islamic countries over the years 2004–2006. The results showed that non-executive directors contributed negatively to cost efficiency. Board size, firm size, and product specialization had a positive effect on the cost efficiency of Takaful insurers. In contrast, the regulatory environment was not statistically significant in terms of improving cost efficiency. Hsu & Petchsakulwong (2010) also verified the impact of corporate governance variables on the technical, allocative, cost, and revenue efficiency performance of public non-life insurance companies in Thailand from 2000 to 2007. The results indicated that the independence of the board and firm size had a positive impact on the efficiency of the performance of Thai non-life insurance companies. Audit committee size, diligence, divergence between voting rights and cash flow rights, board tenure, board age, and board ownership had a negative impact on efficiency performance. However, the relationships between the efficiency of the performance by insurers and the size of their boards, the proportion of financial expertise on an audit committee, and the compensation to the board were unclear. Barros *et al.* (2010) analyzed the effects of deregulation on the efficiency of the Greek insurance industry by employing a two-stage procedure of DEA for the years 1994–2003. The determinants of efficiency demonstrated that life and big insurance companies, on the basis of invested assets, contributed negatively to efficiency. The variable log of market share had a positive impact on efficiency, while the log of equity/invested assets had a negative influence on efficiency. Sinha (2012) suggested that cost efficiency scores were negatively related to the ratio of unit-linked to non-unit-linked products but positively related to the ratio of premiums collected through traditional channels as compared to non-traditional channels in life insurance companies operating in India for the period from 2005–2006 to 2009–2010. Ansah-Adu *et al.* (2012) depicted that market share, firm size, and the ratio of equity to total invested assets were important determinants of an insurance firm's cost efficiency in 30 life and non-life insurance companies in Ghana over the period of 2006–2008. Kwadjo *et al.* (2014) examined the efficiency determinants of the cost efficiency of 16 non-life insurance companies and 14 life insurance companies in Ghana over the period 2006–2008. The results showed that market share, firm size, and the ratio of equity to total invested assets were important determinants of an insurance firm's efficiency. Mathur & Paul (2014) showed that technical reserve and net earnings were positively related to overall technical efficiency in 20 non-life insurance companies operating in India during the financial year 2012–2013. Management expenses had a negative impact on the overall technical efficiency of the insurers. Ilyas & Rajasekaran (2019) found that size and reinsurance had a statistically significant negative relationship with efficiency for Indian non-life insurance companies. It also showed a positive and statistically significant relationship with age and efficiency. Sen (2020) showed that non-discretionary factors, such as age, premium growth, and size of the insurers, along with regulations pertaining to solvency, affected the cost efficiency of Indian life insurers over the period from 2005–2006 to 2015–2016. The variables such as expense ratio, commission ratio, conservation ratio, persistency ratio, and retention ratio established a significant relationship with the different efficiency estimates. The entry of a new insurer negatively affected the cost efficiency of an existing insurer, whereas foreign equity positively improved the cost efficiency. Tuffour *et al.* (2021) explored the determinants of cost and profit efficiency of 12 life insurance companies in Ghana for the period of 2013–2017. The study revealed that the life insurance companies were about 71.2% cost-efficient and 41.7% profit-efficient.

The fixed-effects panel regression results showed that profit and cost efficiency changes had a significant positive impact on firms' return on assets. Sinha (2022) examined the impact of several contextual variables on the profit efficiency scores of 15 general insurance companies for the period 2011–2012 to 2016–2017. The results of pooled ordinary least squares regression showed that solvency ratio had a significant impact on profit efficiency, whereas the impact of return on equity on profit efficiency was contingent based on the model selected.

Thus, the review of literature affirms that different studies choose the variables affecting the efficiency scores of insurance companies randomly. There is a need to bring specificity and precision in identifying variables affecting the efficiency of insurance businesses. Also, studies have not distinguished between life and non-life insurance businesses. The same factors cannot affect the efficiency of both types of insurance businesses, as both sectors do not have similar restrictions and regulations (Barros *et al.*, 2005). Hence, the current study overcomes both of these limitations. It considers only the life insurance companies, and also, the selection of variables is based on a standardized framework, the CARAMEL approach. The study would definitely add value to the existing literature and open up new insights for further research.

3. Database and Research Methodology

3.1. *Sample, time period and data source of the study*

The study is based on all 24 life insurance companies operating in the Indian insurance sector. The efficiency evaluation of the life insurance industry is done over the time period of six years, from 2013–2014 to 2018–2019, as it is the most recent time period in the current decade for which the data are available. The primary source of data for insurance companies is the website of the Insurance Regulatory Development Authority (IRDA), which provides company-wise policyholder accounts, shareholder accounts, and balance sheets. The handbook on Indian insurance statistics, annual reports of insurance companies, and the website of the Life Insurance Council have also been used.

3.2. *Data envelopment analysis (DEA)*

DEA is known as frontier analysis, which was first developed by Charnes *et al.* (1978). Subsequently, it gained the attention of researchers working in different areas, including banking, education, financial institutions, sports, and insurance. DEA is a popular tool for efficiency analysis in various fields, such as banking and finance (Tripe, 2010; Le, 2018; Boubaker *et al.*, 2022) and manufacturing (Becchetti & Sierra, 2003; Wang & Chin, 2010; Vu, 2016; Boubaker *et al.*, 2023). Kaffash *et al.* (2019) reviewed research papers that studied the efficiency of insurance firms and reported that different researchers applied diverse DEA models, such as Simar & Wilson (2007) for bootstrap-DEA, Dao *et al.* (2021) for multi-level DEA, or Ngo & Tsui (2021) for stochastic DEA, to evaluate the efficiency of insurance firms. DEA is a benchmarking technique which permits firms to measure efficiency based on inputs used and outputs produced by them. Some of the studies mentioned that DEA provides better and superior results than parametric approaches (Banker *et al.*, 1988, 1993). DEA applies a linear programming-based technique to assess an empirical production technology frontier. DEA has emerged as a foremost tool owing to its advantages and flexibility (Colwell & Davis, 1992). It calculates the efficiency score of a company by transforming inputs into outputs relative to its best peer group. The best-practice firm defines an envelope surface, which is referred to as an empirical production efficient frontier. Firms that lie on the envelope surface are deemed efficient in DEA, while those units that lie in the interior of the surface are termed inefficient firms. DEA was originally developed for measuring technical efficiency, pure technical efficiency, and scale efficiency, where the prices of inputs and outputs are not required. Afterward, DEA was modified to measure economic efficiency, i.e., cost efficiency, revenue efficiency, and profit efficiency, which requires different input–output combinations as well as their prices (Fried *et al.*, 2008). Panwar *et al.* (2022) stated that there has been a continual development in the theoretical concept of DEA, i.e., from classical models of DEA, it has extended to multi-stage and multi-level DEA models and then to stochastic, fuzzy, and hybrid DEA models. Two-stage DEA is used in this chapter to assess the cost, revenue, and profit efficiency scores of life insurance companies operating in India. Cost efficiency model is an input-oriented model as it minimizes inputs at a given level of output quantities and input prices. On the other hand, the revenue efficiency model is an output-oriented model that maximizes revenue for a given set of input quantities and output prices. Profit efficiency measures how close a company comes to producing maximum profit given the input, output quantities, as well as their prices.

3.3. *Input-output selection*

A critical issue in a non-parametric programming technique is its sensitivity to the selection and number of inputs and outputs, along with number of DMUs to be used, as they can affect the discriminating powers of DEA (Boussofiane *et al.*, 1991). The selection of inputs and outputs is an arbitrary process in DEA (Ariff & Can, 2008; Berger & Humphrey, 1997). Different approaches, namely the intermediate approach, the user cost approach, and the value-added approach, are used for selecting inputs and outputs. The insurance sector is a financial intermediary that borrows funds from shareholders (capital) and policyholders (annuities and insurance policies). It then invests this money in the capital market for transforming it into assets, basically for paying out insurance claims. For such an intermediary financial institution, an intermediation approach is most appropriate (Sinha, 2015; Shinde, 2012; Noronh & Shinde, 2012; Nandi, 2014). Eling & Luhnen (2010) and Kaffash *et al.* (2019) reviewed 95 and 132 articles, respectively, based on the efficiency of the insurance sector. The authors affirmed that labor, or the number of employees/specialists, and equity, or equity capital, are the most frequently used inputs and outputs in the available literature. Thus, after reviewing the literature, the intermediation approach is followed by using two inputs, i.e., the number of agents and the shareholder fund, and two outputs, namely investments and the net premium received along with their prices.

3.4. *Panel data tobit regression model*

In the first stage of analysis, cost, revenue, and profit efficiency scores are calculated, which can be further used in the second stage for checking the consistency of DEA results and for identifying the impact of the explanatory variables. However, the efficiency scores calculated from the first stage of analysis with the help of DEA fall between 0 and 1, thus making the variable a limited dependent variable. Due to the limited nature of the dependent variable, a censored tobit model is preferred (Waisiaturrahma *et al.*, 2020; Le *et al.*, 2022). The tobit model is applied due to the censored nature of the dependent variable as well as the extreme values of the independent variables, which deviate from a normal distribution and are highly skewed in nature. Moreover, the works by previous researchers, such as Donni & Fecher (1997), Diacon (2001), Diacon *et al.* (2002), Sinha (2012), Yakob *et al.* (2014), and Grmanová & Strunz (2017), also suggest that the tobit model is the best model, as it has the ability to handle equations with a restricted threshold. The current study is based on 144 observed values for a period of six years, thus suggesting that the panel tobit regression model is applicable. The fixed-effect tobit model cannot be used, as there is no command for a parametric conditional fixed-effects model that allows the fixed effect to be conditioned out of the likelihood (Stata Press Publications, 2009). Unconditional fixed-effect tobit models may be fit using the tobit model with an individual indicator; however, these estimates will be biased (Stata Press Publications, 2009).

Therefore, to estimate the factors affecting the revenue, cost, and profit efficiency scores, the random-effect panel tobit model is used. This model for the ith company can be specified as

$$y_{it}^* = \beta' x_{it} + u_{it}, \quad i = 1, 2, 3 \ldots, N, \quad t = 1, 2, 3, \ldots, T,$$
$$u_{it} = v_i + \varepsilon_{it}, \tag{1}$$

where, x_{it} is a vector of explanatory variables and the common error term u_{it} could be freely correlated over time. The error term u_{it} can be split into a time-invariant individual random effect (RE) v_i and a time-varying idiosyncratic random error ε_{it}. Thus, the random-effects panel tobit model can be considered in the following form:

$$y_{it}^* = \beta' x_{it} + v_i + \varepsilon_{it}. \tag{2}$$

The observed variables for the left-censored and right-censored observations are

$$y_{it} = \begin{cases} y_{it}^* & \text{If } y_{it}^* > 0 \\ 0 & \text{otherwise} \end{cases}, \tag{3}$$

$$\mathrm{y_{it}} = \begin{cases} y_{it}^* & \text{If } y_{it}^* < 1 \\ 0 & \text{otherwise} \end{cases}. \tag{4}$$

In this study, the second stage of analysis employs the random-effect panel tobit regression model to determine the effect of CARAMEL on various insurance-sector efficiency parameters.

4. The CARAMEL Framework and Hypotheses Development

This section introduces the CARAMEL variables and develops hypotheses for the same:

(1) **Capital adequacy:** Capital adequacy is a key parameter indicating the financial stability of an insurance company. The main purpose of capital is to absorb all risks that arise in the insurance business, such as insurance policy risk, market risk, investment risk, liquidity risk, solvency risk, operational risk, and legal risk. Thus, capital allows insurers to underwrite risks and assists them in promoting stability, safety, and soundness in their insurance business. Life insurers with adequate capital provide confidence of continual financial viability to its customers. Capital adequacy provides a cushion to meet the obligations of a policyholder, as a life insurance company is liable to pay the policy amount under both circumstances, i.e., either the death of the policyholder or the expiry of the time period. Adequate capital covers costs arising from contingent risks in the insurance business and becomes the backbone of earning greater revenues. Thus, insurers must maintain minimum capital so that they can be financially sound to make their payments on time. For the purpose of calculating the capital adequacy ratio of insurance companies, the ratio of **net premium to shareholder fund** is used in the current study. Net premium is an appropriate substitute for calculating the underwriting risk of insurance businesses, as a large volume of risk may cause a threat to the established capital position of a company. The foregoing arguments help us develop the following hypothesis:

H$_1$: Capital adequacy positively affects the cost, revenue, and profit efficiency of life insurance companies.

(2) **Asset quality:** Insurance companies must invest in quality assets. Life insurance companies have to follow the norms set for investments by the IRDA. For instance, the companies are required to invest at least 50% of their funds in government and other approved securities and 15% in infrastructure and housing. This is to confirm that assets match the long-term liabilities

of the policyholders. So, the quality of investment has a direct bearing on the efficiency of Indian life insurance companies. Prudent investment would avoid wastage of resources, thus affecting cost efficiency and leading to better revenue and profit efficiency. In order to measure the asset quality of Indian life insurance companies, the ratio of **total investments to total assets** is used. This ratio measures the percentage of total assets confined in investments, which indicates the extent of the deployment of assets in investments. Hence, with the preceding discussion, the following is hypothesized:

H_2: Asset quality positively affects the cost, revenue, and profit efficiency of life insurance companies.

(3) **Reinsurance and actuarial issues:** As per the regulations, insurance companies must have sufficient capital to pay all probable future claims. This, without doubt, provides protection to the policyholder but limits the amount of business of an insurer. However, the insurer can reduce its liability or responsibility by transferring a part of its liability to another insurer. Reinsurance is such an arrangement in which an insurance company insures itself with another insurance company. This helps it in dispersion of risk. Withholding excessive risks may result in the inability of insurance companies to settle liabilities under insurance contracts. Reinsurance can help insurance companies maintain a lower amount of capital and satisfy insurance authorities that the company is in good financial health and can pay its claims on time. Undoubtedly, reinsurance is a tool for optimizing insurance companies' financial position. It shares the burden of risk/cost and improves revenues and profits. In order to measure the reinsurance and actuarial issues, the **net premium to gross premium ratio** is used. This ratio is also known as the risk retention ratio, which depicts the risk retained by the company. This indicates the risk-bearing capacity of insurance companies. Since a high ratio indicates better risk retention, according to the theoretic concept, the following hypothesis is framed:

H_3: Reinsurance and actuarial issues positively affect the cost, revenue, and profit efficiency of life insurance companies.

(4) **Management soundness:** Management soundness is another vital parameter of the CARAMEL framework. Management

soundness shows the competence and ability of management to organize its resources for the stable and efficient performance of its insurance company. This parameter can be evaluated from the expertise of managers as to how well they are able to manage and minimize their expenses so that they can get the best returns. The expenses of an insurance company primarily include advertising, employee wages, and commissions paid to the sales force. Managing these expenses is one of the most vital requirements, as it indicates an insurance company's efficiency before settling claims on its policies and investment gains or losses. Thus, management soundness is analyzed using three ratios: **operating expenses to gross premium, management expenses to gross premium, and total commission to total premium.** These ratios show what proportion of revenue, i.e., the premium obtained by the insurance company, is used in incurring the related expenses. Sound management would control costs, leading to an increase in revenues and profits. Thus, a lower ratio here indicates better efficiency of the management in managing the operational, management, and commission expenses, which represents management soundness (Mathur & Paul, 2014; Sen, 2020). Thus, the preceding point of view assists in developing the following hypothesis:

H_4: Management soundness measured with the expense ratio negatively affects the cost, revenue, and profit efficiency of life insurance companies.

(5) **Earnings and profitability:** Earning quality refers to the profits earned by an insurance company. Insurance companies mainly have two sources of income: premiums and interests; and dividends and capital gains from various investments. Life insurance companies with higher profits are better able to absorb greater risk. Additionally, profits can be a long-term source of capital for companies. Low profitability may lead to insolvency. Good earnings absorb costs and lead to higher revenues and profits. In order to measure the earning quality of an Indian life insurance company, the ratio of **investment income to investment** is calculated. Investment income is the second major source of revenue for insurance companies. It shows the earnings of the companies through their investments. Hence, with the preceding discussion, the following is hypothesized:

H_5: Earnings and profitability positively affects the cost, revenue, and profit efficiency of life insurance companies.

(6) **Liquidity:** Liquidity refers to the ability of an Indian life insurance company to pay back its liabilities within time. It depicts the capability of a company to achieve a balance between sources and uses of funds. High values of liquidity indicate the ability of an insurer to respond to sudden cash requirements. Life insurance companies need to manage their liquidity properly, as they are required to settle the claims of insured through their short-term assets. They should seriously consider the risk assessment and risk management criteria by balancing their current assets and current liabilities. Optimum levels of liquidity undoubtedly help redeem costs and maintain revenues and profitability. Liquidity management is analyzed by using the current ratio, formulated as **current assets to current liabilities**. This ratio shows whether the company has an adequate amount of current assets to pay its current liabilities. So, taking into consideration the theoretical concept, the following hypothesis is framed:

H_6: Liquidity positively affects the cost, revenue, and profit efficiency of life insurance companies.

(7) **Size**: Size, as measured by the **log of total claims** settled by insurance companies, is taken as a control variable in the regression model. The claims settlement ratio of a life insurance company is a ratio used to measure the percentage of life insurance claims an insurance company has settled during a financial year against the number of claims received plus pending claims from the previous year. A higher claim settlement ratio shows that the insurance company has been successful in providing committed service to its customers. Timely settlement of claims saves on litigation costs, leading to improved revenues and profit efficiency. Thus, the following hypothesis is framed:

H_7: Size positively affects the cost, revenue, and profit efficiency of life insurance companies.

(8) **Age:** The age of an insurance company represents the number of years it has been operating in the insurance sector. With the increase in years of operations, experience and goodwill increase,

which increases efficiency and performance. More years of experience help save on costs and improve revenues and profits. Thus, the following is hypothesized:

H_8: Age has a positive effect on the cost, revenue, and profit efficiency of life insurance companies.

5. Results, Analysis, and Discussion

5.1. *Efficiency of Indian life insurers*

Cost efficiency (CE), revenue efficiency (RE), and profit efficiency (PE) scores for each life insurance company is evaluated using DEA. The overall mean, minimum, maximum, and standard deviation for all life insurance companies are presented in Table 1.

As depicted in Table 1, life insurance companies have mean values of cost, revenue, and profit efficiency at 0.541, 0.411, and 0.577, respectively, during the study time period. The minimum and maximum values of efficiency varies from 0.170 to 0.966 for cost efficiency, 0.097 to 1 for revenue efficiency, and 0.055 to 1 for profit efficiency for all the life insurance companies.

5.2. *Panel tobit regression*

The results of panel tobit regression, assessing the factors affecting the cost, revenue, and profit efficiency of life insurance companies, are presented in Table 2.

In Table 2, the results of panel tobit regression with cost efficiency scores as the dependent variable show that capital adequacy,

Table 1. Results of cost, revenue, and profit efficiency of life insurance companies.

	Mean	**Minimum**	**Maximum**	**Standard deviation**
Cost efficiency	0.541	0.170	0.966	0.233
Revenue efficiency	0.411	0.097	1.000	0.272
Profit efficiency	0.577	0.055	1.000	0.243

Note: The table depicts the mean, minimum, and maximum efficiency score obtained by life insurance companies.

Table 2. Results of panel tobit regression.

CARAMEL (Independent) variables		Definitions of variables	Cost efficiency		Revenue efficiency		Profit efficiency	
			Coef.	$P > z$	Coef.	$P > z$	Coef.	$P > z$
Capital adequacy		Net premium to shareholder fund	0.0038329	0.000	0.0064104	0.000	0.0014446	0.054
Asset quality		Total investment to total assets	0.7245420	0.000	0.9537106	0.001	−0.1400438	0.555
Reinsurance and actuarial issues		Net premium to gross premium	3.2947430	0.002	2.3891890	0.147	1.6146230	0.252
Management soundness		Operating expenses to gross premium	−0.1602447	0.171	0.0488344	0.791	−0.7178859	0.001
		Management expenses to gross premium	−0.7465977	0.005	−0.9421510	0.021	−0.5505833	0.192
		Total commission to total premium	−0.0366379	0.000	−0.0362259	0.000	−0.0848610	0.000
Earning and profitability		Investment income to investments	−0.7536307	0.000	−3.1430740	0.000	−1.0067460	0.011
Liquidity		Current assets to current liabilities and provisions	0.0153835	0.195	0.0480813	0.021	0.0422515	0.039
Control variables	Size	Log of total claims settled	0.0793318	0.003	0.1108917	0.007	0.0587593	0.033
	Age	No. of years since inception	−0.0327159	0.000	−0.0550380	0.000	−0.0105178	0.170
Constant		_cons	−3.0283160	0.005	−2.2197660	0.183	−0.3763862	0.778
Wald chi^2(10)		—	158.1200000	0.000	163.9600000	0.000	220.5600000	0.000
Log likelihood		—	104.9596400	—	36.1501720	—	23.6643810	—

Notes: This table shows the coefficient value and p-value of all CARAMEL variables with cost efficiency, revenue efficiency, and profit efficiency as dependent variables by applying panel tobit regression.

asset quality, and reinsurance and actuarial issues have positive relationships with cost efficiency, which are significant at the 1% level of significance. Management soundness, earnings, and profitability have negative relationships with cost efficiency, which are statistically significant at the 1% level of significance. Among the control variables, size bears a positive while age has a negative relation with cost efficiency, and both are statistically significant at the 1% level of significance. Liquidity reveals a positive relationship with cost efficiency, but it is statistically insignificant.

When the panel tobit regression is run with revenue efficiency scores as the dependent variable, capital adequacy and asset quality disclose positive and significant relationship with revenue efficiency, which is significant at the 1% level of significance. Management soundness has a negative impact on the revenue efficiency and is statistically significant at the 5% and 1% levels of significance. Earning and profitability reveal a negative relationship with revenue efficiency and is statistically significant at the 1% level of significance. Liquidity bears a positive and statistically significant relationship with revenue efficiency, which is significant at the 5% level of significance. Among the control variables, size bears a positive while age has a negative relation with cost efficiency, and both are statistically significant at the 1% level of significance.

When the panel tobit regression is run with the profit efficiency scores of insurance companies taken as the dependent variable, capital adequacy discloses a positive and significant relationship with profit efficiency, which is significant at the 10% level of significance. Management soundness has a negative impact on the profit efficiency, which is statistically significant at the 1% level of significance. Earnings and profitability reveals a negative relationship with profit efficiency and is statistically significant at the 5% level of significance. Liquidity has a positive relationship with profitability, which is statistically significant at the 5% level of significance. Asset quality and reinsurance and actuarial issues have negative and positive relations with profit efficiency, respectively, but they are insignificant. Among the control variables, size bears a positive relationship and is statistically significant at the 5% level of significance. Age reveals a negative relationship with profit efficiency, although it is insignificant. The results of above discussion are presented in a capsule form in Table 3.

Capital adequacy ratio, as depicted by the net premium to shareholder fund, reveals a positive relationship with all the efficiency

Table 3. Summary of expected and actual signs of the explanatory variables.

Independent variables			Expected signs	Signs of efficiency parameters		
Framework		Symbol		Cost	Revenue	Profit
Capital adequacy		Net premium to shareholder fund	+	$+^{*}$	$+^{*}$	$+^{***}$
Asset quality		Total investment to total assets	+	$+^{*}$	$+^{*}$	−
Reinsurance and actuarial issues		Net premium to gross premium	+	$+^{*}$	+	+
Management soundness		Operating expenses to gross premium	−	−	+	$-^{*}$
		Management expenses to gross premium	−	$-^{*}$	$-^{**}$	−
		Total commission to total premium	−	$-^{*}$	$-^{*}$	$-^{*}$
Earning and profitability		Investment income to investments	+	$-^{*}$	$-^{*}$	$-^{**}$
Liquidity		Current assets to current liabilities and provisions	+	+	$+^{**}$	$+^{**}$
Control variables	Size	Log of total claims settled	+	$+^{*}$	$+^{*}$	$+^{**}$
	Age	No. of years since the date of inception	+	$-^{*}$	$-^{*}$	−

Notes: *, **, and *** significant at the 1%, 5%, and 10% levels of significance, respectively. This table shows the expected and actual signs (relationship) of independent variables with cost efficiency, revenue efficiency, and profit efficiency.

parameters in our results. A higher premium earned in relation to shareholder funds helps absorb the heat of rising claims and costs, which makes them cost-efficient. Besides, adequate capital adequacy reduces unexpected losses and has a positive bearing on profit efficiency via the generation of surplus premiums, leading to an affirmative relation with revenue efficiency. Indian life insurance companies are able to collect premiums and settle their claims, leading to a positive impact on all measures of efficiency. The findings coincide with those of Diacon (2001) and Sen (2020), who found that a higher proportion of premiums in insurance business were associated with higher efficiency levels.

Asset quality, measured as the ratio of total investments to total assets, exhibits a significantly positive relationship with cost efficiency and revenue efficiency, indicating that good investments were made by the Indian life insurance companies. Actually, as bound by the statutes, a major proportion of investments by life insurance companies in India are made in safe and secure investment avenues, such as deposits with other financial institutions, government securities, and certificates of deposits. These investments generate stable and consistent returns with minimum costs of investment and without elevated risks. Thus, insurance companies are able to pay their policy amount with bonus to the policyholders. However, since these safer investments are relatively less lucrative, asset quality has a negative association with profit efficiency, though the relation is insignificant.

Reinsurance and actuarial issues, when measured with the net premium to gross premium, have a significant relation only with cost efficiency, which is positive. The results show that, with an increase in risk retention ratio, insurance companies are able to increase cost efficiency. This suggests that the efficiency of companies can be improved through balancing risk retention according to the capability of insurance companies. Our results are in contrast to those of Ilyas & Rajasekaran (2019), who found that reinsurance had a negative and statistically significant relationship with cost efficiency for Indian non-life insurance companies, which indicated that greater dependence on reinsurance for diversification of risk portfolios makes insurers cost-inefficient.

Management soundness, measured as the ratio of operating/management expenses to gross premium and total commission to total premium, reveals a negative and significant impact on cost, revenue, and profit efficiency. The negative sign depicts that lower the expense ratio, better would be the efficiency of the management in managing the operational and management expenses. It seems that Indian life insurance companies are able to balance the expenditure incurred on marketing, advertisement, commissions to highly professional agents, and employees with the benefits received, leading to improved efficiency scores. Our results are supported by Mathur & Paul (2014) and Sen (2020), who found a negative effect of management expenses on the overall technical efficiency of the insurers.

Earnings and profitability, when measured with investment income to investment, have a significantly negative impact on all efficiency scores, as contradictory to the hypothesis framed. Indian life insurance companies seem to invest in very safe channels, not bearing proportionately good returns and hence witnessing a negative impact on efficiency scores. Income from investments does not seem sufficient to generate sufficient revenue to cover the cost of insurance and subsequently generate profits. Nissim (2010) also argues that investment income is highly sensitive to interest rates in both the short and long runs, which can deteriorate the relationship of "E" in CARAMEL with the efficiency scores.

Liquidity reveals a significant and positive relationship with revenue and profit efficiency. This indicates that life insurance companies in India are able to settle their current liabilities without prematurely selling off their long-term investments, leading to better revenue and profit efficiency scores. Our results are supported by Diacon *et al.* (2002), who found a modest relationship between liquidity and efficiency.

Size, as measured by the claim settlement ratio, has a positive and significant impact on all the efficiency measures of insurance companies in the regression model. This highlights a prominent finding that life insurance companies are efficient enough in settling their claims. They do not face unnecessary legal costs for any defaults. They are also able to balance their investments and claims well. Our results are supported by Karim & Jhantasana (2005), who report a positive association between size and efficiency.

Age has a reverse impact, i.e., a negative impact on the efficiency of insurance companies, which is significant for cost and revenue efficiency. Older insurance firms do not use virtual cost-effective modes as compared to the new and young entrants, who tend to save on cost and boost efficiency by using novel modes of business accumulation. However, our results contradict the findings of Ilyas & Rajasekaran (2019), who found a positive and statistically significant relationship between age and efficiency. Sen (2020) found both positive and negative results with different measures of efficiency. Karim & Jhantasana (2005) showed that age had no significant relationship with inefficiency.

6. Conclusion, Implications and Future Scope of the Study

The results of the study emphasize that the acronym CARAMEL validates the impact of vital parameters on the cost, revenue, and profit efficiency of life insurance companies. It can be concluded that, as expected, the quantitative financial CARAMEL indicators of the insurance industry positively impact all the efficiency measures, except earnings and profitability. Overall, the results highlight the fact that insurance companies are focusing on capital adequacy, asset quality, management soundness, and liquidity, which help them improve their efficiency. The results of the study assert that, to be efficient in terms of cost, revenue, and profits, Indian life insurance companies should continuously carry out CARAMEL analysis so that they can take corrective measures for improvement of their financial efficiency scores.

The success story of a life insurance company begins with the deployment of premiums collected from quality assets. Maintaining good quality of their portfolio of investments can assist them in earning profitability and ensuring solvency. Prudent investment also helps them retain liquidity through the convertibility of investments into cash without undue loss of capital for the company as and when required. However, life insurance managers should strategically plan their growth opportunities to achieve sustainable profit margins as well. Managers should also make the insurance business run on contemporary lines by following the latest models of business operation. For instance, virtual business models that help save up on costs and efforts should be adopted so that old insurance companies are able to move hand in hand with their budding competitors in the market.

Though the current chapter has made an attempt to assess efficiency scores in light of the variables of the CARAMEL framework, academicians and researchers should nonetheless try and offer more insights into the efficiency of the insurance sector through future research. Future research work can contribute further by considering various economy-specific variables, along with these company-specific variables. Various risks faced by insurance companies can be taken into consideration as well. Furthermore, some new DEA techniques, such as bootstrap DEA, inverse DEA, and stochastic DEA or hierarchical regression DEA, can be applied to analyze the performance of insurance companies operating in India.

References

Ansah-Adu, K., Andoh, C., & Abor, J. (2012, January). Evaluating the cost efficiency of insurance companies in Ghana. *The Journal of Risk Finance*, 13, 61–76.

Ariff, M. & Can, L. (2008). Cost and profit efficiency of Chinese banks: A non-parametric analysis. *China Economic Review*, 19(2), 260–273.

Banker, R.D., Gadh, V.M., & Gorr, W.L. (1993). A Monte Carlo comparison of two production frontier estimation methods: Corrected ordinary least squares and data envelopment analysis. *European Journal of Operational Research*, 67(3), 332–343.

Banker, R.D., Charnes, A., Cooper, W.W., & Maindiratta, A. (1988). A comparison of DEA and translog estimates of production frontiers using simulated observations from a known technology. In: A. Dogramaci & R. Fare (Eds.), *Applications of Modern Production Theory: Efficiency and Productivity* (pp. 33–55). Boston: Kluwer Academic Publishers.

Barros, C.P., Barroso, N., & Borges, M.R. (2005). Evaluating the efficiency and productivity of insurance companies with a Malmquist index: A case study for Portugal. *The Geneva Papers on Risk and Insurance-Issues and Practice*, 30(2), 244–267.

Barros, C.P., Nektarios, M., & Assaf, A. (2010). Efficiency in the Greek insurance industry. *European Journal of Operational Research*, 205(2), 431–436.

Becchetti, L. & Sierra, J. (2003). Bankruptcy risk and productive efficiency in manufacturing firms. *Journal of Banking & Finance*, 27(11), 2099–2120. https://doi.org/10.1016/S0378-4266(02)00319-9.

Berger, A.N. & Humphrey, D. (1997). Efficiency of financial institutions: International survey and directions for future research. *European Journal of Operational Research*, 98(2), 175–212.

Boubaker, S., Do, D.T., Hammami, H., & Ly, K.C. (2022). The role of bank affiliation in bank efficiency: A fuzzy multi-objective data envelopment analysis approach. *Annals of Operations Research*, 311, 611–639. https://doi.org/10.1007/s10479-020-03817-z.

Boubaker, S., Le, T.D.Q., Ngo, T., & Manita, R. (2023). Predicting the performance of MSMEs: A hybrid DEA-machine learning approach. *Annals of Operations Research*. https://doi.org/10.1007/s10479-023-05230-8(Advance online publication).

Boussofiane, A., Dyson, R.G., & Thanassoulis, E. (1991). Applied data envelopment analysis. *European Journal of Operational Research*, 52(1), 1–15.

Colwell, R.J. & Davis, E.P. (1992). Output and productivity in banking. *The Scandinavian Journal of Economics*, 94, S111–S129.

Cummins, J.D. (1999). Efficiency in the US Life insurance industry: Are insurers minimizing costs and maximizing revenues? In: J.D. Cummins & A.M. Santomero (Eds.), *Changes in the Life Insurance Industry: Efficiency, Technology and Risk Management. Innovations in Financial Markets and Institutions* (Vol. 11). Boston, MA: Springer.

Dao, T.T.T., Mai, X.T.T., Ngo, T., Le, T.D.Q, & Ho, H. (2021). From efficiency analyses to policy implications: A multilevel hierarchical linear model approach. *International Journal of the Economics of Business*, 28(3), 457–470. https://doi.org/10.1080/13571516.2021.1981750.

Diacon, S.R. (2001). The efficiency of UK general insurance companies. CRIS Discussion paper Series. Centre for Risk & Insurance Studies. The University of Nottingham, 3(1), 1–32.

Diacon, S.R., Starkey, K., & O'Brien, C. (2002). Size and efficiency in European long-term insurance companies: An international comparison. *The Geneva Papers on Risk and Insurance. Issues and Practice*, 27(3), 444–466.

Donni, O. & Fecher, F. (1997). Efficiency and productivity of the insurance industry in the OECD countries. *The Geneva Papers on Risk and Insurance-Issues and Practice*, 22(4), 523–535.

Eling, M. & Luhnen, M. (2010). Frontier efficiency methodologies to measure performance in the insurance industry: Overview, systematization, and recent developments. *The Geneva Papers on Risk and Insurance-Issues and Practice*, 35(2), 217–265.

Fried, H.O., Lovell, C.K., Schmidt, S.S., & Schmidt, S.S. (Eds.). (2008). *The Measurement of Productive Efficiency and Productivity Growth*. Oxford, UK: Oxford University Press.

Grmanová, E. & Strunz, H. (2017). Efficiency of insurance companies: Application of DEA and Tobit analyses. *Journal of International Studies*, 10(3), 250–263.

Hsu, W.Y. & Petchsakulwong, P. (2010). The impact of corporate governance on the efficiency performance of the Thai non-life insurance industry. *The Geneva Papers on Risk and Insurance-Issues and Practice*, 35(1), S28–S49.

Ilyas, A.M. & Rajasekaran, S. (2019). An empirical investigation of efficiency and productivity in the Indian non-life insurance market. *Benchmarking: An International Journal*, 26(7), 2343–2371. https://doi.org/10.1108/BIJ-01-2019-0039.

International Association of Insurance Supervisors (2000). https://www.iaisweb.org/.

Kader, H.A., Adams, M., & Hardwick, P. (2010). The cost efficiency of Takaful insurance companies. *The Geneva Papers on Risk and Insurance — Issues and Practice*, 35, 161–181. DOI: 10.1057/g2009.33.

Kaffash, S., Azizi, R., Huang, Y., & Zhu, J. (2020). A survey of data envelopment analysis applications in the insurance industry 1993–2018. *European Journal of Operational Research*, 284(3), 801–813.

Karim, M.Z.A. & Jhantasana, C. (2005). Cost efficiency and profitability in Thailand's life insurance industry: A stochastic cost frontier approach. *International Journal of Applied Econometrics and Quantitative Studies*, 2(4), 19–36.

Kwadjo, A., Charles, A., & Joshua, A. (2012). Evaluating the cost efficiency of insurance companies in Ghana. *Journal of Risk Finance*, 1(3), 61–76.

Le, T.D.Q. (2018). Bank risk, capitalisation and technical efficiency in the Vietnamese banking system. *Australasian Accounting Business & Finance Journal*, 12(3), 42–61. https://doi.org/10.14453/aabfj.v12i3.4.

Le, T.D.Q., Ho, T.N.T., Nguyen, D.T., & Ngo, T. (2022). Intellectual capital — Bank efficiency nexus: Evidence from an emerging market. Cogent Economics & Finance, 10(1), 2127485. https://doi.org/10.1080/23322039.2022.2127485.

Luhnen, M. (2009). Determinants of efficiency and productivity in German property-liability insurance: Evidence for 1995–2006. *The Geneva Papers on Risk and Insurance — Issues and Practice*. 34, 483–505. DOI: 10.1057/gpp.2009.10.

Mathur, T. & Paul, U.K. (2014). Performance appraisal of Indian non-life insurance companies: A DEA approach. *Universal Journal of Management*, 2(5), 173–185.

Meador, J.W., Ryan Jr, H.E., & Schellhorn, C.D. (2000). Product focus versus diversification: Estimates of X-efficiency for the US life insurance industry. *Performance of Financial Institutions: Efficiency, Innovation, Regulation*, 45(9), 175.

Nandi, J.K. (2014). Relative efficiency analysis of selected life insurers in India using data envelopment analysis. *Pacific Business Review International*, 6(8), 69–76.

Ngo, T. & Tsui, K.W.H. (2021). Estimating the confidence intervals for DEA efficiency scores of Asia-Pacific airlines. *Operational Research*. https://doi.org/10.1007/s12351-021-00667-w.

Nissim, D., Analysis and Valuation of Insurance Companies (November 30, 2010). CE—ASA (Center for Excellence in Accounting and Security Analysis) Industry Study No. 2, Available at *SSRN*: https://ssrn.com/abstract=1739204 or http://dx.doi.org/10.2139/ssrn.1739204.

Noronh, M.R. & Shinde, S.R. (2012). A comparative study of cost efficiency of life insurance companies in India. *Ganpat University (Kherva, India) Faculty of Management Studies Journal of Management and Research (GFJMR)*, 4, 1–14.

Panwar, A., Olfati, M., Pant, M., & Snasel, V. (2022). A review on the 40 years of existence of data envelopment analysis models: Historic development and current trends. Archives of Computational Methods in Engineering, 29(7), 5397–5426.

Sen, S. (2020 January). Analysis of cost efficiency of Indian life insurers: A comparison of quantity vs. value based DEA approach. *Asia-Pacific Journal of Risk and Insurance, De Gruyter*, 14(1), 1–25.

Shinde, S. (2012). A comparative study of cost efficiency of life insurance companies in India. *International Journal of Commerce and Business Management*, 5(2), 210–215.

Simar, L. & Wilson, P.W. (2007). Estimation and inference in two-stage, semi-parametric models of production processes. *Journal of Econometrics*, 136, 31–64.

Sinha, R.P. (2012). Are Indian life insurance companies cost efficient? Some recent empirical evidence. *Prajnan*, 41(3), 1–11.

Sinha, R.P. (2022). Profit efficiency of Indian general insurance companies: A ratio-based approach. *Journal of Infrastructure Development*. https://doi.org/10.1177/09749306221100282.

Stata Press Publications (2009). *Stata Longitudinal-data/Panel-data Reference Manual Release 11A*. USA: Stata Press Publication. www.stata.com/manuals13/xt.pdf.

Tripe, D. (2010). Using DEA to investigate bank safety and soundness — Which approach works best? *Journal of Financial Economic Policy*, 2(3), 237–250. http://www.emeraldinsight.com/doi/abs/10.1108/17576381011085449.

Tuffour, J.K., Ofori-Boateng, K., Ohemeng, W., & Akuaku, J.K. (2021). Life insurance companies: Determinants of cost efficiency and profitability. *Journal of Accounting, Business and Management (JABM)*, 28(2), 1–19.

Vu, H.D. (2016). Technical efficiency of FDI firms in the Vietnamese manufacturing sector. *Review of Economic Perspectives*, 16(3), 205–230. https://doi.org/10.1515/revecp-2016-0013.

Wang, Y.-M. & Chin, K.-S. (2009). A new approach for the selection of advanced manufacturing technologies: DEA with double frontiers. *International Journal of Production Research*, 47(23), 6663–6679. https://doi.org/10.1080/00207540802314845.

Wasiaturrahma, S.R., Ajija, S.R., Salama, S.C.U., & Hudaifah, A. (2020). Financial performance of rural banks in Indonesia: A two-stage DEA approach. *Heliyon*, 6(7). https://doi.org/10.1016/j.heliyon.2020.e04390.

Yakob, R., Yusop, Z., Radam, A., & Ismail, N. (2014). Two-stage DEA method in identifying the exogenous factors of insurers' risk and investment management efficiency. *Sains Malaysiana*, 43(9), 1439–1450.

https://doi.org/10.1142/9781800615786_0014

Chapter 14

General Insurer Efficiency and Managerial Ability: A Non-Parametric Approach

Ram Pratap Sinha

Department of Economics,
Government College of Engineering & Leather Technology,
Kolkata, India
rampratapsinha39@gmail.com

Abstract

While it is well known that managerial ability (MA) plays an important role in the performance of business enterprises, such efforts are not directly observable. Following Demerjian *et al.* (2012), the current research study estimates MA in the context of the Indian general insurance sector. The study is based on 120 firm-year observations spread over six years (from 2012–2013 to 2017–2018) and includes 16 private sector general insurers as well as four public sector insurers with diversified businesses. To measure MA, we estimated insurer-wise efficiency using data envelopment analysis and then regressed the efficiency scores on a set of contextual variables which influence insurer performance. In the next stage, we estimated MA from regression residuals (the difference between the observed and fitted values of insurer efficiency). Further, we have used this measure of MA as an independent variable to explain return on equity on the basis of ordinary least squares and general additive model regression, and the relationship is highly significant.

Keywords: Managerial ability, efficiency, data envelopment analysis, directional distance function, GAM regression.

1. Introduction

The performance of a productive firm is commonly measured in terms of its efficiency, which may be conceptualized as the ratio of observed and potential performances. Managerial ability (MA) is one of the key factors which shapes the transformation of enterprise resources into revenue streams, leading to the augmentation of economic value addition of the firm. Thus, MA is an important driver of a firm's efficiency performance. Managerial contributions are crucial in the context of operational and investment decision-making and therefore affect firm performance in the short run as well as in the long run. Strategic management literature has long recognized the role of managerial capability in organizational performance. Stamp (1981) argued that pragmatic and analytic intuition, conceptual analysis, and intuitive analysis can be considered managerial capabilities. Barney (1991) identified the potential of MA to generate a competitive advantage for the concerned firms. Castanias & Helfat (1991, 2001) emphasized the role of the top management of firms in enabling them to generate rents. Helfat & Peteraf (2015) emphasized that MA is a cognitive ability that includes both physical and conceptual abilities. The presence of such capability, therefore, is important for firm performance.

In modern firms, the separation of corporate ownership from management brings to the forefront the problem of evaluating managerial performance. *Inter alia*, there are two consequences of the separation of ownership from control. First, the desires and goals of the firm owner and the firm manager can be very different. Second, managerial efforts are not directly observable. What the stakeholders can observe is the output generated by the company, and it is an outcome of several influential variables, including managerial skills. In fact, the second problem reinforces the first one. This is because top-level managers understand that it is extremely difficult (in most cases) to isolate the impact of managerial effort/skill. This prompts them to take unwarranted risks in many instances. There are numerous research studies that focus on the problem of designing optimal incentive schemes in the presence of asymmetric information.

Given the problems with the measurement of MA, a major part of the empirical literature on firm performance paid attention to firm productivity and efficiency and tried to explain firm performance by estimating the influence of observable environmental variables on observed efficiency. However, considering the importance of MA in the enterprise context, several research studies (Demerjian *et al.*, 2012, 2013) went further and attempted to provide an indirect measure of MA from the efficiency estimates.

The current study estimates and validates MA for the Indian general insurance sector. The estimate of MA is an indirect measure derived from the evaluation of the efficiency of the insurers. For efficiency evaluation, we have applied data envelopment analysis (DEA), followed by the derivation of MA as an indirect estimate generated from regression residuals in the second stage. DEA is a popular tool for efficiency analysis in various fields, such as banking and finance (Tripe, 2010; Le, 2018; Boubaker *et al.*, 2022b) and manufacturing (Becchetti & Sierra, 2003; Wang & Chin, 2009; Vu, 2016; Boubaker *et al.*, 2023).

The motivation behind the current study arose for three reasons. First, the related extant literature focused on the estimation of MA for the entire business sector. However, MA varies substantially across industry and service sectors; therefore, sector-specific measures do carry much importance. Second, there are very few empirical studies of MA in the Indian context. Finally, in the present context, we have provided a more complete measure of efficiency by taking into consideration both the revenue and cost sides of the industry. Given that the existing performance-related literature on the international and Indian insurance sectors is based on the estimation and explanation of efficiency and productivity, the current study adds important value to the literature.

The current study has six sections and proceeds as follows: Section 2 provides a brief overview of the Indian general insurance industry. Section 3 discusses the extant research literature related to the insurance sector and MA. Section 4 describes the methodology. Section 5 is related to the description of variables and data. Section 6 includes results and discussions. Section 7 concludes.

2. Overview of Indian General Insurance Industry

The Indian insurance industry (including the life and non-life insurance segments) experienced significant growth in the current millennium as a consequence of deregulation initiatives by the government. In 1993, the Government of India formed a high-powered committee on the Indian insurance sector to review the then-existing insurance sector scenario, recommend measures for promoting competitive efficiency, and establish a regulatory framework for the sector. In 1999, the Insurance Regulatory and Development Authority (IRDA) was established as the market regulator. This paved the way for deregulation of entry and price control in the insurance sector, which facilitated the entry of private players and deepened the market in terms of scale of operations and product variety. Table 1 provides an overview of the insurance industry growth as measured in terms of changes in insurance penetration and insurance density for the period from 2012–2013 to 2017–18. Insurance penetration refers to the ratio of the insurance premium to the country GDP. Insurance density, on the other hand, refers to the per-capita insurance premium (expressed in US dollars). It may be mentioned that insurance penetration in India was only 0.54 in 1999, which increased to 0.97 by 2018, implying an increase of approximately 80%. During a similar period, insurance density grew from 2.4 to 19. Table 1 also includes the statistics for the United States, China, and the world, and a comparison of the cross-country figures reveals that the Indian general insurance market is quite small and has much scope for further expansion.

Table 1. Growth in general insurance penetration and general insurance density.

	Country	**2013**	**2014**	**2015**	**2016**	**2017**	**2018**
Insurance penetration	India	0.80	0.70	0.70	0.77	0.93	0.97
	United States	4.30	4.30	4.20	4.29	4.28	4.26
	China	1.40	1.50	1.60	1.81	1.89	1.92
	World	2.80	2.70	2.80	2.81	2.80	2.78
Insurance density	India	11	11	12	13.2	18	19
	United States	2,296	2,360	2,377	2,449	2,542	2,672
	China	91	109	128	147.2	159	185
	World	285	294	276	285	297	312

Source: IRDA annual reports.

3. Related Research Literature

3.1. *Performance of the insurance sector*

The growth of the general insurance market across countries due to the initiation of reform measures has attracted increased attention from researchers toward the evaluation of the performance of the industry. Fukuyama & Weber (2001) estimated the efficiency and productivity growth of Japanese non-life insurance companies for the period 1983–1994. Ennsfellner *et al.* (2004) examined the production efficiency of the Austrian insurance market for the period 1994–1999. Choi & Weiss (2005) examined the performance of the US property-liability industry using a SFA approach. Kao & Hwang (2008) applied a relational and independent two-stage DEA model to evaluate the performance of 24 non-life insurance companies in Taiwan. Cummins & Xie (2008) studied the effects of mergers and acquisitions in the US property-liability insurance industry on the productivity and efficiency of the insurers. Barros *et al.* (2010) used a two-stage conditional performance benchmarking for the efficiency evaluation of 71 Greek life and non-life insurance companies. Mahlberg & Url (2010) examined the efficiency and productivity impact of market unification on the German insurance companies for the period 1991–2006. Alhassan & Biekpe (2015) estimated the efficiency, productivity, and returns to scale of South African non-life insurance companies for the period 2007–2012. Ferro & Leon (2017) applied stochastic frontier analysis to estimate the technical efficiency of Argentine non-life insurance companies for the period 2009–2014. In the Indian context, important recent contributions include those from Sinha (2017a, 2017b), Ilyas & Rajasekharan (2019a, 2019b), and Sinha (2021). For an excellent survey of the efficiency literature related to the insurance industry, see Kaffash *et al.* (2020).

3.2. *Estimation of managerial ability*

Demerjian *et al.* (2020) pointed out that the neoclassical model of the firm did not accommodate any separate identity for corporate managers. Initial empirical studies on MA thus considered the efficiency of the observed firm as an indicator of managerial skills/ability. Murthi *et al.* (1996) estimated managerial efficiency by measuring relative marketing and production efficiencies. Relative marketing efficiency refers to the performance based on return on investment and market share relative to five managerial inputs. Relative production efficiency, on the other hand, is measured by the return on investment performance relative to purchase and manufacturing expenses. These efficiency scores were then used to control for managerial skills to estimate the effects of first-mover advantage on market share.

Another strand of research literature estimated the impact of managerial quality effects on firm performance. Barr & Siems (1997) developed two bank failure prediction models which made CAMEL rating-based efficiency estimation and found management quality to be an important determinant of bank failure. Bertrand & Schoer (2003) estimated the impact of managerial (CEO and CFO) fixed effects on corporate policy variables using manager–firm matched panel data. The corporate policy variables incorporated in the study included investment policy, financial policy, organization strategy, and performance. Leverty & Grace (2012) estimated the revenue and cost efficiency of property-liability insurers and estimated the impact of CEO-specific effects on such efficiency.

A third strand of literature is concerned with intellectual capital (Pulic, 2000). The related indicators include capital employed efficiency (= value added from operations relative to the book value of net assets), human capital efficiency (= value added relative to human capital), and structural capital efficiency (= structural capital relative to value added). Kweh *et al.* (2013) considered several individual components of intellectual capital as proxies of managerial skills and estimated the efficiency of Malaysian public-listed software companies in terms of their ability to transform intellectual capital into corporate values (represented by Tobin's Q and return on equity). Nourani *et al.* (2018) estimated intellectual capital efficiency at three levels in the insurance industry for 2005–2012 using a dynamic network DEA model. Neves & Proenca (2021) measured intellectual capital and explored its relationship with the financial performance of Portuguese banks for the period 2009–2016.

The most influential approach concerning MA was introduced by Demerjian *et al.* (2012). Their study is based on managers' relative efficiency in their ability to convert productive resources into revenues. The method suggested by them involves the estimation of MA in a two-stage approach, whereby firm-wise relative efficiencies are calculated in the first stage (by applying DEA) in terms of their observed ability to generate revenues relative to the quantity and mix of resources deployed by them. Firm-level efficiency determination is thus the first part of the whole exercise. Demerjian *et al.* (2012) pointed out that the firm-level efficiency measure estimated in the first stage is determined by several environmental variables, including MA. To isolate the MA component from the other influential variables, it is necessary to eliminate the impact of these other influential variables on firm-efficiency estimates. In the second stage, therefore, they advocated that the effects of firm size, market share, positive free cash flow, the extent of diversification (both product-wise and geographic), the effects of time, industry effects, and firm age need to be removed so that the MA factor could be identified. Since the unexplained part of efficiency may still contain other unidentified environmental variables, it is essential to carry out validation tests at this stage to assess whether the residual measure truly indicates MA. Demerjian *et al.* (2013) examined the relationship between earnings quality and MA. The outcome of the study indicated that managers with a higher level of ability positively influence the quality of firm earnings, as indicated by indicators such as a higher level of earnings, the persistence of earnings, fewer restatements of earnings, and lower errors in the provisions for non-performing assets. Demerjian *et al.* (2020) investigated whether managers with higher abilities enable more stable earnings management. The study found strong evidence that managers with higher ability are significantly more likely to engage in the intentional smoothing of earnings, which is associated with improved operational performance in the future.

Several recent studies followed the approach of Demerjian *et al.* (2012). Cvetkoska *et al.* (2022) estimated MA for the insurance companies of North Macedonia, Croatia, Serbia, and Slovenia and identified its determinants. The study found that CEO duality, board size, board composition, gender diversity, and CEO gender do not have a significant impact on MA, whereas the diversity of nationalities of the CEO revealed a positive and significant impact. Phan (2021) estimated the impact of MA and managerial incentives on firm performance, with specific reference to the influence of MA on the profitability and risk of firms. The study confirmed the presence of a significant and positive relationship between MA and firm leverage. Imeni *et al.* (2021) explored the relationship between MA and earnings classification shifting in Iranian firms based on 1,875 firm-year observations for the period 2005–2019 and applied panel data regression. The results indicate the presence of a positive relationship between MA and the shifting of earnings classification and a negative relationship between MA and real earnings management.

4. Methodology of Estimation

In the current study, we have followed the approach introduced by Demerijian *et al.* (2012) to evaluate MA, subject to some methodological modifications. Accordingly, we have adopted a three-stage estimation process for the measurement and validation of MA. The various stages involved in the process include the application of the directional distance function-based DEA model for the computation of insurer-wise efficiency scores in the first stage and the (regression-based) estimation of the impact of environmental variables on the efficiency scores computed in the first stage. The environmental variables used in the second stage include various other factors (excluding MA) that influence firm performance. Next, we have estimated an indirect estimate of MA as the residual from the application of least squares regression in the second stage. Finally, as a measure of validation of the outcome, we have regressed the return on equity ratio of the observed general insurers on the measure of MA generated earlier. In this stage, we have used general additive model (GAM) regression in addition to the pooled ordinary least squares method for the purpose of regression.

4.1. *Efficiency evaluation using DEA*

The estimation of firm-level efficiency is crucial for deriving estimates of MA. As mentioned, we have applied a non-parametric approach (DEA) for constructing the performance frontier and the estimation of firm-specific performance.

To provide a brief description of the efficiency estimation process, we consider a technology that converts several inputs into multiple outputs. The input vector comprising k inputs $x_1, x_2, \ldots, x_k$ is represented by X, and the output vector including n outputs $y_1, y_2, \ldots, y_1$ is represented by Y. The production possibility set P, which includes feasible production plans of the producer, can be presented as P $= \{(X, Y) \in R^{k+1}, X\,\text{can produce}\,\text{Y}\}$. The production possibility set comprises both output and input sets. The output set O(Y) $\in P$ corresponds to the set of feasible outputs for each input vector, while the input set I(X) includes the input vectors that are feasible under each output vector.

The technology is assumed to exhibit free disposability (of inputs and outputs) and convexity. We may specify them as follows:

(a) **Free disposability:** if $(X, Y) \in P_S$, then $(X_a, Y) \in P_S$, where $X_a \geq X$. Similarly, $(X, Y_a) \in P_S$, provided $Y_a \leq$ Y. The inequalities between the (input and output) vectors are understood component-wise.
(b) **Convexity:** Convexity implies that if (X_a, Y_a) and $(X_b, Y_b) \in P_s$, then $(X_c, Y_c) \in P_s$, where $X_c = \delta X_a + (1 - \delta)X_b$ and $Y_c = \delta Y_a + (1 - \delta)Y_b$ and $\delta \in [0, 1]$.

The process of efficiency estimation is based on the concept of the distance function. Shephard (1953, 1970, 1974) and Farrell (1957) advanced the concept of input and output distance functions. Distance functions relate inefficient observation points to efficient points. The input- and output-oriented distance functions can be formally represented as

$$D^{\text{input}}_{\text{Farrell}} = \text{Min}[\mu : (\theta X, Y) \in P_S], \tag{1}$$

$$D^{\text{output}}_{\text{Farrell}} = \text{Max}[\lambda : (X, \phi Y) \in P_S]. \tag{2}$$

Chambers *et al.* (1996) introduced the directional distance function to provide a new measure of technical efficiency that compares both observed inputs and outputs with the optimal inputs and outputs. The concept of the directional distance function is based on Luenberger's benefit function (1992) and takes into account the potential for increasing outputs and decreasing inputs simultaneously. The directional distance function can be written as

$$D_{DDF} = \text{Max}[\beta : (\{(X - \beta g_x), (Y + \beta g_Y)\} \in P_S], \tag{3}$$

where g_X and g_Y represent the input and direction vectors, respectively, and β is a scalar. The directional distance function provides a complete characterization of the technology, and the input distance functions of Shephard/Farrell are actually special cases of the directional distance function (Färe *et al.*, 2000).

The concept of distance function relates the inefficient point with the best practice points. However, for an empirical estimation of efficiency, a benchmark needs to be constructed, which can serve as the basis for efficiency evaluation. As DEA is a data-driven approach, it enables the construction of the input and output frontiers from the entire set of input and output observations (comprising all the productive units under observation). Thus, we may represent the input and output frontiers as $\sum_{i=1}^{k} \lambda_i X_{ij} (j = 1, 2, \ldots, n)$ and $\sum_{k=1}^{l} \lambda_k Y_{kj} (j = 1, 2, \ldots, n)$, respectively, where there are n firms under observation and λ is a scalar, $\lambda \geq 0$. If we allow returns to scale exhibited by the technology to be variable, we need to add an additional constraint: $\sum \lambda = 1$.

The application of DEA in the first stage yields several advantages over the competing frontier estimation methods. First, we do not require the *a priori* assumption of any particular functional form of the relationship between the inputs and outputs. Second, we are able to accommodate multiple outputs for efficiency evaluation without requiring aggregation. Finally, the assumption of convexity is in conformity with economic theory. It may be noted here that several other important DEA methods are available for the estimation of efficiency, including the double bootstrap approach (Simar *et al.*, 2007, 2015), inverse DEA (Boubaker *et al.*, 2022a), and stochastic DEA (Ngo *et al.*, 2022).

4.2. *Regression methodology for second and third stage*

In the second stage of the estimation process, the ordinary least squares method on efficiency scores has been applied to estimate the impact of environmental variables. In the present context, since the efficiency scores are bounded from both below and above (with the lower and upper bounds being 0 and 1, respectively), we have taken the log of efficiency scores as the dependent variable. The log efficiency scores are then used for regressing on the contextual variables. Once the least squares estimates of the regression parameters are available, we can compute the residuals from the OLS estimate. This residual provides a good measure of MA, provided that the environmental variables do not control for it.

In the final stage, we have used both OLS and GAM regression to estimate the impact of MA (OLS residual) on return on equity. The final stage consists of model validation.

The OLS regression model is

$$\text{Log}(E_{dea}) = \beta_0 + \beta_1 + Z_1 + \beta_2 Z_2 + \cdots + \beta_r Z_r + U. \tag{4}$$

The OLS regression model is preferred over the tobit regression model because, in actuality, the efficiency scores are not generated from a process of data censoring. Instead, the second-stage regression should be carried out by a method that expresses the dependent variable in terms of fractional data (McDonald, 2009).

GAM is a generalized linear model (GLM) in which the dependent variable depends linearly on the unspecified smooth functions of some explanatory variables. GAM regression was originally introduced by Hastie & Tibshirani (1990) by combining the properties of general linear models with additive regression models. If Y denotes the dependent variable and it has r explanatory variables, $Z_1, Z_2, \ldots, Z_r$, which influence the input variables and hence the efficiency scores, then we may write the GAM regression equation as

$$Y = \beta_0 + s(Z_1) + s(Z_2) + \cdots + s(Z_r) + U, \tag{5}$$

where β_0 represents the intercept and $s(Z_i)$ represent the smooth functions of the explanatory variables to be estimated semi-parametrically.

5. Description of Variables and Data

The efficiency evaluation of the observed general insurers requires the specification of the inputs and outputs of the insurers. There are, however, several competing approaches for selecting input and output indicators for the insurance industry. Eling *et al.* (2010) showed that three major types of inputs are used in the insurance industry: labor (agents and office employees), business services, and capital (debt and equity capital). Leverty *et al.* (2010) found three competing views for choosing the outputs: the financial intermediation approach (introduced by Sealey *et al.*, 1977, in the context of the banking sector), the user cost approach (Hancock, 1985, 1991), and the value-added approach (Berger *et al.*, 1987, 1992). The financial intermediation approach considers financial service providers as financial intermediaries who operate between the demanders and suppliers of funds. The user cost approach classifies various asset and liability categories as inputs or outputs, depending on their net contribution to revenue. The value-added approach considers such activities as output indicators (risk pooling and risk bearing, real financial services, and intermediation), which contribute significant value addition as assets while allocating operating costs.

The current study seeks to estimate insurer-wise efficiency, with a specific focus on their ability to generate revenues. Therefore, we need to select the major output variables which contribute to insurer revenue. Thus, we have chosen two outputs: net premium income and income from investments. The outputs represent the flow of income for the insurers. We have considered two inputs: operating expenses (related to the insurance business) and claims incurred, which represent the flow of expenditure for the insurers. Operating expenses are a proxy for labor expenses and business expenses, while claims incurred are a proxy for insurance claims. The selection of inputs and outputs follows Alhassan & Biekpe (2016) and Banker *et al.* (2022).

For the second-stage analysis, we have chosen the following explanatory variables: insurer age, insurer size (log of the total asset), solvency ratio, and ownership dummy (0 for public sector insurers and 1 for private insurers).

To carry out the estimation, we have collected data pertaining to the input, output, and explanatory variables (Table 2) from two sources: IRDA annual reports published by the IRDA for the respective financial years (2012–2013 to 2017–2018) and the Handbook on Indian Insurance Statistics for the relevant years.

Table 2. Inputs, outputs, and environmental variables.

Description	Type of variable
Operating expenses and claims incurred	Input variables
Net premium income and income from investments	Output variables
Insurer age, Insurer size, solvency ratio, and ownership dummy	Environmental variables

6. Results and Discussion

This study includes 12 private sector diversified general insurers, 4 public sector diversified general insurers/ and 4 private sector health insurers. Observations for six financial years (from 2012–2013 to 2017–2018) have been used for the purpose of estimation. Thus, the efficiency evaluation and subsequent analysis have been made on the basis of 120 firm-year observations. We are interested to know the outcome of the three-stage exercise for the estimation of MA and the exploration of the relationship of return on equity with the measure of MA. The results are presented and described in the following.

6.1. *Evaluation of technical efficiency of the observed general insurers*

In the present context, we have assumed that technology is unchanged during the observed period, and accordingly, the current study estimates the efficiency of the in-sample general insurers on the basis of the pooled data. Table 3 presents the mean efficiency scores for the observed general insurers for 2012–2013 to 2017–2018. The table includes the mean efficiency scores for three groups (diversified private, diversified public, and specialized health insurers) as well as for the total number of general insurers under observation. Since we have constructed a global frontier, the efficiency scores are intertemporally comparable. Table 3 indicates that between 2012–2013 and 2015–2016, mean efficiency scores improved secularly but declined in the next two financial years (2016–2017 and 2017–2018). It is further seen that there is not much difference in mean efficiency across the three categories. The insurer-wise efficiency scores are included in Appendix Table A1.

Table 3. Mean efficiency scores of general insurers (2012–2013 to 2017–2018).

Particulars	2013	2014	2015	2016	2017	2018
Private diversified	0.6704	0.6835	0.7117	0.7596	0.7119	0.6980
Private health	0.6673	0.6791	0.7086	0.7540	0.7060	0.6925
Public diversified	0.6743	0.6833	0.7059	0.7640	0.7102	0.6931
Composite	0.6739	0.6832	0.7132	0.7614	0.7114	0.6957

6.2. *Scale efficiency and returns to scale*

Apart from technical efficiency, we have also computed the scale efficiency of the observed general insurers. Table 4 provides the mean scale efficiency scores of the observed general insurers. The mean scale efficiency scores for the aforementioned groups are included in the table in separate rows. The insurer-wise scale efficiency scores are presented in Appendix Table A4. Table 4 clearly shows that the observed private sector general insurers have higher mean scale efficiency than their public sector counterparts, and this might be due to the differences in size across the insurer categories. Further, the private insurers are more proactive in increasing their market share, especially through online channels.

Table 5 provides summary information regarding the type of returns to scale exhibited by the observed insurers. Table 5 shows that most of the general insurers are away from the most productive scale size. Further, the majority of the insurers displayed decreasing returns to scale. This is likely to be caused by the slowdown in investment and economic growth rates in India that has occurred since 2008–2009. Appendix Table A3 provides the general insurer-wise returns to scale for the observed years.

Table 4. Mean scale efficiency scores of general insurers (2012–2013 to 2017–2018).

Particulars	**2013**	**2014**	**2015**	**2016**	**2017**	**2018**
Private diversified	0.9951	0.9886	0.9959	0.9894	0.9682	0.9769
Private health	0.8511	0.9579	0.9743	0.9777	0.9708	0.9684
Public diversified	0.8302	0.9329	0.8651	0.8505	0.8770	0.8641
Composite	0.9333	0.9713	0.9654	0.9593	0.9505	0.9526

Table 5. Returns to scale (RTS) of the in-sample general insurers (2012–2013 to 2017–2018).

Particulars	**2013**	**2014**	**2015**	**2016**	**2017**	**2018**
Constant returns to scale	3	0	5	3	5	4
Increasing returns to scale	14	15	10	11	7	7
Decreasing returns to scale	3	5	5	6	8	9

6.3. *Influence of the environmental variables on insurer efficiency scores*

The efficiency scores generated from the application of the directional distance function model are bound on both sides (below and above). Thus, in the second stage, the log of efficiency is regressed on the environmental variables already mentioned (insurer age, insurer size, solvency ratio, year, and diversification dummy). We have applied pooled OLS, fixed-effects model (FEM), and random effects model (REM). The results are presented in Tables 6–8. They indicate that out of the four explanatory variables, ownership dummy, insurer age, and solvency ratio are statistically significant. However, the effect of insurer size is not significant.

To select the best model, we have applied the F test (for comparing pooled OLS model with FEM), the Breusch–Pagan Lagrange multiplier test (for comparing pooled OLS model with REM), and the

Table 6. Pooled OLS regression of log of efficiency on the environmental variables.

Particulars	**Coefficient**	**Std. error**	***t*-ratio**	***p*-value**
Constant (Intercept)	−1.1212	0.3023	−3.709	0.0015
Ownership index (Dummy)	0.0170	0.0043	3.970	0.0008
Insurer age	0.0091	0.0287	0.3158	0.7556
Solvency ratio	0.0571	0.0289	1.978	0.0626
Log of the total asset	0.3669	0.1589	2.310	0.0323
R-squared	0.3167			
Adjusted R-squared	0.2929			

Table 7. FEM regression of log of efficiency on the environmental variables.

Particulars	**Coefficient**	**Std. error**	***t*-ratio**	***p*-value**
Constant (Intercept)	−1.3754	0.2895	−4.752	∼0.0000
Ownership index (Dummy)	0.5311	0.1555	3.417	0.0009
Insurer age	0.0203	0.0049	4.143	∼0.0000
Solvency ratio	0.0645	0.0316	2.038	0.0443
Log of the total asset	0.0162	0.0313	0.5178	0.6058
Residual variance	0.0253			

Table 8. REM regression of log of efficiency on the environmental variables.

Particulars	**Coefficient**	**Std. error**	***t*-ratio**	***p*-value**
Constant (Intercept)	−1.1793	0.2408	−4.896	∼0.0000
Ownership index (Dummy)	0.4076	0.1303	3.128	0.0022
Insurer age	0.0180	0.0045	4.012	0.0001
Solvency ratio	0.0598	0.0294	2.032	0.0445
Log of the total asset	0.0097	0.0285	0.3395	0.7348

Table 9. Comparison of pooled OLS, FEM, and REM.

Test name	**Test statistic**	***p*-value**
F test	1.4353	0.1288
LM test	0.8229	0.3643
Hausman test	2.9025	0.5743

Hausman test (for comparing REM with FEM). The results (Table 9) indicate that the pooled OLS model is the preferred alternative.

6.4. *Estimation and validation of managerial ability*

In the regression of the log of efficiency scores on the environmental (explanatory) variables, we have included several predictors which influence efficiency performance. However, we did not include any indicators of managerial effort. The residual from the OLS regression, therefore, represents the MA factor. Table 10 provides descriptive statistics for the MA component.

Finally, as a measure of testing the validity of our estimate of MA, we have regressed (using the OLS method) the return on equity of the observed general insurers on the measure of MA, an insurer-specific dummy, and a year dummy. Since return on equity is an important measure of the financial performance of firms, there should be a strong relationship between return on equity and MA. The results of the OLS regression are included in Table 11. It indeed indicates a significant relationship between the two variables (return on equity and MA). We did not consider MA as a dependent variable because the measure of MA may include the influence of some other unknown/unspecified variables.

Table 10. Descriptive statistics of managerial ability.

Particulars	2013	2014	2015	2016	2017	2018	Overall
Mean	−0.0078	−0.0020	0.0080	0.0571	−0.0066	−0.0487	−0.0078
Median	−0.0020	0.0112	0.0112	0.0542	−0.0286	−0.0850	−0.0020
Standard deviation	0.1922	0.1353	0.2249	0.1307	0.1470	0.1165	0.1922
Skewness	0.8355	0.4055	0.5002	0.0866	0.4495	0.6421	0.8355
Kurtosis	3.3358	3.2993	0.0099	−0.5805	0.8584	0.5079	3.3358

Table 11. OLS regression of return on equity on managerial ability.

Particulars	Coefficient	Std. error	t-ratio	p-value of type 1 error
Constant (Intercept)	0.1121	0.0373	3.004	0.0073
OLS residual (Managerial ability)	0.3062	0.0872	3.510	0.0023
Insurer-specific dummy	−0.0118	0.0028	−4.215	0.0005
Year dummy	0.0169	0.0086	1.970	0.0636
R-squared	0.2739			
Adjusted R-squared	0.2551			

Table 11 shows that the relationship of the return on equity and our measure of MA is highly significant. Further, the regression relationship has very high levels of R-squared and adjusted R-squared values.

As an additional measure for testing the validity of the relationship (between return on equity and MA), we have also applied the GAM for the purpose of estimation. Figure 1 provides a graphical relationship between return on equity and MA obtained using GAM.

Table 12 presents the results obtained from the application of GAM. The results are similar to the outcomes obtained from OLS regression. The relationship between MA and return on equity is presented in Figure 1. In case of a linear regression model, the goodness of fit is indicated by the residual sum of squares. However, in the case of GAM, the sum of squares does not make much sense as a measure of the discrepancy between the observations and the fitted values. Instead, we can measure the discrepancy in terms of deviance:

$$D = 2[L(\hat{\beta}_M - L(\hat{\beta}_o)]\phi, \tag{6}$$

Table 12. GAM estimate of the relationship between return on equity and MA.

Approximate significance of parametric coefficients	**Estimate**	**Std. error**	**t-ratio**	**p-value of type 1 error**
Constant (Intercept)	0.1071	0.0394	2.718	0.0076
Insurer-specific dummy	0.0156	0.0079	1.979	0.0502
Year dummy	−0.0109	0.0024	−4.605	0.0000
Approximate significance of the smooth term	edf	Reference edf	F	F(p-value)
Managerial ability	2.268	2.875	6.441	0.0005
R-squared (Adjusted)	0.292			
Generalized cross validation	0.0225			
Scale estimate	0.0215			

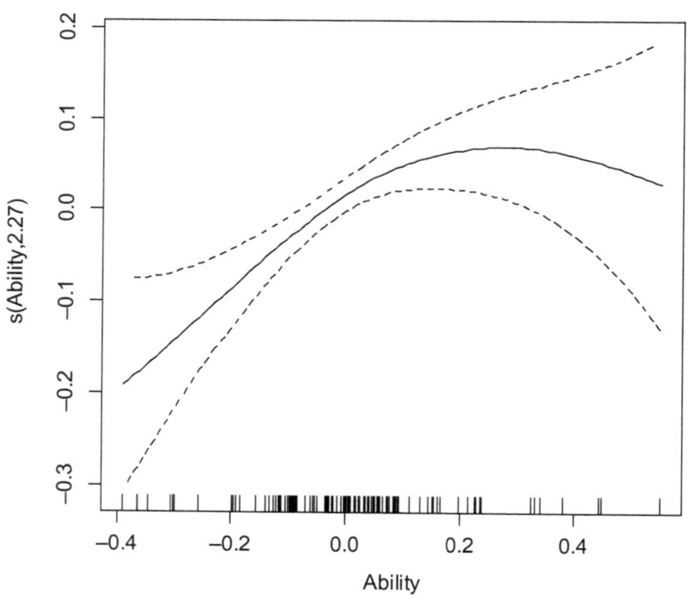

Figure 1. Relationship between return on equity and managerial ability.

where $L(\hat{\beta}_M)$ is the maximum likelihood of the saturated model and $L(\hat{\beta}_O)$ is the maximum likelihood of the observed (fitted) model. Φ is the scale parameter.

Here, model validation is tested by cross-validation of results, which involves out-of-sample testing. The main objective of cross-validation is to test the ability of the chosen model in predicting new data which were not used in the estimation process. The process of cross-validation involves partitioning the data into two complementary subsets: one which is used for performing the analysis, and the other which is used for model testing. Several rounds of cross-validation are usually performed to reduce variability in outcomes. In the present case, we have computed the generalized cross-validation score of the GAM fitted. Smaller values of GCV indicate better model fitting. Overall, the results indicate that the relationship between return on equity and MA is highly significant.

6.5. *Ranking of insurers on the basis of managerial ability*

The estimated results indicate that the MA of the observed general insurers fluctuates over the in-sample years. However, we can rank the general insurers on the basis of mean MA (the mean of MA scores available for the six years under observation). Table 13 provides the ranking of the insurers based on the average MA of the in-sample general insurers. The table shows that out of the 20 observed insurers, 10 (50%) general insurers exhibited a positive mean MA. The remaining insurers exhibited a negative mean MA. Out of the 10 general insurers which showed a positive mean MA, only one insurer is in the public sector, and the remaining nine are privately owned. The year-wise and insurer-wise MA scores are presented in Appendix Table A4.

However, we may want to consider both the mean and standard deviation of the MA of the observed general insurers for the period under consideration. Figure 2 plots the mean MA on the horizontal axis and the standard deviation on the vertical axis. Figure 2 indicates significant variability in respect to MA across the in-sample insurers for the period under consideration.

Table 13. Managerial ability based ranking of general insurers.

Insurer	Mean managerial ability	Standard deviation
Shri Ram General	0.2027	0.1654
Bajaj Allianz	0.1136	0.0819
Tata AIG	0.0817	0.0410
Apollo Munich	0.0714	0.0797
IFFCO Tokio	0.0380	0.2170
New India	0.0226	0.0784
ICICI Lombard	0.0102	0.1085
Oriental	0.0065	0.1228
HDFC Ergo	0.0057	0.2245
Cholamandalam	0.0001	0.0902
Star Health	−0.0198	0.1708
Religare Health	−0.0201	0.1278
Max Bupa	−0.0243	0.2688
National	−0.0265	0.2946
Future Generali	−0.0437	0.0725
SBI General	−0.0467	0.1500
United	−0.0471	0.0731
Reliance General	−0.0506	0.1802
Royal Sundaram General	−0.1428	0.1090
Bharti AXA	−0.1311	0.1469

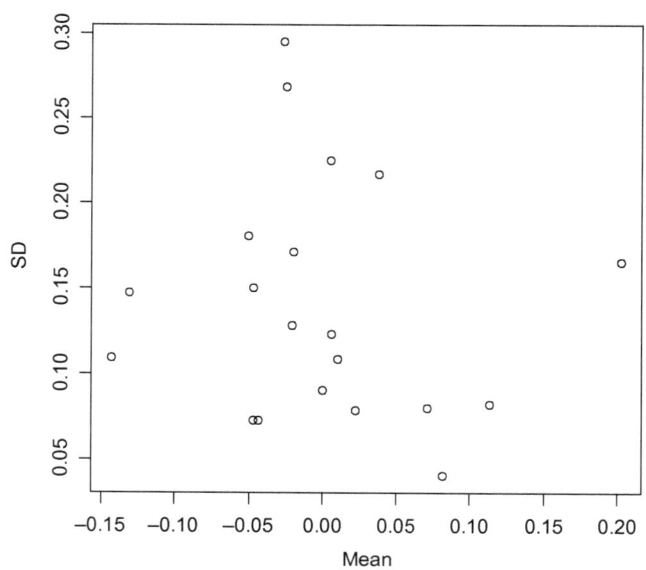

Figure 2. Scatter plot of mean and standard deviation of managerial ability.

7. Concluding Observations

In a competitive insurance market, corporate warfare for survival and business growth is essentially a zero-sum game in which the business gains obtained by market leaders occur at the expense of the other market players. Managerial talent/ability plays a key role in this process by enabling firms to anticipate imminent changes in the business environment and take appropriate measures. In the current study, we have provided a measure of the MA that fluctuates over the in-sample period. This is indicative of the continually changing dynamics of the general insurance market.

This study, however, has several limitations. First, the general insurance market in India is quite small and includes few players in the market when compared to the developed country insurance markets. Second, we have data for only six years. It is essential to consider insurer-wise performance for a longer period (10–15 years) to arrive at a more robust measure of MA. Third, a regression method, such as hierarchical regression (Dao *et al.*, 2021), might provide more insight about regression model selection in the second stage. Finally, we were able to explore the linkage of MA with only one alternative measure of ability (return on equity). Future research studies can consider these improvements as part of their research agenda.

Appendix

Table A1. General insurer-wise technical efficiency (2012–2013 to 2017–2018).

Insurer	2013	2014	2015	2016	2017	2018
Bajaj Allianz	0.7328	0.7714	0.7739	0.8704	0.8314	0.8077
Bharti Axa	0.5269	0.5947	0.7630	0.5548	0.6203	0.6811
Cholamandalam	0.6829	0.6868	0.5592	0.8161	0.6868	0.6398
Future Generali	0.5426	0.6543	0.9850	0.6679	0.5958	0.5902
HDFC Ergo	0.6678	0.6525	0.6135	0.6357	0.6183	0.6058
ICICI Lombard	0.6751	0.6792	0.4865	0.8460	0.7538	0.7276
IFFCO Tokio	0.6997	0.7286	0.6444	0.8791	0.7124	0.6906
Reliance	0.6072	0.6866	0.5258	0.6977	0.6072	0.6205
Royal Sundaram General	0.6043	0.7008	0.6087	0.6918	0.6546	0.6261

(Continued)

Table A1. (*Continued*)

Insurer	2013	2014	2015	2016	2017	2018
SBI General	0.4383	0.5561	1.0000	0.5844	0.6844	0.5831
Shri Ram General	0.6648	0.9907	0.9789	1.0000	0.8566	0.7918
Tata AIG	0.6699	0.7070	0.7690	0.7811	0.6612	0.6603
National	0.6872	0.8020	0.8519	1.0000	0.7458	0.6423
New India	1.0000	0.8318	0.8749	1.0000	1.0000	1.0000
Oriental	0.8256	0.7764	0.8187	0.8560	0.5888	0.7291
United	0.7587	0.7794	0.7598	0.8932	0.6993	0.8042
Apollo Munich	0.6372	0.6381	0.6292	0.6287	0.8112	0.6980
Max Bupa	0.5516	0.5020	0.5370	0.5358	0.8880	0.8141
Religare Health	1.0000	0.4272	0.5174	0.5210	0.5690	0.5482
Star Health	0.4362	0.5049	0.5381	0.7312	0.6536	0.6989

Table A2. General insurer-wise scale efficiency (2012–2013 to 2017–2018).

Insurer	2013	2014	2015	2016	2017	2018
Bajaj Allianz	1.0000	1.0000	1.0000	0.9908	0.9139	0.9519
Bharti Axa	0.9958	0.9966	0.9985	0.9989	0.9956	0.9936
Cholamandalam	0.9935	0.9944	0.9960	0.9964	1.0000	1.0000
Future Generali	0.9886	0.9896	0.9940	0.9945	0.9993	0.9985
HDFC Ergo	0.9960	0.9960	1.0000	1.0000	0.8726	0.9594
ICICI Lombard	1.0000	0.9790	0.9872	0.9000	0.8798	0.8820
IFFCO Tokio	0.9966	0.9968	0.9990	0.9993	1.0000	0.9970
Reliance General	0.9943	0.9948	0.9962	0.9965	1.0000	1.0000
Royal Sundaram General	0.9957	0.9936	0.9966	0.9970	0.9994	0.9984
SBI General	0.9833	0.9830	1.0000	0.9998	0.9990	1.0000
Shri Ram General	0.9989	0.9407	0.9833	1.0000	1.0000	1.0000
Tata AIG	0.9991	0.9990	1.0000	1.0000	0.9596	0.9415
National	0.9284	0.9333	0.8377	0.8274	0.9011	0.9016
New India	1.0000	0.9397	0.8993	0.8681	0.7479	0.7667
Oriental	0.8807	0.9313	0.8267	0.8210	0.9697	0.9371
United	0.5115	0.9272	0.8966	0.8855	0.8894	0.8509
Apollo Munich	0.9338	0.9478	0.9697	0.9866	0.9872	0.9955
Max Bupa	0.9355	0.9714	0.9799	0.9447	0.9046	0.9233
Religare Health	0.5431	0.9183	0.9477	0.9794	0.9915	0.9990
Star Health	0.9920	0.9942	1.0000	0.9999	1.0000	0.9559

Table A3. General insurer-wise returns to scale (2012–2013 to 2017–2018).

Insurer	2013	2014	2015	2016	2017	2018
Bajaj Allianz	Scale efficient	Increasing	Scale efficient	Decreasing	Decreasing	Decreasing
Bharti Axa	Increasing	Increasing	Increasing	Increasing	Increasing	Increasing
Cholamandalam	Increasing	Increasing	Increasing	Increasing	Scale efficient	Scale efficient
Future Generali	Increasing	Increasing	Increasing	Increasing	Increasing	Increasing
HDFC Ergo	Increasing	Increasing	Scale efficient	Scale efficient	Decreasing	Decreasing
ICICI Lombard	Scale efficient	Decreasing	Decreasing	Decreasing	Decreasing	Decreasing
IFFCO Tokio	Increasing	Increasing	Increasing	Increasing	Scale efficient	Increasing
Reliance	Increasing	Increasing	Increasing	Increasing	Scale efficient	Scale efficient
Royal Sundaram	Increasing	Increasing	Increasing	Increasing	Increasing	Increasing
SBI General	Increasing	Increasing	Scale efficient	Increasing	Increasing	Scale efficient
Shri Ram General	Increasing	Increasing	Increasing	Scale efficient	Scale efficient	Scale efficient
Tata AIG	Increasing	Increasing	Scale efficient	Scale efficient	Decreasing	Decreasing
National	Decreasing	Decreasing	Decreasing	Decreasing	Decreasing	Decreasing
New India	Scale efficient	Decreasing	Decreasing	Decreasing	Decreasing	Decreasing
Oriental	Decreasing	Decreasing	Decreasing	Decreasing	Decreasing	Decreasing
United	Decreasing	Decreasing	Decreasing	Decreasing	Decreasing	Decreasing
Apollo Munich	Increasing	Increasing	Increasing	Increasing	Increasing	Increasing
Max Bupa	Increasing	Increasing	Increasing	Increasing	Increasing	Increasing
Religare Health	Increasing	Increasing	Increasing	Increasing	Increasing	Increasing
Star Health	Increasing	Increasing	Scale efficient	Increasing	Constant	Decreasing

Table A4. General insurer-wise managerial ability (2012–2013 to 2017–2018).

Insurer	2013	2014	2015	2016	2017	2018
Bajaj Allianz	0.0724	0.09446	0.08602	0.14509	0.0744	0.0178
Bharti Axa	−0.0948	−0.00190	0.22757	−0.11240	−0.0214	0.0421
Cholamandalam	0.0497	0.02424	−0.19860	0.15983	−0.0337	−0.1222
Future Generali	−0.1201	0.05906	0.44886	0.05022	−0.0926	−0.1172
HDFC Ergo	0.1300	0.08933	0.00668	0.02388	−0.0271	−0.0871
ICICI Lombard	−0.0003	−0.02290	−0.38880	0.15404	0.0025	−0.0484
IFFCO Tokio	0.0354	0.04260	−0.09800	0.19713	−0.0328	−0.0829
Reliance	−0.1259	−0.01380	−0.29920	−0.02990	−0.1959	−0.1922
Royal Sundaram	−0.0913	0.04588	−0.09720	0.03569	−0.0302	−0.1250
SBI General	−0.3630	−0.10550	0.44430	−0.05340	0.0644	−0.1327
Shri Ram General	−0.0033	0.37918	0.33220	0.32388	0.1527	0.0317
Tata AIG	−0.0007	0.03570	0.08608	0.09419	−0.1002	−0.1143
National	−0.0913	0.03288	0.07374	0.23474	−0.1135	−0.2557
New India	0.2157	−0.00240	0.03904	0.16450	0.1523	0.1123
Oriental	0.0914	−0.00800	0.02464	0.05811	−0.3058	−0.1403
United	−0.0568	−0.05400	−0.08780	0.08404	−0.1340	−0.0339
Apollo Munich	0.0606	0.03987	0.01566	0.00978	0.2253	0.0773
Max Bupa	−0.0692	−0.18350	−0.13290	−0.15660	0.3398	0.2368
Religare Health	0.5507	−0.30020	−0.12520	−0.12710	−0.0604	−0.0967
Star Health	−0.3459	−0.19090	−0.19700	−0.11350	0.0041	0.0567

References

Alhassan A.L. & Nicholas, B. (2015). Efficiency, productivity and returns to scale economies in the non-life insurance market in South Africa. *The Geneva Papers on Risk and Insurance-issues and Practice,* 40(3), 493–515.

Alhassan, A.L. & Nicholas, B. (2016). Competition and efficiency in the non-life insurance market in South Africa. *Journal of Economic Studies*, 43(6), 882–909.

Banker, R.D., Amirteimoori, A., & Sinha, R.P. (2022). An integrated data envelopment analysis and generalized additive model for assessing managerial ability with application to the insurance industry. *Decision Analytics Journal*, 4, 100115. https://doi.org/10.1016/j.dajour.2022.100115.

Barney, J.B. (1991). Firm resources and sustained competitive advantage. *Journal of Management*, 17(1), 99–120.

Barr, R. & Siems, T. (1997). Bank failure prediction using DEA to measure management quality. In: R. Barr, R. Helgason & J. Kennington (Eds.), *Interfaces in Computer Science and Operations Research:*

Advances in Metaheuristics, Optimization and Stochastic Modeling Technologies (pp. 341–366). Boston: Kluwer Academic Publishers.

Barros, C.P., Nektarios, M., & Assaf, A. (2010). Efficiency in the Greek insurance industry. *European Journal of Operational Research*, 205(2), 431–436.

Becchetti, L. & Sierra, J. (2003). Bankruptcy risk and productive efficiency in manufacturing firms. *Journal of Banking & Finance*, 27(11), 2099–2120. https://doi.org/10.1016/S0378-4266(02)00319-9.

Berger, A.N. & Humphrey, D. (1992). Measurement and efficiency issues in commercial banking. In: Z. Griliches (Ed.), *Output Measurement in the Service Sectors* (pp. 245–300). Chicago, USA: University of Chicago Press.

Berger, A.N., Hanweck, G., & Humphrey, D. (1987). Competitive viability in banking: Scale, scope and product mix economies. *Journal of Monetary Economics*, 20(3), 501–520.

Boubaker, S., Le, T.D.Q., & Ngo, T. (2022a). Managing bank performance under COVID-19: A novel inverse DEA efficiency approach. *International Transactions in Operational Research*, 30(2023), 2436–2452. https://doi.org/10. 1111/itor.13132.

Boubaker, S., Do, D.T., Hammami, H., & Ly, K.C. (2022b). The role of bank affiliation in bank efficiency: A fuzzy multi-objective data envelopment analysis approach. *Annals of Operations Research*, 311, 611–639. https://doi.org/10.1007/s10479-020-03817-z.

Boubaker, S., Le, T.D.Q., Ngo, T., & Manita, R. (2023). Predicting the performance of MSMEs: A hybrid machine learning approach. *Annals of Operations Research*. https://doi.org/10.1007/s10479-023-05230-8 (Advance online publication).

Castanias, R.P. & Helfat, C.E. (1991). Managerial resources and rents. *Journal of Management*, 17(1), 155–171.

Castanias, R.P. & Helfat, C.E. (2001). The managerial rents model: Theory and empirical analysis. *Journal of Management*, 27(6), 661–678.

Chambers, R.G., Chung, Y., & Fare, R. (1996). Benefit and distance functions. *Journal of Economic Theory*, 70(2), 407–419.

Choi, B.P. & Weiss, M. (2005). An empirical investigation of market structure, efficiency, and performance in property-liability insurance. *Journal of Risk & Insurance*, 72(4), 635–673.

Cvetkoska, V., Eftimov, L., Ivanovsky, I., & Kamenjarska, T. (2022). Measuring the managerial ability in the insurance companies in the Republic of North Macedonia, Croatia, Serbia and Slovenia and identifying its determinants. *International Journal of Banking, Risk and Insurance*, 10(1), 51–67.

Dao, T.T.T., Mai, X.T.T., Ngo, T., Le, T.D.Q., & Ho, H. (2021). From efficiency analyses to policy implications: A multilevel hierarchical linear model approach. *International Journal of the Economics of Business*, 28(3), 457–470.

Demerjian, P., Lev, B., & McVay, S. (2012). Quantifying managerial ability: A new measure and validity tests. *Management Science,* 58(7), 1229–1248.

Demerjian, P., Lewis-Western, M., & McVay, S. (2020). How does intentional earnings smoothing vary with managerial ability? *Journal of Accounting, Auditing & Finance*, 35(2), 1–32.

Demerjian, P., Lev, B., Lewis, M., & McVay, S. (2013). Managerial ability and earnings quality. *The Accounting Review*, 88(2), 463–498.

Eling, M. & Luhnen, M. (2010). Frontier efficiency methodologies to measure performance in the insurance industry: Overview, systematization, and recent developments. *The Geneva Papers on Risk and Insurance — Issues and Practice*, 2010, 35(2), 217–265.

Ennsfellner K.C., Lewis D., & Anderson, R.I. (2004). Production efficiency in the Austrian insurance industry: A Bayesian examination. *Journal of Risk Insurance*, 71(1), 135–159.

Färe, R. & Grosskopf, S. (2000). Theory and application of directional distance functions. *Journal of Productivity Analysis*, 13(2), 93–103.

Farrell, M.J. (1957). The measurement of productive efficiency. *Journal of the Royal Statistical Society, Series A, General*, 120(3), 253–281.

Ferro, G. & León, S.A. (2017). Stochastic frontier analysis of efficiency in Argentina's non-life insurance market. *Geneva Papers on Risk and Insurance-Issues and Practice,* 43(1), 158–174.

Fukuyama, H. & Weber, W.L. (2001). Efficiency and productivity change of non-life insurance companies in Japan. *Pacific Economic Review*, 6(1), 129–146. DOI: 10.1111/1468-0106.00122.

Hancock, D. (1985). The financial firm: Production with monetary and nonmonetary goods, *Journal of Political Economy*, 93(5), 859–880. https://doi.org/10.1086/261339.

Hancock, D. (1991). *A Theory of Production for the Financial Firm.* Boston: Kluwer Academic Publishers.

Hastie, T.J. & Tibshirani, R.J. (1990). *Generalized Additive Models.* New York/Boca Raton: Chapman & Hall/CRC.

Helfat, C.E. & Peteraf, M.A. (2015). Managerial cognitive capabilities and the microfoundations of dynamic capabilities. *Strategic Management Journal*, 36(6), 831–850.

Ilyas, A.M. & Rajasekharan, S. (2019a). An empirical investigation of productivity and efficiency in the Indian non-life insurance market. *Benchmarking: An International Journal.* DOI: 10.1108/BIJ-01-2019-0039.

Ilyas, A.M. & Rajasekharan, S. (2019b). Productivity of the Indian non-life insurance sector: New evidence using Fare-Primont index approach. *International Journal of Productivity and Performance Management.* DOI: 10.1108/IJPPM-04-2019-0147.

Imeni, M., Fallah, M., & Edalatpanah, S.A. (2021). The effect of managerial ability on earnings classification shifting and agency cost of Iranian listed companies. *Discrete Dynamics in Nature and Society.* https://doi.org/10.1155/2021/5565605.

Kao, C. & Hwang, S.N. (2008). Efficiency decomposition in two-stage data envelopment analysis: An application to non-life insurance companies in Taiwan. *European Journal of Operational Research*, 185(1), 418–429.

Kweh, Q.L., Chan, Y.C., & Ting, I.W.K. (2013). Measuring intellectual capital efficiency in the Malaysian software sector. *Journal of Intellectual Capital*, 14(2), 310–324.

Le, T.D.Q. (2018). Bank risk, capitalisation and technical efficiency in the Vietnamese banking system. *Australasian Accounting Business & Finance Journal*, 12(3), 42–61. https://doi.org/10.14453/aabfj.v12i3.4.

Leverty, J.T. & Grace, M.F. (2010). The robustness of output measures in property-liability insurance efficiency studies. *Journal of Banking and Finance*, 34(7), 1510–1524.

Leverty, J.T. & Grace, M.F. (2012). Dupes or incompetents? An examination of management's impact on firm distress. *The Journal of Risk and Insurance*, 79(3), 751–783.

Luenberger, D.G. (1992). Benefit functions and duality. *Journal of Mathematical Economics*, 21(5), 461–481.

Mahlberg, B. & Url, T. (2010). Single market effects on productivity in the German insurance industry. *Journal of Banking & Finance*, 34(7), 1540–1548.

McDonald, J. (2009). Using least squares and Tobit in second stage DEA efficiency analyses. *European Journal of Operational Research*, 197(2), 792–798.

Murthi, B., Srinivasan, K., & Kalyanaram, G. (1996). Controlling for observed and unobserved managerial skills in determining first-mover market share advantage. *Journal of Marketing Research*, 33(3), 329–336.

Neves, E. & Proenca, C. (2021). Intellectual capital and financial performance: Evidence from Portuguese banks, *International Journal of Learning and Intellectual Capital*, 18(1), 93–108.

Ngo, T. & Tsui, K.W.H. (2022). Estimating the confidence intervals for DEA efficiency scores of Asia-Pacific airlines. *Operations Research*, 22(4), 3411–3434. https://doi.org/10.1007/s12351-021-00667-w.

Nourani, M., Chandran, V.G.R., & Kweh, Q.L., & Lu, W.-M. (2018). Measuring human, physical and structural capital efficiency performance of insurance companies. *Social Indicators Research: An International and Interdisciplinary Journal for Quality-of-Life Measurement*, 137(1), 281–315.

Phan, N.H. (2021). Managerial ability, managerial incentives and firm performance: Empirical evidence from Vietnam. *The Journal of Asian Finance, Economics and Business*, 8(4), 193–200.

Pulic, A. (2000). $VAIC^{TM}$ an accounting tool for IC management. *International Journal of Technology Management*, 20(5–8), 702–714.

Shephard, R.W. (1953). *Cost and Productions.* Princeton: Princeton University Press.

Shephard, R.W. (1970). *Theory of Cost and Productions.* Princeton: Princeton University Press.

Shephard, R.W. (1974). *Indirect Production Functions.* Meisenheim am Glan, Germany: Verlag Anton Hain.

Simar, L. & Wilson, P.W. (2007). Estimation and inference in two-stage, semi-parametric models of production processes. *Journal of Econometrics*, 136(1), 31–64.

Simar, L. & Wilson, P.W. (2015). Statistical approaches for non-parametric frontier models: A guided tour. *International Statistical Review*, 83(1), 77–110.

Sinha R.P. (2017a). Dynamic performance benchmarking of Indian general insurance companies. *Journal of Quantitative Economics*, 15(3), 543–564.

Sinha R.P. (2017b). Efficiency-solvency linkage of Indian general insurance companies: A robust non-parametric approach. *Eurasian Economic Review*, 7(3), 353–370.

Stamp, G. (1981). Levels and types of managerial capability. *Journal of Management Studies*, 18(3), 277–298.

Tripe, D. (2010). Using DEA to investigate bank safety and soundness — Which approach works best? *Journal of Financial Economic Policy*, 2(3), 237–250. http://www.emeraldinsight.com/doi/abs/10.1108/17576381011085449.

Vu, H.D. (2016). Technical efficiency of FDI firms in the Vietnamese manufacturing sector. *Review of Economic Perspectives*, 16(3), 205–230. https://doi.org/10.1515/revecp-2016-0013.

Wang, Y.-M. & Chin, K.-S. (2009). A new approach for the selection of advanced manufacturing technologies: DEA with double frontiers. *International Journal of Production Research*, 47(23), 6663–6679. https://doi.org/10.1080/00207540802314845.

Part III

Data Envelopment Analysis (DEA) in Sustainability

https://doi.org/10.1142/9781800615786_0015

Chapter 15

Development of ESG Performance and Sustainability Indicators: A Data Envelopment Analysis Approach with Overlapping Criteria

William Chung[*,‡] **and Yong-Tong Chen**[†,§]

[*]*Department of Management Sciences,*
City University of Hong Kong, Hong Kong
[†]*Department of Information Management,*
School of Government, Beijing Normal University,
Beijing, China
[‡]*mswchung@cityu.edu.hk*
[§]*yongtong.chen@bnu.edu.cn*

Abstract

In developing a firm's environmental, social, and governance (ESG) performance indicator, we note that some criteria belong to more than one dimension. For example, "green energy" can be classified as environmental or social. We call such criteria "overlapping criteria." The main issue is assigning appropriate weighting scores to overlapping and non-overlapping criteria. Another concern is that if there is a negative correlation between criteria, it is inappropriate to construct a composite indicator (CI). To resolve these shortcomings, we propose a two-level approach. The first level involves using data envelopment analysis (DEA) to find three-dimensional CIs. Each pillar of DEA includes the corresponding overlapping and non-overlapping criteria. The second level involves finding the aggregate score of the dimensional CIs by simple weighted average or DEA, depending on the dimension's negative correlation. Due to the nature of DEA, some firms can use a few criteria or dimensions to achieve the best ESG performance, which deviates from the ESG development purpose. To resolve this issue, we propose to impose a set of sustainability constraints so that firms must consider all the criteria in a balanced manner and maintain flexibility to obtain a set of optimal DEA weighting scores.

Keywords: ESG performance indicator, sustainability indicator, overlapping criteria, sustainability constraints, composite indicator, three pillars, negative correlation, tow-level approach, weak sustainability, weighting, firm's efficiency.

1. Introduction

This chapter discusses the issues we encountered while developing sustainability indicators for the United Nations (UN) Sustainable Development Goals (SDGs), see UNSTATS (2020). We found that the issues and the proposed approach can be generalized to not only constructing sustainability indicators for a country but also to the environmental, social, and governance (ESG) performance indicators of firms.

These performance indicators help conduct performance assessments of complex systems. The existing approaches to developing sustainability indicators consist of three steps: (1) selecting criteria related to sustainability performance; (2) classifying criteria based on three dimensions or pillars (economic, social, and environmental); (3) using some mathematical methods to calculate the dimensional indicators, which are aggregated into a sustainability indicator (a composited indicator, CI). The mathematical methods can be a simple calculation of the average of the corresponding criteria and dimensional indicators or a data envelopment analysis (DEA) method. A similar approach can be applied to the ESG performance indicators with ESG dimensions.

One of the motivations of this chapter is that some criteria belong to more than one dimension in the literature. We call such kinds of criteria "overlapping criteria." For example, in developing the sustainability indicator of the 17 SDGs of the UN Summit on Sustainable Development, some goals belong to two dimensions. Allen *et al.* (2017) considered that SDG 6 belongs to "Sustainable and Resilient Societies" and "Human Dignity and Well-being." Not addressing the overlapping criteria issue can lead to impractical decisions in the benchmarking process. A similar observation can be found in the work of Kasirian & Yusuff (2010, 2013). Kasirian & Yusuff (2010) studied the interdependent dimensions in supplier selection via the analytic hierarchy process (AHP) and analytic network process with preemptive goal programming. Kasirian & Yusuff (2013) further addressed the interdependent dimensional criteria within the same dimension. Their results show that disregarding the interdependencies among dimensions and dimensional criteria leads to impractical decisions. To our best knowledge, there is a research gap in that there is no research on developing performance indicators considering overlapping criteria. A similar situation can be found in ESG performance indicators.

The other motivations come from other issues in developing performance indicators, such as the negative correlation of SD/ESG criteria and strong and weak sustainability considerations.

In this chapter, we propose approaches based on DEA to resolve the above issues. The contributions of this study are summarized as follows:

(1) We found a knowledge gap in the literature on developing performance indicators with overlapping criteria.
(2) We propose a two-level approach to resolving the issues of overlapping criteria and the negative correlation via DEA methods.
(3) We propose sustainability constraints to resolve strong and weak sustainability.
(4) We develop a case study of developing sustainability indicators of SDGs for 157 countries, including overlapping criteria, to illustrate the approaches.

We first present the background of ESG indicators and sustainable development (SD) indicators in Sections 2 and 3, respectively. Then, the corresponding indicators' development issues are discussed in Section 4. The proposed methods are given to resolve the issues in Section 5 and are used to develop the sustainability indicator of the SDGs as an application illustration in Section 6. The conclusion is given in Section 7. We propose a two-level approach based on DEA to resolve these issues.

2. ESG Performance Indicator

2.1. *ESG and financial distress*

Sun *et al.* (2014) described that the financial distress process is continuous and dynamic, lasting from a few months to several years, and can ultimately lead to bankruptcy. Hence, financial distress prediction has become a research topic to help firms detect financial difficulties earlier, understand the process of financial distress, and protect firms from bankruptcy. Different prediction models have been proposed based on the literature for predicting firms' financial distress. Balcaen & Ooghe (2006) found that most prediction models are based on either statistical or artificial intelligence approaches. Financial ratios are used to differentiate between non-distressed and distressed firms.

Recently, machine learning techniques have become popular approaches for studying financial distress prediction. For instance, Blascak *et al.* (2021) derived an estimating equation by controlling for prior financial distress. They employed a quantile decomposition strategy to analyze the gender gap along the entire bank card credit limit distribution. Mai *et al.* (2019) developed deep learning models for corporate bankruptcy forecasting using textual disclosures from a comprehensive bankruptcy database of 11,827 US public companies. It is found that the prediction of deep learning models can be further improved when textual data, the traditional accounting-based ratio, and market-based variables are used to develop the models. Huang & Yen (2019) reviewed a set of machine learning approaches to predict the financial distress of the publicly listed Taiwanese firms covering the 2010–2016 sample period. Their review includes supervised approaches (the traditional support vector machine, hybrid associative memory with translation, hybrid GA-fuzzy clustering, and extreme gradient boosting) and unsupervised approaches (the classifier deep belief network and the hybrid DBN–SVM model). Duarte & de Moraes Barboza (2020) provided a review of various academic research on forecasting financial distress with machine learning. Within the scope of credit risk analysis, they found that the demand for identifying and introducing new variables, classifiers, and more assertive methods is constant.

However, Hafiz *et al.* (2015) discussed that financial distress could have devastating effects on all companies' stakeholders, including themselves. Hence, it is believed that financial ratios alone may not be sufficient to understand the process of a firm's financial distress.

To fill this knowledge gap, Citterio (2021) discussed that corporate social responsibility (CSR) has a potential impact on the performance and reputation of companies. However, few pieces of literature consider the relationship between non-financial performance and riskiness. They claimed that no studies consider CSR a predictor of a bank's financial distress. Citterio first assessed the relationship between the ESG score (a proxy for CSR) and bank risk by using an accounting measure (Z-score) and a market-based measure (Merton distance-to-default). Then, fixed-effects regressions were used to develop a model to detect financial distress, which includes the ESG score as one of the predictors. However, to find the firm's efficiency and ESG score, we may employ DEA, a popular technique from the literature described in the following.

2.2. *DEA applications in the firm's efficiency estimations*

The literature shows us many DEA applications in firm's efficiency estimations. For instance, Boubaker *et al.* (2022a) modified the conventional inverse DEA approach algorithm to investigate 49 banks' performance. Using the modified inverse DEA, fewer banks were required to make adjustments to maintain their efficiency levels compared with conventional inverse DEA. Boubaker *et al.* (2021) employed linear regression, semi-parametric, and DEA approaches to investigate the effect of the largest controlling shareholders on French-listed firms' productive efficiency. It is found that a negative association exists between the excess control rights of dominant owners and the firm's productive efficiency. Boubaker *et al.* (2022b) applied a fuzzy multi-objective two-stage DEA technique to study differences in bank efficiency between two types of US commercial banks covering 1994–2018: One type of bank affiliates with single-bank holding companies, and the other with multi-bank holding companies. The results show that the multi-bank type is more efficient than the single-bank type due to the former taking advantage of their parents' resources to enhance their efficiency. Boubaker *et al.* (2020) employed DEA to assess whether foreign currency hedging improves a firm's productive efficiency, considering a sample of French non-financial listed firms covering 2004–2012. It is found that foreign currency hedging has a significant positive effect on efficiency. Le *et al.* (2021) examined the interrelationships between the development of fintech credit and the efficiency of banking systems in 80 countries from 2013 to 2017 using a two-stage framework. In the first stage, using financial ratios, DEA was used to estimate banking efficiency scores, which were then linked with the growth of fintech credit in the second stage, measured by a simultaneous equations model. It is found that fintech credit is more developed in countries with less efficient banking systems.

On the other hand, fintech credit may serve as a wake-up call to the banking system. Ngo & Tsui (2021) combined stochastic frontier analysis and DEA into a single framework to estimate Asia-Pacific airlines' pure inefficiency scores covering 2008–2015. Thaker *et al.* (2021) employed random forest regression in tandem with DEA under a two-phase model to examine corporate governance and the new technical, cost, and profit efficiency of Indian banks covering 2008–2018. Kim (2018) used an ensemble method to develop predictive models for hospitality financial distress, combining particular models' characteristics and advantages. Cheng *et al.* (2019) proposed a purity-based k-nearest neighbor algorithm to improve the performance of the missing value imputation on the application and performance of the classification in financial distress prediction analysis.

2.3. *DEA applications in ESG efficiency estimations*

We can find many DEA applications in ESG efficiency estimations for firms in the literature. For instance, Li *et al.* (2022) employed the bootstrapping DEA approach to estimate the eco-efficiency of China's hotel industry covering 2016–2019. They also extended the bootstrapping procedure of Simar & Wilson (2002) to perform a hypothesis test of returns to scale by considering undesirable by-products, obtaining bias-corrected eco-efficiencies and then testing for comparisons of two means.

Moreover, Kuo *et al.* (2022) employed a two-stage dynamic DEA model to study the sustainability and corporate performances of 29 international transport manufacturing companies covering 2015–2019. Lu *et al.* (2022) used a modified DEA game cross-efficiency approach to examine the longitudinal efficiency of a set of listed retail firms covering 2013–2018. Chang *et al.* (2021) applied

DEA and panel data analysis to study the interactive effects of digital finance and ESG performance in China.

Chen *et al.* (2021) developed a three-step framework by incorporating ESG performance into portfolio optimization, in which two or more attributes and the interactions among the E, S, and G attributes were considered. They constructed a DEA model with quadratic and cubic terms to enhance the evidence of two or more aspects and the interaction between the E, S, and G attributes. These ESG indicators (scores) were combined with financial indicators to select assets based on a cross-efficiency analysis. A portfolio optimization model incorporating ESG indicators with selected assets was constructed to obtain a social responsibility investment strategy. While the ESG performance indicator is used to access the firm's CSR, the SD indicator is mainly employed for national SD performance.

3. Sustainability Development Indicators

The UN (2015) defined SD as the development to "meet the needs of the present without compromising the ability of future generations to meet their own need." It is considered that SD is used to deal with a complex system affected by many criteria. Encouraging the classification of criteria into the three pillars of ecological environment, society, and economy makes analysis more straightforward. Often, SD is presented as aiming to bring the three together in a balanced way, reconciling conflicts.

The UN Conference on Environment and Development (UNCED), held in Rio de Janeiro in 1992, calls on individual countries as well as international governmental and non-governmental organizations to "develop and identify indicators of SD in order to improve the information basis for decision making at all level" (UNCED, 1992). Hence, developing appropriate SD indicators is becoming one of the most popular tasks in the SD literature. Indeed, policymakers face different challenges and issues in realizing SD. An SD indicator is believed to be one of the suitable tools to alleviate this challenge. Surveys of the SD indicator constructions were conducted by Niemeijer & de Groot (2008) and Wilson *et al.* (2007).

Many individuals and groups working on SD seek a way to create harmony between the three pillars. SD indicators are believed to alleviate the difficulties in monitoring progress toward SD using the three pillars' conditions. See Böhringer & Jochem (2007), Mair *et al.* (2018), Swain (2018), Stafford-Smith *et al.* (2017), and Holden *et al.* (2017) for other discussions and applications of the three-pillar model.

4. The Issues of Developing ESG or SD Indicators

In general, the existing approaches to developing SD indicators consist of three steps: (1) selecting criteria related to sustainability performance; (2) classifying criteria into pillar criteria based on the three SD pillars (economic, social, and environmental); (3) using the proposed methods, such as taking an average of the corresponding criteria or the DEA method, to obtain the scores of the SD indicators. However, there is no research work that considers overlapping criteria in the three pillars. A similar approach is employed in ESG performance indicators of the three ESG dimensions: environmental, social, and governance. However, there is a research gap in the issue of overlapping criteria in the three pillars/dimensions in the literature. On top of the overlapping criteria, the negative correlation of SD/ESG criteria and strong and weak sustainability considerations are also addressed in the development of the corresponding indicators.

4.1. *Overlapping criteria in three pillars/dimensions*

In developing SD indicators, the three pillars are often presented as three interconnected rings (Barton, 2000). Figure 1 presents the standard three-pillar view of SD. From the literature, we found that some SD criteria belong to more than one pillar, and we call such criteria "overlapping criteria" (Giddlings *et al.*, 2002). Figure 1, for example, shows overlapping zones among the pillars; some criteria may be the elements of these overlapping zones. The main issue is assigning appropriate weighting scores to overlapping and non-overlapping criteria. Ignoring the overlapping criteria issue can lead to an unrealistic decision in supplier selection. There is a similar observation in the context of sustainable supplier selection problem by Kasirian *et al.* (2010) and Kasirian & Yusuff (2013). It should be noted that Chen *et al.* (2021) addressed the same issue of the interaction between the

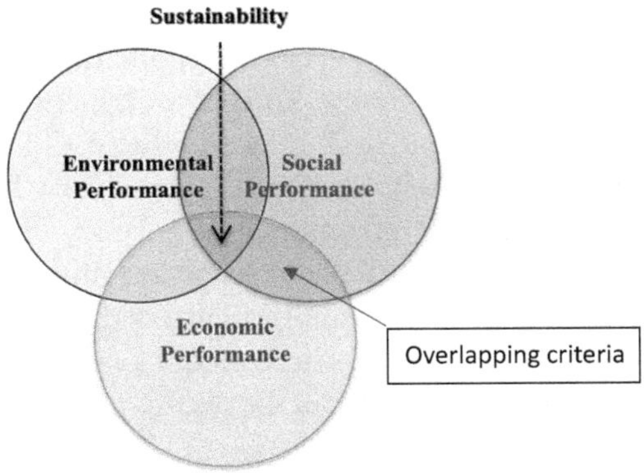

Figure 1. Three-pillar SD with overlapping criteria.

ESG dimensions, which is a similar issue of overlapping criteria in the SD indicator.

4.2. *Negative correlation of SD/ESG criteria*

It is well-known that if SD criteria have a negative correlation, combining them into a composite indicator is inappropriate. A similar shortcoming has been mentioned in the development of the Sustainable Society Index, see the SSI website (2018). Note that ESG indicators also incur problems concerning the negative correlation of criteria.

4.3. *Strong and weak sustainability considerations*

According to Neumayer (2003), *strong sustainable development* is based on the theory of steady-state economics. It is believed that the environmental system contains the economic system, the natural capital is irreplaceable, and there is a limit to material growth, with its policy advocating the distribution and efficiency on a controlled scale. The *weak sustainable development* is based on the neoclassical economics theory, which believes that the economic system contains environmental systems, natural capital can be replaced, and there is no limit to material growth, with its policy advocating maximized efficiency.

Rickels *et al.* (2016) also discussed that using composite indicators comprising indicators for several targets demands an explicit treatment of these trade-offs, some kind of weighting scheme, and an explicit specification of substitution possibilities. The explicit specification of potential substitutions paves the way for a clear distinction between weak and strong sustainability. In theory, weak sustainability allows for unlimited substitution and requires that the (weighted) aggregate of the various indicators does not decline (e.g., Pearce *et al.*, 1989). By contrast, the concept of strong sustainability does not allow for any substitution between the various targets at all.

Similarly, ESG indicators also involve strong and weak sustainability considerations.

5. Methodology

Due to the application of the proposed DEA methods for developing SD indicators, the description of the proposed approach focuses on SD indicator development. It is easy to extend the proposed approach to ESG indicators. Given that three pillar-CIs and SD indicators is a type of CI constructed by DEA, where CI is a composited indicator. This section briefly describes the CI and DEA.

5.1. *Background of CI*

Since CIs are helpful in conducting performance assessments of complex systems, CIs have been used for benchmarking and policy analyses in various fields, such as the economy (i.e., composite leading indicators, CLIs, 2022), environment (i.e., environmental sustainability index, ESI, 2005), and society (i.e., HDI, 2020). Mathematically, CI is a combination of several indicators (also called an index), which are based on sub-indicators without a common meaningful unit of measurement; a straightforward approach to weighting these sub-indicators has not yet been developed (Nardo, 2008).

Saisana & Tarantola (2002) and Nardo (2008) listed the major steps in constructing CIs. One of these steps is the weighting and aggregation of sub-indicators. However, assigning the "appropriate" weights to sub-indicators is a crucial step that may pose difficulties for decision-makers. In particular, equal weighting and subjective weight selection imply that objective evidence to support any subjective choice of weighting values may not be sufficient.

Aside from equal weighting, several weighting and aggregation methods were applied to construct CIs in the literature. According to Nardo (2008), some commonly used methods include principal component analysis, the analytic hierarchy process, conjoint analysis, and DEA. These methods can appropriately weight and aggregate a set of sub-indicators into a CI. However, the first three methods require prior or external information to determine the weights of sub-indicators.

5.2. *DEA for constructing CIs*

DEA was developed by Charnes *et al.* (1978) and Banker *et al.* (1984). It is a non-parametric approach mainly used to evaluate the technical efficiency of decision-making units (DMUs). An essential feature of DEA is that it does not need prior information to weigh sub-indicators (i.e., criteria). Instead, it can endogenously optimize the weights for DMUs, which is the country in question in the case of the current study.

Cherchye *et al.* (2004), Cherchye & Kuosmanen (2004), and Despotis (2005a, 2005b) almost simultaneously developed the following DEA model to construct a CI. For each country, $i(=1,\ldots,n)$, the following linear programming is solved to find its efficiency in terms of the performance level of SD, where y_{ij} is the value of criteria j $(=1,\ldots,J)$ of DMU i (or k) and θ_i is the (relative) performance of DMU i:

(DEA-CI)

$$\theta_i = \max_{u_{ij}} \sum_{j}^{J} u_{ij} y_{ij}, \tag{1}$$

$$\text{s.t.:} \quad \sum_{j}^{J} u_{ij} y_{kj} \leq 1 \quad \forall k, \tag{2}$$

$$u_{ij} \geq 0 \quad \forall j. \tag{3}$$

Countries that achieve a score of $\theta_i = 1$ correspond to "efficient countries" in the DEA terminology. If the score is $\theta_i < 1$, this country i is considered "inefficient." However, "efficiency" has no special meaning in the current study because no type of transformation of input to output is assumed (Despotis, 2005a).

Cherchye *et al.* (2007) discussed that model (DEA-CI) provides the CI context from its invariance to measurement units, which implies that a normalization stage can be skipped. Moreover, it fills the information gap in the "right" set of weights (u_{ij}) by generating flexible "benefit-of-the-doubt" (BoD) weights for each evaluated country, resulting in the model being called the BoD model for CI construction. Model (DEA-CI) has been applied to several areas; see Filippetti & Peyrache (2011), Murias *et al.* (2012), Tofallis (2013), Saeidi *et al.* (2013), Mizobuchi (2014), Yang *et al.* (2014), and Charles & Díaz (2017).

The current study utilizes model (DEA-CI) to obtain a rating score representing each pillar's sustainable performance. That is, after solving the above model (DEA-CI) $\boldsymbol{n}$ times for $\boldsymbol{n}$ countries, $\theta = (\theta_1, \theta_2, \ldots, \theta_n)$ is obtained, which shows the sustainability level of all countries in one of the three pillars. Hence, three sets of θ are developed. From model (DEA), the closer the value θ_i is to 1, the more efficient the corresponding DMU. Thus, in model (DEA-CI), a higher value of θ_i of the country implies a better performance level.

5.3. *Two-level approach*

We propose a two-level approach with the DEA method to resolve the first two issues (see Section 4.1 on overlapping criteria and Section 4.2 on the negative correlation of criteria). Moreover, we also propose that using a kind of sustainability constraint with DEA models can enhance the SD indicator results' robustness and flexibility, enabling modelers to consider strong and weak sustainability (see Section 4.3). We also find that the above issues and the proposed two-level approach can be applied to a firm's ESG performance indicators.

The approach consists of two levels: the first level is to use DEA to find three pillar-CIs: environmental-CI (Envi-CI), social-CI (Soci-CI), and economic-CI (Econ-CI). The second level uses a simple DEA method with sustainability constraints.

5.3.1. *For overlapping criteria*

At the first level, the overlapping criteria will be included in the corresponding pillar-DEA model. For example, if Criteria_A is an overlapping criterion belonging to the economic and environmental pillars, Criteria_A will be one of the criteria in economic-DEA and environmental-DEA accordingly. The resulting weighting scores of Criteria_A would be different in different dimensions. At the first level, the problem of overlapping criteria (Section 4.1) can be resolved accordingly.

5.3.2. *For the negative correlation of pillars-CIs*

At the second level, we apply the DEA method with three pillar-CIs. Consequently, the weighting scores of the overlapping criteria will be considered repeatedly without loss of fairness to both the overlapping and non-overlapping criteria. Note that if there is a negative correlation between the three pillar-CIs, the simple weighted average method cannot be applied at the second level. Hence, employing the DEA method at the second level can resolve the negative correlation among the three pillar-CIs (see Section 4.2).

5.3.3. *For the strong and weak sustainability*

To resolve the strong and weak sustainability issues (see Section 4.3), we impose a set of sustainability constraints in the corresponding DEA models. It is believed that different countries have different optimal conditions to secure an integrated implementation of the three pillars of SD. Hence, countries should be flexible to have different weighting schemes of substitution possibilities. A DEA model with sustainability constraints can provide such kind of flexibility.

5.3.4. *Sustainability constraints*

To achieve the above-mentioned flexibility, a set of sustainability constraints is introduced into model (DEA-CI), as shown in the following:

$$u_{ij}y_{kj} \Big/ \sum_{j}^{J} u_{ij}y_{kj} \leq p, \tag{4}$$

where p is the percentage of the flexibility of using other criteria to substitute the criterion in question. For example, if a pillar-CI consists of five criteria,

$$p = 25\% = 1/(\text{Number of criteria} - 1). \quad (5)$$

That is, we enforce the countries using at least four out of five criteria to obtain their DEA scores (see Wong & Beasley, 1990). These constraints also imply that the individual countries cannot heavily rely on some criteria to obtain the efficient score of $\theta_i = 1$, which is an inappropriate result in the context of sustainability studies. Indeed, countries should improve all the criteria in a balanced manner and maintain flexibility to obtain a set of optimal weighting scores from the DEA analysis. Hence, these constraints are named "sustainability constraints" in our DEA models.

6. An Application of SDG Sustainability Indicators

In 2015, world leaders met at the UN Summit on Sustainable Development in New York. They formally adopted 17 **SDGs** consisting of 169 specific targets as a global SD framework. More descriptions and discussions about the principles of SDGs can be found in the works of Simon *et al.* (2016), Griggs *et al.* (2014), Nilsson *et al.* (2013), and Mair *et al.* (2018).

Rickels *et al.* (2016) argued that a CI of SDGs could provide essential additional information and help facilitate the assessment of SD in general and in the SDG context. They considered that the current indicator framework of the SDGs, with its broad set of individual indicators, prevents straightforward assessment of synergies and trade-offs between the various indicators, targets, and goals, thus heightening the significance of policy guidance in achieving SD. By analyzing SDG 14 (Ocean) for the European Union coastal states, they demonstrated how the (complementary) inclusion of CIs that aggregate the individual indicators by applying a generalized mean could enhance their analysis.

Indeed, few people would be able to remember all 17 SDGs with 169 targets, and it is believed that it would be hard to create accountability around any of them. Moreover, Hák *et al.* (2016) discussed the need for relevant indicators of SDGs. It is considered that the SDGs and their targets have laid out a policy framework. However, the indicators may be ambiguous without thorough expert and scientific follow-up on their operationalization. See Allen *et al.* (2016) for a review of modeling tools for studying national SDGs and Burford *et al.* (2016) for the review of indicators in SDGs.

6.1. *The 17 overlapping SDGs and three pillars*

Mair *et al.* (2018) considered that the 17 SDGs bring together environmental, social, and economic concerns. Swain (2018) inspected which pillars are the most effective for achieving SD through SD measurement. Stafford-Smith *et al.* (2017) considered that the SDGs seek to profoundly link the goals' social, economic, and environmental aspects. Indeed, the SDGs also consider that "for sustainable development to be achieved, it is crucial to harmonize three core elements: economic growth, social inclusion, and environmental protection. These elements are interconnected, and all are crucial for the well-being of individuals and societies" (UN, 2015).

Le Blanc (2015) showed the interrelationships between the 17 SDGs through the network of SDGs and targets. Sena *et al.* (2016) reviewed the SDGs and targets. The 17 SDGs were classified into three SD pillars, and the relationship between the targets on water, drought (desertification), and health was highlighted. Sena *et al.* proposed a conceptual framework that shows the interrelationships between the 17 SDGs, identifying with a greater or smaller degree of intensity those that are key to understanding and acting on the subject of drought from the point of view of health and human well-being. Sena *et al.* (2016) are among those authors who demonstrated interrelationships between the 17 SDGs by classifying them into three pillars. Similar interrelationship properties can be found in the work of Allen *et al.* (2017), in which an SDG was classified into more than one dimension. For example, SDG 6 belongs to "Sustainable and Resilient Societies" and "Human Dignity and Well-being." Nunes *et al.* (2016) also presented some overlapping SDGs belonging to some criteria of the pillar of "Health and Well-being." See Pongiglione (2015) and Nilsson *et al.* (2016) for a similar discussion. Hence, we have overlapping and non-overlapping SDGs in the three-pillar framework, and the corresponding distribution (Sena *et al.*, 2016) is presented in Table 1. The overlapping SDGs are SDGs 3, 6, 7, 11, 13, and 17. Note that SDG 17 belongs to three pillars.

Table 1. Descriptions of SDGs and the corresponding pillars.

Pillar	Goal	Descriptions
Econ	1	End poverty in all its forms everywhere
Soci	2	End hunger, achieve food security and improved nutrition and promote sustainable agriculture
Soci–Econ–Envi	3	Ensure healthy lives and promote well-being for all at all ages
Soci	4	Ensure inclusive and quality education for all and promote lifelong learning
Soci	5	Achieve gender equality and empower all women and girls
Econ–Envi	6	Ensure access to water and sanitation for all
Econ–Envi	7	Ensure access to affordable, reliable, sustainable and modern energy for all
Econ	8	Promote inclusive and sustainable economic growth, employment and decent work for all
Econ	9	Build resilient infrastructure, promote sustainable industrialization and foster innovation
Soci	10	Reduce inequality within and among countries
Soci–Econ	11	Make cities inclusive, safe, resilient and sustainable
Econ	12	Ensure sustainable consumption and production patterns
Soci–Envi	13	Take urgent action to combat climate change and its impacts
Envi	14	Conserve and sustainably use the oceans, seas and marine resources
Envi	15	Sustainably manage forests, combat desertification, halt and reverse land degradation, halt biodiversity loss
Soci	16	Promote just, peaceful and inclusive societies
Soci–Econ–Envi	17	Revitalize the global partnership for sustainable development

6.2. *Data of SDGs*

The data are obtained from the SDG Index and Dashboards (Schmidt-Traub *et al.*, 2017; SDG Index, 2022), which were introduced as analytical tools for assessing countries' baselines for the SDGs that can be applied by researchers in cross-disciplinary analyses required for implementation. The Index and Dashboards synthesize available country-level data for all 17 goals and, for each country, estimate the size of the gap toward achieving the SDGs. A total of 157 countries have sufficient data for all 17 goals, which are used in the current chapter.

In the database, three SDGs have no data for some countries: two countries in SDG 1 (end poverty in all its forms everywhere), 13 countries in SDG 10 (reduce inequality within and among countries), and 39 countries in SDG 14 (conserve and sustainably use the oceans, seas, and marine resources). For SDG 1, zero is used for "no data." For SDG 14, the average value is used. For SDG 10, zero is used, except for New Zealand, using the average value.

6.3. *Results and discussion*

We use model (DEA-CI) with the sustainability constraints to obtain the corresponding DEA scores from SDI Index and Dashboards for 157 countries. The resulting DEA scores are reported in Table A1 of the Appendix. The second to fourth columns present the pillar-indexes and the first-level results. The fifth column is the second-level results, the SD indicator. Note that the country's order is based on the descending order of the SD indicator. The following sections present a summary of the results.

6.3.1. *Summary of the first-level results*

Table 2 provides the average scores, as well as the minimum scores, of the 157 countries in the Envi, Soci, and Econ pillars. On average, the relative performance of the environmental pillar (0.802) is better than that of the social (0.744) and economic (0.726) pillars.

Table 3 summarizes the top five best and worst countries according to the DEA score of three pillar-indexes. The table highlights that **Denmark**, **Finland**, **Norway**, and **Sweden** are

Table 2. Maximum, minimum, average, and median of DEA results.

	Envi-CI	**Soci-CI**	**Econ-CI**
Max	1.000	1.000	1.000
Average	0.802	0.744	0.726
Min	0.524	0.000	0.000
Median	0.828	0.748	0.803

Table 3. Top 10 best and worst countries according to three pillar-indexes.

	Envi-index		**Soci-index**		**Econ-index**	
Ranking	**Country**	**Score**	**Country**	**Score**	**Country**	**Score**
1	**Denmark**	1.000	**Denmark**	1.000	**Sweden**	1.000
2	**Norway**	1.000	**Sweden**	1.000	**Norway**	0.972
3	Croatia	1.000	Iceland	1.000	**Denmark**	0.970
4	**Finland**	1.000	**Norway**	1.000	Japan	0.968
5	**Sweden**	1.000	**Finland**	0.978	Cuba	0.968
6	Slovenia	0.989	Germany	0.960	**Finland**	0.961
7	Estonia	0.982	Belgium	0.955	France	0.948
8	France	0.978	Slovenia	0.950	Germany	0.948
9	Romania	0.971	Switzerland	0.937	Netherlands	0.945
10	Malta	0.970	New Zealand	0.930	Australia	0.942
148	Angola	0.622	Mauritania	0.537	Turkmenistan	0.348
149	Afghanistan	0.620	Liberia	0.532	CDR**	0.345
150	Madagascar	0.603	Mali	0.525	Malawi	0.343
151	Liberia	0.587	CDR**	0.516	Sierra Leone	0.255
152	Sierra Leone	0.570	Nigeria	0.509	Liberia	0.252
153	Singapore	0.569	Afghanistan	0.497	Niger	0.168
154	Chad	0.564	Niger	0.482	Madagascar	0.163
155	CAR*	0.549	Chad	0.426	Chad	0.115
156	Togo	0.543	CAR*	0.359	CAR*	0.105
157	CDR**	0.524	Saudi Arabia	0.000	Burundi	0.000

Note: *CAR: Central African Republic; **CDR: Democratic Republic of the Congo.

in the top 10 best countries that appear in all three pillar-indexes. On the other hand, the Central African Republic, Chad, the Democratic Republic of the Congo, and Liberia are the worst countries in all three pillar indexes.

6.3.2. *Sustainability indicator of SDGs*

As discussed in previous sections, if there is a negative correlation existing between the three pillar-indexes, developing an aggregated indicator by adding up three SD dimensions with different weighting schemes is unsuitable. Table 4 shows the positive correlation results of the three pillar-indexes from the results of the DEA analysis. Hence, we can use different weighting schemes to determine the aggregated SD indicator of SDGs.

Instead of using different weighting schemes, we use model (DEA-CI) with sustainability constraints for the three pillar-indexes to find the aggregate SD indicator (sustainability indicator of SDGs) shown in the last column of Table A1 in the Appendix. That is the second level of the proposed approach. Table 5 provides the best and worst five countries in terms of the SD indicator, with an average of 0.791. It is consistent with Table 5 that Denmark, Finland,

Table 4. Correlation results of three pillar-indexes.

	Envi	**Soci**	**Econ**
Envi	1.000	—	—
Soci	0.777	1.000	—
Econ	0.794	0.770	1.000

Table 5. Top best and worst five countries in SD indicator.

	Country	**Envi**	**Soci**	**Econ**	**SD indicator**
1	Denmark	1.000	1.000	0.970	1.000
2	Sweden	1.000	1.000	1.000	1.000
3	Norway	1.000	1.000	0.972	1.000
4	Finland	1.000	0.978	0.961	0.989
5	Slovenia	0.989	0.950	0.906	0.969
153	Afghanistan	0.620	0.497	0.494	0.552
154	Niger	0.630	0.482	0.168	0.546
155	Congo, Dem. Rep.	0.524	0.516	0.345	0.520
156	Chad	0.564	0.426	0.115	0.485
157	Central African Republic	0.549	0.359	0.105	0.435
				Avg.	0.791

Norway, and Sweden are the top 10 best countries. The Central African Republic, Chad, and the Democratic Republic of the Congo are the worst in all three pillar indexes.

7. Conclusions

Generally, the popular approach to constructing SD indicators for countries follows three steps: (1) selecting criteria related to sustainability performance; (2) classifying criteria into three-pillar (or dimensional) criteria based on three SD pillars (environmental, social, and economic); (3) using DEA to measure the pillars performance of countries as the sustainability indicator. A similar approach is used for the firm's ESG performance measures.

We found a research gap in the literature: the consideration of overlapping criteria in three pillars (or dimensions). The overlapping criteria can be classified into more than one of three sustainable dimensions (environmental, social, and economic). Such overlapping criteria can also be found in the firm's ESG performance measures. For example, "green energy" can be classified as either economic or environmental. Other issues are (1) the shortcoming of a negative correlation between criteria and pillars and (2) the strong and weak sustainability considerations.

We propose a two-level approach with the DEA method to resolve the overlapping criteria issue and the negative correlation of criteria. Moreover, we propose using a weight constraint in the DEA models (sustainability constraint), which enhances the robustness and flexibility of the SD indicator results and enables modelers to consider strong and weak sustainability. We also find that the above issues and the proposed two-level approach can be applied to a firm's ESG performance indicators. Note that the sustainability constraints "enforce" DMUs to consider all the criteria in a balanced manner and maintain flexibility to obtain optimal weighting scores from the DEA analysis. The approach consists of two levels. The first level uses DEA to find three pillar-CIs: environmental-CI (Envi-CI), social-CI (Soci-CI), and economic-CI (Econ-CI). The second level uses a DEA method with sustainability constraints.

Finally, we apply the two-level approach to find the sustainability indicator of the UN SDGs. After the first-level approach, we have three pillar-CIs with no negative correlation. We further use the second-level DEA method to find the sustainability indicators for each country. Denmark, Sweden, and Norway are among the best.

Appendix

Table A1. Pillar indexes and SD indicator by country.

Country	Envi	Soci	Econ	SD indicator
Sweden	1.000	1.000	1.000	1.000
Denmark	1.000	1.000	0.970	1.000
Norway	1.000	1.000	0.972	1.000
Finland	1.000	0.978	0.961	0.989
Slovenia	0.989	0.950	0.906	0.969
France	0.978	0.921	0.948	0.963
Iceland	0.828	1.000	0.921	0.959
Germany	0.955	0.960	0.948	0.958
Japan	0.945	0.917	0.968	0.957
Cuba	0.945	0.911	0.968	0.956
Croatia	1.000	0.876	0.910	0.953
The Czech Republic	0.950	0.910	0.931	0.940
Estonia	0.982	0.886	0.900	0.939
The Netherlands	0.902	0.929	0.945	0.937
Belgium	0.918	0.955	0.912	0.936
Malta	0.970	0.901	0.876	0.934
Austria	0.917	0.928	0.939	0.933
United Kingdom	0.915	0.922	0.941	0.931
New Zealand	0.900	0.930	0.933	0.931
Switzerland	0.921	0.937	0.923	0.930
Ireland	0.930	0.927	0.904	0.928
Canada	0.872	0.916	0.935	0.925
Spain	0.908	0.893	0.930	0.919
Romania	0.971	0.824	0.858	0.911
Italy	0.929	0.861	0.893	0.911
Australia	0.873	0.880	0.942	0.910
Hungary	0.931	0.875	0.886	0.908
Latvia	0.940	0.841	0.869	0.903
Uruguay	0.883	0.854	0.920	0.901
Chile	0.898	0.850	0.904	0.901
Belarus	0.900	0.870	0.896	0.898
Argentina	0.898	0.831	0.898	0.898
Portugal	0.893	0.891	0.901	0.897
Brazil	0.922	0.783	0.869	0.895
Bulgaria	0.939	0.811	0.852	0.893
Suriname	0.939	0.777	0.851	0.893

(Continued)

Table A1. (*Continued*)

Country	Envi	Soci	Econ	SD indicator
Poland	0.893	0.870	0.892	0.893
Moldova	0.900	0.885	0.856	0.893
Greece	0.936	0.839	0.852	0.892
Korea, Rep.	0.853	0.838	0.933	0.891
Ukraine	0.908	0.861	0.872	0.890
Mexico	0.885	0.808	0.881	0.883
Ecuador	0.889	0.809	0.878	0.883
Costa Rica	0.926	0.842	0.762	0.882
Venezuela, RB	0.910	0.720	0.856	0.882
Russian Federation	0.910	0.698	0.855	0.881
Serbia	0.892	0.867	0.835	0.879
Slovak Republic	0.896	0.861	0.856	0.878
Uzbekistan	0.862	0.889	0.802	0.875
Trinidad and Tobago	0.849	0.777	0.899	0.873
Algeria	0.885	0.784	0.856	0.870
Dominican Republic	0.900	0.748	0.839	0.869
Cyprus	0.886	0.834	0.778	0.859
Turkey	0.851	0.774	0.864	0.857
Macedonia, FYR	0.902	0.801	0.815	0.856
Luxembourg	0.802	0.888	0.825	0.855
Israel	0.756	0.830	0.879	0.854
Malaysia	0.825	0.804	0.884	0.854
United States	0.785	0.807	0.905	0.853
Kyrgyz Republic	0.857	0.847	0.778	0.852
Thailand	0.844	0.763	0.860	0.852
Jamaica	0.830	0.797	0.872	0.850
Lithuania	0.854	0.845	0.841	0.850
Tunisia	0.875	0.788	0.825	0.849
Belize	0.825	0.746	0.870	0.847
Albania	0.863	0.781	0.830	0.846
Montenegro	0.803	0.840	0.852	0.846
Morocco	0.870	0.769	0.818	0.843
Armenia	0.858	0.827	0.807	0.843
Colombia	0.845	0.771	0.837	0.841
Bosnia and Herzegovina	0.864	0.816	0.808	0.839
Gabon	0.912	0.708	0.771	0.835
Georgia	0.855	0.814	0.814	0.834
Paraguay	0.840	0.794	0.816	0.828
Singapore	0.569	0.744	0.929	0.826
Azerbaijan	0.841	0.791	0.805	0.823
Bolivia	0.855	0.731	0.791	0.822

Table A1. (*Continued*)

Country	Envi	Soci	Econ	SD indicator
El Salvador	0.829	0.736	0.807	0.818
Peru	0.847	0.770	0.790	0.817
Lebanon	0.803	0.722	0.830	0.816
Vietnam	0.822	0.808	0.810	0.816
Jordan	0.780	0.807	0.816	0.811
Saudi Arabia	0.813	0.000	0.806	0.810
Kazakhstan	0.781	0.823	0.795	0.809
Barbados	0.744	0.778	0.836	0.806
Tajikistan	0.824	0.788	0.752	0.806
Honduras	0.837	0.743	0.776	0.805
Guyana	0.838	0.716	0.770	0.803
South Africa	0.844	0.759	0.712	0.799
Nicaragua	0.826	0.750	0.774	0.799
Bhutan	0.838	0.698	0.762	0.798
Sri Lanka	0.837	0.730	0.762	0.798
Iran, Islamic Rep.	0.803	0.729	0.780	0.792
Egypt, Arab Rep.	0.844	0.702	0.743	0.790
Mauritius	0.704	0.742	0.824	0.781
China	0.752	0.748	0.803	0.777
United Arab Emirates	0.744	0.681	0.802	0.772
Panama	0.790	0.662	0.736	0.762
Philippines	0.779	0.727	0.744	0.761
Oman	0.759	0.695	0.757	0.758
Nepal	0.793	0.720	0.612	0.755
Namibia	0.839	0.678	0.644	0.750
Mongolia	0.748	0.748	0.679	0.748
Bahrain	0.647	0.704	0.779	0.739
Botswana	0.770	0.706	0.687	0.737
Timor-Leste	0.774	0.693	0.597	0.731
Burundi	0.780	0.686	0.000	0.730
Guatemala	0.708	0.676	0.747	0.727
Qatar	0.684	0.722	0.724	0.723
Indonesia	0.730	0.704	0.716	0.723
Kuwait	0.751	0.633	0.692	0.720
Lao PDR	0.719	0.717	0.661	0.718
Rwanda	0.784	0.660	0.493	0.717
Ghana	0.755	0.675	0.659	0.713
Syrian Arab Republic	0.758	0.672	0.553	0.712
Senegal	0.705	0.690	0.663	0.698
Myanmar	0.770	0.628	0.583	0.692
Turkmenistan	0.667	0.708	0.348	0.687

(*Continued*)

Table A1. (*Continued*)

Country	Envi	Soci	Econ	SD indicator
Zimbabwe	0.703	0.665	0.576	0.684
Zambia	0.747	0.628	0.512	0.682
Iraq	0.631	0.668	0.697	0.682
Cambodia	0.624	0.700	0.646	0.672
Lesotho	0.698	0.645	0.469	0.670
Swaziland	0.667	0.625	0.670	0.669
Djibouti	0.640	0.590	0.689	0.664
Bangladesh	0.695	0.635	0.450	0.664
Uganda	0.705	0.626	0.435	0.663
Pakistan	0.720	0.558	0.613	0.662
India	0.684	0.609	0.638	0.660
Tanzania	0.669	0.641	0.457	0.655
Kenya	0.650	0.658	0.596	0.654
Cameroon	0.718	0.601	0.498	0.654
Malawi	0.722	0.594	0.343	0.651
Ethiopia	0.662	0.627	0.552	0.644
Yemen, Rep.	0.725	0.571	0.535	0.639
Gambia, The	0.721	0.570	0.409	0.637
Mozambique	0.671	0.603	0.423	0.635
Congo, Rep.	0.658	0.611	0.502	0.633
Burkina Faso	0.719	0.563	0.368	0.632
Cote d'Ivoire	0.644	0.577	0.559	0.609
Benin	0.625	0.581	0.386	0.602
Angola	0.622	0.579	0.506	0.600
Sudan	0.622	0.578	0.560	0.599
Guinea	0.647	0.556	0.355	0.598
Haiti	0.640	0.559	0.431	0.597
Mauritania	0.669	0.537	0.517	0.595
Sierra Leone	0.570	0.616	0.255	0.592
Mali	0.635	0.525	0.372	0.575
Madagascar	0.603	0.547	0.163	0.574
Togo	0.543	0.602	0.417	0.571
Nigeria	0.639	0.509	0.461	0.567
Liberia	0.587	0.532	0.252	0.558
Afghanistan	0.620	0.497	0.494	0.552
Niger	0.630	0.482	0.168	0.546
Congo, Dem. Rep.	0.524	0.516	0.345	0.520
Chad	0.564	0.426	0.115	0.485
Central African Republic	0.549	0.359	0.105	0.435

References

Allen, C., Metternicht, G., & Wiedmann, T. (2016). National pathways to the Sustainable Development Goals (SDGs): A comparative review of scenario modelling tools. *Environmental Science & Policy*, 66, 199–207.

Allen, C., Nejdawi, R., El-Baba, J., Hamati, K., Metternicht, G., & Wiedmann, T. (2017). Indicator-based assessments of progress towards the sustainable development goals (SDGs): A case study from the Arab region. *Sustainability Science*, 12(6), 975–989.

Balcaen, S. & Ooghe, H. (2006). Thirty-five years of studies on business failure: An overview of the classic statistical methodologies and their related problems. *The British Accounting Review*, 38(1), 63–93.

Banker, R.D., Charnes, A., & Cooper, W.W. (1984). Some models for estimating technical and scale inefficiencies in data envelopment analysis. *Management Science*, 30(9), 1078–1092.

Barton, H. (2000). Conflicting perceptions of the neighbourhood. In: H. Barton (Ed.), *Sustainable Communities* (pp. 3–18). London: Earthscan.

Blascak, N. & Tranfaglia, A. (2021). *Decomposing Gender Differences in Bankcard Credit Limits*. Finance and Economics Discussion Series 2021-072. Washington: Board of Governors of the Federal Reserve System. https://doi.org/10.17016/FEDS.2021.072.

Böhringer, C. & Jochem, P.E. (2007). Measuring the immeasurable — A survey of sustainability indices. *Ecological Economics*, 63(1), 1–8.

Boubaker, S., Le, T.D.Q., & Ngo, T. (2022a). Managing bank performance under COVID-19: A novel inverse DEA efficiency approach. *International Transactions in Operational Research*, 30(2023), 2436–2452. https://doi.org/10.1111/itor.13132.

Boubaker, S., Manita, R., & Mefteh-Wali, S. (2020). Foreign currency hedging and firm productive efficiency. *Annals of Operations Research*, 313, 833–854. https://doi.org/10.1007/s10479-020-03730-5.

Boubaker, S., Manita, R., & Rouatbi, W. (2021). Large shareholders, control contestability and firm productive efficiency. *Annals of Operations Research*, 296(1), 591–614.

Boubaker, S., Do, D.T., Hammami, H., & Ly, K.C. (2022b). The role of bank affiliation in bank efficiency: A fuzzy multi-objective data envelopment analysis approach. *Annals of Operations Research*, 311, 611–639. https://doi.org/10.1007/s10479-020-03817-z.

Burford, G., Tamás, P., & Harder, M.K. (2016). Can we improve indicator design for complex sustainable development goals? A comparison of a values-based and conventional approach. *Sustainability*, 8(9), 861.

Chang, K., Cheng, X., Wang, Y., Liu, Q., & Hu, J. (2021). The impacts of ESG performance and digital finance on corporate financing efficiency in China. *Applied Economics Letters*, 30(40), 516–523.

Charles, V. & Díaz, G. (2017). A non-radial DEA index for Peruvian regional competitiveness. *Social Indicators Research*, 134(2), 747–770.

Charnes, A., Cooper, W.W., & Rhodes, E. (1978). Measuring the efficiency of decision making units. *European Journal of Operational Research*, 2(6), 429–444.

Chen, L., Zhang, L., Huang, J., Xiao, H., & Zhou, Z. (2021). Social responsibility portfolio optimization incorporating ESG criteria. *Journal of Management Science and Engineering*, 6(1), 75–85.

Cheng, C.H., Chan, C.P., & Sheu, Y.J. (2019). A novel purity-based k nearest neighbors imputation method and its application in financial distress prediction. *Engineering Applications of Artificial Intelligence*, 81, 283–299.

Cherchye, L. & Kuosmanen, T. (2004). Benchmarking sustainable development: A synthetic meta-index approach (No. 2004/28). Research Paper, UNU-WIDER, United Nations University (UNU).

Cherchye, L., Moesen, W., & Puyenbroeck, T. (2004). Legitimately diverse, yet comparable: On synthesizing social inclusion performance in the EU. *JCMS: Journal of Common Market Studies*, 42(5), 919–955.

Cherchye, L., Moesen, W., Rogge, N., & Van Puyenbroeck, T. (2007). An introduction to 'benefit of the doubt' composited indicators. *Social Indicators Research*, 82(1), 111–145.

Citterio, A. (2021). The role of ESG in predicting bank financial distress: Cross-country evidence. In *6th Annual International Workshop on Financial System Architecture & Stability (IWFSAS)*, August 30–31 2021. https://iwfsas.org/iwfsas2021/wp-content/uploads/2021/08/50_The-role-of-ESG-in-predicting-bank-financial-distress-cross-country-evidence.pdf.

CLI (2022). Composite leading indicators, organisation for economic co-operation and development. OECD.Stat. https://www.oecd.org/sdd/leading-indicators/oecd-composite-leading-indicators-clis.htm#:~:text=The%20OECD%20Composite%20Leading%20Indicators,and%20short%2Dterm%20economic%20situation.

Despotis, D.K. (2005a). A reassessment of the human development index via data envelopment analysis. *Journal of the Operational Research Society*, 56(8), 969–980.

Despotis, D.K. (2005b). Measuring human development via data envelopment analysis: The case of Asia and the Pacific. *Omega*, 33(5), 385–390.

Duarte, D.L. & de Moraes Barboza, F.L. (2020). Forecasting financial distress with machine learning — A review. *Future Studies Research Journal: Trends and Strategies*, 12(3), 528–574.

ESI (2005). Environmental sustainability index. Socioeconomic Data and Applications Center. http://sedac.ciesin.columbia.edu/data/collection/esi/.

Filippetti, A. & Peyrache, A. (2011). The patterns of technological capabilities of countries: A dual approach using composite indicators and data envelopment analysis. *World Development*, 39(7), 1108–1121.

Griggs, D., Smith, M.S., Rockström, J., Öhman, M.C., Gaffney, O., Glaser, G., Kanie, N., Noble, I., Steffen, W., & Shyamsundar, P. (2014). An integrated framework for sustainable development goals. *Ecology and Society*, 19(4). https://openresearch-repository.anu.edu.au/bitstream/1885/13423/1/Griggs%20et%20al%20An%20Integrated%20Framework%202014.pdf.

Hafiz, A., Lukumon, O., Muhammad, B., Olugbenga, A., Hakeem, O., & Saheed, A. (2015, March). Bankruptcy prediction of construction businesses: Towards a big data analytics approach. In *2015 IEEE First International Conference on Big Data Computing Service and Applications* (pp. 347–352), 30 March–2 April 2015, Redwood City, California, USA. IEEE.

Hák, T., Janoušková, S., & Moldan, B. (2016). Sustainable development goals: A need for relevant indicators. *Ecological Indicators*, 60, 565–573.

HDI (2020). Human Development Index (HDI), human development reports. United Nations Development Programme. http://hdr.undp.org/en/content/human-development-index-hdi.

Holden, E., Linnerud, K., & Banister, D. (2017). The imperatives of sustainable development. *Sustainable Development*, 25(3), 213–226.

Huang, Y.P. & Yen, M.F. (2019). A new perspective of performance comparison among machine learning algorithms for financial distress prediction. *Applied Soft Computing*, *83*, 105663.

Kasirian, M.N. & Yusuff, R.M. (2013). An integration of a hybrid modified TOPSIS with a PGP model for the supplier selection with interdependent criteria. *International Journal of Production Research*, 51(4), 1037–1054. https://doi.org/10.1080/00207543.2012.663107.

Kasirian, M.N., Yusuff, R.M., Ismail, M.Y. (2010). Application of AHP and ANP in supplier selection process-a case in an automotive company. *International Journal of Management Science and Engineering Management*, 5(2), 125–135.

Kim, S.Y. (2018). Predicting hospitality financial distress with ensemble models: The case of US hotels, restaurants, and amusement and recreation. *Service Business*, 12(3), 483–503.

Kuo, K.C., Yu, H.Y., Lu, W.M., & Le, T.T. (2022). Sustainability and corporate performance: Moderating role of environmental, social, and governance investments in the transportation sector. *Sustainability*, 14(7), 4095.

Le, T.D.Q., Ho, T.H., Nguyen, D.T., & Ngo, T. (2021). Fintech credit and bank efficiency: International evidence. *International Journal of Financial Studies*, 9(3), 44. https://www.mdpi.com/2227-7072/9/3/44.

Le Blanc, D. (2015). Towards integration at last? The sustainable development goals as a network of targets. *Sustainable Development*, 23(3), 176–187.

Li, Y., Liu, A.C., Yu, Y.Y., Zhang, Y., Zhan, Y., & Lin, W.C. (2022). Bootstrapped DEA and clustering analysis of eco-efficiency in China's hotel industry. *Sustainability*, 14(5), 2925.

Lu, W.M., Kuo, K.C., & Tran, T.H. (2022). Impacts of positive and negative corporate social responsibility on multinational enterprises in the global retail industry: DEA game cross-efficiency approach. *Journal of the Operational Research Society*, 1–16.

Mai, F., Tian, S., Lee, C., & Ma, L. (2019). Deep learning models for bankruptcy prediction using textual disclosures. *European Journal of Operational Research*, 274(2), 743–758.

Mair, S., Jones, A., Ward, J., Christie, I., Druckman, A., & Lyon, F. (2018). A critical review of the role of indicators in implementing the sustainable development goals. In: *Handbook of Sustainability Science and Research* (pp. 41–56). Cham: Springer.

Mizobuchi, H. (2014). Measuring world better life frontier: A composite indicator for OECD better life index. *Social Indicators Research*, 118(3), 987–1007.

Murias, P., Novello, S., & Martinez, F. (2012). The regions of economic well-being in Italy and Spain. *Regional Studies*, 46(6), 793–816.

Nardo, M. (2008). *Handbook on Constructing Composite Indicators-Methodology and User Guide*. Paris, France: OECD Publishing.

Neumayer, E. (2003). *Weak Versus Strong Sustainability: Exploring the Limits of Two Opposing Paradigms*. Northampton, USA: Edward Elgar Publishing.

Ngo, T. & Tsui, K.W.H. (2021). Estimating the confidence intervals for DEA efficiency scores of Asia-Pacific airlines. *Operational Research*. https://doi.org/10.1007/s12351-021-00667-w.

Niemeijer, D. & de Groot, R.S. (2008). A conceptual framework for selecting environmental indicator sets. *Ecological Indicators*, 8(1), 14–25.

Nilsson, M., Griggs, D., & Visbeck, M. (2016). Map the interactions between sustainable development goals: Mans Nilsson, Dave Griggs and Martin Visbeck present a simple way of rating relationships between the targets to highlight priorities for integrated policy. *Nature*, 534(7607), 320–323.

Nilsson, M., Lucas, P., & Yoshida, T. (2013). Towards an integrated framework for SDGs: Ultimate and enabling goals for the case of energy. *Sustainability*, 5(10), 4124–4151.

Nunes, A.R., Lee, K., & O'Riordan, T. (2016). The importance of an integrating framework for achieving the sustainable development goals: The example of health and well-being. *BMJ Global Health*, 1(3), e000068.

Pearce, D.W., Markandya, A., & Barbier, E.B. (1989). *Blueprint for a Green Economy Earthcan.* Abingdon, UK: Earthscan.

Pongiglione, F. (2015). The need for a priority structure for the sustainable development goals. *Journal of Global Ethics*, 11(1), 37–42.

Rickels, W., Dovern, J., Hoffmann, J., Quaas, M.F., Schmidt, J.O., & Visbeck, M. (2016). Indicators for monitoring sustainable development goals: An application to oceanic development in the European Union. *Earth's Future*, 4(5), 252–267.

Saeidi, R., Amin, G., Raissi, S., & Gattoufi, S. (2013). An efficient DEA method for ranking woven fabric defects in textile manufacturing. *International Journal of Advanced Manufacturing Technology*, 68, 349–354.

Saisana, M. & Tarantola, S. (2002). State-of-the-art report on current methodologies and practices for composite indicator development. European Commission, Joint Research Centre, Institute for the Protection and the Security of the Citizen, Technological and Economic Risk Management Unit.

Schmidt-Traub, G., Kroll, C., Teksoz, K., Durand-Delacre, D., & Sachs, J.D. (2017). National baselines for the sustainable development goals assessed in the SDG index and dashboards. *Nature Geoscience*, 10(8), 547.

SDG Index (2022). Sustainable development report 2022. http://sdgindex.org/.

Sena, A., Freitas, C.M.D., Barcellos, C., Ramalho, W., & Corvalan, C. (2016). Measuring the invisible: Analysis of the sustainable development goals in relation to populations exposed to drought. *Ciencia & Saude Coletiva*, 21(3), 671–684.

Simar, L. & Wilson, P.W. (2002). Non-parametric tests of returns to scale. *European Journal of Operational Research*, 139(1), 115–132.

Simon, D., Arfvidsson, H., Anand, G., Bazaz, A., Fenna, G., Foster, K., Jain, G., Hansson, S., Evans, L.M., Moodley, N., Nyambuga, C.,

Oloko, M., Ombara, D.C., Patel, Z., Perry, B., Primo, N., Revi, A., Van Niekerk, B., Wharton, A., & Wright, C. (2016). Developing and testing the urban sustainable development goals targets and indicators — A five-city study. *Environment and Urbanization*, 28(1), 49–63. https://doi.org/10.1177/0956247815619865.

SSI website (2018). Aggregation, calculation methodology. https://ssi.with-koeln.de/documents/summary-ssi-2018.pdf.

Stafford-Smith, M., Griggs, D., Gaffney, O., Ullah, F., Reyers, B., Kanie, N., Stigson, B., Shrivastava, P., Leach, M., & O'Connell, D. (2017). Integration: The key to implementing the sustainable development goals. *Sustainability Science*, 12(6), 911–919.

Sun, J., Li, H., Huang, Q.H., & He, K.Y. (2014). Predicting financial distress and corporate failure: A review from the state-of-the-art definitions, modeling, sampling, and featuring approaches. *Knowledge-Based Systems*, 57, 41–56.

Swain, R.B. (2018). A critical analysis of the sustainable development goals. In: *Handbook of Sustainability Science and Research* (pp. 341–355). Cham: Springer.

Thaker, K., Charles, V., Pant, A., & Gherman, T. (2021). A DEA and random forest regression approach to studying bank efficiency and corporate governance. *Journal of the Operational Research Society*, 1–28. https://doi.org/10.1080/01605682.2021.1907239.

Tofallis, C. (2013). An automatic-democratic approach to weight setting for the new human development index. *Journal of Population Economics*, 26(4), 1325–1345.

UN (2015). The sustainable development agenda. United Nations. http://www.un.org/sustainabledevelopment/development-agenda/.

UNCED (1992). Report of the United Nations conference on environment and development. https://www.un.org/esa/dsd/agenda21/res_agenda21_40.shtml.

UNSTATS (2020). Global indicator framework for the sustainable development goals and targets of the 2030 agenda for sustainable development. https://unstats.un.org/sdgs/indicators/indicators-list/.

Wilson, J., Tyedmers, P., & Pelot, R. (2007). Contrasting and comparing sustainable development indicator metrics. *Ecological Indicators*, 7(2), 299–314.

Wong, Y.H. & Beasley, J.E. (1990). Restricting weight flexibility in data envelopment analysis. *Journal of the Operational Research Society*, 41(9), 829–835.

Yang, F.C., Kao, R.H., Chen, Y.T., Ho, Y.F., Cho, C.C., & Huang, S.W. (2017). A common weight approach to construct composite indicators: The evaluation of fourteen emerging markets. *Social Indicators Research*, 137, 463–479.

https://doi.org/10.1142/9781800615786_0016

Chapter 16

Applications of Data Envelopment Analysis to Evaluate Resource Mobilization and Regional Integration Outcomes: Recent Trends and Future Challenges

Dominik Naeher[*,‡] **and Raghavan Narayanan**[†,§]

[*]*University of Goettingen, Goettingen, Germany*
[†]*European Bank for Reconstruction and Development, London, UK*
[‡]*dnaeher@uni-goettingen.de*
[§]*narayanr@ebrd.com*

Abstract

The literature that uses data envelopment analysis (DEA) for policy analysis is thriving and has expanded from studies evaluating efficiency at the micro level to macroeconomic applications of DEA, assessing efficiency across different domains at the country, regional, and global levels. This chapter reviews recent trends in this body of literature, focusing on applications of DEA in the contexts of national resource mobilization and regional economic integration. These applications have been particularly influential in informing policymakers in the international development community, as several recent World Bank reports involving insights obtained from DEA demonstrate. In addition, this literature has developed a new approach to not only compare achieved absolute levels of outcomes but also to track relative performance by accounting for each country's and region's individual economic structure and associated capacity. Despite recent progress on some of the methodological issues facing the application of DEA, several challenges remain. The chapter discusses possible avenues for future research to address these challenges, including refinements on existing techniques for DEA and exploring new ways of combining DEA with other methods, such as artificial intelligence and machine learning techniques.

Keywords: Data envelopment analysis, efficiency, non-parametric frontier, resource mobilization, private capital flows, revenue potential, taxation, regional integration, international trade, machine learning, evidence-based policymaking.

1. Introduction

Empirical applications of data envelopment analysis (DEA) initially focused on evaluating efficiency at the micro level, considering decision-making units (DMUs) such as firms, hospitals, and schools, among others. More recent research has expanded the application of DEA to macroeconomic contexts, and there is now a well-established body of research that uses DEA to assess efficiency at the country, regional, and global level. For example, this is evident in the increasing number of studies applying DEA to assess public sector efficiency (Gupta & Verhoeven, 2001; Afonso *et al.*, 2005; Herrera & Pang, 2005), domestic and foreign resource mobilization (Bogetić *et al.*, 2022; Naeher & Narayanan, 2022), national transportation and commerce systems (Rashidi & Cullinane, 2019; Wang *et al.*, 2020), and regional economic integration (Naeher, 2015; Naeher & Narayanan, 2020).

This chapter contributes to the literature by reviewing recent trends in this body of research and discussing current challenges as well as possible avenues for future research to address these challenges. In doing so, we focus on applications of DEA to two important macroeconomic contexts, namely resource mobilization and regional integration. This focus is motivated by two observations. First, these two strands of literature have been particularly influential in informing policymakers within the international development community, as two recent evaluation studies conducted by the World Bank (2018, 2020), which prominently feature applications of DEA, demonstrate. In particular, these studies highlight that applying DEA to macroeconomic contexts can provide useful, data-driven evidence as input to policy advice, including for assessing achievements in targeting areas with certain needs and characteristics and informing decisions about future policy interventions.

Second, we argue that applications of DEA to these macroeconomic contexts have helped to fill an important gap in the academic literature by providing a new approach for researchers to not only compare achieved absolute levels of outcomes (e.g., revenue-to-GDP ratios in the context of taxation, or intra-regional trade shares in the context of economic integration), but also to assess achieved outcomes relative to each country's and region's individual economic structure and associated capacity. Specifically, using DEA to account for differences in economic structure across countries and regions provides a way to track relative performance over time and assess untapped potential to further improve performance.

At the same time, various challenges remain which future research in this field must address to improve upon the currently used methodology. We therefore also contribute to the literature by discussing possible avenues for future research in this direction, including both refinements on existing techniques for DEA and exploring new ways of combining DEA with other methods, such as artificial intelligence and machine-learning techniques.

The chapter is structured as follows. Section 2 provides a brief overview of the selection of input and output variables and methods used in the literature for applying DEA at the macroeconomic level. Section 3 discusses the results of recent studies which use DEA to evaluate national resource mobilization and regional integration outcomes. Section 4 concludes by discussing the limitations and challenges facing the covered body of literature and suggesting possible avenues for future research to address these challenges.

2. Variable Selection and Methods

Studies applying DEA at the macroeconomic level typically focus on a particular topic that largely determines the choice of input and output variable(s) included in the analysis. Especially for cross-country studies with a global scope, the selection of variables is often driven by the availability of data on empirical macro indicators, which tends to be limited for samples going beyond OECD countries. For instance, Gupta & Verhoeven (2001) and Herrera & Pang (2005) estimate and compare the efficiency of government spending on health and education across samples that include many low-income countries. The output variables are thus chosen to represent a set of standard health and education indicators with wide data availability, such as life expectancy, infant mortality, school enrollment, and adult literacy rate. As an input variable, government per-capita spending on health and education, measured in purchasing power parity (PPP) terms, is used.

A common approach in this literature is to aggregate information on several input (or output) variables into a single composite index. By normalizing the resulting index, one benefit of this approach is that the results are easy to interpret. For example, Bogetić *et al.* (2022) and Naeher & Narayanan (2022) construct composite indices capturing the enabling environment for resource mobilization in each country, which are then used as input variables in a DEA. The included indicators are based on a structured literature review and selected to represent those factors identified in the literature as driving forces behind domestic and international resource flows. The same approach has also been used for the output variable in DEA. For example, Naeher & Narayanan (2020) study efficiency and untapped potential in regional economic integration by constructing a composite regional integration index, which is then used as the output variable in a DEA.

Another feature shared by many studies in this body of literature is that various different model specifications are estimated in parallel within the same study, instead of focusing on a single (best) specification. For example, the study by Herrera & Pang (2005) considers both input-oriented and output-oriented models, as well as both DEA and free disposal hull (FDH) approaches, and then compares the results obtained across the different specifications. The main goal behind this approach is to verify the robustness of the obtained results. At the same time, it may also be argued that there is no established 'best practice' in model selection within this literature yet, with convergence on a single preferred approach still lacking.

More generally, economic studies using DEA (or other non- and semi-parametric methods such as stochastic frontier analysis and artificial neural networks) often view themselves as needing to defend their choice of method against critics that favor parametric (regression-based) approaches. Given the strong focus on causal inference in modern economics, this is perhaps no big surprise. On the other hand, it is important to note that the research questions addressed by studies using DEA in macroeconomic contexts are typically of a purely descriptive nature and thus do not require structural approaches that allow for causal identification.

Using DEA to evaluate efficiency at the macro level has been argued to offer several benefits compared to alternative methods for performance assessments, such as regression-based approaches. Most importantly, DEA calculates the efficiency score for a given DMU based on the performance of a subset of efficient units in the sample with similar input factors as the considered DMU (i.e., the pair of units which define the segment of the frontier against which the performance of the respective DMU is assessed). This differs from regression-based approaches, which rely on comparisons relative to the average performance across all units in the sample. If the considered DMUs are very heterogenous — as is typically the case in macroeconomic contexts where countries from different parts of the world with different economic structures and stages of development are compared — then DEA has the advantage that the efficiency estimates for any given DMU are not determined by other units with fundamentally different characteristics and prospects but only by those units that feature comparable levels of the considered input factors. In addition, applying no-parametric DEA typically requires fewer assumptions about the structure of the data than parametric methods such as regression analysis. The results generated by DEA therefore tend to be easier to interpret and are not subject to the same econometric challenges that regression-based methods typically face, which makes DEA particularly suitable for policy-oriented analyses.

3. Applications of DEA to Resource Mobilization and Regional Integration

In 2015, the global development community adopted the 2030 Sustainable Development Agenda and the Sustainable Development Goals (SDGs) that underpin it. Countries made commitments at the 21st Conference of the Parties to the United Nations Framework Convention on Climate Change. In July 2015, the third international conference on Financing for Development in Addis Ababa recognized that the financial resources needed to achieve the SDGs far exceeded the financial flows at the national level. These mandates have created a critical role for multilateral institutions to analyze national resource mobilization opportunities in depth, including creating conditions to attract private capital at scale, strengthening government revenue generation capacity, and benefiting from deeper regional cooperation and integration.

Despite the importance of resource mobilization and economic integration to both national policymakers and international development organizations, there is currently only limited data-driven evidence available that allows decisionmakers to quantify and compare achievements across countries, map achieved progress against stated goals, and guide decision-making. For example, the indicators used to assess countries' performance in attracting private capital flows are typically based exclusively on achieved outcomes (such as foreign direct investment (FDI) inflows relative to GDP), which do not account for the different economic conditions and constraints faced by different countries. Similarly, efforts by governments to raise tax revenues are most commonly assessed based on revenue-to-GDP ratios as the main performance indicator. Given that poorer countries tend to feature less favorable infrastructure and economic conditions for generating private or public finances, it is not surprising that low-income countries (LICs) are typically found to perform poorly along these indicators.

It is important to note, however, that the results obtained based on these indicators provide little insight on how *efficient* countries are in mobilizing resources, i.e., how well each country performs relative to its own potential, as determined by its economic structure and constraints. We now turn to the discussion of three recent studies that address this gap in the literature by using DEA to evaluate countries' relative efficiencies in mobilizing resources and fostering regional economic integration, relative to their enabling domestic environments.

3.1. *Private capital mobilization*

The study by Naeher & Narayanan (2022) applies DEA to assess countries' achievements in attracting private capital relative to the quality of each country's domestic enabling environment, i.e., relative to the economic conditions and constraints identified in the literature that determine each country's potential capital flows. The analysis is tailored to the context of private capital mobilization and is conducted along the following three steps. First, data on the achieved levels of private capital flows for 135 countries are obtained, covering all income groups and geographical regions. The considered types of capital flows are FDI, portfolio equity investment, and private sector borrowing, and each of these indicators is measured relative to a country's GDP. The three indicators are then combined into a single composite measure (the DEA output variable) using standard normalization and aggregation techniques.

Second, a set of input variables is identified to capture countries' domestic economic enabling environment for private capital mobilization (also referred to as "pull factors"; see Fernandez-Arias, 1996; Ghosh *et al.*, 2014; Reinhart *et al.*, 2016). The selection of these variables is based on a structured literature review and results in 14 empirical indicators covering seven dimensions of countries' domestic enabling environment: (1) market-related factors such as capital return and growth potential, (2) institutional and regulatory quality, (3) openness to trade, (4) economic and political stability, (5) infrastructure development, (6) financial development, and (7) natural and human resources. To facilitate the analysis and interpretation of the results, the 14 indicators are aggregated into a single composite index (the DEA input index), which can be thought of as a proxy variable capturing the enabling environment for private capital mobilization in each country.

Finally, the DEA input index and output variable are combined to construct the production possibility frontier for private capital mobilization. The performance of each country in attracting private capital is then assessed relative to the estimated frontier based on the efficiency scores calculated via an output-oriented DEA. The objective of this analysis is to identify those countries that generate relatively high levels of private capital compared to other countries with similar economic conditions and constraints and those countries that apparently fall short of their potential.

In the context of private capital mobilization, the results obtained from the DEA can be interpreted in the following way. The frontier defines the maximum possible level of capital flows for any given level of the considered input factors, based on what other countries

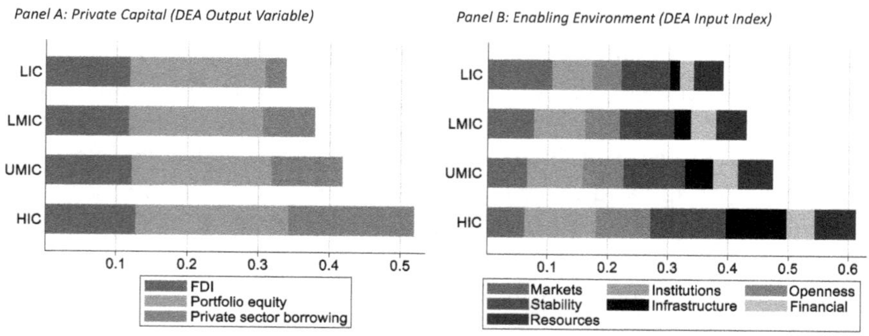

Figure 1. Private capital flows and enabling environment by income group, 2015–2018.

Note: This figure shows normalized and aggregated values of private capital flows (Panel A) and enabling factors (Panel B), with higher values indicating larger flows (relative to a country's GDP) and better enabling environments. The sample consists of 135 countries.

Source: Authors' illustration based on the results in Naeher & Narayanan (2022).

in the sample are achieving. The relative performance of a country can thus be assessed based on the distance to the estimated frontier. In essence, this provides an estimate of the capital each country should be able to generate based on what other countries with similar economic characteristics are achieving. Countries that generate more capital with the same economic characteristics (or countries that generate the same level of capital with less favorable economic characteristics) are thus considered to be more efficient than others.

Figure 1 shows the normalized measure of aggregate private capital flows (DEA output variable) and the composite proxy measure of each country's enabling environment (DEA input index) for the period 2015–2018 by income group. Clearly, countries with lower income levels tend to mobilize less private capital relative to GDP than countries belonging to higher income groups (Panel A). At the same time, countries with lower income levels also tend to feature less favorable economic conditions for generating private capital flows (Panel B). Together, this implies that poorer countries are not necessarily less efficient in mobilizing private capital than richer economies when considering the limited resources available to them.

Indeed, the efficiency scores obtained from DEA reveal that many LICs which do not rank highly according to achieved levels of private capital flows relative to GDP are in fact performing on or very close to the efficient frontier. This indicates that these countries are good performers given their available resources and the constraints facing them. In particular, this finding complements the insights from existing studies in this literature by providing a different perspective on countries' achievements that accounts for differences in countries' fundamental economic structures and constraints. In addition, the frontier analysis provides quantitative estimates of the untapped mobilization potential of each country, i.e., the magnitude by which each country should, in principle, be able to increase its private capital flows given its available economic resources and conditions. By repeating the analysis for different time periods (2007–2010, 2011–2014, and 2015–2018), it is furthermore possible to track the performance of countries (and regions) over time. Specifically, the results show that the countries in Sub-Saharan Africa experienced the strongest increases in DEA efficiency scores, on average, among all regions between 2007 and 2018. This indicates that various (low-income) countries in this region were catching up with the global frontier.

3.2. *Government revenue mobilization*

The study by Bogetić *et al.* (2022) applies DEA to assess governments' achievements in raising domestic revenue. The applied methodology is similar to the one in Naeher & Narayanan (2022) and follows along the same three major steps as described above. First, the output variable is specified as total government revenue as a share of GDP. Data on this variable is available for 182 countries, and the analysis focuses on 118 low- and middle- income countries in the period 2008–2019 (high-income countries are excluded). The variables considered as input factors in the DEA are selected based on the economic literature on the determinants of government revenue potential (see the survey in Gupta, 2007) and include (1) per-capita income levels, (2) sectoral composition of output, (3) international trade relative to GDP, and (4) the age dependency ratio (capturing a country's demographic composition). As before, the indicators are aggregated into a single composite measure (the DEA input index), which can be thought of as a proxy variable capturing the conditions determining each country's economic revenue potential.

Finally, the DEA input index and output variable are combined to construct the production possibility frontier for government revenues. In the context of revenue mobilization, the efficiency scores obtained from an output-oriented DEA can be thought of as capturing the degree to which each country is raising revenue up to its own potential (or still features untapped potential to further increase revenue).

According to the results, the 118 low- and middle-income countries in the sample achieved about 62% of their revenue potential, on average, in the period 2016–2019. The analysis also shows that there is considerable variation in efficiency scores, both across income groups and among countries belonging to the same income group. This suggests that accounting for differences in fundamental economic conditions via DEA can lead to important additional insights about the revenue potential of countries at different income levels. Specifically, a simple comparison of revenue-to-GDP ratios across countries suggests that LICs perform considerably worse than countries in higher income groups. This is illustrated in Figure 2 (Panel A): Except for one outlier (South Sudan), revenue levels in LICs show moderate variation, ranging from 2% of GDP (Somalia) to 23% of GDP (Mozambique), while the revenue levels in the other three income groups tend to be much higher. In contrast, the efficiency scores of LICs cover almost the entire possible range of DEA scores from zero to one (see Figure 2, Panel B). The DEA results thus reveal that there are many LICs that, in fact, perform better than countries from higher income groups given the economic resources and conditions available to them. Hence, looking only at absolute revenue-to-GDP ratios will likely be misleading for drawing conclusions about how much more revenues a country can potentially raise, given its economic structure. This puts the efforts of less-developed countries to increase their revenue-to-GDP ratios (IMF, 2018) in perspective, as it may be much harder for these countries to raise their revenue levels to the same levels as observed among countries in higher income groups.

The results also indicate that countries with low or high efficiency in revenue mobilization are not concentrated in a particular income group or geographical region. Rather, all income groups and world regions include countries with high and low efficiency scores. This suggests that if future efforts aimed at improving revenue mobilization were to be targeted at countries considering their revenue

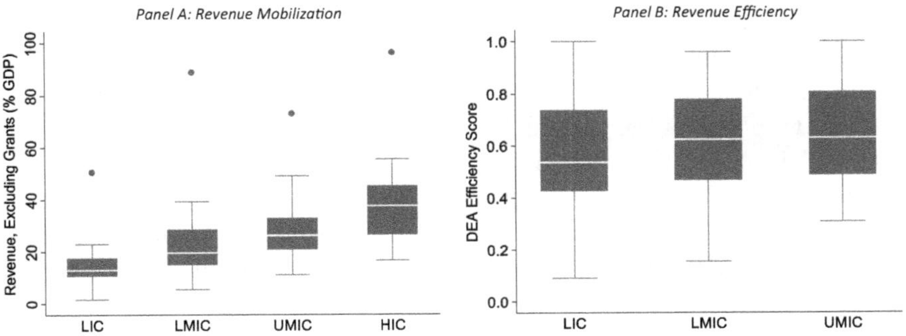

Figure 2. Revenue-to-GDP Ratios and DEA Scores by Income Group, 2016–2019.

Note: This figure shows box plots of government revenues (Panel A) and efficiency scores estimated via DEA (Panel B), with higher values indicating higher revenues (relative to a country's GDP) and better performance, respectively. The boxes represent the range of values between the 25th and 75th percentiles (including the median), with the ends of the whiskers representing the lower and upper adjacent values. The sample in Panel A consists of 182 countries. The sample in Panel B consists of 118 countries (high-income countries are excluded).

Source: Authors' illustration based on the results in Bogetić *et al.* (2022).

efficiency as captured by the DEA scores, this would not necessarily end up benefiting countries belonging to a particular income group or geographical region in a disproportional way. In other words, efficient government revenue mobilization appears to be a challenge across all income groups and regions; contrary to previous results based solely on comparing revenue-to-GDP ratios, it is not the exclusive province of low-income countries.

3.3. *Regional economic integration*

Regional integration is frequently argued to constitute a key strategy for reducing conflict and fostering an environment conducive to economic growth and development (UNESCAP, 2016; EU, 2021). In the economic literature on regional integration, different dimensions and indicators of integration are considered, ranging from studies focusing exclusively on trade, migration, or financial market integration to multidimensional composite measures capturing even regional value chains, peace and security aspects, and digital-economy integration (Naeher, 2015; Huh & Park, 2018; UNESCAP, 2020).

The study by Naeher & Narayanan (2020) follows the latter approach and measures regional integration using a composite regional integration (CRI) index that is constructed by aggregating information on 11 empirical indicators covering five dimensions of regional integration: (1) trade integration, (2) financial integration, (3) regional investment and production networks, (4) movement of people, and (5) peace and security. With the exception of dimension (6), all included indicators capture intraregional shares based on bilateral (dyadic) data rather than national (country-level) data.

The CRI index is then used as the output variable in an output-oriented DEA designed to estimate the empirical magnitudes of untapped regional integration potential across geographical subregions of the world. This is done by setting the CRI index in relation to a composite proxy measure of the enabling environment for regional integration (the DEA input index), capturing enabling factors such as trade openness, cross-border infrastructure, and business regulation environment. Specifically, the analysis first constructs the empirical production possibility frontier for regional integration outcomes, where the frontier specifies the potential level of integration (quantified by the CRI index) as a function of currently available resources and enabling factors needed to foster economic integration in each subregion. The frontier is then used to compare each subregion's achieved level of integration relative to its integration potential to obtain an estimate of the subregion's untapped integration potential.

The key assumption underlying this approach is that subregions that feature similar levels of enabling factors for integration should, in principle, be able to achieve similar levels of regional integration outcomes. Estimated efficiency scores below one can thus be interpreted as untapped potential in regional integration. This provides a way to identify those subregions that are achieving relatively high levels of integration compared to other subregions facing similar enabling environments, and those subregions that still feature large untapped integration potential.

The considered sample comprises 193 economies, grouped into 19 subregions spanning all geographical regions of the world (subregions are based on the classification of the UN, 2017). Figure 3 plots the CRI index over the DEA input index and shows the resulting production possibility frontier for regional integration (dotted line) in the year 2017. The frontier turns out to be defined by four subregions: Middle Africa (at the lower end of the enabling environment), Southeast Asia, East Asia, and Western Europe (upper end). Globally, average regional integration levels across all subregions are estimated to be at 60% of the estimated potential. The subregions in Europe achieve the highest value, around 67% on average, of their integration potential. In Asia, subregions achieve about 60% of their integration potential, on average. The subregions of the Americas and Africa are found to achieve 57% of their integration potential.

According to the constructed CRI index, Western Europe features the highest level of regional integration. This is consistent with the view that Europe (and the members of the European Union in particular) features one of the highest levels of economic integration in the world. At the same time, East Asia follows closely behind, and there are several dimensions of regional integration, such as trade and regional investment and production networks, for which both East Asia and Southeast Asia achieve outcomes that are comparable to those achieved by European subregions.

As can be observed from Figure 3, Northern Europe, North America, and South America all feature similar values of the CRI index as Southeast Asia. However, only the latter is found to be efficient. In contrast, Northern Europe, North America, and South America are each located at a considerable distance from the frontier, indicating that these subregions still feature large untapped integration potential. This highlights the value added from performing this DEA as compared to a mere comparison of achieved levels of integration (which would have suggested that Northern Europe, North America, South America, and Southeast Asia are all performing very similarly).

The subregions with the lowest values of the CRI index are Middle Africa, Northern Africa, Central Asia, and Central America. Among these, however, Middle Africa is in fact efficient (given the much weaker enabling environment), while Northern Africa, Central Asia, and Central America each feature large untapped integration potential. Notice that this also suggests that all world regions include subregions with large untapped integration potential.

There are several ways in which these findings are useful from a policy perspective. First, they may assist organizations with a mandate to foster regional integration to assess their achievements in

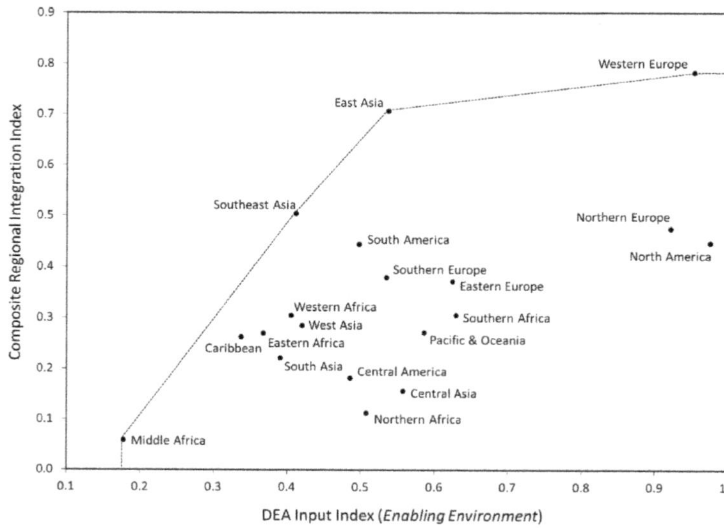

Figure 3. Regional integration frontier.

Note: This figure shows the production possibility frontier for regional integration (dotted line) across geographical regions. The DEA output variable is a Composite Regional Integration Index, with higher values indicating stronger regional economic integration. The DEA input variable is a composite proxy measure of enabling factors, with higher values indicating better enabling environments.

Source: Authors' illustration based on the results in Naeher & Narayanan (2020).

targeting subregions with certain needs and characteristics, such as subregions that perform relatively well despite low absolute levels of integration or subregions that are still far away from reaching their full integration potential. Moreover, the analysis provides insights about which enabling factors (e.g., trade-related factors, physical infrastructure, or private sector regulations) are particularly strong or weak in certain subregions. For example, these insights may be used to evaluate the relevance and appropriateness of interventions aimed at increasing integration in any given subregion and prioritize different types of interventions in future programs.

4. Conclusion and Avenues for Future Research

We have argued that the recent expansion of applications of DEA to macroeconomic contexts, such as national resource mobilization and regional economic integration, has generated added value both to the international scientific community and to policymakers working in these contexts. Most importantly, this body of literature demonstrates the suitability of DEA to generate results that complement the insights obtained from comparisons of achieved outcomes, by providing a method to account for differences in economic structure and input factors across countries. Moreover, we have discussed examples illustrating that DEA offers several advantages over other methods for performance assessment (including regression-based approaches), including that DEA (a) calculates efficiency based on relevant "peer" DMUs with similar levels of input factors rather than relying on comparisons relative to the average performance across all units in the sample, (b) allows for a simultaneous analysis of multiple inputs and outputs, (c) requires fewer assumptions about the structure of the data than parametric methods, and (d) tends to be easier to interpret.

At the same time, various challenges remain which are yet to be resolved. The rest of this chapter, therefore, focuses on discussing some of these challenges and how ongoing and future research may be able to address them.

One important issue facing any application of DEA is the selection of input and output variables. These choices are crucial as they directly affect the obtained efficiency scores, and considering different inputs may, in general, lead to different conclusions. This issue is exacerbated by the fact that, due to its non-parametric nature, DEA provides little feedback on the relevance and adequacy of the considered combinations of input and output variables. In fact, implementing DEA will produce efficiency scores for any combination of inputs and outputs, irrespective of how these have been selected. Applications of DEA are thus always at risk of considering inadequate combinations of input and output variables and must provide justification for the choices made.

In the literature, this issue is typically addressed in two ways. First, the selection of variables can be based on previously existing evidence on the factors determining the outputs of interest (e.g., based on a structured literature review). Since DEA does not provide insights about (causal) determinants, in practice the evidence used to inform the selection of inputs in DEA often relies on evidence obtained from regression-based studies that examine the determinants of the considered output variables. This also highlights the limitations arising from the lack of causal identification facing applications of non-parametric frontier analysis more generally. While, for the purpose of efficiency analysis, DEA offers several benefits compared to the traditional econometric approach based on regression analysis, it does not provide the same tools required to identify causal relationships in observational data (Thanassoulis, 1993; Sickles & Zelenyuk, 2019). The results obtained from DEA are therefore generally of a descriptive nature and rely critically on the criteria applied by researchers in choosing the considered input and output variables. Unfortunately, so far, no unified, generally accepted approach to defining these criteria exists.

A second way in which the literature has implicitly addressed the issue of model selection is by performing comprehensive robustness checks to assess the sensitivity of the obtained results to changes in the chosen input variables. For example, Afonso *et al.* (2005) and Naeher & Narayanan (2020) report results for different weighting and aggregation schemes, including for alternative specifications obtained by iteratively excluding individual input variables from the analysis. While such robustness checks can provide useful insights, they only represent an imperfect solution as they do not typically consider all possible alternative specifications, and even if that were the case, it would not be clear which specification would be more adequate. We therefore conclude that more research and scientific debate is needed to inform the variable choices in future applications of DEA and move closer to a widely accepted framework for DEA model selection criteria.

Another important challenge for studies applying DEA, and for studies focused on macroeconomic contexts in particular, is to explain variations in the efficiency scores across countries estimated via DEA. For example, the study by Bogetić *et al.* (2022) applies DEA to estimate the empirical production possibility frontier for government revenues as a function of countries' domestic economic structures and constraints, and then uses the frontier to assess each country's economic revenue potential (i.e., the level of revenue, relative to GDP, that each country should be able to mobilize given its economic structure and constraints). Once the frontier and associated efficiency scores have been obtained, the question of interest at the second stage then is which factors can explain the observed variation in efficiency scores across countries, i.e., why do some countries manage to generate more revenues than other countries with comparable economic conditions? Unlike the mere quantification of efficiency in the first stage of the analysis, this question is essentially of a causal nature. Many studies therefore follow a two-stage approach in which the second stage consists of using the DEA scores obtained from the DEA in the first stage as the dependent variable in a subsequent regression analysis.

However, there are well-documented econometric challenges that can lead to biased results when using the efficiency scores obtained from frontier analyses as a dependent variable in a regression analysis (Simar & Wilson, 2007). In fact, Simar & Wilson (2011) argue that there are only two statistical models available in which second-stage regressions lead to consistent estimates. The first one is the truncated regression approach proposed by Simar & Wilson (2007), and the second one is a model based on ordinary least squares (OLS) proposed by Banker & Natarajan (2007). However, second-stage OLS estimation is consistent only under a narrow set of assumptions on the data-generating process. Simar & Wilson (2011) therefore argue that bootstrap methods are required to obtain an adequate tool for inference in the second stage. Other approaches proposed in the literature to address these issues include stochastic frontier analysis (Battese & Coelli, 1995; Karakaplan & Kutlu, 2017) and combinations of DEA with stochastic models (Ngo & Tsui, 2021; Tsionas, 2021). In addition, Dao *et al.* (2021) argue that a hierarchical regression approach can provide better results by offering higher statistical power.

Future research applying DEA at the macroeconomic level may benefit from using these recently developed methodological advancements to analyze determinants of efficiency in regression frameworks while at the same time preserving the benefits of non-parametric approaches such as DEA at the first stage. The same applies to other methodological refinements of DEA, such as the inverse DEA approach, which focuses on computing the magnitudes by which the inputs of a given DMU need to be reduced to keep efficiency fixed following a decline in output (Boubaker *et al.*, 2022).

Finally, another promising direction for future research is to investigate ways to combine DEA with recently developed methods in the literature on artificial intelligence and machine learning. For example, in the context of regional economic integration, some studies apply machine-learning techniques such as network clustering algorithms, hierarchical clustering algorithms, and evolutionary algorithms to evaluate the compositions of regional agreements, such as monetary unions (Benassy-Quere & Coupet, 2005; Tsangarides & Qureshi, 2008) and customs unions (De Lombaerde *et al.*, 2021, 2022a), and related research questions (see the survey in De Lombaerde *et al.*, 2023). The key idea behind these approaches is to start with a sample of individual countries and then subsequently merge those countries that are most integrated with each other according to specified criteria (e.g., the criteria implied by the Optimum Currency Area theory). The landscape of country clusters emerging at later steps of this iterative process is then used to assess the extent to which actual groupings of countries (e.g., real-world customs unions) are aligned with the structure of 'natural markets' observable from the data or appear to be driven by other (e.g., political) factors (De Lombaerde *et al.*, 2022b).

Future research may be able to combine DEA with machine-learning techniques in order to generate new ways for studying (macro)economic phenomena. For example, in the context of regional integration, it may be interesting to perform assessments of (untapped) integration potential not only for existing regional organizations, but also for those regional clusters of countries emerging from the machine-learning algorithms described above. In particular, this would provide a new, data-driven benchmark for assessing the overall efficiency of the global system of regionalization against which the current landscape of regional agreements could be evaluated.

Overall, it can be concluded that the economic literature applying DEA is growing in various directions, including a trend to study macroeconomic phenomena as well, which previously remained unconsidered by applications of DEA. At the same time, there appears to be ample scope for further contributions to the field of DEA, both in refining existing techniques for DEA and by exploring new ways of combining DEA with other methods.

References

Afonso, A., Schuknecht, L., & Tanzi, V. (2005). Public sector efficiency: An international comparison. *Public Choice*, 123(3–4), 321–347.

Banker, R.D. & Natarajan, R. (2007). Evaluating contextual variables affecting productivity using data envelopment analysis. *Operations Research*, 56(1), 48–58.

Battese, G.E. & Coelli, T.J. (1995). A model for technical inefficiency effects in a stochastic frontier production function for panel data. *Empirical Economics*, 20(2), 325–332.

Benassy-Quere, A. & Coupet, M. (2005). On the adequacy of monetary arrangements in sub-Saharan Africa. *World Economy*, 28(3), 349–373.

Bogetić, Ž., Naeher, D., & Narayanan, R. (2022). Measuring untapped revenue potential in developing countries: Cross-country frontier and panel data analysis. *Journal of Public Finance and Public Choice.* https://doi.org/10.1332/251569121X16613592544147.

Boubaker, S., Le, T.D.Q., & Ngo, T. (2022). Managing bank performance under COVID-19: A novel inverse DEA efficiency approach. *International Transactions in Operational Research*, 30(2023), 2436–2452. https://doi.org/10.1111/itor.13132.

Dao, T.T., Mai, X.T., Ngo, T., Le, T.D.Q., & Ho, H. (2021). From efficiency analyses to policy implications: A multilevel hierarchical linear model approach. *International Journal of the Economics of Business*, 28(3), 457–470.

De Lombaerde, P., Naeher, D., & Saber, T. (2021). Regional integration clusters and optimum customs unions: A machine learning approach. *Journal of Economic Integration*, 36(2), 262–281.

De Lombaerde, P., Naeher, D., & Saber, T. (2022a). On the optimal size and composition of customs unions: An evolutionary approach. *Computational Economics.* https://doi.org/10.1007/s10614-022-10307-w.

De Lombaerde, P., Naeher, D., & Saber, T. (2022b). Evaluating accession decisions in customs unions: A dynamic machine learning approach. Mimeo. University of Goettingen.

De Lombaerde, P., Naeher, D., Vo, H.T., & Saber, T. (2023). Regional economic integration and machine learning: Policy insights from the review of literature. *Journal of Policy Modeling*, 45(5), 1077–1097.

EU (2021). *The European Union and Regional Economic Integration: Creating Collective Public Goods — Past, Present and Future.* Strasbourg: European Parliamentary Research Service.

Fernandez-Arias, E. (1996). The new wave of private capital inflows: Push or pull? *Journal of Development Economics*, 48(2), 389–418.

Ghosh, A.R., Qureshi, M., Kim, J., & Zalduendo, J. (2014). Surges. *Journal of International Economics*, 92(2), 266–285.

Gupta, A.S. (2007). Determinants of tax revenue efforts in developing countries. IMF Working Paper, No. 07/184. Washington, DC: International Monetary Fund.

Gupta, S. & Verhoeven, M. (2001). The efficiency of government expenditure: Experiences from Africa. *Journal of Policy Modelling*, 23, 433–467.

Herrera, S. & Pang, G. (2005). Efficiency of public spending in developing countries: An efficiency frontier approach. World Bank Policy Research Working Paper, No. 3645. Washington, DC: World Bank.

Huh, H.S. & Park, C.Y. (2018). Asia-Pacific regional integration index: Construction, interpretation, and comparison. *Journal of Asian Economics*, 54, 22–38.

IMF (2018). *Fiscal Monitor: Capitalizing on Good Times*. Washington, DC: International Monetary Fund.

Karakaplan, M.U. & Kutlu, L. (2017). Endogeneity in panel stochastic frontier models: An application to the Japanese cotton spinning industry. *Applied Economics*, 49(59), 5935–5939.

Naeher, D. (2015). An empirical estimation of Asia's untapped regional integration potential using data envelopment analysis. *Asian Development Review*, 32(2), 178–195.

Naeher, D. & Narayanan, R. (2020). Untapped regional integration potential: A global frontier analysis. *Journal of International Trade & Economic Development*, 29(6), 722–747.

Naeher, D. & Narayanan, R. (2022). Attracting private capital for development: Are poorer countries less efficient? *International Economics and Economic Policy*. https://doi.org/10.1007/s10368-022-00549-2.

Ngo, T. & Tsui, K.W.H. (2021). Estimating the confidence intervals for DEA efficiency scores of Asia-Pacific airlines. *Operational Research*, 22(4), 3411–3434.

Rashidi, K. & Cullinane, K.P. (2019). Evaluating the sustainability of national logistics performance using data envelopment analysis. *Transport Policy*, 74, 35–46.

Reinhart, C.M., Reinhart, V., & Trebesch, C. (2016). Global cycles: Capital flows, commodities, and sovereign default. *American Economic Review*, 106(5), 574–580.

Sickles, R.C. & Zelenyuk, V. (2019). *Measurement of Productivity and Efficiency: Theory and Practice*. Cambridge: Cambridge University Press.

Simar, L. & Wilson, P. (2007). Estimation and inference in two-stage, semiparametric models of production processes. *Journal of Econometrics*, 136, 31–64.

Simar, L. & Wilson, P. (2011). Two-stage DEA: Caveat emptor. *Journal of Productivity Analysis*, 36, 205–218.

Thanassoulis, E. (1993). A comparison of regression analysis and data envelopment analysis as alternative methods for performance assessments. *Journal of the Operational Research Society*, 44(11), 1129–1144.

Tsangarides, C.G. & Qureshi, M.S. (2008). Monetary union membership in West Africa: A cluster analysis. *World Development*, 36(7), 1261–1279.

Tsionas, M.G. (2021). Optimal combinations of stochastic frontier and data envelopment analysis models. *European Journal of Operational Research*, 294(2), 790–800.

UN (2017). *World Population Prospects: The 2017 Revision*. New York: United Nations Population Division.

UNESCAP (2016). Enhancing regional economic cooperation and integration in Asia and the Pacific. Note by the Secretariat: E/ESCAP/72/5. United Nations Economic and Social Commission for Asia and the Pacific, Bangkok.

UNESCAP (2020). *Regional Integration for Sustainable Development in Asia and the Pacific: ESCAP Digital and Sustainable Regional Integration Index and indicator Framework*. Bangkok: United Nations ESCAP.

Wang, T., Kang, J.W., & Valentine, V.F. (2020). A holistic analysis of national E-commerce and logistics development. *Maritime Economics & Logistics*, 22, 500–513.

World Bank (2018). *Two to Tango: An Evaluation of World Bank Group Support for Fostering Regional Integration*. Washington, DC: World Bank.

World Bank (2020). *World Bank Group Approaches to Mobilize Private Capital for Development: An Independent Evaluation*. Washington, DC: World Bank.

https://doi.org/10.1142/9781800615786_0017

Chapter 17

Efficiency Analysis of Brazilian Regions for the Sustainable Expansion of Food Production Based on Satellite Data: A Data Envelopment Analysis Approach

Paulo Nocera Alves Junior[*,‡]
and Damaris Chieregato Vicentin[†,§]

[*]*Universidad Católica del Norte (UCN),*
Escuela de Ingeniería de Coquimbo (EIC), Coquimbo, Chile
[†]*School of Applied Science,*
State University of Campinas (Unicamp),
Limeira, Brazil
[‡]*paulo.alves@ce.ucn.cl*
[§]*damaris_sv@hotmail.com*

Abstract

An essential topic for policymakers is the environmental and socioeconomic impact of food supply chains. The world population is growing; consequently, food consumption is expected to increase, and Brazil is one of the important countries in global food production with the potential to supply food to meet this higher food demand in the world. Considering this context, the aim of this chapter is to propose a double-frontier data envelopment analysis (DEA) model to benchmark Brazilian regions for the sustainable expansion of food production. Thus, for this purpose, a slack-based measure (SBM) DEA model combined with the composite index double-frontier method is developed for efficiency analysis of the Brazilian regions, deemed decision-making units (DMUs), based on satellite data from the Brazilian Annual Land Use and Land Cover Mapping

Project (MAPBIOMAS) and the Greenhouse Gas Emission and Removal Estimating System (SEEG). The data are split into negative factors (NFs), undesirable outputs (UOs), or inputs (e.g., irrigation and CO_2 equivalent emissions related to the agricultural sector), and positive factors (PFs) or outputs (e.g., food productivity or pasture area). This chapter uses averaged data from 2017 to 2021 for 14 of the 26 (plus the Federal District) Brazilian states. As for the results, Santa Catarina is in the first position of the ranking (first recommendation for expansion of agricultural production). In terms of regions, the northeast has two states in the efficient frontier and none in the inverted frontier, while the southeast has none in the efficient frontier and two in the inverted frontier. The other regions have states in both frontiers. These results would help provide policymakers with strategies for future investment to expand food production. For future research, it is suggested to apply the proposed model using data from other countries and develop stochastic DEA models, besides investigating the efficiency of future scenarios.

Keywords: Data envelopment analysis (DEA), slack-based measure (SBM), double-frontier, sustainability, satellite data, food production, agricultural frontier expansion, CO_2 equivalent emissions, irrigation, and pasture area.

1. Introduction

Food production can be fundamental for the development of a country; however, it needs to consider a series of social and environmental challenges, such as the global loss of forests, global emissions of greenhouse gases (FAO, The State of Agricultural Commodity Markets, 2018; Bijleveld *et al.*, 2015), decline in biodiversity, cases of land grabbing, and forced labor (Zalles *et al.*, 2019). According to studies by Ermgassen *et al.* (2020), there is a risk of deforestation inherent in countries with greater production and export of food, such as Brazil.

In addition to these problems, Tian *et al.* (2021) demonstrate that in the future, the population growth rate will increase, so it is important that countries like Brazil, which export a lot of food, increase their production rate while improving their efficiency in the use of water and fertilizers. The authors also state that the imbalance of supply and demand for food could cause hunger, malnutrition, and environmental degradation, especially in countries with financial distress.

To compare the performance of Brazilian regions for the sustainable expansion of food production requires a more comprehensive approach. Therefore, different data envelopment analysis (DEA) models have been proposed by researchers in the literature, as can be seen in the network DEA models from Alves Junior *et al.* (2021b) and Sun *et al.* (2022) or the dual-step procedure with network equilibrium model (NEM) and DEA from Branco *et al.* (2021).

Other recent models applied to different contexts could also be applied to the subject of expansion of food production, including comparative analyses between analytical hierarchy processes (AHP) and DEA (Rashidi, 2020), fuzzy multi-objective DEA method (Boubaker *et al.*, 2022b), comparing firm productive efficiency using regression, semi-parametric model of Olley and Pakes, and DEA approaches (Boubaker *et al.*, 2021) to estimating the confidence intervals for DEA efficiency scores (Ngo & Tsui, 2021), DEA with random forest regression approach (Thaker *et al.*, 2021), combination of stochastic frontier and DEA (Tsionas, 2021), and the inverse DEA efficiency approach (Boubaker *et al.*, 2022a) to measure the performance of DMUs.

A double-frontier DEA model with a slack-based measure (SBM) can contribute to the efficiency analysis, which identifies a possibility to include optimist and pessimistic views in a composite index (CI) analysis, introducing sustainability factors to improve them (Khoshroo *et al.*, 2022; Alves Junior *et al.*, 2022a, 2022c; Mai *et al.*, 2023).

Hence, a double-frontier DEA model using the SBM was proposed for efficiency analysis of the Brazilian regions, deemed Decision-Making Units (DMUs), based on data from the Brazilian Institute for Geography and Statistics (IBGE) and satellite data from the Brazilian Annual Land Use and Land Cover Mapping Project (MAPBIOMAS) and the Greenhouse Gas Emission and Removal Estimating System (SEEG).

The results of this research contribute to the expansion of the agricultural frontier by recommending the improvement of production in efficient areas found by the application of the model proposed in this chapter, i.e., in regions that require less area to be under irrigation and discourage deforestation by incentivizing the recovery of degraded pasture areas to expand production.

The chapter is divided into the following sections, besides this introduction: In Section 2, a synthesis of the literature review about the DEA, SBM, and the sustainable expansion of food production is made, which bases the proposal. In Section 3, the scientific methodology used is described. In Section 4, the results of the double-frontier

DEA model proposed for the sustainable expansion of food production are discussed. Finally, Section 5 concludes the chapter.

2. Literature Review

In the context of evaluating the performance of agricultural areas, some authors applied DEA models to evaluate these areas considered DMUs in terms of converting inputs (e.g., capital, labor, and land) into outputs (agricultural production), besides evaluating the impacts of interventions (Kalirajan *et al.*, 1996; Fulginiti & Perrin, 1997; Nin Pratt & Yu, 2008; Toma *et al.*, 2015).

Due to increasing concerns about the possibility of agriculture causing damage to the environment, other authors applied DEA models to measure the sustainability-related efficiency of agricultural supply chains, thus considering the production to satisfy human needs while reducing environmental impacts (DeSimone & Popoff, 2000; Masuda, 2016; Mwambo *et al.*, 2020).

Another possible application of DEA models related to agriculture is the assessment of irrigation performance in food-producing areas (Kibirige *et al.*, 2019; Yilmaz *et al.*, 2009; Raju & Kumar, 2006; Rodrigues-Díaz *et al.*, 2004).

Also, recent studies developed double-frontier DEA models or combined DEA models with other methodologies and evaluated the interaction among different systems, such as the one between the use of agricultural land (land use) and the delivery of agricultural products (transportation).

In this context, Khoshroo *et al.* (2022) developed a double-frontier DEA-based Malmquist productivity index to analyze the energy of DMUs in the agriculture sector, considering undesirable pollutant outputs. Alves Junior *et al.* (2022a) developed a double-frontier DEA model to evaluate the efficiency of regions to produce sugarcane destined for ethanol demand for implementing vehicle fleet renovation in the South American market. Alves Junior *et al.* (2022c) developed a double-frontier DEA model to evaluate the impact of the COVID-19 pandemic on the green-fuel supply chain of the ethanol sector. Alves Junior *et al.* (2021b) also developed a double-frontier DEA model for hierarchical network systems and applied it to support the bidding decisions of smallholder farmers in public calls in Brazil.

Regarding combining models, Alves Junior *et al.* (2021a) analyzed the efficiency of green transport corridors in exporting Brazilian agricultural bulk by applying a dual-step procedure of NEM and DEA, resulting in potential corridors to guide infrastructure investments and address public policies. Branco *et al.* (2021) analyzed agriculture land use and transportation networks by applying a dual-step procedure of NEM and DEA, resulting in suitable areas for future locations of soybean and corn production in Brazil. Further, Alves Junior *et al.* (2022b) focused on the development of a modeling framework to measure the FEW Nexus performance in agricultural production and transportation, among other systems, considering a dual-step procedure combining DEA and NEM. This resulted in the analysis of a sustainable agricultural frontier expansion under the FEW Nexus context.

In addition to these studies, some important studies on double-frontier DEA were developed, such as those by Paradi *et al.* (2004), who used DEA and worst-practice DEA in credit risk evaluation; Wang & Chin (2009), who proposed a new approach for the selection of advanced manufacturing technologies; and Azizi (2014), who established a DEA efficiency analysis using a DEA approach with double frontier.

3. Methodology

Due to the importance of countries like Brazil that export a lot of food, as is the case with Brazilian soybean production for exportation, this chapter aims to determine the efficient states where production could be expanded first.

The analysis considers the most productive Brazilian states. A double-frontier DEA model is used to measure the efficiency of the Brazilian soybean producing states. In the analysis, four measures related to the Brazilian soybean producing states were considered: (1) productivity, (2) area under irrigation, (3) pasture area, and (4) CO_2 equivalent emission due to deforestation.

First, in DEA, inefficiency is defined in frontier analysis as a deviation from the efficient frontier, and it is measured by following the non-parametric approach with multiple outputs and inputs (Tran & Ngo, 2014).

Charnes *et al.* (1978) were the pioneers in developing the first DEA model, also known as the CCR model, under a constant returns to scale (CRS) assumption. This model aimed to measure the efficiency of DMUs. In simple words, efficiency can be the sum of weighted outputs divided by the sum of weighted inputs. The fractional programming model in multiplier form is given in Equations (1–3) (Charnes *et al.*, 1978):

$$\text{Maximize } \theta = \frac{\sum_{j=1}^{n} u_j y_{j0}}{\sum_{i=1}^{m} v_i x_{i0}}, \tag{1}$$

$$\text{s.t.: } \frac{\sum_{j=1}^{n} u_j y_{jk}}{\sum_{i=1}^{m} v_i x_{ik}} \leq 1; \quad k = 1, 2, \ldots, z, \tag{2}$$

$$u_j, v_i \geq 0, \tag{3}$$

where θ is the efficiency, u_j is the weight of the jth output, v_i is the weight of the ith input, x_{i0} is the amount of ith input used by DMU_0 (under analysis), y_{j0} is the amount of the jth output produced by DMU_0 (under analysis), x_{ik} is the amount of the ith input used by the kth DMU, y_{jk} is the amount of the jth output produced by the kth DMU, m is the number of inputs used in the model, n is the number of outputs considered in the model, and z is the number of DMUs in our analysis.

According to Coelli *et al.* (2005) and Cooper *et al.* (2006), it is possible to impose $\sum_{i=1}^{m} v_i x_{i0} = 1$ to avoid an infinite number of solutions and create the input-oriented linear CCR as follows in Equations (4–7):

$$\text{Maximize } \theta = \sum_{j=1}^{n} u_j y_{j0}, \tag{4}$$

$$\text{s.t.: } \sum_{i=1}^{m} v_i x_{i0} = 1, \tag{5}$$

$$\sum_{j=1}^{n} u_j y_{jk} - \sum_{i=1}^{m} v_i x_{ik} \leq 0; \quad k = 1, 2, \ldots, z, \tag{6}$$

$$u_j, v_i \geq 0. \tag{7}$$

The input-oriented CCR in the envelopment form is as follows in Equations (8–11) (Cooper *et al.*, 2006):

$$\text{Minimize } \theta$$

$$\text{s.t.: } \sum_{k=1}^{z} \lambda_k x_{ik} + S_i^- - \theta x_{i0} = 0, \quad i = 1, 2, \ldots, m, \tag{8}$$

$$\sum_{k=1}^{z} \lambda_k y_{jk} - S_j^+ = y_{j0}, \quad j = 1, 2, \ldots, n, \tag{9}$$

$$\lambda_k \geq 0, \quad S_i^- \geq 0, \quad \text{and} \quad S_j^+ \geq 0, \tag{10}$$

where S_i^- is the slack variable for the ith input, S_j^+ is the slack variable for the jth output, and Λ_k is the contribution of the kth DMU to DMU_0 (under analysis). The other parameters and variables are the same as before.

After the CCR model's development, other DEA models were proposed with specific features. For example, Banker *et al.* (1984) developed a variable returns to scale (VRS), known as BCC. The model is similar to CCR, but BCC includes a constraint given in Equation (11) (Banker *et al.*, 1984):

$$\sum_{k=1}^{z} \lambda_k = 1. \tag{11}$$

These basic models are either input-oriented or output-oriented; on the other hand, the SBM developed by Tone (2001) uses slacks, considering improvements in both orientations (inputs and outputs) simultaneously, and it is possible to address CRS or VRS assumptions.

In contrast to the CCR and BCC models, the SBM is an additive model that is unit-invariant, non-oriented, and non-radial. The objective function can simultaneously improve both inputs and outputs without necessarily requiring them to be uniform or in equal proportion. The model computes the maximum possible improvement in each dimension (Thrall, 1996).

According to Liu *et al.* (2010), besides the conventional interpretation without considering undesirable measures, DMUs aim to minimize inputs and undesirable outputs as negative factors (Rentizelas *et al.*, 2019; Seiford & Zhu, 2002) while maximizing outputs and undesirable inputs as positive factors (Liu *et al.*, 2015). In this sense, productivity and pasture area are used as positive factors, i.e., the behavior of an output as "the higher, the better," because a higher productivity indicates more production using less area, while pasture area indicates suitable areas for future locations of soybean production without deforestation. Conversely, irrigation and CO_2 equivalent emissions due to deforestation are used as negative factors, i.e., the behavior of an input as "the lower, the better," because using less irrigation to produce soybean indicates less usage of water and energy to produce the same amount of soybean (e.g., using more natural rainfall water), while CO_2 equivalent emissions due to deforestation indicate no suitable areas for future locations of soybean production due to its environmental impact to discourage deforestation.

This chapter uses averaged data from 2017 to 2021 for 14 of the 26 (plus the Federal District) Brazilian states. From the 26 Brazilian states plus the Federal District (i.e., 27 regions), only 21 states produced soybeans from 2017 to 2021, i.e., Rio Grande do Norte, Paraíba, Pernambuco, Sergipe, Espírito Santo, and Rio de Janeiro were excluded from the analysis. From those 21 states, four had pasture areas only in the Amazonia biome, so Acre, Amazonas, Roraima, and Amapá were excluded from the analysis. From the other 17 states, three had no irrigation data, i.e., Rondônia, Ceará, and Alagoas were excluded from the analysis, resulting in the 14 states considered in this chapter.

The soybean productivity data were obtained from the Brazilian Institute for Geography and Statistics (IBGE, 2020). The data on pasture area and area under irrigation were obtained from the Brazilian Annual Land Use and Land Cover Mapping Project (MAPBIOMAS, 2022). And the CO_2 equivalent emissions due to deforestation were obtained from the Greenhouse Gas Emission and Removal Estimating System (SEEG, 2022).

In this study, we applied a double-frontier DEA model from the framework of Alves Junior *et al.* (2022c) and based on the SBM model from Tone (2001), and the variables related to the studies were obtained from Alves Junior *et al.* (2021a), Branco *et al.* (2021), and Alves Junior *et al.* (2022b).

The SBM model (Tone, 2001) allows DMUs to simultaneously maximize their outputs and minimize their inputs (Cook *et al.*, 2014).

The SBM model under the VRS assumption can be seen in Equations (12–17) (Tone, 2001):

$$\text{Minimize } \tau = t - \left(\frac{1}{m}\right)\left(\sum_{i=1}^{m} \frac{S_i^-}{x_{i0}}\right), \tag{12}$$

$$\text{s.t.: } t + \left(\frac{1}{n}\right)\left(\sum_{j=1}^{n} \frac{S_j^+}{y_{j0}}\right) = 1, \tag{13}$$

$$\sum_{k=1}^{z} \Lambda_k x_{ik} + S_i^- - t x_{i0} = 0 \quad i = 1, 2, \ldots, m, \tag{14}$$

$$\sum_{k=1}^{z} \Lambda_k y_{jk} - S_j^+ - t y_{j0} = 0 \quad j = 1, 2, \ldots, n, \tag{15}$$

$$\sum_{k=1}^{z} \Lambda_k - t = 0, \tag{16}$$

$$\Lambda_k \geq 0,\ S_i^- \geq 0,\ S_j^+ \geq 0, \quad \text{and} \quad t > 0, \tag{17}$$

where τ is the efficiency level, t is the model linearization variable, S_i^- is the slack variable for the ith input, S_j^+ is the slack variable for the jth output, Λ_k is the contribution of the kth DMU to DMU_0 (under analysis), x_{i0} is the amount of the ith input used by DMU_0 (under analysis), y_{j0} is the amount of the jth output produced by DMU_0 (under analysis), x_{ik} is the amount of the ith input used by the kth DMU, y_{jk} is the amount of the jth output produced by the kth DMU, m is the number of inputs used in the model, n is the number of outputs considered in the model, and z is the number of DMUs in our analysis.

For the double-frontier approach, we considered the CI suggested by Leta *et al.* (2005) and Alves Junior *et al.* (2022c), and the inverted efficiency model is as follows in Equations (18–23):

$$\text{Minimize } \tau^{\text{inverted}} = t - \left(\frac{1}{n}\right)\left(\sum_{j=1}^{n} \frac{S_j^+}{y_{j0}}\right), \tag{18}$$

$$\text{s.t.:}\quad t + \left(\frac{1}{m}\right)\left(\sum_{i=1}^{m} \frac{S_i^-}{x_{i0}}\right) = 1, \tag{19}$$

$$\sum_{k=1}^{z} \Lambda_k x_{ik} - S_i^- - t x_{i0} = 0 \quad i = 1, 2, \ldots, m, \tag{20}$$

$$\sum_{k=1}^{z} \Lambda_k y_{jk} + S_j^+ - t y_{j0} = 0 \quad j = 1, 2, \ldots, n, \tag{21}$$

$$\sum_{k=1}^{z} \Lambda_k - t = 0, \tag{22}$$

$$\Lambda_k \geq 0,\ S_i^- \geq 0,\ S_j^+ \geq 0, \quad \text{and} \quad t > 0. \tag{23}$$

According to Leta *et al.* (2005) and Alves Junior *et al.* (2022c), the CI can be seen in Equation (24), and it is the standardized arithmetic average value between the standard efficiency in Equations (12–17) and the inverted one (handling outputs as inputs, and vice versa) in Equations (18–23) used to compute the scores:

$$\tau_k^{\text{composite}} = \frac{[\tau_k^{\text{standard}} + (1 - \tau_k^{\text{inverted}})]/2}{\max\{[\tau_k^{\text{standard}} + (1 - \tau_k^{\text{inverted}})]/2]\}}, \quad k = 1, 2, 3, \ldots, z, \tag{24}$$

where τ_k^{inverted} is the inverted efficiency score estimated for the kth DMU and τ_k^{standard} is the standard efficiency score calculated for the kth DMU.

This approach indicates the most efficient states in terms of the solution that results in more productivity but using less area under irrigation and, at the same time, with more pasture area but less CO_2 equivalent emissions due to deforestation. According to Stabile *et al.* (2020) and Branco *et al.* (2021), it is possible to increase production and simultaneously decrease deforestation by using only suitable pasture to expand the cropland for production. For example, in Brazil, it is possible to achieve the expected production in 2050 using only pasture areas without deforestation. So, our approach is in accordance with the literature, considering these variables to discourage deforestation while expanding the agricultural frontier.

4. Results and Discussion

First, descriptive statistics were calculated using the analyzed data, as shown in Table 1. It is possible to observe that the variable with less variability is productivity, and the one with more variability is area under irrigation (ha). The variables were organized into two categories: outputs and inputs. The input variables are area under irrigation (ha) and CO_2e emissions due to deforestation (10^6 t), while the output variables are productivity (t/ha) and pasture area (10^3 ha).

Posteriorly, the double-frontier SBM model represented by Equations (12–24) was applied. The results from the efficiency analysis for the 14 states can be seen in Table 2. It shows the efficiency and ranking of the states in expanding the agricultural production frontier.

According to the results of this chapter presented in Table 2, nine states are part of the efficient frontier, but only five are benchmarks for inefficient DMUs (Maranhão, São Paulo, Paraná, Rio Grande do Sul, and Mato Grosso), and three of the nine states are part of both frontiers (Pará, Minas Gerais, and Distrito Federal). From the other five states, two are part of the inverted frontier only (São Paulo, and Rio Grande do Sul) and three are not part of any frontier (Maranhão, Paraná, and Mato Grosso).

Considering the CI (double frontier), Santa Catarina from the southern (S) region is the best state to expand the agricultural

Table 1. Descriptive statistics of the data.

Variable	Average	Standard deviation	Max	Min	Variable
Productivity (t/ha)	3.34	0.24	3.66	2.98	Output
Pasture area (10^3ha)	6,200.45	6,692.58	19,706.03	81.68	PF treated as output
Area under irrigation (ha)	76,945.32	102,265.21	337,734.39	129.82	Input
CO_2e emission due to deforestation (10^6t)	45.66	78.28	289.33	0.14	UO or NF treated as input

Table 2. Efficiency, CI, and ranking of the states.

Region	State	Abbreviation	Efficiency	Inv. ef	1-Inv. ef	Average	CI	Ranking
N	Pará	PA	1.0000	1.0000	0.0000	0.5000	0.5013	11
N	Tocantins	TO	1.0000	0.0963	0.9037	0.9518	0.9543	4
NE	Maranhão	MA	0.5199	0.0709	0.9291	0.7245	0.7264	8
NE	Piauí	PI	1.0000	0.0200	0.9800	0.9900	0.9925	2
NE	Bahia	BA	1.0000	0.1573	0.8427	0.9214	0.9237	6
SE	Minas Gerais	MG	1.0000	1.0000	0.0000	0.5000	0.5013	10
SE	São Paulo	SP	0.3583	1.0000	0.0000	0.1792	0.1796	12
S	Paraná	PR	0.6338	0.0491	0.9509	0.7923	0.7944	7
S	Santa Catarina	SC	1.0000	0.0051	0.9949	0.9975	1.0000	1
S	Rio Grande do Sul	RS	0.0201	1.0000	0.0000	0.0101	0.0101	14
CW	Mato Grosso do Sul	MS	1.0000	0.0569	0.9431	0.9715	0.9740	3
CW	Mato Grosso	MT	0.1011	0.7790	0.2210	0.1611	0.1615	13
CW	Goiás	GO	1.0000	0.1471	0.8529	0.9264	0.9288	5
CW	Distrito Federal	DF	1.0000	1.0000	0.0000	0.5000	0.5013	9

production frontier (CI = 1). The first six states that are part of the efficient frontier (best practices) only have a final CI >0.9000: one is from the northern (N) region (Tocantins), two from the northeast (NE) region (Piauí and Bahia), and two are from the center-west (CW) region (Mato Grosso do Sul and Goiás), but none of them are from the southeast (SE) region.

Regarding the inverted frontier (worst practices), from the last six positions of the ranking based on the CI, three areas are "inverted benchmarks" to other DMUs: Rio Grande do Sul from the S region, Pará from the N region, and Minas Gerais from the SE region, while São Paulo from the SE region and Distrito Federal from the CW region are part of the inverted frontier too, but none of them are from the NE region.

So, in terms of regions (considering more than one state), NE has two states in the efficient frontier and none in the inverted frontier, while SE has none in the efficient frontier and two in the inverted frontier, and the other regions have states in both frontiers.

Despite Santa Catarina in the S region being the first in the rank, it is important to note that Santa Catarina is the second state that uses less irrigation and has a relatively high productivity and low CO_2e emissions; however, it is in the 11th position among the 14 states when considering pasture area for the expansion. Therefore, considering the expected soybean production for 2050 from Branco *et al.* (2021), other states need to expand the production besides Santa Catarina. The second position is Piauí in the NE region, but it is 10th among the 14 states when considering pasture area for expansion. Along with the Tocantins in the N region, they are two of the four states from the MATOPIBA region. On the other hand, in the CW region, Mato Grosso do Sul (MS) and Goiás have a lot of pasture area to expand soybean production, despite the high usage of irrigation, but the productivity and low CO_2e emissions compensate for it.

In terms of benchmarks, Table 3 is a summary of benchmarks (the best measures for efficiency) and inverted benchmarks (the worst measures for efficiency).

Santa Catarina (S) is the best benchmark for states from the NE (Maranhão) and S (Paraná), and the third one for the Mato Grosso (CW), using the outputs productivity (t/ha) and Pasture area (10^3ha) with the inputs area under irrigation (ha) and CO_2e emission due to deforestation (10^6t) to be considered as the most sustainable state. States from CW (Mato Grosso do Sul and Distrito

Table 3. Benchmarks (best practices) and inverted benchmarks (worst practices).

Region	State	Part of the efficient frontier	Benchmarks for inefficient DMUs	Inverted bench-marks for the DMUs	Part of the inverted frontier
N	PA	Yes	—	TO, MA, BA, PR, MS, MT, GO	Yes
N	TO	Yes	MA	—	No
NE	MA	No	—	—	No
NE	PI	Yes	MT	—	No
NE	BA	Yes	—	—	No
SE	MG	Yes	—	BA, MS, MT, GO	Yes
SE	SP	No	—	—	Yes
S	PR	No	—	—	No
S	SC	Yes	MA, PR, MT	—	No
S	RS	No	—	TO, MA, PI, PR, SC	Yes
CW	MS	Yes	SP, PR, RS, MT	—	No
CW	MT	No	—	—	No
CW	GO	Yes	—	—	No
CW	DF	Yes	SP, PR	—	Yes

Federal) are the best benchmarks for states from CW (Mato Grosso), SE (São Paulo), and S (Rio Grande do Sul), as well as Paraná (S). Piauí (NE) is the benchmark for Mato Grosso (CW). Tocantins (N) is the benchmark for Maranhão (NE).

In summary, aiming to improve production in efficient areas and in accordance with the results and data analysis, policymakers are recommended to incentivize soybean production in regions with high productivity (as in Santa Catarina and Mato Grosso do Sul) and in those that need less area under irrigation (as in Santa Catarina, Piauí, and Tocantins). They should incentivize the recovery of degraded pasture areas to expand production but discourage deforestation (as in Santa Catarina, Goiás, Piauí, and Mato Grosso do Sul), besides promoting technological evolution, e.g., to improve seed quality for better productivity and reduce environmental impact, while improving the positive factors.

5. Conclusion

We developed a double-frontier DEA model based on the SBM to analyze and rank the soybean-producing states, considering two positive factors (productivity and pasture area to expand the agricultural frontier) and two negative factors (area under irrigation and CO_2e emissions due to deforestation that need to be discouraged). It was applied to averaged data from 2017 to 2021 for 14 Brazilian soybean producing states, including the Federal District (Pará, Tocantins, Maranhão, Piauí, Bahia, Minas Gerais, São Paulo, Paraná, Santa Catarina, Rio Grande do Sul, Mato Grosso do Sul, Mato Grosso, Goiás, and Distrito Federal).

The results showed that the most efficient state (Santa Catarina) had relatively low pasture area to expand production, so other states (Mato Grosso, Mato Grosso do Sul and Distrito Federal, São Paulo, Rio Grande do Sul, Paraná, Piauí, and Tocantins) need to be considered as well. The first five positions in the ranking had a CI above 0.9. These five states (Santa Catarina, Piauí, Mato Grosso do Sul, Tocantins, and Goiás) were distributed among four regions (S, NE, CW, and N). SE was the only region with all states resulting in a CI lower than 0.6000, with two states among the last five states in the ranking.

The recommendations for policymakers were related to improved production in efficient areas, i.e., in regions that need less area under irrigation and discourage deforestation, by incentivizing the recovery of degraded pasture areas to expand production, besides promoting technological evolution, e.g., to improve seed quality for better productivity, and reduce environmental impact, while improving positive factors.

A limitation of this study is the use of data only from one country. Hence, for future research, it is suggested to apply the proposed model using data from other countries. Also, it is suggested to develop models to investigate the efficiency of future scenarios, in terms of applications, and the development of other models, such as stochastic DEA models, in terms of methods.

References

Alves Junior, P.N., Melo, I.C., Okamura, F.T., & Cruz Júnior, J.C. (2022a). Analysis of the renewable energy generation capability for attending a national renovation fleet through ethanol-cell electric vehicles in a South American Market. In: M. Fathi, E. Zio & P.M. Pardalos (Eds.), *Handbook of Smart Energy Systems*. Cham: Springer. https://doi.org/10.1007/978-3-030-72322-4_127-1.

Alves Junior, P.N., Branco, J.E.H., Bartholomeu, D.B., Cruz Junior, J.C., & Caixeta-Filho, J.V. (2022b). Methods for evaluating food-energy-water nexus: Data envelopment analysis and network equilibrium model approaches. In: L.L.B. Lazaro, L.L. Giatti, L.S. Valente de Macedo & J.A. Puppim de Oliveira (Eds.), *Water-Energy-Food Nexus and Climate Change in Cities*. Sustainable Development Goals Series. Cham: Springer. https://doi.org/10.1007/978-3-031-05472-3_10.

Alves Junior, P.N., Melo, I.C., Branco, J.E.H., Bartholomeu, D.B., & Caixeta-Filho, J.V. (2021a). Which Green Transport Corridors (GTC) are efficient? A dual-step approach using Network Equilibrium Model (NEM) and Data Envelopment Analysis (DEA). *Journal of Marine Science and Engineering*, 9, 247. https://doi.org/10.3390/jmse9030247.

Alves Junior, P.N., Melo, I.C., Santos, R.M., Rocha, F.V. & Caixeta-Filho, J.V. (2022c). How did COVID-19 affect green-fuel supply chain? — A performance analysis of Brazilian ethanol sector. *Research in Transportation Economics*, 93, 101137, https://doi.org/10.1016/j.retrec.2021.101137.

Alves Junior, P.N., Melo, I.C., Yamanaka, L., Severino, M.R., & Rentizelas, A. (2021b). Supporting the bidding decisions of smallholder farmers in public calls in Brazil. *Agriculture*, 12, 48. https://doi.org/10.3390/agriculture12010048.

Azizi, H. (2014). DEA efficiency analysis: A DEA approach with double frontiers. *International Journal of Systems Science*, 45(11), 2289–2300. https://doi.org/10.1080/00207721.2013.768715.

Bijleveld, M., Naber, N., & Odegard, I. (2015). Food commodity footprints: Global GHG footprints and water scarcity footprints in agriculture. https://www.cedelft.eu/en/publications/1766/food-commodity-footprints-global-ghg-footprints-and-water-scarcity-footprints-in-agriculture (Accessed 2 February 2020).

Boubaker, S., Le, T.D.Q., & Ngo, T. (2022a). Managing bank performance under COVID-19: A novel inverse DEA efficiency approach. *International Transactions in Operational Research*, 30(2023), 2436–2452. https://doi.org/10.1111/itor.13132.

Boubaker, S., Manita, R., & Rouatbi, W. (2021). Large shareholders, control contestability and firm productive efficiency. *Annals of Operations Research*, 296(1), 591–614.

Boubaker, S., Do, D.T., Hammami, H., & Ly, K.C. (2022b). The role of bank affiliation in bank efficiency: A fuzzy multi-objective data envelopment analysis approach. *Annals of Operations Research*, 311, 611–639. https://doi.org/10.1007/s10479-020-03817-z.

Branco, J.E.H., Bartholomeu, D.B., Alves Junior, P.N., & Caixeta-Filho, J.V. (2021). Mutual analyses of agriculture land use and transportation networks: The future location of soybean and corn production in Brazil. *Agricultural Systems*. https://doi.org/10.1016/j.agsy.2021.103264.

Brazilian Institute of Geography and Statistics (IBGE) (2020). *Produção Agrícola Municipal*, Rio de Janeiro. https://www.ibge.gov.br/estatisticas/economicas/agricultura-e-pecuaria/9117-producao-agricola-municipal-culturas-temporarias-e-permanentes.html?edicao=31675. (Accessed 10 January 2022).

Charnes, A, Cooper, W.W., & Rhodes E. (1978). Measuring the efficiency of decision making units. *European Journal of Operational Research*, 2(6), 429–444.

Coelli, T.J., Rao, D.S.P., O'Donnell, C.J., & Battese, G.E. (2005). *An Introduction to Efficiency and Productivity Analysis* (2nd edn.). New York: Springer.

Cook, W.D., Tone, K., & Zhu, J. (2014). *Data Envelopment Analysis: Prior to Choosing a Model* (Vol. 44, pp. 1–4). UK: Omega.

Cooper, W.W., Seiford, L.M., & Tone, K. (2006). *Data Envelopment Analysis: A Comprehensive Text with Models, Applications, References, and DEA-Solver Software* (2nd edn.). New York: Springer.

DeSimone, L.D. & Popoff, F. (2000). *Eco-efficiency: The Business Link to Sustainable Development.* Massachusetts: The MIT Press.

Ermgassen, Z.E.K.H.J., *et al.* (2020). Using supply chain data to monitor zero deforestation commitments: An assessment of progress in the Brazilian soy sector. *Environmental Research Letters*, 15, 035003. https:/doi.org/10.1088/1748-9326/ab6497.

FAO, The State of Agricultural Commodity Markets (2018). *Agricultural Trade, Climate Change and Food Security.* Rome: FAO.

Fulginiti, L.E. & Perrin, R.K. (1997). LDC agriculture: Nonparametric Malmquist productivity indexes. *Journal of Development Economics*, 53, 373–390. https://doi.org/10.1016/S0304-3878(97)00022-9.

Kalirajan, K.P., Obwona, M.B., & Zhao, S. (1996). A decomposition of total factor productivity growth: The case of Chinese agricultural growth before and after reforms. *The American Journal of Agricultural Economics* 78, 331–338. https://doi.org/10.2307/1243706.

Khoshroo, A., Izadikhah, M., & Emrouznejad, A. (2022). Total factor energy productivity considering undesirable pollutant outputs: A new double frontier based malmquist productivity index. *Energy*, 258, 124819. https://doi.org/10.1016/j.energy.2022.124819.

Kibirige, D., Singh, A.S., & Rugube, L.M. (2019). Small-scale irrigation and production efficiency among vegetable farmers in the Eastern Cape Province of South Africa: The DEA approach. *Journal of Agricultural Studies* 7, 149. https://doi.org/10.5296/jas.v7i1.14564.

Leta, F.R., Soares de Mello, J.C.C.B., Gomes, E.G., & Meza, L.A. (2005). Métodos de melhora de ordenação em DEA aplicados à avaliação estática de tornos mecânicos, Investig. *Operacional*, 25(2), 229–242.

Liu, W., Zhou, Z., Ma, C., Li, D., & Shen, W. (2015). Two-stage DEA models with undesirable input-intermediate-outputs. *Omega*, 56, 74–87.

Liu, W.B., Meng, W., Li, X.X., & Zhang, D.Q. (2010). DEA models with undesirable inputs and outputs, *Annals of Operations Research*, 173(1), 177–194.

Mai, X.T.T., Nguyen, H.T.N., Ngo, T., Le, T.D.Q., & Nguyen, L.P. (2023). Efficiency of the Islamic banking sector: Evidence from two-stage DEA double frontiers analysis. *International Journal of Financial Studies*, 11(1), 32. https://www.mdpi.com/2227-7072/11/1/32.

MAPBIOMAS — Brazilian Annual Land Use and Land Cover Mapping Project (2022). Cobertura e Transições Bioma & Estados (Coleção 7). http://mapbiomas.org/ (Accessed 10 January 2022).

Masuda, K. (2016). Measuring eco-efficiency of wheat production in Japan: A combined application of life cycle assessment and data envelopment analysis. *The Journal of Cleaner Production*, 126, 373–381. https://doi.org/10.1016/j.jclepro.2016.03.090.

Mwambo, F.M., Fürstb, C., Nyarkoc, B.K., Borgemeistera, C., & Martiusd, C. (2020). Maize production and environmental costs: Resource evaluation and strategic land use planning for food security in northern Ghana by means of coupled energy and data envelopment analysis. *Land Use Policy*, 95, 104490.

Ngo, T. & Tsui, K.W.H. (2021). Estimating the confidence intervals for DEA efficiency scores of Asia-Pacific airlines. *Operational Research*. https://doi.org/10.1007/s12351-021-00667-w.

Nin, Pratt, A. & Yu, B., (2008). *An Updated Look at the Recovery of Agricultural Productivity in Sub-Saharan Africa*, IFPRI Discussion Paper 00787, Washington, D.C. International Food Policy Research Institute.

Paradi, J.C., Asmild, M., & Simak, P.C. (2004). Using DEA and worst practice DEA in credit risk evaluation. *Journal of Productivity Analysis*, 21(2), 153–165.

Raju, K.S. & Kumar, D.N. (2006). Ranking irrigation planning alternatives using data envelopment analysis. *Water Resource Management*, 20, 553–566. https://doi.org/10.1007/s11269-006-3090-5.

Rashidi, K. (2020). AHP versus DEA: A comparative analysis for the gradual improvement of unsustainable suppliers. *Benchmarking: An International Journal*, 27, 2283–2321.

Rentizelas, A., Melo, I.C., Alves Junior, P.N., Campoli, J.S., & Rebelatto, D.A.N. (2019). Multi-criteria efficiency assessment of international biomass supply chain pathways using data envelopment analysis. *Journal of Cleaner Production*, 237, 117690.

SEEG — Greenhouse Gas Emission and Removal Estimating System (2022). Tabela Geral das Emissões.

Seiford, L.M. & Zhu, J. (2002). Modeling undesirable factors in efficiency evaluation. *European Journal of Operational Research*, 142(1), 16–20.

Sun, B., Wang, H., Ortiz, J., Huang, J., Zhao, C., & Wang, Z. (2022). A decomposed data analysis approach to assessing city sustainable development performance: A network DEA model with a slack-based measure. *Sustainability*, 14(17), 11037. https://doi.org/10.3390/su14 1711037.

Thaker, K., Charles, V., Pant, A., & Gherman, T. (2021). A DEA and random forest regression approach to studying bank efficiency and corporate governance. *Journal of the Operational Research Society*, 1–28. https://doi.org/10.1080/01605682.2021.1907239.

Thrall, R.M. (1996). Duality, classification, and slacks in DEA. *Annals of Operations Research*, 66, 23–125.

Tian, X., Engel, B.A., Qian, H., Hua, E., Sun, S., & Wang, Y. (2021). Will reaching the maximum achievable yield potential meet future global food demand? *Journal of Cleaner Production*, 294. https://doi.org/10.1016/j.jclepro.2021.126285.

Toma, E., Dobre, C., Dona, I., & Cofas, E. (2015). DEA applicability in assessment of agriculture efficiency on areas with similar geographically patterns. *Agriculture and Agricultural Science Procedia*, 6, 704–711. https://doi.org/10.1016/j.aaspro.2015.08.127.

Tone, K. (2001). A slacks-based measure of efficiency in data envelopment analysis. *European Journal of Operational Research*, 130(3), 498–509. DOI: 10.1016/S0377-2217(99)00407-5.

Tran, D.H. & Ngo, D.T. (2014). Performance of the Vietnamese automobile industry: A measurement using DEA. *Asian Journal of Business and Management*, 2(3), 184–191.

Tsionas, M.G. (2021). Optimal combinations of stochastic frontier and data envelopment analysis models. *European Journal of Operational Research*. https://doi.org/https://doi.org/10.1016/j.ejor.2021.02.003.

Wang, Y.-M. & Chin, K.-S. (2009). A new approach for the selection of advanced manufacturing technologies: DEA with double frontiers. *International Journal of Production Research*, 47(23), 6663–6679. https://doi.org/10.1080/00207540802314845.

Yilmaz, B., Yurdusev, M.A., & Harmancioglu, N.B. (2009). The assessment of irrigation efficiency in Buyuk Menderes Basin. *Water Resource Management*, 23, 1081–1095.

https://doi.org/10.1142/9781800615786_0018

Chapter 18

Organizational Performance Assessment Using Data Envelopment Analysis: A Case of a State University in Turkey

Sıdıka Ece Yılmaz[*,†] **and İzzettin Özdemir**[‡]

[*]*Career Planning Application and Research Center, Adana Alparslan Türkeş Science and Technology University, Turkey*

[†]*eyilmaz@atu.edu.tr*

[‡]*ozdemir.izzettin@hotmail.com*

Abstract

Every organization, for-profit or non-profit, has a different standard of performance. Performance is a critical factor in determining whether an organization will succeed in a highly competitive market. Due to the increasing number of institutions, educational institutions are becoming significant organizations whose performance must be scrutinized. As a matter of fact, the chapter's goal is to assess the performance of a state university's 11 faculties in Turkey based on their activities at the end of the academic year 2020–2021. Data envelopment analysis has been used as a performance measuring tool to evaluate the existing situation with consideration of both input and output variables. This study's input variables include the number of classrooms, the budget, and academic staff members. The numbers of students, graduates, and articles have all been utilized as output variables. They are used in this study to assess the efficiency levels of academic departments. Eventually, inefficient academic units were identified, and faculty performance comparisons were accomplished. Finally, the potential causes were discussed comprehensively.

Keywords: Data envelopment analysis, organizational performance, organizational efficiency, performance measurement, performance assessment, performance evaluation.

1. Introduction

Universities are crucial in developing a skilled labor force for emerging nations. For this reason, educational institutions must run properly. Resources must be properly allocated to ensure efficiency. Effective structuring and planning, as well as the allocation of resources appropriately within the university, are fundamental for university performance. Since resources are limited, the efficiency of allocations depends heavily on this factor. However, there are now more universities than ever before. The efficiency of these universities is in doubt given these circumstances.

Turkey has the highest number of university students by population in Europe.[1] The number of universities, which was 52 before 2006, has increased to 208 in 2022. The number of students has grown, along with the quick expansion of institutions in Turkey in recent years. For instance, 6,80,599 students attended foundation universities and vocational high schools and 7,616,360 students enrolled in state universities in the 2021–2022 academic year in Turkey.[2] However, the growing number of universities and students raises concerns about their efficiency.

A performance measurement tool is needed to demonstrate efficiency and measure performance across the university. Data envelopment analysis (DEA) is one of the methods used to evaluate the performance of universities. As it allows for the handling of multiple inputs and outputs without making presumptions regarding the monetary values of inputs and outputs, DEA is used as an effective performance measurement tool in non-profit organizations, such as schools, hospitals, and universities (Kuah & Wong, 2011).

[1]https://tr.euronews.com/2021/11/16/avrupa-da-nufusa-gore-en-cok-universite-ogrencisi-turkiye-de-grafik (Accessed 11 July 2022).

[2]https://istatistik.yok.gov.tr/ (Accessed 11 July 2022).

The technique can identify which organizational unit has the best practice (Albright & Winston, 2007). DEA is a popular tool for efficiency analysis in various fields, such as banking and finance (Tripe, 2010; Le, 2018; Boubaker *et al.*, 2022b) and manufacturing (Becchetti & Sierra, 2003; Wang & Chin, 2010; Vu, 2016; Boubaker *et al.*, 2023).

As it enables university administrators to assess academic units using the same set of parameters, the DEA method has also been favored in the chapter. Additionally, DEA offers significant outputs regarding the units that need to be given priority when allocating resources in order to make them productive going forward (Moreno & Tadepalli, 2002). As a result, eligible university administrators may assess the chapter's findings and evaluate how they correspond to those at their own institutions.

DEA has already been commonly used to assess the efficiency of universities (Tomkins & Green, 1988; Beasley, 1995; Moreno & Tadepalli, 2002; Abbott & Doucouliagos, 2003; Gökşen *et al.*, 2015). Several studies using this method measured the efficiency of a specific unit or department (Lopes & Lanzer, 2002; Moreno & Tadepalli, 2002) within the university, while others examined the overall efficiency of the institution (Rhodes ve Southwick, 1986, Flegg *et al.*, 2004; Dikmen, 2008; Ulucan, 2011).

The efficiency status is an intriguing issue, given the increase in the number of universities in Turkey. As a matter of fact, it is critical for the education sector to conduct studies on university performance. This chapter contributes to the validity of the DEA method, enriches the literature by using it as a performance measurement tool in educational institutions, and contributes to the existing literature as a study investigating a university in Turkey and university performances in a different region at the faculty level.

In this chapter, the activities of faculty at a state university have been evaluated, and the university's performance has been measured using DEA. According to the analyses, seven faculties have efficient performance, while four faculties have inefficient performance. The chapter includes a literature review, method, results and findings, and conclusion sections.

2. Literature Review

2.1. *Performance measurement and data envelopment analysis*

The process of assessing the efficiency and effectiveness of activities is referred to as performance measurement (Neely *et al.*, 1995). Organizational performance not only reflects concrete data regarding the goals but also represents organizational efficiency and whether the organization is using existing resources efficiently (Dobák, 1996). Performance assessment is an essential part of higher education institutions' self-regulation ability, as it plays a critical role in the preparation, planning, and feedback of managerial decisions on resource distribution. (Lábas, 2014).

Charnes, Cooper, and Rhodes developed DEA in 1978 in order to assess the efficiency of public schools in the United States (Charnes *et al.*, 1978). DEA is a mathematical method used to obtain a productivity index for decision-making units (DMUs) when multiple inputs and outputs are available at the same time (Donthu & Yoo, 1998). It is a technique that advises managers and decision-makers on how to improve the efficiency of inefficient DMUs (Reisman, 2004). The efficiency of any decision unit is defined as the largest ratio of the weighted sum of outputs derived from the output/input ratio to the sum of weighted inputs, provided that similar ratios for each decision unit are less than or equal to one (Charnes *et al.*, 1978).

Charnes *et al.* (1978) presented the CCR model, which provides DMU efficiency measurement for constant returns to scale (CRS), in which all DMUs perform at their optimal scale. Afterward, Banker *et al.* (1984) developed the BCC model, which refers to a variable returns to scale (VRS) efficiency measurement model. This model allows for dividing efficiency into technical and scale efficiencies in DEA. In contrast to the BCC model, which produces variable output for each input, the CCR model yields a comparable rise in output for each input. Moreover, these models are separated into two groups: input- and output-oriented DEA models. The best input combination to yield the best output composition is sought after by the input-oriented DEA model. Conversely, the output-oriented DEA model's objective is to maximize output while using the least amount of input (Charnes *et al.*, 1978). Overall efficiency is divided into two components in the form of multiplication (Ganley & Cubbin, 1992), and technical efficiency is divided into local technical efficiency and scale efficiency components (Cooper *et al.*, 2000).

The DEA method is now exhibiting a variety of trends. For instance, Simar & Wilson (2007) proposed two bootstrap procedures for the two-stage efficiency estimation problem. This method is currently used in many studies. In this trend, double bootstrap regression estimation is used to assess the relevance of explanatory variables after DEA in the initial phase (López-Penabad *et al.*, 2020). Another trend is inverse DEA based on the reduced efficiency method. In this method, only the groups with low activity scores were focused on. Thus, it is aimed at reducing the computational load of the analysis (Boubaker *et al.*, 2022a). In fact, this method was developed by Zhang & Cui (1999) as an inverse DEA based on output reductions to evaluate the investment projects of the Chinese State Economic Information System. Hammami *et al.* (2020) developed a Euclidean common set of weights. It is stated that this method is more realistic in terms of weight determination. With this method, both efficient and inefficient firms can be ranked. In addition, it provides the opportunity to sort according to the credit ratings of banks. It is obvious that various DEA trends are currently developing.

2.2. *Performance measurement in higher education*

DEA enables the evaluation of relative activities between universities as well as between university departments or courses (Kuah & Wong, 2011). In the DEA carried out in higher education institutions, evaluation entails gathering information about students, staff, and the functioning of educational organizations (Astin, 1993).

DEA calculates an academic unit's efficiency by comparing its inputs and outputs with all of the parameters studied. If a department's outputs are optimal (maximum possible) in comparison to its inputs and outputs of all comparable parameters, it is considered efficient (efficiency = 1). After obtaining results, if there is an inefficient department unit, DEA can assist that department in determining how to better allocate resources and thus improve its efficiency (Moreno & Tadepalli, 2002).

The performance of academic departments at Brazil's Federal University of Santa Catarina was evaluated using DEA. It was discovered that 15 of the 58 departments evaluated in the results performed poorly (Lopes & Lanzer, 2002). The efficiency of 42 academic departments at a state university was evaluated using DEA in another study. *Faculty salaries, staff salaries, operational budget, equipment budget,* and *building space allocated to each academic unit* were the inputs, whereas the outputs were *the number of graduates, undergraduates, full-time equivalents produced, student credit hours generated,* and *the amount of grants awarded.* According to the findings, 22 of the departments were relatively productive (Moreno & Tadepalli, 2002).

Research showed that four faculties were found to be inefficient after examining the efficiency of eight faculties at Cumhuriyet University. Inputs included *the number of academic staff, administrative staff, personnel, service procurement,* and *consumption expenses,* while outputs included *the number of students, student fees, projects,* and *postgraduate students.* The analysis revealed that four faculties, namely the faculties of Medicine, Dentistry, Fine Arts, and Theology, were inefficient when compared to other faculties (Kutlar & Kartal, 2004). Between 2004 and 2006, another study looked at the relative technical efficiency of academic departments at the Islamic University of Gaza. *Operating expenses, credit hours,* and *training resources* were inputs, while *graduates, promotions,* and *public service activities* were outputs. According to the findings, 10 departments out of 30 were found to be efficient (Agha *et al.*, 2011).

In a study conducted with 24 foundation universities in 2006 using the CCR model, it was discovered that 15 universities were relatively efficient and nine universities (Bahçeşehir, Çağ, Doğuş, Haliç, İstanbul Bilgi, Okan, Ufuk, and Yeditepe Universities) were relatively inefficient in total, according to their relative total efficiency values (Özden, 2008). In a study conducted with the BCC model in 50 state universities with data from 2008, 15 universities were found to be inefficient (Ulucan, 2011). In a study in which the CCR model was applied to 23 foundation universities for 2012, it was found that six universities were efficient in total and 11 universities were technically efficient, according to the BCC model (Bal, 2013). This study shows that the efficiency of foundation universities has decreased over the years in Turkey. Another study evaluated the performance of 48 universities out of 108 state universities using DEA, and it showed that 21 universities were efficient with the assumption of CRS and 28 universities were efficient with the assumption of VRS (Kırda, 2014).

A study determined the performance levels of departments at Dokuz Eylül University. The input variables were *the university's outdoor-indoor area, the number of academic staff,* and *the number of administrative staff*, while the output variables were *the numbers of publications* and *graduate students.* It showed that eight out of 26 DMUs were found to be efficient in both BCC and CCR analyses. Three DMUs were found to be inefficient in the CCR analysis, while 15 DMUs were not efficient in both BCC and CCR analyses (Gökşen *et al.*, 2015). In another study investigating the efficiency of Gaziantep University's faculties based on the 2014–2015 academic year, the inputs were *the number of academic personnel, the budget used,* and *the capacity of the educational area*, and the outputs were *the number of students, the number of graduate students, the number of projects,* and *the number of publications.* According to the findings, 11 out of 12 faculties were qualified as active (Doğan *et al.*, 2016).

The 2014–2015 academic year was used to assess the training efficiency of 43 Turkish state universities. As input variables, *total expenditures, the number of instructors, and the number of lecturers* were used. Output variables included *the number of supported public and infrastructure projects, citations, publications in the SCI, SSCI, and AHCI indices, total undergraduate students,* and *graduate students.* According to data development analysis, the following universities are efficient: *Gebze Technical University, Anadolu University, Middle East Technical University, Istanbul Technical University, Istanbul University, Marmara University, Hacettepe University, Gazi University, Ankara University,* and *Ege University.* In general, 23% of universities are efficient (Türkan & Özel, 2017). Another study used data envelopment to examine how efficiently 100 Turkish state universities used their resources in 2013. The CCR model found nine universities to be efficient, while the BCC model found 27 universities to be efficient (Arslan & Güven, 2018).

The activities of 18 faculties at Erciyes University were investigated in 2016. While *the number of academic staff, general budget expenses,* and *students* were defined as inputs, *the number of graduates, projects,* and *publications* were defined as outputs.

Table 1. Inputs and outputs regarding efficiency analysis in higher education.

Author	Input variables	Output variables
Bessent *et al.* (1983)	Revenue from the state government Number of students completing a program Employer satisfaction with the training of students	Student contact hours Number of full-time equivalent instructors Square feet of facilities for each program Direct instructional expenditures
Tomkins & Green (1988)	Number of full-time employees Personnel costs Operating costs Other costs	Number of university students Number of PhD students Total income Number of publications
Beasley (1990)	Research income Expenditure was treated	Undergraduate student numbers Postgraduate student numbers Research ratings
Johnes & Johnes (1993)	Research income	Research output
Sinuany-Stern *et al.* (1994)	Operating costs Salaries	Research grants Publications Graduate students Contact hours
Beasley (1995)	Research income Personnel costs Operating costs	Number of undergraduate students and postgraduate students Number of publications that take part in indexes
Moreno & Tadepalli (2002)	Faculty salaries Staff salaries Operational budget Equipment budget Building space allocated to each academic unit	The number of graduates The number of undergraduates Full-time equivalents produced Student credit hours generated The amount of grants awarded
Abbott & Doucouliagos (2003)	Number of academic staff Number of non-academic staff Operating costs Fixed assets	Number of students Number of graduate students with an associate degree undergraduate and postgraduate degree Amount of research

Table 1. (*Continued*)

Author	Input variables	Output variables
Flegg *et al.* (2004)	Number of faculty members Number of undergraduate students Number of postgraduate students Total expenses	Research and consultancy income Number of graduate students with an undergraduate degree Number of graduate students with a postgraduate degree
Warning (2004)	Personnel costs Other costs	Number of publications that take part in indexes Number of students
Kutlar & Kartal (2004)	Number of academic staff Number of administrative staff Personnel, service procurement, and consumption expenses Acreage	Number of students Student fees Projects Number of postgraduate students
Baysal *et al.* (2005)	Personnel costs Other current expenditures Investment expenses Transfers Number of faculty members	Number of undergraduate students Number of postgraduate students Number of PhD students Number of publications
Babacan *et al.* (2007)	General budget expenditures Expenditure out of budget Number of professors Number of associate professors Number of assistant professor Number of assistant instructors Number of administrative staff	Number of publications that take part in indexes University income Number of undergraduate students Number of postgraduate students Number of graduate students with a postgraduate degree
Kurtlar & Babacan (2008)	General budget expenditures Expenditure out of budget Number of professors Number of associate professors Number of assistant professor Number of assistant instructors Number of administrative staff	Number of publications that take part in indexes University income Number of undergraduate students Number of postgraduate students Number of graduate students with a postgraduate degree

(*Continued*)

Table 1. (*Continued*)

Author	Input variables	Output variables
Dikmen (2008)	The number of academic staff The number of administrative staff Budget allowances	Number of undergraduate Students Number of graduate Students Number of domestic and international Publications
Agha *et al.* (2011)	Operating expenses Credit hours Training resources	Number of graduates Promotions Public service activities
Bal (2013)	The number of faculty members Other numbers of academic staff	Number of students/Faculty number rate The sum of SCI, SSCI, AHCI indexed articles and citations
Gökşen *et al.*, (2015)	The university's outdoor-indoor area The number of academic staff The number of administrative staff	Number of publications Number of graduate students
Doğan *et al.* (2016)	The number of academic personnel The budget used The capacity of the educational area	The number of students The number of graduate students The number of projects The number of publications
Türkan & Özel (2017)	Total expenditure The number of instructors The number of lecturers	The number of supported public and infrastructure Projects The number of citations The number of SCI, SSCI, AHCI indexed publications Total number of undergraduate students Total number of graduate students
Çınaroğlu *et al.* (2018)	The number of academic staff The general budget expenses The number of students	The number of graduates The number of projects The number of publications
Ergülen *et al.* (2021)	The number of academic staff The number of administrative staff The number of students	Total expense Number of graduates Academic grade point average

Note: This table shows the inputs and outputs for efficiency analysis in previous higher education studies.
Source: Adapted from Avkiran (2001) and Özden (2008) and updated with other studies.

It has been determined that a total of 12 faculties were technically active, including the Dental, Pharmacy, Education, Science, Aviation and Space Sciences, Communication, Engineering, Health Sciences, Seyrani Agriculture, Medicine, Türkan Tuncer Tourism, and Veterinary Faculties. In terms of scale efficiency, seven faculties were found to be efficient: Dentistry, Education, Aviation and Space Sciences, Communication, Health Sciences, Seyrani Agriculture, and Veterinary Medicine (Çınaroğlu *et al.*, 2018).

The efficiency of 20 faculties was examined in a study based on the Selçuk University 2018 annual report and using the CCR model. The input variables were the *number of academic staff, administrative staff*, and *students*, and the output variables were the *total expense, number of graduates*, and *academic grade point average*. Ten faculties were found to be efficient in the DEA method (Ergülen *et al.*, 2021). Inputs and outputs regarding efficiency analysis in higher education, which are generally used in the studies, are shown in Table 1.

As formerly indicated, Turkey has the most universities in 2022. Because of this, research on university performance analysis could contribute significant knowledge to the literature during this time. In addition, specific variables that were not previously preferred were identified in this chapter. As an illustration, the number of classrooms as input and the number of articles as output were examined with the DEA method. Compared to previous studies, the chapter focused on a few particular variables that hadn't been addressed before. As a result, it offers a distinction in terms of literature.

3. Method

DEA is used as a tool to evaluate the performance of faculties at a state university. The input-oriented weighted CCR model was applied, and the solver plug-in of the Excel software package provided by Microsoft Office was used.

DEA is a method that searches for the most appropriate combination of inputs to produce an output combination in the most efficient way. The optimal weights to maximize the proportion of a weighted sum of outputs to a weighted sum of inputs are determined using DEA for the selected DMU. The input-oriented weighted CCR model with linear data envelopment facilitates the calculations.

It is converted into programming. The objective function, shown in Equation (1), aims to maximize the efficiency of faculties:

$$E_k = \text{Max}\left(\sum_{r=1}^{p} u_r \gamma_{rk}\right). \tag{1}$$

In linear programming, the objective function does not consist of a denominator. Therefore, the denominator of the objective function is set to 1. As seen in Equation (2), this equality is added to the model as a constraint:

$$\left(\sum_{i-1}^{m} v_i x_{ik}\right) = 1, \tag{2}$$

where the decision unit is $\boldsymbol{k}$, $\boldsymbol{u_r}$ denotes the unit of weight given to the output, and $\boldsymbol{v_i}$ is the unit of weight given to the input. The weight values must satisfy the inequality $\boldsymbol{u_r}, \boldsymbol{v_i} \geq \boldsymbol{\varepsilon}$. Here, $\boldsymbol{\varepsilon}$ is specified as a sufficiently small positive number.

We have one more constraint to calculate the efficiency of the output value. In this constraint, the weighted sum of outputs must not be greater than the weighted sum of inputs. The second restriction in the model is shown in Equation (3):

$$\left(\sum_{r=1}^{P} u_r Y_{rj}\right) - \left(\sum_{1=1}^{m} v_i X_{ij}\right) \leq 0. \tag{3}$$

Consider a set of n DMUs, with each DMU $\boldsymbol{j}(j = 1, \ldots, n)$ using m inputs $\boldsymbol{X}_{\boldsymbol{ij}}(i = 1, \ldots, m)$ to produce k outputs $\boldsymbol{Y}_{\boldsymbol{rj}}(r = 1, \ldots, p)$.

3.1. *Decision-making units*

In the study, 11 faculties were identified as DMUs. The completeness of the data was considered when selecting faculties as the DMUs. These faculties are Engineering and Architecture, Dentistry, Education, Faculty of Arts and Sciences, Economics and Administrative Sciences, Theology, Health Sciences, Sports Sciences, Fisheries, Maritime, and Agriculture for the academic year 2020–2021. The 2020–2021 academic year covers the fall semester of 2020 and the spring semester of 2021. The data used in this study were evaluated using the data collected at the end of the 2021 spring semester.

The study followed the $\boldsymbol{n \geq m + p + 1}$ principle, where $\boldsymbol{m}$ is the number of inputs, $\boldsymbol{p}$ is the number of outputs, and $\boldsymbol{n}$ is the number of DMUs (Behdioğlu & Özcan, 2009; Boussofiane *et al.*, 1991). Sherman reported that the number of DMUs should be greater than the sum of the number of inputs and outputs (Sherman, 1984). As a result, 11 DMUs are suitable based on the input and output numbers used in the study, as $11 > 3 + 3 + 1$.

3.2. *Input and output selection*

Annual reports are regularly published by universities. Therefore, for the research, a state university with a current annual report for each faculty was chosen as the study's sample. In order to reliably measure system efficiency with DEA, the input and output variables that best express the system process should be chosen (Özden, 2008). In this study, the number of classrooms, representing the total number of classrooms and laboratories, the budget allocated to faculties for the year 2020–2021, the number of academic staff working in the faculty were used as inputs, and the number of students, the number of graduates, and the number of articles published by faculties were used as outputs, as shown in Table 2.

The number of classrooms selected as an input has not been specifically used before in the literature. However, similar variables, such as the capacity of the educational area (Doğan *et al.*, 2016) and building space allocated to each academic unit (Moreno & Tadepalli, 2002), have been included in the literature. The budget allocated to faculties has been used in previous studies (Dikmen, 2008; Moreno & Tadepalli, 2002). The number of academic staff working in the faculty has been tested as input in other studies (Abbott & Doucouliagos, 2003; Ergülen *et al.*, 2021). The number of students among the outputs selected in the study has been examined in previous studies (Doğan *et al.*, 2016; Türkan & Özel, 2017). The number of graduates has been chosen as an output many times before (Agha *et al.*, 2011; Gökşen, 2015). Finally, the number of articles published by faculties has not been specifically tested in the literature before. However, similar variables, such as the number of publications (Çınaroğlu *et al.*, 2018) and the number of SCI-, SSCI-, and AHCI-indexed publications (Türkan & Özel, 2017), have been included in the literature. The reason for choosing these variables is generally because they are

Table 2. Inputs and outputs.

	Outputs			Inputs		
Faculties	Students	Graduates	Articles	Class-room	Budget	Academic staff
Engineering and Architecture F.	1,050	162	150	69	8,032,000	85
Dentistry F.	422	48	23	15	6,210,700	69
Education F.	1,675	454	86	64	14,096,500	91
Science and Literature F.	1,500	150	153	117	18,959,800	132
Economics and Administrative Sciences F.	1,998	436	62	65	10,863,900	81
Theology F.	1,959	314	34	86	10,548,954	65
Health Sciences F.	415	92	11	25	1,899,500	13
Sports Sciences F.	378	49	18	4	1,539,800	14
Fisheries F.	55	5	56	27	6,090,300	36
Maritime F.	358	66	13	42	2,314,200	19
Agriculture F.	119	15	17	21	2,304,600	15

Note: This table presents the values of the input and output variables of 11 faculties as a result of the DEA method.

popularly used by researchers or have not been specifically tested before. As a result, the performance analysis literature has evaluated these identified variables repeatedly, either identically or similarly. This demonstrates that the DEA method can be applied to variables.

4. Results and Findings

The input-oriented weighted CCR model was employed in this study to assess the efficiency of the state university's faculties, and efficiency results were generated among these units. The solver plug-in of the Excel software package provided by Microsoft Office was applied for the analysis. The information pertaining to the 11 faculties used in the efficiency analysis is displayed in Table 3.

In the faculties of engineering and architecture, dentistry, education, economics and administrative sciences, theology, health sciences, and sports sciences, the inputs and outputs are balanced.

Table 3. Faculty efficiency scores.

Faculties	**Efficiency scores (%)**	**Efficiency results**
Faculty of Engineering and Architecture	100.00	Efficient
Faculty of Dentistry	100.00	Efficient
Faculty of Education	100.00	Efficient
Faculty of Arts and Sciences	70.58	Inefficient
Faculty of Economics and Administrative Sciences	100.00	Efficient
Faculty of Theology	100.00	Efficient
Faculty of Health Sciences	100.00	Efficient
Sports Science Faculty	100.00	Efficient
Faculty of Fisheries	93.24	Inefficient
Maritime Faculty	68.30	Inefficient
Faculty of Agriculture	64.22	Inefficient

Note: This table shows the efficiency score and efficiency results of 11 faculties as a result of the DEA method.

These faculties have been observed to function efficiently. In exchange for the available inputs, the Faculty of Arts and Sciences is 70.58% efficient, the Fisheries Faculty is 93.24% efficient, the Maritime Faculty is 68.30% efficient, and the Agriculture Faculty is 64.22% efficient. These findings reveal that these faculties did not work efficiently.

In terms of effect, it can be considered that the university is moderately successful at the faculty level. This university appears to need development work and effort for the Faculty of Arts and Sciences, the Fisheries Faculty, the Maritime Faculty, and the Agriculture Faculty. While four faculties were not found to be efficient in the study, only one faculty was not found to be inefficient in a study comparing the efficiency of Gaziantep University faculties in Turkey in 2016 (Dogan *et al.*, 2016). Among the reasons for this situation may be the differences in the budgets transferred to the universities according to the years and the insufficient or different numbers of academicians, students, graduates, and publications in the determined DPUs in terms of university scale. In a study examining 20 faculties of Selcuk University, 10 faculties were found to be inefficient (Ergülen *et al.*, 2021). According to this, it can be stated that the current sample is more successful than the Selcuk University sample. However, this situation reflects only the overall picture. After all, each faculty has its own success criteria. Sample size, budget, number of academics, and classroom size are different. For this reason, instead of expressing it distinctly as successful or unsuccessful, it can be stated that the university is moderately successful in numbers.

5. Conclusion

The DEA method (the input-oriented weighted CCR model) was used in this chapter to assess the efficiency of faculties at a state university in Turkey. The DMUs of the research were 11 faculty, and the study considered the 2020–2021 academic year. Three input and output parameters were chosen to assess the efficiency of faculties. The number of classrooms, the budget allocated to faculties, and the number of academic staff were used as inputs, while the outputs were the number of students, the number of graduates, and the number of articles published by faculties. According to the results, the faculties of engineering and architecture, dentistry, education, economics, administrative sciences, theology, health sciences, and sports sciences were efficient, while the faculty of arts and sciences, the fisheries faculty, the maritime faculty, and the agriculture faculty were inefficient.

Currently, the increasing number of universities draws attention to the performance levels of universities. Therefore, more such studies are needed. Future studies can be conducted utilizing new methodologies by selecting regionally fixed faculties or with new DEA methods.

References

Abbott, M. & Doucouliagos, C. (2003). The efficiency of Australian Universities: A data envelopment analysis. *Economics of Education Review*, 22(1), 89–97.

Agha, S.R., Kuhail, I., Abdul Nabi, N., Salem, M., & Ghanim, A. (2011). Assessment of academic departments efficiency using data envelopment analysis. *Journal of Industrial Engineering and Management*, 4(2), 301–325.

Albright, S.C. & Winston, W.L. (2007). *Management Science Modeling* (Revised 3rd edn., Thomson, pp. 184–192). Boston: Cengage Learning, Inc.

Arslan, A.E. & Güven, Ö.Z. (2018). Veri Zarflama Analizi İle Üniversite Etkinliklerinin Belirlenmesine Yönelik Bir Çalışma: Türkiye Örneği. *Uluslararası Afro-Avrasya Araştırmaları Dergisi*, 3(6), 86–105.

Astin, A.W. (1993). *Assessment for Excellence.* Phoenix, AZ: Oryx Press.

Avkiran, N.K. (2001). Investigating technical and scale efficiencies of Australian universities through data envelopment analysis. *Socio-economic Planning Sciences*, 35(1), 57–80.

Babacan, A., Kartal, M., & Bircan, M.H. (2007). Cumhuriyet Üniversitesi'nin Etkinliğinin Kamu Üniversiteleri ile Karşılaştırılması: Bir VZA Tekniği Uygulaması. *C.U. İktisadi ve İdari Bilimler Fakültesi Dergisi*, 8(2), 97–114.

Bal, V. (2013). Vakıf üniversitelerinde veri zarflama analizi ile etkinlik belirlenmesi. *Manas Sosyal Araştırmalar Dergisi*, 2(1), 1–20.

Banker, R.D., Charnes, A., & Cooper, W.W. (1984). Some models for estimating technical and scale inefficiencies in data envelopment analysis. *Management Science*, 30(9), 1078–1092.

Baysal, M.E. Alçılar, B. Çerçioğlu, H., & Toklu, B. (2005). Türkiye'deki Devlet Üniversitelerinin 2004 Yılı Performanslarının, Veri Zarflama Analizi Yöntemiyle Belirlenip Buna Göre 2005 Yılı Bütçe Tahsislerinin Yapılması. *Sakarya Üniversitesi Fen Bilimleri Enstitüsü Dergisi*, 9(1), 67–73.

Beasley, J.E. (1990). Comparing university departments. *Omega-International Journal*, 18(2), 171–183.

Beasley, J.E. (1995). Determining teaching and research efficiencies. *Journal of The Operational Research Society*, 46(4), 441–452.

Becchetti, L. & Sierra, J. (2003). Bankruptcy risk and productive efficiency in manufacturing firms. *Journal of Banking & Finance*, 27(11), 2099–2120.

Behdioğlu, S. & Özcan, G. (2009). Veri zarflama analizi ve bankacılık sektöründe bir uygulama, Süleyman Demirel Üniversitesi İktisadi ve İdari Bilimler Fakültesi Dergisi, 14(3), 304 (2009).

Bessent, A.M., Bessent, E.W., Charnes, A., Cooper, W.W., & Thorogood, N.C. (1983). Evaluation of educational program proposals by means of DEA. *Educational Administration Quarterly*, 19(2), 82–107.

Boubaker, S., Le, T.D.Q., & Ngo, T. (2022a). Managing bank performance under COVID-19: A novel inverse DEA efficiency approach. *International Transactions in Operational Research*, 30(2023), 2436–2452. https://doi.org/10.1111/itor.13132.

Boubaker, S., Do, D.T., Hammami, H., & Ly, K.C. (2022b). The role of bank affiliation in bank efficiency: A fuzzy multi-objective data envelopment analysis approach. *Annals of Operations Research*, 311, 611–639. https://doi.org/10.1007/s10479-020-03817-z.

Boubaker, S., Le, T.D.Q., Ngo, T., & Manita, R. (2023). Predicting the performance of MSMEs: A hybrid DEA-machine learning approach. *Annals of Operations Research.* https://link.springer.com/article/10.1007/s10479-023-05230-8.

Boussofiane, A., Dyson, R., & Rhodes, E. (1991). Applied DEA. *European Journal of Operational Research*, 52(1), 1–15.

Charnes, A., Cooper, W.W., & Rhodes, E. (1978). Measuring the efficiency of decision making units. *European Journal of Operations Research*, 2, 429–444.

Çinaroğlu, E., Doruk, N., & Tunahan, A.V.C.I. (2018). Erciyes Üniversitesi fakültelerinin veri zarflama analizi yöntemiyle etkinlik analizi. *Uluslararası Yönetim İktisat ve İşletme Dergisi*, 14(4), 1025–1043.

Dikmen, F.C. (2008). Veri Zarflama Analizi ile Üniversitelerin Etkinliğinin Ölçülmesi. Kocaeli Üniversitesi, İktisadi ve İdari Bilimler Fakültesi Dergisi, 3–6, 1–22.

Doğan, A., Soylemez, İ., & Özcan, U. (2016). Measuring effectiveness using data envelopment analysis: A case of university. *The Eurasia Proceedings of Educational & Social Sciences (EPESS)*, 5(2), 70–75.

Donthu, N. & Yoo, B. (1998). Retail productivity assessment using data envelopment analysis. *Journal of Retailing*, 74(1), 89–105.

Ergülen, A., Ünal, Z., & Harmankaya, İ. (2021). Data envelopment analysis and efficiency analysis of higher education institutions: Example of Selçuk University. *Konuralp Journal of Mathematics*, 9(1), 201–208.

Flegg, T., Allen, D.O., Field, K., & Thurlow, T.W. (2004, December). Measuring the efficiency of British universities: A multi-period data envelopment analysis. *Education Economics*, 12(3), 231–249.

Gökşen, Y., Doğan, O., & Özkarabacak, B. (2015). A data envelopment analysis application for measuring efficiency of university departments. *Procedia Economics and Finance*, 19, 226–237.

Hammami, H., Ngo, T., Tripe, D., & Vo, D.T. (2020). Ranking with a Euclidean common set of weights in data envelopment analysis: With application to the Eurozone banking sector. *Annals of Operations Research*, 311(2), 675–694.

Johnes, G. & Johnes, J. (1993). Measuring the research performance of UK economics departments: An application of data envelopment analysis. *Oxford Economic Papers*, 45, 332–347.

Kırda, K. (2014). Veri zarflama analizi ile Türkiye'deki devlet üniversitelerinin performanslarının değerlendirilmesi. *QUO VADIS: International Social Sciences Conference*, 15–17 Ekim 2014, Artvin, Türkiye.

Kuah, C.T. & Wong, K.Y. (2011). Efficiency assessment of universities through data envelopment analysis. *Procedia Computer Science*, 3, 499–506.

Kutlar, A. & Babacan, M. (2008). Türkiye'deki Kamu Üniversitelerinde CCR Etkinliği-Ölçek Etkinliği Analizi: DEA Tekniği Uygulaması. Kocaeli *Üniversitesi Sosyal Bilimler Enstitüsü Dergisi*, 15(1), 148–172.

Kutlar, M. & Kartal, M.H. (2004). Cumhuriyet Üniversitesinin Verimlilik Analizi: Fakülteler Düzeyinde Veri Zarflama Yöntemiyle Bir Uygulama. *Kocaeli Üniversitesi Sosyal Bilimler Enstitüsü Dergisi*, 8(2), 49–79.

Lábas, I. (2014). Analysis of the financial management of the University of Debrecen. *Annals of the University of Oradea Economic Science*, XXIII(1), 763–772.

Le, T.D.Q. (2018). Bank risk, capitalisation and technical efficiency in the Vietnamese banking system. *Australasian Accounting, Business and Finance Journal*, 12(3), 41–61.

Lopes, A.L.M. & Lanzer, E.A. (2002). Data envelopment analysis-DEA and fuzzy sets to assess the performance of academic departments: A case study at Federal University of Santa Catarina-UFSC. *Pesquisa Operacional*, 22, 217–230.

López-Penabad, M.C., Maside-Sanfiz, J.M., Torrelles Manent, J., & Iglesias-Casal, A. (2020). Application of the DEA double bootstrap to analyze efficiency in Galician Sheltered Workshops. *Sustainability*, 12(16), 6625.

Moreno, A.A. & Tadepalli, R. (2002). Assessing academic department efficiency at a public university. *Managerial and Decision Economics*, 23(7), 385–397.

Neely, A., Gregory, M., & Platts, K. (1995). Performance measurement system design. *International Journal of Operations & Production Management*, 15(4), 80–116.

Özden, Ü. (2008). Veri zarflama analizi (VZA) ile Türkiye'deki vakıf üniversitelerinin etkinliğinin ölçülmesi. *İstanbul Üniversitesi İşletme Fakültesi Dergisi*, 37(2), 167–185.

Reisman, A. (2004). What Can Or/Ms Academics Learn from and About Data Envelopment Analysis? Lots!, United States, Available at *SSRN*: https://ssrn.com/abstract=529542.

Rhodes, E.L., & Southwick, L. (1986). *Determinants of Efficiency in Public and Private Universities*. Department of Economics, University of South Carolina, Columbia, South Carolina, United States.

Sherman, H.D. (1984). Hospital efficiency measurement and evaluation. *Medical Care*, 22(10), 922–938.

Simar, L. & Wilson, P.W. (2007). Estimation and inference in two-stage, semi-parametric models of production processes. *Journal of Econometrics*, 136(1), 31–64.

Sinuany-Stern, Z., Mehrez, A., & Barboy, A. (1994). Academic departments efficiency via DEA. *Computers & Operations Research*, 21(5), 543–556.

Tomkins, C. & Green, R. (1988). An experiment in the use of data envelopment analysis for evaluating the efficiency of UK university departments of accounting. *Financial Accountability & Management*, 4(2), 147–164.

Tripe, D. (2010). Using DEA to investigate bank safety and soundness–which approach works best? *Journal of Financial Economic Policy*, 2(3), 237–250.

Türkan, S., & Özel, G. (2017). Efficiency of state universities in Turkey during the 2014–2015 academic year and determination of factors affecting efficiency. *TED Education and Science*, 42(191), 307–322.

Ulucan, A. (2011). Measuring the efficiency of Turkish universities using 10 measure- specific data envelopment analysis. *Sosyoekonomi*, 14(14), 181–196.

Vu, H.D. (2016). Technical efficiency of FDI firms in the Vietnamese manufacturing sector. *Review of Economic Perspectives*, 16(3), 205–230.

Wang, Y.M. & Chin, K.S. (2009). A new approach for the selection of advanced manufacturing technologies: DEA with double frontiers. *International Journal of Production Research*, 47(23), 6663–6679.

Warning, S. (2004). Performance differences in German higher education: Empirical analysis of strategic groups. *Review of Industrial Organization*, 24, 393–408.

Zhang, X.S. & Cui, J.C. (1999). A project evaluation system in the state economic information system of china an operations research practice in public sectors. *International Transactions in Operational Research*, 6(5), 441–452.

https://doi.org/10.1142/9781800615786_0019

Chapter 19

Productivity Growth and Technical Efficiency Changes in Rice Farms: An Input-Oriented Study Using Biosolid

Carlos Alberto Zúniga-González

School of Agricultural and Veterinary Sciences, Department of Agroecology, Bioeconomy and Climate Change Research Center, National Autonomous University of Nicaragua, León, Nicaragua

czuniga@ct.unanleon.edu.ni

Abstract

This chapter focuses on evaluating the efficiency of applying Biosolid Xolotlan (BX) on farms in western Nicaragua as a bio-input business. It measures efficiency and productivity for each farm engaged in bio-input operations. The data collected for the measurement process were gathered from each farm. The evaluation employed the DEA approach alongside the Malmquist index methodology to assess impact, while stochastic frontier methods were utilized to estimate econometric frontiers. The results indicate that the contribution of wage laborers (Li) and investment cost (Ki) positively influences farms that implemented BX. Additionally, the findings reveal an average technical efficiency of 48% in BX utilization, which had a positive impact on implementing farms. Farms that achieved higher production levels (frontiers) stood out: farm 9 at 88%, farm 5 at 86%, and farm 3 at 79%. Ultimately, it was discovered that the investment made by producers to enhance soil nutrient levels using BX for bioremediation of degraded soils significantly improved production efficiency.

Keywords: DEA, bio-business, eco-intensification, bioremediation, scale efficiency, Malmquist, technical efficiency, Biosolid Xolotlan, rice farm.

1. Introduction

This chapter focuses on measuring the productivity and efficiency of Biosolid Xolotlan (BX), previously known as treatable wastewater (Wolters *et al.*, 2022; Harley-Nyang *et al.*, 2022), applied on agricultural production farms in western Nicaragua. The interest of the Nicaraguan Aqueduct Company and Sanitary Sewers (ENACAL) for validating its product, "Biosolido Xolotlan," has motivated this type of research (Ziajahromi & Leusch, 2022). Consequently, this study measures the productive sector eco-intensification in the bioeconomy (Zúniga-González *et al.*, 2014) as a result of the application of BX. However, bad agricultural practices and climate variability have consequences for the impoverishment of soils due to anthropic processes in an ecosystem. There is evidence of soil degradation in Nicaragua, one of which is the way producers prepare the land, causing the problem of burning as a custom transferred by tradition. Although it is true that there are savings in the investment of labor, there is also a loss of nitrogen (Machín, 2006).

Deforestation is another of the actions that cause soil degradation and therefore affect the efficiency and productivity of agricultural production subsystems (Zúniga-González, 2011; Thanker *et al.*, 2021). Altamirano & Delgado (2019) conclude that there is a risk factor in the types of permeable and sandy loam soils for the contamination of aquifers by soluble contaminants (UNAN León, 2006; Guerrero *et al.*, 2012).

Another serious problem is solid waste, which causes damage, such as contamination of water resources and soil, as well as being harmful to health, air, wildlife, landscapes, and the aesthetics of cities. According to the Nicaraguan Environmental Plan (PAANIC), of the total number of municipalities in the country, 56% have a regular garbage collection system, while 87% lack sanitary authorization (Guerrero *et al.*, 2012; Perez *et al.*, 2012).

Garcia *et al.* (2011) present some indicators that give us an idea of the extent of contamination. For example, in Nicaragua, the contamination (garbage) was 5,938 m^3 per day, of which 75% is domiciliary and 25% is from commerce, industry, and hospitals. About 49% is added for garbage collection.

In the study areas where cotton has historically been cultivated, including the municipalities of El Viejo, Chinandega, Chichigalpa, León, and Tisma, organs were found to be chlorinated with contaminants, such as DDT, DDD, Endrin, Toxaphene, DDE, and Aldrín, above the permitted concentrations (Guerrero *et al.*, 2012).

The contribution of this study is to validate BX. The impact of the use of biosolids was examined by measuring it in terms of efficiency and the technology used by the evaluated farms (Lin *et al.,* 2022; Harley-Nyang *et al.*, 2022). We identify which farm was the most efficient, where the biosolid was more effective in terms of technology as well as in terms of capacity for the use and management of the validated product.

It is a common practice to use the frontier analysis approach to evaluate and examine the efficiency and performance of the agricultural sector by estimating a multi-dimensional "best practice" frontier for all the farms analyzed that used BX and then comparing those farms with that frontier (Akimov & Simshauser, 2020; Lam *et al.*, 2020; Rattanachai *et al.*, 2012). Data envelopment analysis (DEA) and stochastic frontier analysis (SFA) are the two most popular tools used for frontier analysis, with DEA being more popular in the agricultural (biosolid) efficiency literature because of its flexibility within the complex setting of the agricultural sector and its relevance to small samples (Gandiglio *et al.*, 2017; Theodoridis & Psychoudakis, 2008; Cobanoglu, 2013). More specifically, DEA evaluates the efficiency of the farms by using a given set of inputs to produce a maximum set of outputs (output-oriented DEA) or by using the minimum set of inputs to produce a given set of outputs (input-oriented DEA), or a combination of both (Lam *et al.*, 2020; Heifler *et al.,* 2018; Judy *et al.*, 2015; Clarke & Cummins, 2015). Other applications include banking (Ngo & Le, 2019; Hammami *et al.*, 2020; Boubaker *et al.*, 2021). From a practical perspective, however, one may raise the question of how much input (or output) needs to be consumed (or produced) if a farm wants to reach a certain efficiency level. This is the main question of the traditional DEA with BX as input, which was proposed and developed by Zhou *et al.* (2012) and

Adler & Volta (2016). Given the impact of low yields due to soil degradation, many farms have seen their inputs and especially their outputs decrease, such as low yields per hectare and high input costs, so the issue of maintaining such good efficiency has been raised and becomes important. To date, most DEA studies (e.g., Bravo-Ureta *et al.*, 2007; Dios-Palomares *et al.*, 2015) have focused on traditional inputs and some environmentally harmful chemicals; few have examined the impacts of such factors on soil bioremediation.

The rest of the chapter is organized as follows: Section 2 reviews the literature on BX's efficiency and the use of DEA in the agricultural sector. Section 3 explains the relevant methodologies, including DEA, and the methodology used for the quantification of the indicator parameters of the nutritional status of the soil. Section 4 introduces the data and reports and discusses the empirical findings. Section 5 concludes the study and suggests some directions for future research.

2. Literature Review

Through the literature review, many cases of biosolids applied to agricultural production were found. Soto & Meléndez (2004) reported that in Costa Rica and Nicaragua, banana and coffee plantations have been using organic and conventional fertilizers. The large- and medium-sized ones acquire them in the market, opening possibilities for the marketing channels for this item, while the small ones make them on their own farms.

The evaluation of the technical-environmental performance of systems, which includes the evaluation of the treatment levels achieved by the treatment processes (for example, measured as a reduction in volatile solids, the energy balance, and the final disposal of waste), is essential to compare the efficiency of the systems and, even more importantly, how their products, such as BX, can affect their impacts. However, this has not yet been investigated in previous research studies. To fill this research gap, this study proposes an analysis involving multi-criteria decision-makers using BX integrated into their inputs and a data envelopment analysis (DEA) approach to assess the profitability of their farms (Lam *et al.*, 2020).

Gonzáles *et al.* (2017) and Deago (2020) studied the applications of biosolids, where, despite increased soil salinity, the biosolid structure improved, increasing permeability. In the Dominican Republic, Duran (2020), and in the United States, Ozores & Mendez (2010) analyzed the application of biosolids in the production of vegetables, as in soils it can increase plant growth and produce comparable yields with less inorganic amounts compared to synthetic commercial fertilizers. They also investigated how to use waste from urban water treatment plants in agriculture. Alba & Humanez (2020) studied compost generated from biosolids under laboratory conditions. Their conclusion was that the favorable environment for the humus of the sludge can germinate seeds, which allows it to be converted into raw material for agriculture. Krishnamurthy *et al.* (2007) applied different types of doses of sewage sludge in organic fertilizers for corn and the initial development of *Pinus douglasiana.* García Sánchez *et al.* (2020) made a proposal for the coagulation of domestic wastewater. Albán & Molano (2017) considered bio-business as a profitable alternative, and Rodríguez (2021) highlights how, in pandemic conditions, it is important to rethink this type of business based on wealth and resources (biodiversity).

Floravanti's thesis (2003) draws attention to the proposed use of efficient microorganisms (EM) to eliminate putrefied malodors. The almost complete reduction of total and fecal coliforms favors treatment with EM. Sepulveda *et al.* (2012) applied biosolids and assessed the properties of soil during its temporal evolution in the Araucania region, and the recovery of soils was addressed by analyzing the case of Inceptisol soil under degraded conditions.

3. Methodology

3.1. *Methodology used for the quantification of the indicator parameters of the nutritional status of soil*

In the entire process, the soil samples underwent several stages, which included: sampling, transportation, and preparation, including drying, grinding and sieving, packaging, coding and storage, and laboratory analysis (physical and chemical).

3.1.1. *Evaluation and interpretation of the results*

The volume was estimated using the area and depth, as given in Equation (1):

$$V = A\,(m^2) \times h\,(m). \tag{1}$$

The mass is given by the volume and bulk density of the soil, as shown in Equation (2):

$$m = V \times \rho. \tag{2}$$

Once the content of each of the parameters present in the soils was estimated, the biosolid dose to be applied was determined using the following formula in Equation (3):

$$\text{Dosis} = \text{mpo}\frac{\text{Crop consumption} - \text{Soil contribution}}{\text{Relative Fertilizer Efficiency}}. \tag{3}$$

3.2. *Methodology for measuring the technical efficiency of bio-businesses*

3.2.1. *Enveloping data analysis methodology in bio-businesses*

Kumbhakar & Lovell (2000) explain the DEA methodology as an estimate of bio-businesses (firms) located on the border and their behavior. This frontier is explained as the maximum product obtained by the application of BX, including the treatment of water as energy generated (Ding *et al.*, 2020).

The database includes the application of the BX as an estimate of the bio-businesses for farms of 11 rice producers in the western area of Nicaragua in the period 2013–2017. The data consider the producers in terms of space and time, which allowed us to organize a data panel of the information about the farms studied (BB). The information makes it possible to identify the producer of BB per year over time.

Coelli *et al.* (2005) consider technical efficiency as determined by the following cost minimization programming (LP; Equation (4)):

$$ET_n = \min_{\theta n \lambda i} \theta_n, \tag{4}$$

$$\text{s.t.: } -q_n + \sum_{i=1}^{j} \lambda_i q_i \geq v,$$

$$\theta_n X_{ij}^* - \sum_{i=1}^{j} \lambda_i \chi_j \geq 0,$$

$$\sum_{i=1}^{j} \lambda_{i=1},$$

$$\lambda_i \geq 0,$$

where, ET_n is the technical efficiency for the farm o (BB), n is represented, $i = 1, \ldots, I$ represents the number of farms with BB, $j = 1, \ldots, J$ represents the amount of inputs, q_i is the amount produced on the farm i, q_n is the amount produced on the farm n, χ_{ij} is the input obtained from j where BX applied on the farm is included i, χ_{nj} is the input obtained j used on the farm o (BB) n, λi are weighted by farm $y\, \theta n$ is a scalar vector less than one, where the technical efficiency is defined for the farm n. LP1 (Linear Programming 1 or Equation 1) presented is repeated for each of the farms considered in the analysis.

Coelli *et al.* (2005) explain that $\sum_{i=1}^{j} \lambda_{i=1}$ is a safe constraint, where TE is estimated assuming VRS, otherwise it would be CRS. Farrel (1957) indicates that efficiency reduces its radius area, which is interpreted as working with the input level without affecting the price level (Wilson, 2006).

Cost efficiency is defined by economic efficiency. This is estimated in two stages. The linear programming problem (LP) solved first is shown in the following:

$$MC_n = \min_{\lambda i \chi j} W_{nj} X_{nj}, \tag{5}$$

$$\text{s.t.:} \quad -q_n + \sum_{i=1}^{j} \lambda_i q_i \geq 0,$$

$$X_{nj} - \sum_{i=1}^{j} \lambda_i \chi_j \geq 0,$$

$$\sum_{i=1}^{j} \lambda_{i=1},$$

$$\lambda_i \geq 0,$$

where, MC_n represent minimizing costs on the farm n, W_{nj} represents the price of input i on farm n, and X^*_{ij} input level j that minimizes the cost on farm n. Calculating the economic efficiency is the second step, as given by

$$\mathrm{EE}_n = \frac{\sum_{j=1}^{j} W_{nj} X^*_{nj}}{\sum_{j=1}^{j} W_{nj} X_{nj}}. \tag{6}$$

In the first step, the numerator is the minimum cost obtained on farm n and the denominator is the observed cost on farm n. This value is always less than or equal to one. As a result of the quotient between economic and technical efficiencies, the allocation efficiency is estimated:

$$\mathrm{AE}_n = \mathrm{EE}/\mathrm{ET}. \tag{7}$$

Figure 1 presents the assumption of decision-making units (DMUs) that refer to two inputs x_1 and x_2 expressed as q of the product unit. For the DMU A, the quotient is between 0A′ and 0A. It can be defined graphically as the technical efficiency or, in the case of the DMU C, by the ratio 0C′/0C. The distance 0C″ represents the ratio of 0C and is regarded as a measure of cost efficiency within economic efficiency. The quotient 0C″/0C′ measures allocation efficiency. An inefficiency in the allocation of factors in Case B represents a technically efficient DMU but basically presents economic inefficiencies. Case D which represents a technically and economically efficient DMU.

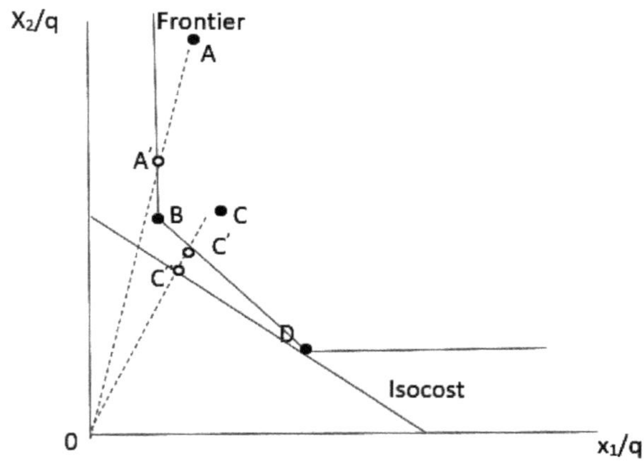

Figure 1. Efficiency measurement.

The TE at constant returns to scale (CRS) divided by the TE at variable returns to scale (VRS) gives the quotient of the efficiency of scale (ES). The following formula is applied to each farm:

$$ES = ET_{CRS}/ET_{VRS}. \tag{8}$$

Values less than one are considered for ES, in such a way that the farm functions at CRS if ES $= 1$ and is at increasing (IRS) or diminishing (DRS) returns to scale if ES < 1. Changing the condition from $\sum_{i=1}^{j} \lambda_i = 1$ to $\sum_{i=1}^{j} \lambda_i \leq 1$ in Equation (1) leads to the reformulation of the problem. As a result, for a variable returns to scale (VRS) farm, the newly estimated technical efficiency (ET) becomes non-constant returns to scale. Hence, the farm operates under variable returns to scale (VRS) if it's. Subsequently, the constant returns to scale (CRS) equals ET_{VRS}. Is different from?

Coelli *et al.* (2005) graphically represents this situation in Figure 2, where the type of scale inefficiency considered is the CRS frontier. Point B represents the frontier determined by ET_{NIRS}, which differs from the one determined by ET_{VRS}. In DMU B, they are increasing; therefore, the returns to scale grow. Point D represents that the returns to scale will decrease.

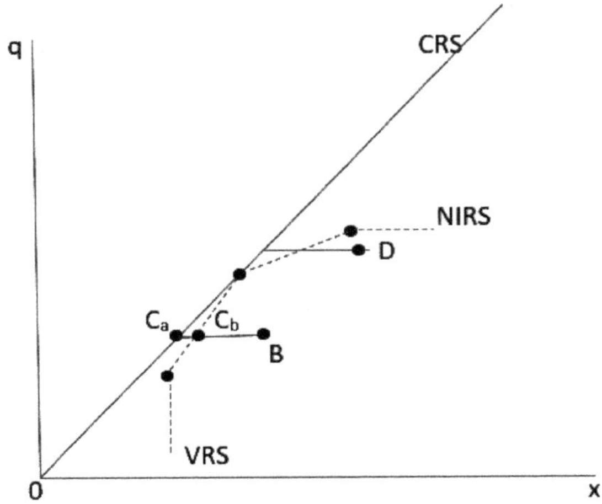

Figure 2. Measuring efficiencies of scale (Coelli *et al.*, 2005).

3.3. *Panel data, DEA, and the Malmquist index*

In this chapter, we considered panel data with DEA-like linear programs, and an input-based Malmquist index was estimated to measure productivity change. This productivity was split into technical change and technical efficiency change.

Some authors define productivity as the division between BX and the elements used. We refer to partial productivity when a single factor is considered; on the other hand, considering all the inputs is called total factor productivity (TFP). Farrell (1957) defines technical efficiency (TE) by considering the company that uses the minimum resources; in our case, it would be using BX to achieve a certain level of product (rice). Daraio & Simar (2007) compare the relationship between inputs and outputs to measure efficiency, which defines the production frontier. Lovell (1993) defines it as that productivity that we observe and the one obtained as good or optimal.

Farrell (1957) provides the concepts of allocation efficiency and economic efficiency. The first measures how efficient a company is in combining its inputs at a market price. The second, according to Farrell, is akin to the ratio between technical and allocation efficiencies.

On the other hand, it is noted that efficiency is related to scale efficiencies (SE) defined in the Coob–Douglas production function (Zellner & Dreze, 1966), which refers to scale growths. A bio-business, farm, or company works at CRS when its function is homogenous of grade 1, fulfilling $f(\lambda x) = \lambda f(x)$, for all $\lambda \geq 0$. If $f(\lambda x) < \lambda f(x)$, for all $\lambda > 1$, it presents DRS. In the opposite case, it would imply an error during measurement.

According to Grosskopf (1993, p. 183), input-oriented Malmquist indices can be defined in a similar way as the output-oriented ones, so Färe *et al.* (1997) define it as

$$m_0(y_{t+1}, x_{t+1}, y_t, x_t) = \left[\frac{d_0^t(x_{t+1}, y_{t+1})}{d_0^t(x_t, y_t)} x \frac{d_0^{t+1}(x_{t+1}, y_{t+1})}{d_0^{t+1}(x_t, y_t)}\right]^{1/2}. \quad (9)$$

The agricultural production point (x_{t+1}, y_{t+1}) represents productivity relative to the agrarian production point (x_t, y_t). A positive TFP growth from period t to period $t + 1$ is when its value is greater than one. Then, TFP indices are based on the geometric mean of two outputs or two inputs, as is the case in this chapter. Period t

technology uses one index, and the other period uses $t + 1$ technology. To solve Equation (9), we must calculate the four component distance functions, which will involve four LP problems.

A CRS output-oriented LP technology is assumed to calculate $d_0^t(x_t, y_t)$, i.e.,

$$[d_0^t(x_t, y_t)]^{-1} = \min_{\emptyset\lambda}\phi, \tag{10}$$

$$\begin{aligned} \text{s.t.:} \quad -\phi y_{it} + Y_t\lambda &\geq 0,\\ x_{it} - X_t\lambda &\geq 0,\\ \lambda &\geq 0. \end{aligned}$$

The remaining three LP problems are simple variants of this:

$$[d_0^{t+1}(x_{t+1}, y_{t+1})]^{-1} = \min_{\emptyset\lambda}\phi, \tag{11}$$

$$\begin{aligned} \text{s.t.:} \quad -\phi y_{i,t+1} + Y_{t+1}\lambda &\geq 0,\\ x_{i,t+1} - X_{t+1}\lambda &\geq 0,\\ \lambda &\geq 0; \end{aligned}$$

$$[d_0^t(x_{t+1}, y_{t+1})]^{-1} = \min_{\emptyset\lambda}\phi, \tag{12}$$

$$\begin{aligned} \text{s.t.:} \quad -\phi y_{i,t+1} + Y_t\lambda &\geq 0,\\ x_{i,t+1} - X_t\lambda &\geq 0,\\ \lambda &\geq 0; \end{aligned}$$

$$[d_0^{t+1}(x_t, y_t)]^{-1} = \min_{\emptyset\lambda}\phi, \tag{13}$$

$$\begin{aligned} \text{s.t.:} \quad -\phi y_{it} + Y_{t+1}\lambda &\geq 0,\\ x_{it} - X_{t+1} + \lambda &\geq 0,\\ \lambda &\geq 0. \end{aligned}$$

It is important to consider that in LPs (12) and (13), the parameter ϕ need not be ≥ 1, where the agricultural points are compared to technology from different time periods. In LP (12), agricultural point from period t¡1 is compared to technology in period t. When the value $\Phi < 1$ is possible, then technical progress occurs. In this study, we have $N = 11$ farms, so we need to calculate $N \times (3T-2)$, where $T = 5$ (i.e., the period 2013–2017), this would provide $11 \times (3 \times 5 - 2) = 143$ LPs.

3.4. *Data*

DEA (Boubaker *et al.*, 2020; Ngo & Tsui, 2021; Le *et al.*, 2021) was considered based on the authors' recommendations, where the BX variables were used as an alternative in the bio-businesses. So, the procedure proposes a linear programming (LP) where there are no censored variables but instead truncates the population by 1 to estimate the impact of the technical efficiency of a bio-business in rice production. If the soils are recovered, they are no longer used as polluting inputs, and the profitability of bio-businesses increases (Thanker *et al.*, 2021; Tsionas, 2021).

In this chapter, the variables used are as follows:

Y_i represents the output quantities obtained in each productive cycle measured in *qq* of rice.
L_i represents the variable cost of each agricultural production cycle per farm. This cost includes the inputs used in the production process, including the amount of BX estimated in Table 1.
K_i represents cost of investment or recovery of capital.

4. Results and Discussion

The results of estimating the Total Factor Productivity (TFP) indexes, measuring the impact of using BX as a bioremediator according to Wandscheer (2015), Molano (2005), and Rodríguez (2021), highlight the significance of bio-business in agricultural farms. They underscore its relevance for rice producers. Table 1 provides a summary of the soil's nutrient levels and the necessary BX dosage required for application.

Table 2 presents the statistical description of the data used in the Malmquist indices, observing that 55 data points were used, organized in a data panel for 11 farms that applied BX in the rice field during five productive cycles in different areas of western Nicaragua (Chen *et al.*, 2022; Zúniga-González, 2020; Zúniga-González *et al.*, 2022). It is noted that, on average, during the rice production cycle (2013–2017), the producers obtained 275.6 qq per mz (194.13 qq X ha). The production costs, including the labor costs, for the use of BX on average were 96.83 cordobas ($3.14), and the cost of capital recovery per BX represented 12.65 corbodas ($0.41). The dispersion of these data with respect to the mean shows that the rice production

Table 1. Summary of physical chemical elements contained in the soil, soil demand per crop, and the suggested amount of biosolids.

	N		P		K		Ca		Mg		S		DOSE
Ref.	CS lb / Mz	DS lb / Mz	CS lb / Mz	DS lb / Mz	CS lb / Mz	DS lb / Mz	CS lb / Mz	DS lb / Mz	CS lb / Mz	DS lb / Mz	CS lb / Mz	DS lb / Mz	qq BS /Mz
—	618.3	—	337.5	—	705.5	—	7457.9	—	372.9	—	82.4	—	—
1555	146.8	−471.5	183.9	−153.6	606.1	−99.4	5143.2	−2314.7	132.8	−240.1	213.8	131.4	23.1
1556	247.3	−371.0	133.3	−204.2	692.7	86.6	4822.8	−2635.1	193.4	60.6	42.5	−39.9	26.4
1557	332.3	−286.0	99.6	−237.9	484.2	−208.4	4179.2	−3278.7	104.9	−88.6	24.5	−57.9	32.8
1558	266.6	−351.7	69.2	−268.3	426.5	−57.7	3215.2	−4242.7	65.3	−39.6	47.7	−34.7	42.4
1559	255.0	−363.3	89.1	−248.4	812.0	385.5	6173.0	−1284.9	239.8	174.6	105.7	23.3	12.8
1560	235.7	−382.6	121.5	−216.0	1314.8	502.8	6430.2	−1027.7	259.2	19.4	16.4	−66.0	10.3
1561	208.7	−409.6	43.9	−293.6	1109.6	−205.2	10931.3	3473.4	741.9	482.7	30.9	−51.5	−34.7
1562	180.8	−437.5	93.1	−244.4	615.7	−493.9	5787.2	−1670.7	201.4	−540.5	88.9	6.5	16.7
1563	139.1	−479.2	104.1	−233.4	869.7	254.0	6558.8	−899.1	289.4	87.9	51.3	−31.1	9.0
1564	235.7	−382.6	92.8	−244.7	375.2	−494.5	3215.1	−4242.8	51.5	−237.9	113.3	30.9	42.4
1565	208.7	−409.6	119.8	−217.7	500.3	125.1	3215.1	−4242.8	65.4	13.9	34.4	−48.0	42.4
1566	336.2	−282.1	50.6	−286.9	561.2	60.9	3215.1	−4242.8	68.2	2.8	88.5	6.1	42.4
1569	92.7	−525.6	172.8	−164.7	564.4	3.2	2572.1	−4885.8	72.7	4.5	22.7	−59.7	48.9
1570	132.9	−485.4	218.7	−118.8	610.6	46.2	2572.1	−4885.8	75.7	3.0	110.0	27.6	48.9

Notes: 1 lb = 0.4535 kg; 1 Mz = 0.7044 ha; CS: Soil Content; Ref.: Laboratory code per sample.

Table 2. Statistics panel data production cycles 2013–2017.

Statisticians	(Y)	(L)	(K)
Mean	275.60	96.63	12.65
Standard deviation	179.67	45.76	9.86
Minimum	56.98	27.00	2.00
Maximum	734.12	190.90	34.93

during the average of the years studied was not high, so it was estimated that the SD for (Y) was 179.67 qq X mz (139.24 X ha), while the production costs were 45.75 cordobas ($1.49) and the cost of recovery of the capital invested in the BX was 9.86 cordobas ($0.32).

The efficiency results are in the range investigated by other authors who have measured efficiency in Nicaragua, such as Dios-Palomares *et al.* (2015), González (2011), López-González *et al.* (2015), Abdulai (2001), Bravo-Ureta (2007), Milan & Zúniga-González (2021), Zúniga-González (2020), and Battese & Coelli (1995), indicate that the number of variables studied explains the specified model directly.

4.1. *TFP estimation and efficiencies*

The results of the DEA and TFP calculations are summarized in this section. Given that there are five annual observations in 11 farms in the western region of Nicaragua (Coelli, 1996; Zúniga-González, 2020, 2022; Battese and Corra, 1997), the calculations involved solving $11 \times (5 \times 3 - 2) = 143\,\text{LP}$ iterations for the optimal solution of the problem (Coelli, 1996). Table 3 shows the changes in technical efficiency (effch), where we consider the capacity of the producer to handle the technology of BX, the change in technology (techch) referred to the physical-chemical composition of BX, the change in pure efficiency (pech), and the change in efficiency at scale (sech) referred to the size of the farm. The change in total factor productivity (tfpch) was referred to BX used as input at the means or dose indicated or estimated by the soil laboratory, as indicated in Table 1. Consequently, it was observed that the change in total factor productivity was 30% below the frontier, mainly explained by the change in technical efficiency (0.18) and efficiency at scale (0.18). Changes in pure efficiency were not significant or indifferent. The change in technology was below the border (31%).

Table 3. Mean technical efficiency change, technology change, and TFP change in farms that applied BX during the 2013–2017 production cycles.

Firm	effch	techch	pech	sech	tfpch
1	1.080	0.682	1	1.080	0.737
2	1.016	0.669	1	1.016	0.680
3	1.007	0.669	1	1.007	0.674
4	1.012	0.669	1	1.012	0.677
5	1.012	0.669	1	1.012	0.677
6	1.023	0.713	1	1.023	0.729
7	1.022	0.694	1	1.022	0.710
8	1.020	0.665	1	1.020	0.678
9	1.011	0.667	1	1.011	0.674
10	1.000	0.817	1	1.000	0.817
11	1.000	0.669	1	1.000	0.669
Mean	1.018	0.688	1	1.018	0.701

Table 4. Annual mean technical efficiency, technology change, and TFP change in farms that applied BX during the 2013–2017 production cycles.

Year	effch	techch	pech	sech	tfpch
2014	1.012	0.532	1	1.012	0.538
2015	1.031	0.674	1	1.031	0.695
2016	0.995	0.772	1	0.995	0.769
2017	1.035	0.810	1	1.035	0.839
Mean	1.018	0.688	1	1.018	0.701

Table 4 shows the changes in effch, techch, pech, sech, and ptf during the period 2013–2017 and explains that during the year 2016, there was a decrease in technical efficiency and the size of the areas that used BX was reduced, although in the following year, the tendency to increase the levels of both technical efficiency and technological change was resumed. This tendency was increasing with the interannual changes in BX technology and the capacity of the producers to manage and improve its application in the productive process of rice.

Table 5. Annual mean changes of technology (BX) in CRS and the variation of returns to scale of technical efficiency in farms that applied BX during the 2013–2017 production cycles.

	CRS TE rel to tech in yr			
Year	$t-1$	t	$t+1$	**VRS TE**
2013	0.000	0.554	1.096	1
2014	0.340	0.561	0.857	1
2015	0.415	0.575	0.759	1
2016	0.464	0.572	0.719	1
2017	0.481	0.581	0.000	1

Table 5 shows the interannual changes in technology (BX) under CRS and the variation of returns to scale of technical efficiency. It can be noted that although at low levels, a growing trend in the use of technology (BX) in CRS is observed, that is, although below the distance from the border, the efforts of producers to improve the way in which the application of BX on the farms under study, on the other hand, as the production area increased, the variation in yields at the technical efficiency scale shows that they remain on the border. These results imply that the size variable is fundamental to the productivity of BX since, as the areas increase, the productivity and technical efficiencies increase.

On the other hand, it is very important to consider the application of DEA models to eco-intensification approaches, considering the trend of DEA development, such as a censored regression (tobit) by Watkins *et al.* (2014), DEA and SFA software by Coelli (2005), Boris *et al.* (2007), Boubaker *et al.* (2021, 2022a, 2022b), Zúniga-González & Jaramillo-Villanueva (2022), and Zúniga-González *et al.* (2022), and bootstrap DEA by Simar & Wilson (2007).

5. Conclusion

With this chapter, it was possible to validate the productivity and technical efficiency of the use of BX as a soil bioremediation, making it profitable for the producers who used the amounts demanded by the soil laboratory. The main conclusion in the validation of BX reflects the need to improve soils in the western zone, constituting a bio-business with BX as an input validated by rice producers, but which represents an opportunity for producers of other items, enabling bio-businesses from available biodiversity resources.

BX is validated by the proposed technical efficiency model and the TPF index in addition. The model suggests the expansion of use to other items, where producers could improve their yields and the quality of their items using this bioremediator, as a business necessity to advance in a model that considers the availability of biodiversity resources (Carraresi & Bröring, 2021). The TFP indices show an increasing tendency to reduce the distance from the frontier. Both the BX technology and the technical efficiency of the producers to improve each year in the management of the application of BX to their crops showed an increasing interannual trend closer to the distance from the border.

References

Abdulai, A. & Eberlin, R. (2001). Technical efficiency during economic reform in Nicaragua: Evidence from farm household survey data. *Economic System*, 25, 113–125.

Adler, N. & Volta, N. (2016). Accounting for externalities and disposability: A directional economic environmental distance function. *European Journal of Operational Research*, 250(1), 314–327.

Akimov, A. & Simshauser, P. (2020). Performance measurement in Australian water utilities. Current state and future directions. *Australian Journal of Public Administration*, 79(1), 111–142.

Alba Pinto, M.F. & Humanez Poveda, L.F. (2020). Propuesta para la obtención de compost partiendo de los biosólidos generados en la operación de Laboratorios Esko Ltda (Bachelor's thesis, Fundación Universidad de América).

Albán, I.A., Pulido, M.Á., & Molano, J.H. (2017). GRUPO DE BIONEGOCIOS. Bionegocios, un sendero para el desarrollo. *Revista Colombiana de Biotecnología*, 35–37.

Altamirano, M. & Delgado, V. (2019). Contaminación natural por arsénico en las aguas subterráneas de la comunidad rural "La Fuente", para sugerir y promover el uso de fuentes alternativas de agua segura municipio La Paz Centro, León, Nicaragua. *Revista Torreón Universitario*, 8(23), 58–72.

Battese, G.E. & Coelli, T.J. (1995). A model for technical inefficiency effects in a stochastic frontier production function for panel data. *Empirical Economics*, (20), 325–332. https://doi.org/10.1007/BF01205442.

Battese, G.E. & Corra, G.S. (1997). Estimation of a production frontier model: With application to the pastoral zone of Eastern Australia. *Australian Journal of Agricultural Economics*, 21, 169–179. https://doi.org/10.1111/j.1467-8489.1977.tb00204.x.

Boubaker, S., Le, T.D.Q., & Ngo, T. (2022a). Managing bank performance under COVID-19: A novel inverse DEA efficiency approach. *International Transactions in Operational Research*, 30(2023), 2436–2452. https://doi.org/10.1111/itor.13132.

Boubaker, S., Manita, R., & Mefteh-Wali, S. (2020). Foreign currency hedging and firm productive efficiency. *Annals of Operations Research*, 1–22.

Boubaker, S., Manita, R., & Rouatbi, W. (2021). Large shareholders, control contestability and firm productive efficiency. *Annals of Operations Research*, 296(1), 591–614.

Boubaker, S., Do, D.T., Hammami, H., & Ly, K.C. (2022b). The role of bank affiliation in bank efficiency: A fuzzy multi-objective data envelopment analysis approach. *Annals of Operations Research*, 311, 611–639. https://doi.org/10.1007/s10479-020-03817-z.

Bravo-Ureta, B.E., Solís, D., López, V.H.M., Maripani, J.F., Thiam, A., & Rivas, T. (2007). Technical efficiency in farming: A meta-regression analysis. *Journal of Productivity Analysis*, 27(1), 57–72.

Carraresi, L. & Bröring, S. (2021). How does business model redesign foster resilience in emerging circular value chains? *Journal of Cleaner Production*, 289, 125823. https://doi.org/10.1016/j.jclepro.2021.125823.

Cheng, M., Lu, Y., Zhu, H., & Xiao, J. (2022). Measuring CO2 emissions performance of China's construction industry: A global Malmquist index analysis. *Environmental Impact Assessment Review*, 92, 106673.

Clarke, R.M. & Cummins, E. (2015). Evaluation of "classic" and emerging contaminants resulting from the application of biosolids to agricultural lands: A review. *Human and Ecological Risk Assessment: An International Journal*, 21(2), 492–513.

Cobanoglu, F. (2013). Measuring the technical efficiency of cotton farms in Turkey using stochastic frontier and data envelopment analysis. *Outlook on Agriculture*, 42(2), 125–131.

Coelli, T. (1996). A guide to DEAP version 2.1: A data envelopment analysis (computer) program. Centre for Efficiency and Productivity Analysis, University of New England, Australia (Vol. 96, No. 8, pp. 1–49).

Coelli, T., Prasada Rao, D.S., O'Donnell, C.J., & Battese, G.E. (2005). *An Introduction to Efficiency and Productivity Analysis*. New York: Springer, p. 338.

Daraio, C. & Simar, L. (2007). *Advanced Robust and Nonparametric Methods in Efficiency Analysis, Methodology and Applications* (p. 248). New York: Springer. (Studies in Productivity and Efficiency).

Deago, E. (2020). Digestividad anaeróbica de biosólidos de aguas residuales usando el sistema Oxitop para estimar la producción de biogás. *I+D Tecnológico*, 16(2), 51–58. https://doi.org/10.33412/idt.v16.2.2831.

Ding, T., Wu, H., Jia, J., Wei, Y., & Liang, L. (2020). Regional assessment of water-energy nexus in China's industrial sector: An interactive meta-frontier DEA approach. *Journal of Cleaner Production*, 244, 118797. https://doi.org/10.1016/j.jclepro.2019.118797.

Dios-Palomares, R. Alcaide, D., Diz, J., Jurado, M., Prieto, A., Morantes, M., & Zúniga-González, C.A. (2015). 7. Analysis of the efficiency of farming systems in Latin America and the Caribbean considering environmental issues. *Revista Cientifica-Facultad de Ciencias Veterinarias*, 25(1). https://publons.com/publon/3106827/.

Durán, A.D.J.M., Núñez, V.A.R., & Orgaz-Agüera, F. (2020). Use of biosolids from urban water treatment plants in agriculture. Case study in the Dominican Republic. *DELOS: Sustainable Local Development*, 13(37), 12.

Färe, R., Grosskopf, S., & Norris, M. (1997). Productivity growth, technical progress, and efficiency change in industrialized countries: Reply. *The American Economic Review*, 87(5), 1040–1044.

Farrell, M.J. (1957). The measurement of productive efficiency. *Journal of the Royal Statistical Society. Series A (General),* 120(3), 253–290. https://doi.org/10.2307/2343100.

Fioravanti Basombrío, M.T. & Vega Araya, N. (2003). Tesis: Eficiencia de los microorganismos eficaces (EM) en la estabilización de lodos sépticos para su reuso agrícola. http://usi.earth.ac.cr/glas/sp/pdf/2000100.pdf.

Gandiglio, M., Lanzini, A., Soto, A., Leone, P., & Santarelli, M. (2017). Enhancing the energy efficiency of wastewater treatment plants through co-digestion and fuel cell systems. *Frontiers in Environmental Science*, 5, 70.

García Sánchez, S.S. & Palacios Unda, R.E. (2020). Propuesta de mejoramiento en el proceso de coagulación de las aguas residuales domesticas en la PTAR del municipio de Puerto Gaitán Meta (Bachelor's thesis, Fundación Universidad de América). Diponible en https://hdl.handle.net/20.500.11839/8200 (visitado 4/8/21).

García-Torres, R., Rios-Leal, E., Martínez-Toledo, Á., Ramos-Morales, F.R., Cruz-Sanchez, J.S., & Cuevas-Díaz, M.D.C. (2011). Uso de cachaza y bagazo de caña de azúcar en la remoción de hidrocarburos en suelo contaminado. *Revista internacional de Contaminación Ambiental*, 27(1), 31–39.

González, C.A.Z. (2011). Technical efficiency of organic fertilizer in small farms of Nicaragua: 1998–2005. *African Journal of Business Management*, 5(3), 967–973. https://publons.com/publon/11272633/.

González-Flores, E., Ramos-Barragán, J.E., Tornero-Campante, M.A., & Murillo-Murillo, M. (2017). Evaluación de dosis de biosólidos urbanos en maíz bajo condiciones de invernadero. *Revista mexicana de Ciencias Agrícolas*, 8(1), 119–132.

Grosskopf, S. (1993). Efficiency and productivity. In: T. Coelli (Ed.), *A Guide to DEAP Version 2.1: a Data Envelopment Analysis (Computer) Program* (pp. 1–49). Australia: Centre for Efficiency and Productivity Analysis, University of New England (Vol. 96, No. 8).

Guerrero Guido, P.L., Hernández Flores, K.D., & Hernández Hernández, R.J. (2012). Manual de procedimientos sobre el Manejo de Residuos Peligrosos para establecimientos de asistencia sanitaria e industria farmacéutica de Nicaragua 2012. Tesis. Repositorio UNAN Leon, http://riul.unanleon.edu.ni:8080/jspui/bitstream/123456789/5714/1/222158.pdf.

Harley-Nyang, D., Memon, F.A., Jones, N., & Galloway, T. (2022). Investigation and analysis of microplastics in sewage sludge and biosolids: A case study from one wastewater treatment works in the UK. *Science of the Total Environment*, 823, 153735.

Heifler, O., Carmeli, C., & Carmeli, I. (2018). Enhanced optoelectronics by oriented multilayers of photosystem I proteins in dry hybrid bio-solid devices. *The Journal of Physical Chemistry C*, 122(21), 11550–11556.

Judy, J.D., McNear Jr, D.H., Chen, C., Lewis, R.W., Tsyusko, O.V., Bertsch, P.M., Rao, W., Stegemeier, J., Lowry, G.V., McGrath, S.P., Durenkamp, M. and Unrine, J.M. (2015). Nanomaterials in biosolids inhibit nodulation, shift microbial community composition, and result in increased metal uptake relative to bulk/dissolved metals. *Environmental Science & Technology*, 49(14), 8751–8758.

Krishnamurthy, L., Zamora Natera, F., Hernández Alvárez, E., Rodríguez Macías, R., Salcedo Pérez, E., & Vázquez Alarcón, A. (2007). Evaluación de lodos residuales como abono orgánico en suelos volcánicos de uso agrícola y forestal en Jalisco, México. *Revista de ciencia y tecnología de América*, 32(2), 115–120.

Kumbhakar, S.C. & Lovell, C.A.K. (2000). *Stochastic Frontier Analysis* (p. 333). Cambridge: Cambridge University Press. https://doi.org/10.1017/CBO9781139174411.

Lam, C.M., Hsu, S.C., Alvarado, V., & Li, W.M. (2020). Integrated life-cycle data envelopment analysis for techno-environmental performance evaluation on sludge-to-energy systems. *Applied Energy*, 266, 114867.

Le, T.D.Q., Ho, T.H., Nguyen, D.T., & Ngo, T. (2021). Fintech credit and bank efficiency: International evidence. *International Journal of Financial Studies*, 9(3), 44. https://www.mdpi.com/2227-7072/9/3/44.

Lin, Z., Price, G.W., Burton, D.L., & Clark, O.G. (2022). Effects on soil nitrogen and plant production from land applying three types of biosolids to an agricultural field for three consecutive years. *Soil and Tillage Research*, 223, 105458.

López-González, Á.S., Zúniga-González, C.A., López, M.R., Quirós-Madrigal, O.J., Colón-García, A.P., Navas-Calderón, J., & Rangel-Cura, R.A. (2015). Estado del arte de la medición de la productividad y la eficiencia técnica en América Latina: Caso Nicaragua. *Revista Iberoamericana de Bioeconomía y Cambio Climático*, 1(2), 76–100.

Lovell, C.A.K. (1993). Production frontiers and productive efficiency. In: H.O. Fried, C.A.K. Lovell & S.S. Schmidt (Eds.), *The Measurement of Productive Efficiency*. New York: Oxford University Press. p. 440.

Luana, L. & Morone, P. (2021). Holistic approach in the evaluation of the sustainability of bio-based products: An integrated assessment tool. *Sustainable Production and Consumption*, 28, 911–924. https://doi.org/10.1016/j.spc.2021.07.006.

Milán-Pérez, J.A. & Zúniga-González, C.A. (2021). Necesidades de investigación y transferencia de tecnologías sobre cambio climático en Nicaragua: Una oportunidad en la Bioeconomía. *Revista Iberoamericana De Bioeconomía y Cambio Climático*, 7(13), 1518–1543. https://doi.org/10.5377/ribcc.v7i13.11270.

Molano, J. (2005). Qué son los bionegocios y su entorno. Presentación del Grupo de Bionegocios — Gobernación del Huila, Neiva, Colombia.

Ngo, T. & Tsui, K.W.H. (2021). Estimating the confidence intervals for DEA efficiency scores of Asia-Pacific airlines. *Operational Research*, https://doi.org/10.1007/s12351-021-00667-w.

Ozores-Hampton, M. and Mendez, J. (2010). Uso de biosólidos en producción de hortalizas. Publicación del Departamento de Horticultural Sciences. Universidad de la Florida, Institute of Food and Agricultural Sciences Extension, Florida.

Pérez Rodríguez, L.J. & Chávez Porras, Á. (2012). Tesis: Evaluación del uso de sustratos preparados con lodos residuales en la germinación de semillas de zanahoria Daucus Carota l. Universidad militar Nueva Granada. Facultad de Ingeniería, dirección de posgrados, especialización en planeación ambiental y manejo integral de los recursos naturales, Bogotá.

Rattanachai, A., Srilaong, V., Kanlayanarat, S., Wasusri, T., & Tanprasert, K. (2012, February). Comparative execution efficiency of growers in Thailand's baby corn supply chain by data envelopment analysis. In Southeast Asia Symposium on Quality Management in Postharvest Systems and Asia Pacific Symposium on Postharvest Quality 989 (pp. 231–234).

Rodriguez, R.R.V. (2021). Ante una pandemia global existen oportunidades para una producción y consumo responsable en el Perú. *Visionarios en ciencia y tecnología*, 6(1), 167–185.

Sepúlveda, A., Carolina, I.H., & Francisco, E.M. (2012). Efecto de la aplicación de biosólidos sobre la calidad del suelo: Evolución temporal de suelo degradado, región de la Araucanía, Chile. http://repositoriodigital.uct.cl/bitstream/handle/10925/1003/art%C3%ADculo_Sep%C3%BAlveda%20et%20al_.pdf?sequence=1&isAllowed=y.

Simar, L. & Wilson, P.W. (2007). Estimation and inference in two-stage, semi-parametric models of production processes. *Journal of Econometrics*, 136, 31–64. https://doi.org/10.1016/j.jeconom.2005.07.009.

Soto, G. & Meléndez, G. (2004). Cómo medir la calidad de los abonos orgánicos. Repositorio institucional de la Universidad de Costa Rica http://hdl.handle.net/10669/318.

Thaker, K., Charles, V., Pant, A., & Gherman, T. (2021). A DEA and random forest regression approach to studying bank efficiency and corporate governance. *Journal of the Operational Research Society*, 1–28. https://doi.org/10.1080/01605682.2021.1907239.

Theodoridis, A.M. & Psychoudakis, A. (2008). Efficiency measurement in Greek dairy farms: Stochastic frontier vs. data envelopment analysis. *International Journal of Economic Sciences and Applied Research*, 1(2), 53–67.

Tsionas, M.G. (2021). Optimal combinations of stochastic frontier and data envelopment analysis models. *European Journal of Operational Research*. https://doi.org/https://doi.org/10.1016/j.ejor.2021.02.003.

UNAN León (2006). Laboratorio de Microbiología, Facultad de Ciencias. Informe del diagnóstico preliminar de la calidad del agua de consumo en las comunidades del sector rural noreste del municipio de León.

Wandscheer, C.B. (2015). Desenvolvimento de bionegócios e conhecimento tradicional: uma análise de suas influências nas populações tradicionais no Brasil. *Revista Brasileira de Planejamento e Desenvolvimento*, 4(2), 50–68.

Watkins, K.B., Hristovska, T., Mazzanti, R., Wilson, C.E., & Schmidt, L. (2014). Measurement of technical, allocative, economic, and scale efficiency of rice production in Arkansas using data envelopment analysis. *Journal of Agricultural and Applied Economics*. 46(1), 89–106. https://doi.org/10.1017/S1074070800000651.

Wilson, P. (2006). FEAR: A Software Package for Frontier Efficiency Analysis with R [En línea]. Consultado 1 julio 2014. Disponible en: http://www.clemson.edu/economics/faculty/wilson/software/fear/Paper/fear.pdf.

Wolters, B., Hauschild, K., Blau, K., Mulder, I., Heyde, B.J., Sørensen, S.J., Siemens, J., Jechalke, S., Smalla, K., & Nesme, J. (2022). Biosolids for safe land application: Does wastewater treatment plant size matters when considering antibiotics, pollutants, microbiome, mobile genetic elements and associated resistance genes? *Environmental Microbiology*, 24(3), 1573–1589.

Zellner, A., Kmenta, J., & Dreze, J. (1966). Specification and estimation of Cobb-Douglas production function models. *Econometrica: Journal of the Econometric Society*, 34(4), 784–795.

Zhou P., Ang B., & Wang, H. (2012). Energy and CO_2 emission performance in electricity generation: A non-radial directional distance function approach. *European Journal of Operational Research*, 221(3), 625–635.

Ziajahromi, S. & Leusch, F.D. (2022). Systematic assessment of data quality and quality assurance/quality control (QA/QC) of current research on microplastics in biosolids and agricultural soils. *Environmental Pollution*, 294, 118629.

Zúniga-González, C.A. (2011). Impact of deforestation on the sustainable local development of households: Nicaragua, 1998–2005. *Encuentro*, 88, 101–119. https://doi.org/10.5377/encuentro.v43i88.502.

Zúniga-González, C.A. (2020). Total factor productivity growth in agriculture: Malmquist index analysis of 14 countries, 1979–2008. *Revista Electrónica De Investigación En Ciencias Económicas*, 8(16), 68–97. https://doi.org/10.5377/reice.v8i16.10661.

Zúniga-González, C.A. (2022). Data for: Application of biosolido Xolotlan as a soil bioremediation. figshare. Book. https://doi.org/10.6084/m9.figshare.21625889.v1.

Zúniga-González, C.A. & Jaramillo-Villanueva, J.L. (2022). Frontier model of the environmental inefficiency effects on livestock bioeconomy [version 1; peer review: awaiting peer review]. *F1000Research*, 11, 1382. https://doi.org/10.12688/f1000research.128071.1.

Zúniga-González, C.A., Moreno-Mayorga, L.F., & Quiroz-Medina, C.R. (2022). Estudio de la eficiencia técnica en escuelas de campo de Nicaragua. *Revista Tecnología en Marcha*, 128–140. https://doi.org/10.18845/tm.v35i3.5696.

Zúniga-González, C.A., Durán Zarabozo, O., Dios Palomares, R., Sol Sánchez, A., Guzman Moreno, M.A., Quiros, O., & Montoya Gaviria, G.D.J. (2014). Estado del arte de la bioeconomía y el cambio climático (No. 1133-2016-92457, pp. 20–329). DOI: 10.22004/ag.econ.168356. https://ageconsearch.umn.edu/record/168356.

https://doi.org/10.1142/9781800615786_bmatter

Index

C

D

E

K

L

O

P

Q

R

Printed in the USA
CPSIA information can be obtained
at www.ICGtesting.com
LVHW011610240824
789147LV00002B/92

9 781800 61577